W9-AHC-038

# INFORMATION TECHNOLOGY PROJECT MANAGEMENT

# INFORMATION TECHNOLOGY PROJECT MANAGEMENT

**REVISED Sixth Edition**

**Kathy Schwalbe, Ph.D., PMP**
*Augsburg College*

Australia • Brazil • Japan • Korea • Mexico • Singapore • Spain • United Kingdom • United States

**Information Technology Project Management, REVISED Sixth Edition**

**Kathy Schwalbe**


Executive Vice President and Publisher: Jonathan Hulbert

Executive Vice President of Editorial, Business: Jack Calhoun

Publisher: Joe Sabatino

Senior Acquisitions Editor: Charles McCormick, Jr.

Senior Product Manager: Kate Mason

Development Editor: Deb Kaufmann

Editorial Assistant: Nora Heink

Marketing Director: Keri Witman

Marketing Manager: Adam Marsh

Senior Marketing Communications Manager: Libby Shipp

Marketing Coordinator: Suellen Ruttkay

Content Project Manager: Suganya Selvaraj

Media Editor: Chris Valentine

Senior Art Director: Stacy Jenkins Shirley

Cover Designer: Craig Ramsdell

Cover Image: ©Getty Images/Digital Vision

Manufacturing Coordinator: Julio Esperas

Compositor: PreMediaGlobal



© 2011 Course Technology, Cengage Learning

ALL RIGHTS RESERVED. No part of this work covered by the copyright herein may be reproduced, transmitted, stored or used in any form or by any means graphic, electronic, or mechanical, including but not limited to photocopying, recording, scanning, digitizing, taping, Web distribution, information networks, or information storage and retrieval systems, except as permitted under Section 107 or 108 of the 1976 United States Copyright Act, without the prior written permission of the publisher.

For product information and technology assistance, contact us at **Cengage Learning Customer & Sales Support, 1-800-354-9706.**

For permission to use material from this text or product, submit all requests online at **cengage.com/permissions.**

Further permissions questions can be emailed to **permissionrequest@cengage.com.**

Microsoft and the Office logo are either registered trademarks or trademarks of Microsoft Corporation in the United States and/or other countries. Course Technology, a part of Cengage Learning, is an independent entity from the Microsoft Corporation, and not affiliated with Microsoft in any manner.

Some of the product names and company names used in this book have been used for identification purposes only and may be trademarks or registered trademarks of their respective manufacturers and sellers.

Information pertaining to Northwest Airlines was used with their express permission. No part of it may be reproduced or used in any form without prior written permission from Course Technology.

The material that is reprinted from the *PMBOK® Guide, Fourth Edition* (©2008 Project Management Institute, Inc., all rights reserved) is used with permission of the Project Management Institute, Inc., Four Campus Boulevard, Newtown Square, PA 19073-2399, USA. Phone: (610) 356-4600, Fax: (610) 356-4647. Project Management Institute (PMI) is the world's leading project management association with over 270,000 members worldwide. For further information, contact PMI Headquarters at (610) 356-4600 or visit the Web site at www.pmi.org.

PMI, PMP, and PMBOK are registered marks of the Project Management Institute, Inc.



Library of Congress Control Number: 2010929569

**Student Edition:**
ISBN-13: 978-1-111-22175-1
ISBN-10: 1-111-22175-8

**Instructor's Edition:**
ISBN-13: 978-0-538-48268-4
ISBN-10: 0-538-48268-0

**Course Technology**
20 Channel Center Street
Boston, MA 02210
USA

Course Technology, a part of Cengage Learning, reserves the right to revise this publication and make changes from time to time in its content without notice.

Cengage Learning is a leading provider of customized learning solutions with office locations around the globe, including Singapore, the United Kingdom, Australia, Mexico, Brazil and Japan. Locate your local office at: **www.cengage.com/global**

Cengage Learning products are represented in Canada by Nelson Education, Ltd.

To learn more about Course Technology, visit **www.cengage.com/coursetechnology**

Purchase any of our products at your local college store or at our preferred online store **www.cengagebrain.com**

Printed in the United States of America
2 3 4 5 6 7 14 13 12 11

*For Dan, Anne, Bobby, and Scott*

# BRIEF CONTENTS

# TABLE OF CONTENTS

# PREFACE

The future of many organizations depends on their ability to harness the power of information technology, and good project managers continue to be in high demand. Colleges have responded to this need by establishing courses in project management and making them part of the information technology, management, engineering, and other curriculum. Corporations are investing in continuing education to help develop effective project managers and project teams. This text provides a much-needed framework for teaching courses in project management, especially those that emphasize managing information technology projects. The first five editions of this text were extremely well received by people in academia and the workplace. The Sixth Edition builds on the strengths of the previous editions and adds new, important information and features.

It's impossible to read a newspaper, magazine, or Web page without hearing about the impact of information technology on our society. Information is traveling faster and being shared by more individuals than ever before. You can buy just about anything online, surf the Web on a mobile phone, or use a wireless Internet connection at your local coffee shop. Companies have linked their many systems together to help them fill orders on time and better serve their customers. Software companies are continually developing new products to help streamline our work and get better results. When technology works well, it is almost invisible. But did it ever occur to you to ask, "Who makes these complex technologies and systems happen?"

Because you're reading this text, you must have an interest in the "behind-the-scenes" aspects of technology. If I've done my job well, as you read you'll begin to see the many innovations society is currently experiencing as the result of thousands of successful information technology projects. In this text, you'll read about IT projects around the world that went well, including Mittal Steel Poland's Implementation of SAP project that unified IT systems to improve business and financial processes; Dell Earth and other green computing projects that save energy and millions of dollars; and Six Sigma projects such as the project to improve case load management at Baptist St. Anthony's Hospital in Amarillo, Texas; the systems infrastructure project at the Boots Company in the United Kingdom that is taking advantage of supplier competition to cut costs and improve services; Kuala Lumpur's state-of-the-art Integrated Transport Information System (ITIS) project; and many more. Of course, not all projects are successful. Factors such as time, money, and unrealistic expectations, among many others, can sabotage a promising effort if it is not properly managed. In this text, you'll also learn from the mistakes made on many projects that were not successful. I have written this book in an effort to educate you, tomorrow's project managers, about what will help make a project succeed—and what can make it fail. You'll also see how projects are used in everyday media, such as television and film, and how companies use best practices in project management. Many readers tell me how much they enjoy reading these real-world examples in the What Went Right?, What Went Wrong?, Media Snapshot, and Best Practice features. As practitioners know, there is no "one size fits all" solution to managing projects. By seeing how different organizations successfully implement project management, you can help your organization do the same.

Although project management has been an established field for many years, managing information technology projects requires ideas and information that go beyond standard project management. For example, many information technology projects fail because of a lack of user input, incomplete and changing requirements, and a lack of executive support. This book includes suggestions on dealing with these issues. New technologies can also aid in managing information technology projects, and examples of using software to assist in project management are included throughout the book.

*Information Technology Project Management, REVISED Sixth Edition,* is still the only textbook to apply all nine project management knowledge areas—project integration, scope, time, cost, quality, human resource, communications, risk, and procurement management—and all five process groups—initiating, planning, executing, monitoring and controlling, and closing—to information technology projects. This text builds on the *PMBOK® Guide, Fourth Edition,* an American National Standard, to provide a solid framework and context for managing information technology projects. It also includes an appendix, *Guide to Using Microsoft Project 2010,* which many readers find invaluable. A second appendix provides advice on earning and maintaining Project Management Professional (PMP) certification from the Project Management Institute (PMI) as well as information on other certification programs, such as CompTIA's Project+ certification. A third appendix provides new case studies and information on using simulation software to help readers apply their project management skills.

*Information Technology Project Management, REVISED Sixth Edition,* provides practical lessons in project management for students and practitioners alike. By weaving together theory and practice, this text presents an understandable, integrated view of the many concepts, skills, tools, and techniques involved in information technology project management. The comprehensive design of the text provides a strong foundation for students and practitioners in project management.

## New to the REVISED Sixth Edition

Building on the success of the previous editions, *Information Technology Project Management, REVISED Sixth Edition,* introduces a uniquely effective combination of features. The main changes made to the REVISED Sixth Edition only involve Appendix A. We know that faculty cannot update texts every single year, so this revision only provides you the option of teaching your students with the latest edition of Microsoft Project, Project 2010. The Beta release has been out for several months, and the final product should be available in summer 2010.

Appendix A has been thoroughly updated based on Microsoft Project 2010. There are many updates in Project 2010. In addition to adopting the Ribbon interface, Project 2010 provides a manual scheduling option, a simple Timeline feature, and a Team Planner view to easily assign people to tasks and reduce overallocations.

The main changes between the Sixth Edition and the Fifth Edition include the following:

- Several changes were made to synchronize the Sixth Edition with the *PMBOK® Guide, Fourth Edition,* which PMI published in December 2008. Several processes have changed, a few have been deleted, and a few have been added. For example, project scope management now includes a process for collecting requirements, which produces requirements documentation, a requirements management plan, and a requirements traceability matrix as outputs. This text describes this and other new processes and provides more details and examples of their outputs.

- Appendix C, Additional Running Cases, provides two new cases and information about using Fissure's simulation software. One of the new cases focuses on green computing projects, and the other involves finding or creating video clips related to project management. There is also a running case at the end of each knowledge area chapter, and the old cases from the Fifth Edition text are available on the new companion (premium) Web site. Several additional exercises are also provided at the end of chapters.
- A new Jeopardy-like game is provided on the companion (premium) Web site to help students study important concepts from each chapter in a fun and engaging way.
- A new companion (premium) Web site for the Sixth Edition (*www.cengage.com/mis/schwalbe*) provides you with access to informative links from the end notes, lecture notes, interactive quizzes, templates, additional running cases, suggested readings, podcasts, the new Jeopardy-like game, and many other items to enhance your learning.

### ACCESSING THE COMPANION (PREMIUM) WEB SITE

To access the companion (premium) Web site, open a Web browser and go to *www.cengage.com/login*. Locate your companion (premium) access card in the front of each new book purchase, and click "Create My Account" to begin the registration process. If you've purchased a used book, please search for ***Information Technology Project Management, Sixth Edition*** at *www.CengageBrain.com* where you can purchase instant access.

- Updated examples are provided throughout the text. You'll notice several new examples in the Sixth Edition that explain recent events in managing real information technology projects. Several of the What Went Right?, What Went Wrong?, Media Snapshot, and Best Practice examples have been updated to keep you up-to-date. Additional examples and results of new studies are also included throughout the text, with appropriate citations.
- User feedback is incorporated. Based on feedback from reviewers, students, instructors, practitioners, and translators (this book has been translated into Chinese, Japanese, Russian, and Czech), you'll see several additional changes to help clarify information.

## Approach

Many people have been practicing some form of project management with little or no formal study in this area. New books and articles are being written each year as we discover more about the field of project management, and project management software continues to advance. Because the project management field and the technology industry change rapidly, you cannot assume that what worked even a few years ago is still the best approach today. This text provides up-to-date information on how good project management and effective use of software can help you manage projects, especially information technology projects. Five distinct features of this text include its relationship to the Project Management Body of Knowledge, its detailed guide for using Microsoft Project 2010, its value in preparing for Project Management Professional and other certification exams, its inclusion of running case studies and online templates, and its companion (premium) Web site. You can also purchase a special bundling of this text that includes simulation software by Fissure, or you can order the Fissure simulation separately.

### Based on the *PMBOK® Guide, Fourth Edition*

The Project Management Institute (PMI) created the Guide to the Project Management Body of Knowledge (the *PMBOK® Guide*) as a framework and starting point for understanding project management. It includes an introduction to project management, brief descriptions of all nine project management knowledge areas, and a glossary of terms. The *PMBOK® Guide* is, however, just that—a guide. This text uses the *PMBOK® Guide, Fourth Edition-like* (December 2008) *as a foundation, but goes beyond it by providing more details, highlighting additional topics, and providing a real-world context for project management.* Information Technology Project Management, Sixth Edition, *explains project management specifically as it applies to managing information technology projects in the twenty-first century. It includes several unique features to bring you the excitement of this dynamic field (for more information on features, see the section entitled "Pedagogical Features").*

### Contains a Detailed Guide on How to Use Microsoft Project 2010

Software has advanced tremendously in recent years, and it is important for project managers and their teams to use software to help manage information technology projects. Each copy of *Information Technology Project Management, REVISED Sixth Edition,* includes a detailed guide in Appendix A on using the leading project management software on the market—Microsoft Project 2010. Examples using Project and other software tools are integrated throughout the text, not as an afterthought. Appendix A, Guide to Using Microsoft Project 2010, teaches you in a systematic way to use this powerful software to help in project scope, time, cost, human resource, and communications management.

### Resource for PMP and Other Certification Exams

Professional certification is an important factor in recognizing and ensuring quality in a profession. PMI provides certification as a Project Management Professional (PMP), and this text is an excellent resource for studying for the certification exam. This text will also help you pass other certification exams, such as CompTIA's Project+ exam. Having experience working on projects does not mean you can easily pass the PMP or other certification exams.

I like to tell my students a story about taking a driver's license test after moving to Minnesota. I had been driving very safely and without accidents for over 16 years, so I thought I could just walk in and take the test. I was impressed by the sophisticated computer system used to administer the test. The questions were displayed on a large touch-screen monitor, often along with an image or video to illustrate different traffic signs or driving situations. I became concerned when I found I had no idea how to answer several questions, and I was perplexed when the test seemed to stop and a message displayed saying, "Please see the person at the service counter." This was a polite way of saying I had failed the test! After controlling my embarrassment, I picked up one of the Minnesota driving test brochures, studied it for an hour or two that night, and successfully passed the test the next day.

The point of this story is that it is important to study information from the organization that creates the test and not be overconfident that your experience is enough. Because this text is based on PMI's *PMBOK® Guide, Fourth Edition,* it provides a valuable reference for studying for PMP certification. It is also an excellent reference for CompTIA's Project+ exam. I have earned both of those certifications and kept them in mind when writing this text.

### Provides Exercises, Running Cases, Templates, Sample Documents, and Optional Simulation Software

Based on feedback from readers, the Sixth Edition continues to provide challenging exercises and running cases to help students apply concepts in each chapter. There are over 50 templates, examples of real project documents, and optional simulation software developed by Fissure, a PMI Registered Education Provider, that you can use to actively practice your skills in managing a project. All of these features help the subject matter come alive and have more meaning.

### Includes a Companion (Premium) Web site

A companion (premium) Web site provides you with a one-stop location to access informative links and tools to enhance your learning. Similar to other companion (premium) Web sites provided by Course Technology, this site will be a valuable resource as you view lecture notes, templates, interactive quizzes, podcasts, student files for Project 2010, important articles, references, and more. You can also link to the author's site to see real class syllabi, samples of student projects, and other helpful links.

## Organization and Content

*Information Technology Project Management, REVISED Sixth Edition,* is organized into three main sections to provide a framework for project management, a detailed description of each project management knowledge area, and three appendices to provide practical information for applying project management. The first three chapters form the first section, which introduces the project management framework and sets the stage for the remaining chapters.

Chapters 4 through 12 form the second section of the text, which describes each of the project management knowledge areas—project integration, scope, time, cost, quality, human resource, communications, risk, and procurement management—in the context of information technology projects. An entire chapter is dedicated to each knowledge area. Each knowledge area chapter includes sections that map to their major processes as described in the *PMBOK® Guide, Fourth Edition.* For example, the chapter on project quality management includes sections on planning quality, performing quality assurance, and performing quality control. Additional sections highlight other important concepts related to each knowledge area, such as Six Sigma, testing, maturity models, and using software to assist in project quality management. Each chapter also includes detailed examples of key project management tools and techniques as applied to information technology projects. For example, the chapter on project integration management includes samples of various project-selection documents, such as net present value analyses, ROI calculations, payback analyses, and weighted scoring models. The project scope management chapter includes a sample project charter, a project scope statement, and several work breakdown structures for information technology projects.

Appendices A through C form the third section of the text, which provides practical information to help you apply project management skills on real or practice projects. By following the detailed, step-by-step guide in Appendix A, which includes more than 60 screen shots, you will learn how to use Project 2010. Appendix B summarizes what you need to know to earn PMP or other certifications related to project management. Appendix C provides additional running cases and information on using simulation software to help you practice your new skills.

## Pedagogical Features

Several pedagogical features are included in this text to enhance presentation of the materials so that you can more easily understand the concepts and apply them. Throughout the text, emphasis is placed on applying concepts to current, real-world information technology project management.

### Learning Objectives, Chapter Summaries, Discussion Questions, Exercises, Quick Quizzes, Running Cases, and Companion (Premium) Web site

Learning Objectives, Chapter Summaries, Quick Quizzes, Discussion Questions, Exercises, Running Cases, and the companion (premium) Web site are designed to function as integrated study tools. Learning Objectives reflect what you should be able to accomplish after completing each chapter. Chapter Summaries highlight key concepts you should master. The Discussion Questions help guide critical thinking about those key concepts. Quick Quizzes test knowledge of essential chapter concepts and include an answer key. Exercises provide opportunities to practice important techniques, as do the Running Cases. The companion (premium) Web site provides several study aids, such as podcasts, the new Jeopardy-like game, and interactive quizzes for each chapter, which are different from the Quick Quizzes in the text.

### Opening Case and Case Wrap-Up

To set the stage, each chapter begins with an opening case related to the material presented in that chapter. These "real-life" case scenarios (most based on the author's experiences) spark student interest and introduce important concepts in a real-world context. As project management concepts and techniques are discussed, they are applied to the opening case and other similar scenarios. Each chapter then closes with a case wrap-up—with some ending successfully and some, realistically, failing—to further illustrate the real world of project management.

### What Went Right? and What Went Wrong?

Failures, as much as successes, can be valuable learning experiences. Each chapter of the text includes one or more examples of real information technology projects that went right as well as examples of projects that went wrong. These examples further illustrate the importance of mastering key concepts in each chapter.

### Media Snapshot

The world is full of projects. Several televisions shows, movies, newspapers, Web sites, and other media highlight project results, good and bad. Relating project management concepts to all types of projects highlighted in the media will help you understand and see the importance of this growing field. Why not get people excited about studying project management by showing them how to recognize project management concepts in popular television shows, movies, or other media?

### Best Practice

Every chapter includes an example of a best practice related to topics in that chapter. For example, Chapter 1 describes best practices written by Robert Butrick, author of *The Project Workout,* from the *Ultimate Business Library's Best Practice* book. He suggests that organizations ensure their projects are driven by their strategy and engage project stakeholders.

### Key Terms

The fields of information technology and project management both include many unique terms that are vital to creating a workable language when the two fields are combined. Key terms are displayed in bold face and are defined the first time they appear. Definitions of key terms are provided in alphabetical order at the end of each chapter and in a glossary at the end of the text.

### Application Software

Learning becomes much more dynamic with hands-on practice using the top project management software tool in the industry, Microsoft Project 2010, as well as other tools, such as spreadsheet software and the Internet. Each chapter offers you many opportunities to get hands-on experience and build new software skills. This text is written from the point of view that reading about something only gets you so far—to really understand project management, you have to do it for yourself. In addition to the exercises and running cases found at the end of each chapter and in Appendix C, several challenging exercises are provided at the end of Appendix A, Guide to Using Microsoft Project 2010.

## SUPPLEMENTS

The following supplemental materials are available when this text is used in a classroom setting. All of the teaching tools available with this text are provided to the instructor on a single CD-ROM.

- **Electronic Instructor's Manual** The Instructor's Manual that accompanies this textbook includes additional instructional material to assist in class preparation, including suggestions for lecture topics and additional discussion questions.
- **ExamView®** This textbook is accompanied by ExamView, a powerful testing software package that allows instructors to create and administer printed,

computer (LAN-based), and Internet exams. ExamView includes hundreds of questions that correspond to the topics covered in this text, enabling students to generate detailed study guides that include page references for further review. The computer-based and Internet testing components allow students to take exams at their computers, and also save the instructor time by grading each exam automatically.

- **PowerPoint Presentations** This text comes with Microsoft PowerPoint slides for each chapter. These are included as a teaching aid for classroom presentation, to make available to students on the network for chapter review, or to be printed for classroom distribution. Instructors can add their own slides for additional topics they introduce to the class.
- **Solution Files** Solutions to end-of-chapter questions can be found on the Instructor Resource CD-ROM and may also be found on the Course Technology Web site at *www.cengage.com/mis/schwalbe*. The solutions are password-protected.
- **Distance Learning** Course Technology is proud to present online courses in WebCT and Blackboard, to provide the most complete and dynamic learning experience possible. When you add online content to one of your courses, you're adding a lot: self tests, links, glossaries, and, most of all, a gateway to the twenty-first century's most important information resource. We hope you will make the most of your course, both online and offline. For more information on how to bring distance learning to your course, contact your Course Technology sales representative.

## ACKNOWLEDGMENTS

I never would have taken on this project—writing this book, the first, second, third, fourth, fifth, and sixth edition—without the help of many people. I thank the staff at Course Technology for their dedication and hard work in helping me produce this book and in doing such an excellent job of marketing it. Kate Mason (formerly Hennessy), Deb Kaufmann, Matthew Hutchinson, Patrick Franzen, and many more people did a great job in planning and executing all of the work involved in producing this book.

I thank my many colleagues and experts in the field who contributed information to this book. David Jones, Rachel Hollstadt, Cliff Sprague, Michael Branch, Barb Most, Jodi Curtis, Rita Mulcahy, Karen Boucher, Bill Munroe, Tess Galati, Joan Knutson, Neal Whitten, Brenda Taylor, Quentin Fleming, Jesse Freese, Nick Matteucci, Nick Erndt, Dragan Milosevic, Bob Borlink, Arvid Lee, Kathy Christenson, Peeter Kivestu, and many other people who provided excellent materials included in the Sixth Edition of this book. I really enjoy the network of project managers, authors, and consultants in this field who are very passionate about improving the theory and practice of project management.

I also thank my students and colleagues at Augsburg College and the University of Minnesota for providing feedback on the earlier editions of this book. I received many valuable comments from them on ways to improve the text and structure of my courses. I learn something new about project management and teaching all the time by interacting with students, faculty, and staff.

I also thank faculty reviewers for providing excellent feedback for me in writing this edition: Brian Ameling, Limestone College; Michel Avital, University of Amsterdam; Al Fundaburk, Bloomsburg University; Suleyman Guleyupoglu, University of Phoenix; Aurore Kamssu, Tennessee State University; Angela Lemons, North Carolina A&T State University; Alan L. Matthews, Travecca Nazarene University; Thyra Nelson, Macon State College; Samir Shah, Pennsylvania State University—York; and Andrew Urbaczewski, University of Michigan-Dearborn. I also wish to thank the many reviewers of the earlier editions of this text. I also thank the many other instructors and readers who have contacted me directly with praise as well as suggestions for improving this text. I really appreciate the feedback and do my best to incorporate as much as I can.

Most of all, I am grateful to my family. Without their support, I never could have written this book. My wonderful husband, Dan, has always supported me in my career, and he helps me keep up-to-date with software development since he is a lead architect for ComSquared Systems, Inc. Our three children, Anne, Bobby, and Scott, actually think it's cool that their mom writes books and speaks at conferences. They also see me managing projects all the time. Anne, now 25, teases me for being the only quilter she knows who treats each quilt as a project. (Maybe that's why I get so many done!) Our children all understand the main reason why I write—I have a passion for educating future leaders of the world, including them.

As always, I am eager to receive your feedback on this book. Please send comments to me at schwalbe@augsburg.edu.

Kathy Schwalbe, Ph.D., PMP
Professor, Department of Business Administration
Augsburg College

## ABOUT THE AUTHOR

Kathy Schwalbe is a Professor in the Department of Business Administration at Augsburg College in Minneapolis, where she teaches courses in project management, problem solving for business, systems analysis and design, information systems projects, and electronic commerce. Kathy was also an adjunct faculty member at the University of Minnesota, where she taught a graduate-level course in project management in the engineering department. She also provides training and consulting services to several organizations and speaks at several conferences. Kathy worked for ten years in industry before entering academia in 1991. She was an Air Force officer, systems analyst, project manager, senior engineer, and information technology consultant. Kathy is an active member of PMI, having served as the Student Chapter Liaison for the Minnesota chapter of PMI, VP of Education for the Minnesota chapter, Director of Communications and Editor of the Information Systems Specific Interest Group (ISSIG) Review, and member of PMI's test-writing team. Kathy earned her Ph.D. in Higher Education at the University of Minnesota, her MBA at Northeastern University's High Technology MBA program, and her B.S. in mathematics at the University of Notre Dame.

# INFORMATION TECHNOLOGY PROJECT MANAGEMENT

CHAPTER 1

# INTRODUCTION TO PROJECT MANAGEMENT

## LEARNING OBJECTIVES

**After reading this chapter, you will be able to:**

- Understand the growing need for better project management, especially for information technology projects
- Explain what a project is, provide examples of information technology projects, list various attributes of projects, and describe the triple constraint of project management
- Describe project management and discuss key elements of the project management framework, including project stakeholders, the project management knowledge areas, common tools and techniques, and project success
- Discuss the relationship between project, program, and portfolio management and the contributions they each make to enterprise success
- Understand the role of the project manager by describing what project managers do, what skills they need, and what the career field is like for information technology project managers
- Describe the project management profession, including its history, the role of professional organizations like the Project Management Institute (PMI), the importance of certification and ethics, and the advancement of project management software

## OPENING CASE

Anne Roberts, the Director of the Project Management Office for a large retail chain, stood in front of 500 people in the large corporate auditorium to explain the company's new strategies. She was also broadcasting to thousands of other employees, suppliers, and stockholders throughout the world using live video via the Internet. The company had come a long way in implementing new information systems to improve inventory control, sell products using the Web, streamline the sales and distribution processes, and improve customer service. However, the stock price was down, the nation's economy was weak, and people were anxious to hear about the company's new strategies.

Anne began to address the audience, "Good morning. As many of you know, our CEO promoted me to this position as Director of the Project Management Office two years ago. Since then, we have completed many projects, including the advanced data networks project. That project enabled us to provide persistent broadband between headquarters and our retail stores throughout the world, allowing us to make timely decisions and continue our growth strategy. Our customers love that they can return items to any store, and any sales clerk can look up past sales information. Local store managers can make timely decisions using up-to-date information. Of course, we've had some project failures, too, and we need to continually assess our portfolio of projects to meet business needs. Two big IT initiatives this coming year include meeting new green IT regulations and providing enhanced online collaboration tools for our employees, suppliers, and customers. Our challenge is to work even smarter to decide what projects will most benefit the company, how we can continue to leverage the power of information technology to support our business, and how we can exploit our human capital to successfully plan and execute those projects. If we succeed, we'll continue to be a world-class corporation."

"And if we fail?" someone asked from the audience.

"Let's just say that failure is not an option," Anne replied.

## INTRODUCTION

Many people and organizations today have a new—or renewed—interest in project management. Until the 1980s, project management primarily focused on providing schedule and resource data to top management in the military, computer, and construction industries. Today's project management involves much more, and people in every industry and every country manage projects. New technologies have become a significant factor in many businesses. Computer hardware, software, networks, and the use of interdisciplinary and global work teams have radically changed the work environment. The following statistics demonstrate the significance of project management in today's society, especially for projects involving information technology (IT). Note that IT projects involve using hardware, software, and/or networks to create a product, service, or result.

- Total global spending on technology goods, services, and staff was projected to reach $2.4 trillion in 2008, an 8 percent increase from 2007. IT purchases in the U.S. grew less than 3 percent, while the rest of the Americas expanded in local currencies at 6-percent rates. Asia Pacific and the oil-exporting areas of Eastern Europe, the Middle East, and Africa were the main engines of growth.[1]

- In the U.S. the size of the IT workforce topped 4 million workers for the first time in 2008. Unemployment rates in many information technology occupations were among the lowest in the labor force at only 2.3 percent. Demand for talent is high, and several organizations throughout the world cannot grow as desired due to difficulties in hiring and recruiting the people they need.[2]
- In 2007 the total compensation for the average senior project manager in U.S. dollars was $104,776 per year in the United States, $111,412 in Australia, and $120,364 in the United Kingdom. The average total compensation of a program manager was $122,825 in the United States, $133,718 in Australia, and $165,489 in the United Kingdom. The average total compensation for a Project Management Office (PMO) Director was $134,422 in the United States, $125,197 in Australia, and $210,392 in the United Kingdom. This survey was based on self-reported data from more than 5,500 practitioners in 19 countries.[3]
- The number of people earning their Project Management Professional (PMP) certification continues to increase each year.
- A research report showed that the U.S. spends $2.3 trillion on projects every year, an amount equal to 25 percent of the nation's gross domestic product. The world as a whole spends nearly $10 trillion of its $40.7 trillion gross product on projects of all kinds. More than 16 million people regard project management as their profession.[4]

Today's companies, governments, and nonprofit organizations are recognizing that to be successful, they need to be conversant with and use modern project management techniques. Individuals are realizing that to remain competitive in the workplace, they must develop skills to become good project team members and project managers. They also realize that many of the concepts of project management will help them in their everyday lives as they work with people and technology on a day-to-day basis.

## WHAT WENT WRONG?

In 1995, the Standish Group published an often-quoted study entitled "The CHAOS Report." This consulting firm surveyed 365 information technology executive managers in the United States who managed more than 8,380 information technology application projects. As the title of the study suggests, the projects were in a state of chaos. U.S. companies spent more than $250 billion each year in the early 1990s on approximately 175,000 information technology application development projects. Examples of these projects included creating a new database for a state department of motor vehicles, developing a new system for car rental and hotel reservations, and implementing a client-server architecture for the banking industry. The study reported that the overall success rate of information technology projects was *only* 16.2 percent. The surveyors defined success as meeting project goals on time and on budget. The study also found that more than 31 percent of information technology projects were canceled before completion, costing U.S. companies and

*continued*

government agencies more than $81 billion. The study authors were adamant about the need for better project management in the information technology industry. They explained, "Software development projects are in chaos, and we can no longer imitate the three monkeys—hear no failures, see no failures, speak no failures."[5]

In a more recent study, PricewaterhouseCoopers surveyed 200 companies from 30 different countries about their project management maturity and found that *over half of all projects fail.* They also found that only 2.5 percent of corporations consistently meet their targets for scope, time, and cost goals for all types of project.[6]

Although several researchers question the methodology of such studies, their popularity has prompted managers throughout the world to examine their practices in managing projects. Many organizations assert that using project management provides advantages, such as:

- Better control of financial, physical, and human resources
- Improved customer relations
- Shorter development times
- Lower costs and improved productivity
- Higher quality and increased reliability
- Higher profit margins
- Better internal coordination
- Positive impact on meeting strategic goals
- Higher worker morale

This chapter introduces projects and project management, explains how projects fit into programs and portfolio management, discusses the role of the project manager, and provides important background information on this growing profession. Although project management applies to many different industries and types of projects, this text focuses on applying project management to information technology projects.

## WHAT IS A PROJECT?

To discuss project management, it is important to understand the concept of a project. A **project** is "a temporary endeavor undertaken to create a unique product, service, or result."[7] Operations, on the other hand, is work done in organizations to sustain the business. Projects are different from operations in that they end when their objectives have been reached or the project has been terminated.

### Examples of Information Technology Projects

Projects can be large or small and involve one person or thousands of people. They can be done in one day or take years to complete. As described earlier, information technology projects involve using hardware, software, and/or networks to create a product, service, or result. Examples of information technology projects include the following:

- A technician replaces ten laptops for a small department
- A small software development team adds a new feature to an internal software application for the finance department

- A college campus upgrades its technology infrastructure to provide wireless Internet access across the whole campus
- A cross-functional taskforce in a company decides what Voice-over-Internet-Protocol (VoIP) system to purchase and how it will be implemented
- A company develops a new system to increase sales force productivity and customer relationship management
- A television network implements a system to allow viewers to vote for contestants and provide other feedback on programs
- The automobile industry develops a Web site to streamline procurement
- A government group develops a system to track child immunizations
- A large group of volunteers from organizations throughout the world develops standards for environmentally friendly or green IT

Gartner, Inc., a prestigious consulting firm, identified the top ten strategic technologies for 2008. A few of these technologies include the following:

- *Green IT:* Simply defined, **green IT** or **green computing** involves developing and using computer resources in an efficient way to improve economic viability, social responsibility, and environmental impact. For example, government regulations now encourage organizations and IT departments to use low-emission building materials, recycle computing equipment, and use alternative energy and other green technologies.
- *Unified communications:* The majority of organizations are expected to migrate from PBX (private branch exchange) to IP (Internet protocol) telephony in the next three years.
- *Business process modeling:* Enterprise and process architects, senior developers, and business process analysts must work together to help organizations effectively use IT to improve processes. Business process modeling (BPM) suites are expected to fill a critical role as a compliment to service-oriented architecture (SOA).
- *Virtualization 2.0:* **Virtualization** hides the physical characteristics of computing resources from their users, such as making a single server, operating system, application, or storage device appear to function as multiple virtual resources. Virtualization technologies can improve IT resource management and increase flexibility for adapting to changing requirements and workloads. Virtualization 2.0 adds automation technologies so that resource efficiency can improve dramatically.
- *Social software:* Most students and professionals today use online social networking sites such as MySpace, Facebook, LinkedIn, and YouTube to collaborate with others. Organizations will increasingly use social software technologies to augment traditional collaboration.[8]

As you can see, a wide variety of projects use information technologies, and organizations rely on them for their success.

## MEDIA SNAPSHOT

Nicholas Carr published his exposé "IT Doesn't Matter" in the May 2003 issue of *Harvard Business Review,* a topic he expanded on in the following year with his book *Does IT Matter? Information Technology and the Corrosion of Competitive Advantage.*[9] Both sparked heated debates on the value of information technology in today's society. Carr suggested that information technology has followed a pattern similar to earlier infrastructure technologies like railroads and electric power. As availability increased and costs decreased, information technology has become a commodity; therefore, Carr argued, it can no longer provide companies with a competitive advantage. In 2006, *Baseline* magazine published the article, "Where I.T. Matters: How 10 Technologies Transformed 10 Industries" as a retort to Carr's ideas. Below are a few of the technologies and industries that have made IT an important part of their business strategy. (Visit *www.nicholasgcarr.com* to see more recent work by Carr, including a free 2008 eBook, *IT in 2018: From Turing's Machine to the Computing Cloud.*)

- *VoIP:* VoIP has totally transformed the telecommunications industry and broadband Internet access. Phone companies do not have a lock on dial tones anymore; you can make a phone call through a cable TV provider or over any Internet channel for less than the cost of ordinary phone service. These technologies, along with regulatory changes, have forced major phone companies to be more competitive to keep and attract customers. VoIP is more efficient and less expensive than traditional phone networks. The ramp-up to VoIP is expected to happen quickly. Research firm IDC estimates that U.S. subscribers to residential VoIP services will grow from 3 million in 2005 to 27 million by the end of 2009.
- *Global Positioning Systems (GPS) and Business Intelligence:* "Farming is the oldest known human activity," says Michael Swanson, an agricultural economist at Wells Fargo bank, the largest lender to U.S. farmers. "You'd think that after 10,000 years there'd be nothing left to improve. Not true." How have GPS and Business Intelligence changed the farming industry? In 1950, American farmers planted 83 million acres of corn, which produced 38 bushels per acre. In 2004, farmers planted 81 million acres of corn, which produced 160 bushels. That means that 2.5 percent fewer acres produced more than four times as much corn. Swanson estimates that if farmers did not use the technology they do today, they would have had to plant 320 million acres of corn last year to meet demand: "We'd be planting parking lots and backyards," Swanson joked.
- *Digital Supply Chain:* The entertainment industry's distribution system has changed dramatically due to new information technologies. "The great promise of digital technology is that consumers will be able to choose how they want to consume content," says Kevin Tsujihara, president of Warner Home Entertainment Group, a new department formed to handle the digital delivery of entertainment to consumers. Before Warner underwent a major digital transformation, they were only able to process one or two pictures at a time. "Today, we have the capability of taking upward of 10 simultaneous motion picture projects and working on them in this environment. The creation of these digital masters obviously is important in that we can make a transformation to whatever channel we need to get to the consumer."[10]

## Project Attributes

As you can see, projects come in all shapes and sizes. The following attributes help to define a project further:

- *A project has a unique purpose.* Every project should have a well-defined objective. For example, Anne Roberts, the Director of the Project Management Office in the opening case, might sponsor an information technology collaboration project to develop a list and initial analysis of potential information technology projects that might improve operations for the company. The unique purpose of this project would be to create a collaborative report with ideas from people throughout the company. The results would provide the basis for further discussions and projects. As in this example, projects result in a unique product, service, or result.
- *A project is temporary.* A project has a definite beginning and a definite end. In the information technology collaboration project, Anne might form a team of people to work immediately on the project, and then expect a report and an executive presentation of the results in one month.
- *A project is developed using progressive elaboration.* Projects are often defined broadly when they begin, and as time passes, the specific details of the project become clearer. Therefore, projects should be developed in increments. A project team should develop initial plans and then update them with more detail based on new information. For example, suppose a few people submitted ideas for the information technology collaboration project, but they did not clearly address how the ideas would support the business strategy of improving operations. The project team might decide to prepare a questionnaire for people to fill in as they submit their ideas to improve the quality of the inputs.
- *A project requires resources, often from various areas.* Resources include people, hardware, software, and other assets. Many projects cross departmental or other boundaries to achieve their unique purposes. For the information technology collaboration project, people from information technology, marketing, sales, distribution, and other areas of the company would need to work together to develop ideas. The company might also hire outside consultants to provide input. Once the project team has selected key projects for implementation, they will probably require additional resources. And to meet new project objectives, people from other companies—product suppliers and consulting companies—may be added. Resources, however, are limited and must be used effectively to meet project and other corporate goals.
- *A project should have a primary customer or sponsor.* Most projects have many interested parties or stakeholders, but someone must take the primary role of sponsorship. The **project sponsor** usually provides the direction and funding for the project. In this case, Anne Roberts would be the sponsor for the information technology collaboration project. Once further information technology projects are selected, however, the sponsors for those projects would be senior managers in charge of the main parts of the company affected by the projects. For example, if the vice president of sales initiates a project to

improve direct product sales using the Internet, he or she might be the project sponsor.

- *A project involves uncertainty.* Because every project is unique, it is sometimes difficult to define its objectives clearly, estimate how long it will take to complete, or determine how much it will cost. External factors also cause uncertainty, such as a supplier going out of business or a project team member needing unplanned time off. This uncertainty is one of the main reasons project management is so challenging, especially on projects involving new technologies.

An effective **project manager** is crucial to a project's success. Project managers work with the project sponsors, the project team, and the other people involved in a project to meet project goals.

## The Triple Constraint

Every project is constrained in different ways by its scope, time, and cost goals. These limitations are sometimes referred to in project management as the **triple constraint**. To create a successful project, a project manager must consider scope, time, and cost and balance these three often-competing goals. He or she must consider the following:

- *Scope:* What work will be done as part of the project? What unique product, service, or result does the customer or sponsor expect from the project? How will the scope be verified?
- *Time:* How long should it take to complete the project? What is the project's schedule? How will the team track actual schedule performance? Who can approve changes to the schedule?
- *Cost:* What should it cost to complete the project? What is the project's budget? How will costs be tracked? Who can authorize changes to the budget?

Figure 1-1 illustrates the three dimensions of the triple constraint. Each area—scope, time, and cost—has a target at the beginning of the project. For example, the information technology collaboration project might have an initial scope of producing a 40- to 50-page report and a one-hour presentation on about 30 potential information technology projects. The project manager might further define project scope to include providing a description of each potential project, an investigation of what other companies have implemented for similar projects, a rough time and cost estimate, and assessments of the risk and potential payoff as high, medium, or low. The initial time estimate for this project might be one month, and the cost estimate might be $45,000–$50,000. These expectations provide the targets for the scope, time, and cost dimensions of the project. Note that the scope and cost goals in this example include ranges—the report can be between 40- to 50-pages long and the project can cost between $45,000 and $50,000. Because projects involve uncertainty and limited resources, projects rarely finish according to discrete scope, time, and cost goals originally planned. Instead of discrete target goals, it is often more realistic to set a range of goals such as spending between $45,000 and $50,000 and having the length of the report between 40 and 50 pages. These goals might mean hitting the target, but not the bull's eye.

Managing the triple constraint involves making trade-offs between scope, time, and cost goals for a project. For example, you might need to increase the budget for a project to meet

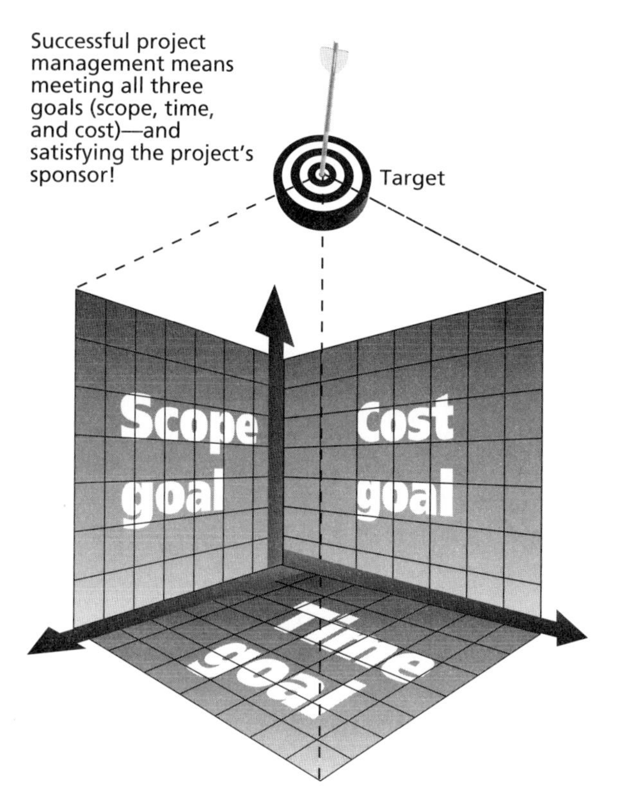


**FIGURE 1-1** The triple constraint of project management

scope and time goals. Alternatively, you might have to reduce the scope of a project to meet time and cost goals. Experienced project managers know that you must decide which aspect of the triple constraint is most important. If time is most important, you must often change the initial scope and/or cost goals to meet the schedule. If scope goals are most important, you may need to adjust time and/or cost goals.

For example, to generate project ideas, suppose the project manager for the information technology collaboration project sent an e-mail survey to all employees, as planned. The initial time and cost estimate may have been one week and $5,000 to collect ideas based on this e-mail survey. Now, suppose the e-mail survey generated only a few good project ideas, and the scope goal was to collect at least 30 good ideas. Should the project team use a different method like focus groups or interviews to collect ideas? Even though it was not in the initial scope, time, or cost estimates, it would really help the project. Since good ideas are crucial to project success, it would make sense to inform the project sponsor that you want to make adjustments.

Although the triple constraint describes how the basic elements of a project—scope, time, and cost—interrelate, other elements can also play significant roles. Quality is often a key factor in projects, as is customer or sponsor satisfaction. Some people, in fact, refer to the *quadruple constraint* of project management, which includes quality as well as scope, time, and cost. Others believe that quality considerations, including customer satisfaction, must be inherent in setting the scope, time, and cost goals of a project. A project team may meet scope, time, and cost goals but fail to meet quality standards or satisfy their sponsor, if they have not adequately addressed these concerns. For example, Anne Roberts may receive a 50-page report describing 30 potential information technology projects and hear a presentation on the findings of the report. The project team may have completed the work on time and within the cost constraint, but the quality may have been unacceptable. Anne's view of an executive presentation may be very different from the project team's view. The project manager should be communicating with the sponsor throughout the project to make sure the project meets his or her expectations.

How can you avoid the problems that occur when you meet scope, time, and cost goals, but lose sight of quality or customer satisfaction? The answer is *good project management, which includes more than meeting the triple constraint.*

## WHAT IS PROJECT MANAGEMENT?

**Project management** is "the application of knowledge, skills, tools and techniques to project activities to meet project requirements."[11] Project managers must not only strive to meet specific scope, time, cost, and quality goals of projects, they must also facilitate the entire process to meet the needs and expectations of the people involved in or affected by project activities.

Figure 1-2 illustrates a framework to help you understand project management. Key elements of this framework include the project stakeholders, project management knowledge areas, project management tools and techniques, and the contribution of successful projects to the enterprise.

### Project Stakeholders

**Stakeholders** are the people involved in or affected by project activities and include the project sponsor, project team, support staff, customers, users, suppliers, and even

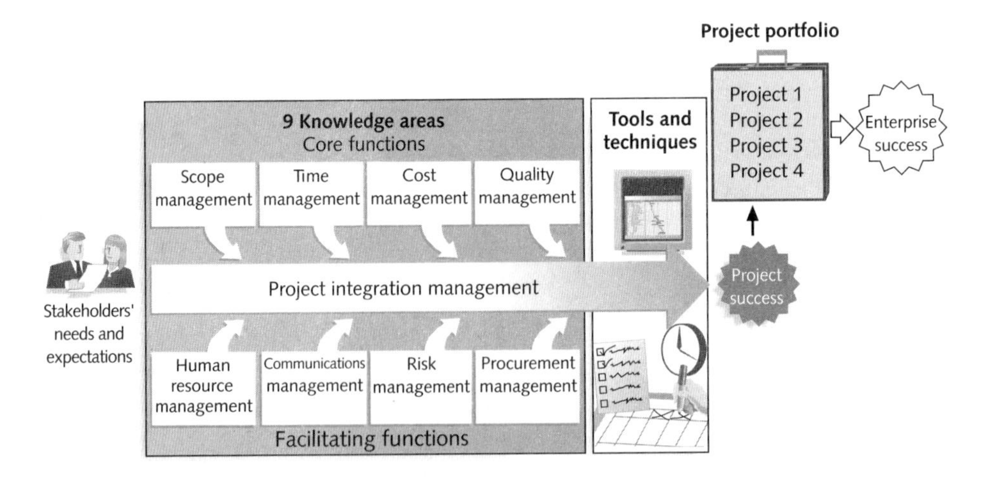


**FIGURE 1-2** Project management framework

opponents of the project. These stakeholders often have very different needs and expectations. For example, building a new house is a well-known example of a project. There are several stakeholders involved in a home construction project.

- The project sponsors would be the potential new homeowners. They would be the people paying for the house and could be on a very tight budget, so they would expect the contractor to provide accurate estimates of the costs involved in building the house. They would also need a realistic idea of when they could move in and what type of home they could afford given their budget constraints. The new homeowners would have to make important decisions to keep the costs of the house within their budget. Can they afford to finish the basement right away? If they can afford to finish the basement, will it affect the projected move-in date? In this example, the project sponsors are also the customers and users for the product, which is the house.
- The project manager in this example would normally be the general contractor responsible for building the house. He or she needs to work with all the project stakeholders to meet their needs and expectations.
- The project team for building the house would include several construction workers, electricians, carpenters, and so on. These stakeholders would need to know exactly what work they must do and when they need to do it. They would need to know if the required materials and equipment will be at the construction site or if they are expected to provide the materials and equipment. Their work would need to be coordinated since there are many interrelated factors involved. For example, the carpenter cannot put in kitchen cabinets until the walls are completed.
- Support staff might include the buyers' employers, the general contractor's administrative assistant, and other people who support other stakeholders. The buyers' employers might expect their employees to still complete their work but allow some flexibility so they can visit the building site or take phone calls related to building the house. The contractor's administrative assistant would support the project by coordinating meetings between the buyers, the contractor, suppliers, and so on.
- Building a house requires many suppliers. The suppliers would provide the wood, windows, flooring materials, appliances, and so on. Suppliers would expect exact details on what items they need to provide, where and when to deliver those items, and so on.
- There may or may not be opponents of a project. In this example, there might be a neighbor who opposes the project because the workers are making so much noise that she cannot concentrate on her work at home, or the noise might wake her sleeping children. She might interrupt the workers to voice her complaints or even file a formal complaint. Or, the neighborhood might have association rules concerning new home design and construction. If the homeowners did not follow these rules, they might have to halt construction due to legal issues.

As you can see from this example, there are many different stakeholders on projects, and they often have different interests. Stakeholders' needs and expectations are important in the beginning and throughout the life of a project. Successful project managers develop good relationships with project stakeholders to understand and meet their needs and expectations.

## Project Management Knowledge Areas

**Project management knowledge areas** describe the key competencies that project managers must develop. The center of Figure 1-2 shows the nine knowledge areas of project management. The four core knowledge areas of project management include project scope, time, cost, and quality management. These are core knowledge areas because they lead to specific project objectives.

- Project scope management involves defining and managing all the work required to complete the project successfully.
- Project time management includes estimating how long it will take to complete the work, developing an acceptable project schedule, and ensuring timely completion of the project.
- Project cost management consists of preparing and managing the budget for the project.
- Project quality management ensures that the project will satisfy the stated or implied needs for which it was undertaken.

The four facilitating knowledge areas of project management are human resource, communications, risk, and procurement management. These are called facilitating knowledge areas because they are the processes through which the project objectives are achieved.

- Project human resource management is concerned with making effective use of the people involved with the project.
- Project communications management involves generating, collecting, disseminating, and storing project information.
- Project risk management includes identifying, analyzing, and responding to risks related to the project.
- Project procurement management involves acquiring or procuring goods and services for a project from outside the performing organization.

Project integration management, the ninth knowledge area, is an overarching function that affects and is affected by all of the other knowledge areas. Project managers must have knowledge and skills in all nine of these areas. This text includes an entire chapter on each of these knowledge areas because all of them are crucial to project success.

## Project Management Tools and Techniques

Thomas Carlyle, a famous historian and author, stated, "Man is a tool-using animal. Without tools he is nothing, with tools he is all." As the world continues to become more complex, it is even more important for people to develop and use tools, especially for managing important projects. **Project management tools and techniques** assist project managers and their teams in carrying out work in all nine knowledge areas. For example, some popular time-management tools and techniques include Gantt charts, project network diagrams, and critical path analysis. Table 1-1 lists some commonly used tools and techniques by knowledge area. You will learn more about these and other tools and techniques throughout this text.

**TABLE 1-1** Common project management tools and techniques by knowledge area



| Knowledge area/category | Tools and techniques |
|---|---|
| Integration management | Project selection methods, project management methodologies, stakeholder analyses, project charters, project management plans, **project management software, change requests,** change control boards, project review meetings, **lessons-learned reports** |
| Scope management | **Scope statements, work breakdown structures,** statements of work, **requirements analyses,** scope management plans, scope verification techniques, and scope change controls |
| Time management | **Gantt charts,** project network diagrams, critical path analysis, crashing, fast tracking, schedule performance measurements |
| Cost management | Net present value, return on investment, payback analysis, earned value management, project portfolio management, cost estimates, cost management plans, cost baselines |
| Quality management | Quality metrics, checklists, quality control charts, Pareto diagrams, fishbone diagrams, maturity models, statistical methods |
| Human resource management | Motivation techniques, empathic listening, responsibility assignment matrices, project organizational charts, resource histograms, team building exercises |
| Communications management | Communications management plans, **kick-off meetings,** conflict management, communications media selection, status and **progress reports,** virtual communications, templates, project Web sites |
| Risk management | Risk management plans, risk registers, probability/impact matrices, risk rankings |
| Procurement management | Make-or-buy analyses, contracts, requests for proposals or quotes, source selections, supplier evaluation matrices |

A 2006 survey of 753 project and program managers was conducted to rate several project management tools. Respondents were asked to rate tools on a scale of 1–5 (low to high) based on the extent of their use and the potential of the tools to help improve project success. "Super tools" were defined as those that had high use and high potential for improving project success. These super tools included software for task scheduling (such as project management software), scope statements, requirement analyses, and lessons-learned reports. Tools that are already extensively used and have been found to improve project importance include progress reports, kick-off meetings, Gantt charts, and change requests. These super tools are bolded in Table 1-1.[12] Of course, different tools can be more effective in different situations. It is crucial for project managers and their team members to determine which tools will be most useful for their particular projects.

## WHAT WENT RIGHT?

Follow-up studies by the Standish Group (see the previously quoted CHAOS study in the What Went Wrong? passage) showed some improvement in the statistics for information technology projects in the past decade:

- The number of successful IT projects has more than doubled, from 16 percent in 1994 to 35 percent in 2006.
- The number of failed projects decreased from 31 percent in 1994 to 19 percent in 2006.
- The United States spent more money on IT projects in 2006 than 1994 ($346 billion and $250 billion, respectively), but the amount of money wasted on challenged projects (those that did not meet scope, time, or cost goals, but were completed) and failed projects was down to $53 billion in 2006 compared to $140 billion in 1994.[13]

The good news is that project managers are learning how to succeed more often; the bad news is that it is still very difficult to lead successful IT projects. "The reasons for the increase in successful projects vary. First, the average cost of a project has been more than cut in half. Better tools have been created to monitor and control progress and better skilled project managers with better management processes are being used. The fact that there are processes is significant in itself."[14]

Despite its advantages, project management is not a silver bullet that guarantees success on all projects. Project management is a very broad, often complex discipline. What works on one project may not work on another, so it is essential for project managers to continue to develop their knowledge and skills in managing projects. It is also important to learn from the mistakes and successes of others.

### Project Success

How do you define the success or failure of a project? There are several ways to define project success. The list that follows outlines a few common criteria for measuring the success of a project using the example of upgrading 500 desktop computers within three months for $300,000:

1. *The project met scope, time, and cost goals.* If all 500 computers were upgraded and met other scope requirements, the work was completed in three months or less, and the cost was $300,000 or less, you could consider it a

successful project based on this criterion. The Standish Group studies used this definition of success. Several people question this simple definition of project success and the methods used for collecting the data. (See the references by Glass on the companion Web site for this text to read more about this debate.)

2. *The project satisfied the customer/sponsor.* Even if the project met initial scope, time, and cost goals, the users of the computers or their managers (the main customers or sponsors in this example) might not be satisfied. Perhaps the project manager or team members never returned calls or were rude. Perhaps users had their daily work disrupted during the upgrades or had to work extra hours due to the upgrades. If the customers were not happy with important aspects of the project, it would be deemed a failure. Conversely, a project might not meet initial scope, time, and cost goals, but the customer could still be very satisfied. Perhaps the project team took longer and spent more money than planned, but they were very polite and helped the users and managers solve several work-related problems. Many organizations implement a customer satisfaction rating system for projects to measure project success instead of only tracking scope, time, and cost performance.
3. *The results of the project met its main objective, such as making or saving a certain amount of money, providing a good return on investment, or simply making the sponsors happy.* Even if the project cost more than estimated, took longer to complete, and the project team was hard to work with, if the users were happy with the upgraded computers it would be a successful project, based on this criterion. As another example, suppose the sponsor really approved the upgrade project to provide a good return on investment by speeding up work and therefore generating more profits. If those goals were met, the sponsor would deem the project a success, regardless of other factors involved.

Why do some IT projects succeed and others fail? Table 1-2 summarizes the results of the 2001 CHAOS study, describing, in order of importance, what factors contribute most to the success of information technology projects. The study lists executive support as the most important factor, overtaking user involvement, which was ranked first in earlier studies. Also note that several other success factors can be strongly influenced by executives such as encouraging user involvement, providing clear business objectives, assigning an experienced project manager, using a standard software infrastructure, and following a formal methodology. Other success factors are related to good project scope and time management such as having a minimized scope, firm basic requirements, and reliable estimates. In fact, experienced project managers, who can often help influence all of these factors to improve the probability of project success, led 97 percent of successful projects.

It is interesting to compare success factors for information technology projects in the U.S. with those in other countries. A 2004 study summarizes the results of a survey of 247 information systems project practitioners in mainland China. One of the study's key findings is that relationship management is viewed as a top success factor for information systems in China, while it is not mentioned in U.S. studies. The study also suggested that having competent team members is less important in China than in the U.S. The Chinese, like the Americans, included top management support, user involvement, and a competent project manager as vital to project success.[15]

**TABLE 1-2** What helps projects succeed?

| |
|---|
| 1. Executive support |
| 2. User involvement |
| 3. Experienced project manager |
| 4. Clear business objectives |
| 5. Minimized scope |
| 6. Standard software infrastructure |
| 7. Firm basic requirements |
| 8. Formal methodology |
| 9. Reliable estimates |
| 10. Other criteria, such as small milestones, proper planning, competent staff, and ownership |

The Standish Group, "Extreme CHAOS," (2001).

It is also important to look beyond individual project success rates and focus on how organizations as a whole can improve project performance. Research comparing companies that excel in project delivery—the "winners"—from those that do not found four significant best practices. The winners:

1. *Use an integrated toolbox.* Companies that consistently succeed in managing projects clearly define what needs to be done in a project, by whom, when, and how. They use an integrated toolbox, including project management tools, methods, and techniques. They carefully select tools, align them with project and business goals, link them to metrics, and provide them to project managers to deliver positive results.
2. *Grow project leaders.* The winners know that strong project managers—referred to as project leaders—are crucial to project success. They also know that a good project leader needs to be a business leader as well, with strong interpersonal and intrapersonal skills. Companies that excel in project management often grow their project leaders internally, providing them with career opportunities, training, and mentoring.
3. *Develop a streamlined project delivery process.* Winning companies have examined every step in the project delivery process, analyzed fluctuations in workloads, searched for ways to reduce variation, and eliminated bottlenecks to create a repeatable delivery process. All projects go through clear stages and clearly define key milestones. All project leaders use a shared road map, focusing on key business aspects of their projects while integrating goals across all parts of the organization.

4. *Measure project health using metrics.* Companies that excel in project delivery use performance metrics to quantify progress. They focus on a handful of important measurements and apply them to all projects. Metrics often include customer satisfaction, return on investment, and percentage of schedule buffer consumed.[16]

Project managers play an important role in making projects, and therefore organizations, successful. Project managers work with the project sponsors, the project team, and the other stakeholders involved in a project to meet project goals. They also work with the sponsor to define success for that particular project. Good project managers do not assume that their definition of success is the same as the sponsors'. They take the time to understand their sponsors' expectations and then track project performance based on important success criteria.

## PROGRAM AND PROJECT PORTFOLIO MANAGEMENT

As mentioned earlier, about one-quarter of the world's gross domestic product is spent on projects. Projects make up a significant portion of work in most business organizations or enterprises, and successfully managing those projects is crucial to enterprise success. Two important concepts that help projects meet enterprise goals are the use of programs and project portfolio management.

### Programs

A **program** is "a group of related projects managed in a coordinated way to obtain benefits and control not available from managing them individually."[17] As you can imagine, it is often more economical to group projects together to help streamline management, staffing, purchasing, and other work. The following are examples of common programs in the IT field.

- *Infrastructure:* An IT department often has a program for IT infrastructure projects. Under this program, there could be several projects, such as providing more wireless Internet access, upgrading hardware and software, and developing and maintaining corporate standards for IT.
- *Applications development:* Under this program, there could be several projects, such as updating an enterprise resource planning (ERP) system, purchasing a new off-the-shelf billing system, or developing a new capability for a customer relationship management system.
- *User support:* In addition to the many operational tasks related to user support, many IT departments have several projects to support users. For example, there could be a project to provide a better e-mail system or one to develop technical training for users.

A **program manager** provides leadership and direction for the project managers heading the projects within a program. Program managers also coordinate the efforts of project teams, functional groups, suppliers, and operations staff supporting the projects to ensure that project products and processes are implemented to maximize benefits. Program managers are responsible for more than the delivery of project results; they are change agents responsible for the success of products and processes produced by those projects. For example, the popular video game *Rock Band*™ lists the program manager and team first under the credits section for the game.

Program managers often have review meetings with all their project managers to share important information and coordinate important aspects of each project. Many program managers worked as project managers earlier in their careers, and they enjoy sharing their wisdom and expertise with their project managers. Effective program managers recognize that managing a program is much more complex than managing a single project. They recognize that technical and project management skills are not enough—program managers must also possess strong business knowledge, leadership capabilities, and communication skills.

## Project Portfolio Management

In many organizations, project managers also support an emerging business strategy of **project portfolio management** (also called just **portfolio management** in this text), in which organizations group and manage projects and programs as a portfolio of investments that contribute to the entire enterprise's success. Portfolio managers help their organizations make wise investment decisions by helping to select and analyze projects from a strategic perspective. Portfolio managers may or may not have previous experience as project or program managers. It is most important that they have strong financial and analytical skills and understand how projects and programs can contribute to meeting strategic goals.

Figure 1-3 illustrates the differences between project management and project portfolio management. Notice that the main distinction is a focus on meeting tactical or strategic goals. Tactical goals are generally more specific and short-term than strategic goals, which emphasize long-term goals for an organization. Individual projects often address tactical goals, whereas portfolio management addresses strategic goals. Project management addresses questions like "Are we carrying out projects well?", "Are projects on time and budget?", and "Do project stakeholders know what they should be doing?"

Portfolio management addresses questions like "Are we working on the right projects?", "Are we investing in the right areas?", and "Do we have the right resources to be competitive?" Pacific Edge Software's product manager, Eric Burke, defines project portfolio management as "the continuous process of selecting and managing the optimum set of project initiatives that deliver maximum business value."[18]

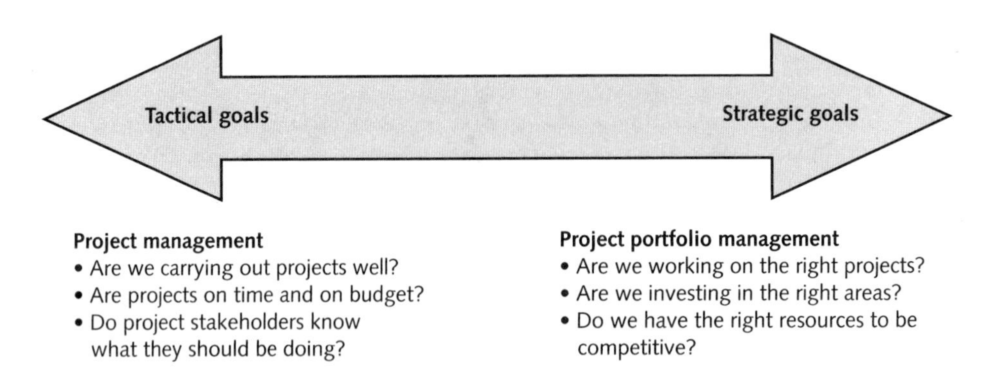


**FIGURE 1-3** Project management compared to project portfolio management

Many organizations use a more disciplined approach to portfolio management by developing guidelines and software tools to assist in project portfolio management. The Project Management Institute (described later in this chapter) first published the *Organizational Project Management Maturity Model (OPM3) Knowledge Foundation* in 2003,[19] which describes the importance not only of managing individual projects or programs well, but the importance of following organizational project management to align projects, programs, and portfolios with strategic goals. OPM3 is a standard that organizations can use to measure their organizational project management maturity against a comprehensive set of best practices.

## BEST PRACTICE

A **best practice** is "an optimal way recognized by industry to achieve a stated goal or objective."[20] Rosabeth Moss Kanter, a Professor at Harvard Business School and well-known author and consultant, says that visionary leaders know "the best practice secret: Stretching to learn from the best of the best in any sector can make a big vision more likely to succeed."[21] Kanter also emphasizes the need to have measurable standards for best practices. An organization can measure performance against its own past, against peers, and even better, against potential. Kanter suggests that organizations need to continue to reach for higher standards. She suggests the following exercise regime for business leaders who want to adapt best practices in an intelligent way to help their own organizations:

- Reach high. Stretch. Raise standards and aspirations. Find the best of the best and then use it as inspiration for reaching full potential.
- Help everyone in your organization become a professional. Empower people to manage themselves through benchmarks and standards based on best practice exchange.
- Look everywhere. Go far afield. Think of the whole world as your laboratory for learning.

Robert Butrick, author of *The Project Workout,* wrote an article on best practices in project management for the *Ultimate Business Library's Best Practice* book. He suggests that organizations need to follow basic principles of project management, including these two mentioned earlier in this chapter:

- Make sure your projects are driven by your strategy. Be able to demonstrate how each project you undertake fits your business strategy, and screen out unwanted projects as soon as possible.
- Engage your stakeholders. Ignoring stakeholders often leads to project failure. Be sure to engage stakeholders at all stages of a project, and encourage teamwork and commitment at all times.[22]

As you can imagine, project portfolio management is not an easy task. Figure 1-4 illustrates one approach for project portfolio management where one large portfolio exists for the entire organization. This allows top management to view and manage all projects at an enterprise level. Sections of that portfolio are then broken down to improve the management of projects in each sector. For example, a company might have the main

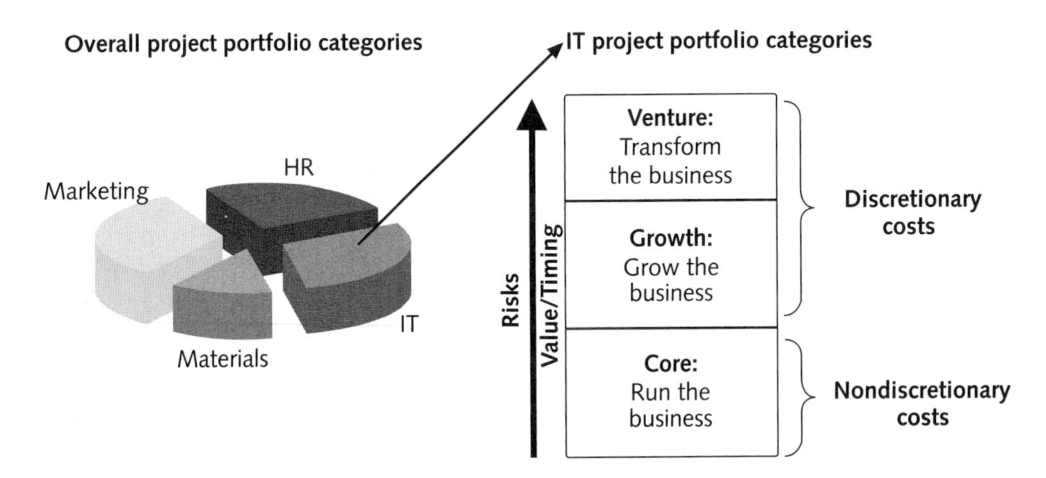


**FIGURE 1-4** Sample project portfolio approach

portfolio categories as shown in the left part of Figure 1-4—marketing, materials, IT, and human resources (HR)—and divide each of those categories further to address their unique concerns. The right part of this figure shows how the IT projects could be categorized in more detail to assist in their management. In this example, there are three basic IT project portfolio categories:

- *Venture:* Projects in this category help transform the business. For example, the large retail chain described in the opening case might have an IT project to provide kiosks in stores and similar functionality on the Internet where customers and suppliers could quickly provide feedback on products or services. This project could help transform the business by developing closer partnerships with customers and suppliers.
- *Growth:* Projects in this category would help the company grow in terms of revenues. For example, a company might have an IT project to provide information on their corporate Web site in a new language, such as Chinese or Japanese. This capability could help them grow their business in those countries.
- *Core:* Projects in this category must be accomplished to run the business. For example, an IT project to provide computers for new employees would fall under this category.

Note on the right part of Figure 1-4 that the Core category of IT projects is labeled as nondiscretionary costs. This means that the company has no choice in whether to fund these projects; they must fund them to stay in business. Projects that fall under the Venture or Growth category are discretionary costs because the company can use its own discretion or judgment in deciding whether or not to fund them. Notice the arrow in the center of Figure 1-4 labeled Risks, Value/Timing. This arrow indicates that the risks, value, and timing of projects normally increase as you move from Core to Growth to Venture projects. However, some core projects can also be high risk, have high value, and require good timing. As you can see, many factors are involved in portfolio management.

Many organizations use specialized software to organize and analyze all types of project data into project portfolios. **Enterprise** or **portfolio project management software** integrates information from multiple projects to show the status of active, approved, and future projects across an entire organization. For example, Figure 1-5 provides a sample screen from portfolio management software provided by Planview. The charts and text in the upper half of the screen show the number and percentage of projects in this project portfolio that are on target and in trouble in terms of schedule and cost variance. The bottom half of the screen lists the names of individual projects, percent complete, schedule variance, cost variance, budget variance, and risk percentage. The last section in this chapter provides more information on project management software.

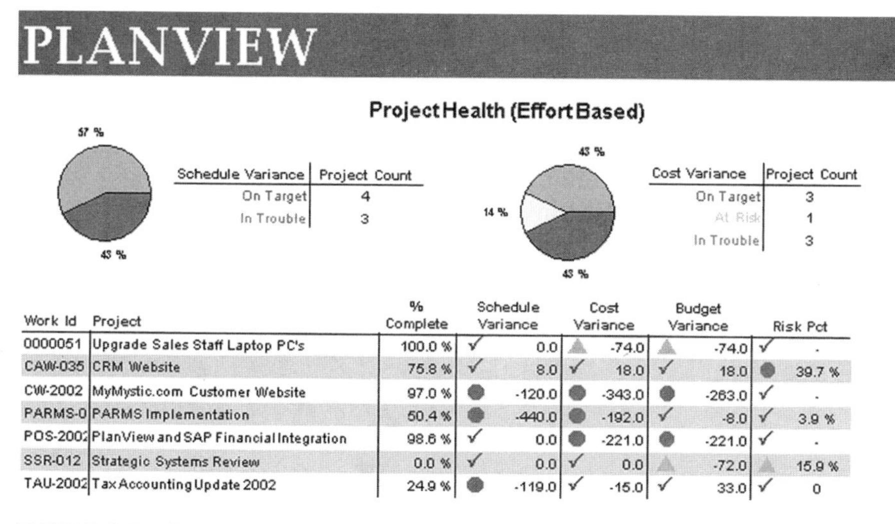


| Work Id | Project | % Complete | Schedule Variance | Cost Variance | Budget Variance | Risk Pct |
|---|---|---|---|---|---|---|
| 0000051 | Upgrade Sales Staff Laptop PC's | 100.0 % | ✓ 0.0 | -74.0 | -74.0 | ✓ . |
| CAW-035 | CRM Website | 75.8 % | ✓ 8.0 | ✓ 18.0 | ✓ 18.0 | 39.7 % |
| CW-2002 | MyMystic.com Customer Website | 97.0 % | -120.0 | -343.0 | -263.0 | ✓ . |
| PARMS-0 | PARMS Implementation | 50.4 % | -440.0 | -192.0 | ✓ -8.0 | ✓ 3.9 % |
| POS-2002 | PlanView and SAP Financial Integration | 98.6 % | ✓ 0.0 | -221.0 | -221.0 | ✓ . |
| SSR-012 | Strategic Systems Review | 0.0 % | ✓ 0.0 | ✓ 0.0 | -72.0 | 15.9 % |
| TAU-2002 | Tax Accounting Update 2002 | 24.9 % | -119.0 | ✓ -15.0 | ✓ 33.0 | ✓ 0 |

**FIGURE 1-5** Sample project portfolio management screen showing project health

## THE ROLE OF THE PROJECT MANAGER

You have already read that project managers must work closely with the other stakeholders on a project, especially the sponsor and project team. They are also more effective if they are familiar with the nine project management knowledge areas and the various tools and techniques related to project management. Experienced project managers help projects succeed. But what do project managers do exactly? What skills do they really need to do a good job? The next section provides brief answers to these questions, and the rest of this book gives more insight into the role of the project manager. Even if you never become a project manager, you will probably be part of a project team, and it is important for team members to help their project managers.

### Project Manager Job Description

A project manager can have many different job descriptions, which can vary tremendously based on the organization and the project. For example, Monster.com includes thousands of job listings for project managers. They even have a job category for project/program managers. Here are a few edited postings:

- *Project manager for a consulting firm:* Plans, schedules, and controls activities to fulfill identified objectives applying technical, theoretical, and managerial skills to satisfy project requirements. Coordinates and integrates team and individual efforts and builds positive professional relationships with clients and associates.
- *IT project manager for a financial services firm:* Manages, prioritizes, develops, and implements information technology solutions to meet business needs. Prepares and executes project plans using project management software following a standard methodology. Establishes cross-functional end-user teams defining and implementing projects on time and within budget. Acts as a liaison between third-party service providers and end-users to develop and implement technology solutions. Participates in vendor contract development and budget management. Provides post implementation support.
- *IT project manager for a nonprofit consulting firm:* Responsibilities include business analysis, requirements gathering, project planning, budget estimating, development, testing, and implementation. Responsible for working with various resource providers to ensure development is completed in a timely, high-quality, and cost-effective manner.

The job description for a project manager can vary by industry and by organization, but there are similar tasks that most project managers perform regardless of these differences. In fact, project management is a skill needed in every major information technology field, from database administrator to network specialist to technical writer.

## Suggested Skills for Project Managers

In an interview with two chief information officers (CIOs), John Oliver of True North Communications, Inc. and George Nassef of *Hotjobs.com*, both men agreed that the most important project management skills seem to depend on the uniqueness of the project and the people involved.[23] Project managers need to have a wide variety of skills and be able to decide which particular skills are more important in different situations. As you can imagine, good project managers should have many skills. *A Guide to the Project Management Body of Knowledge*—the *PMBOK® Guide*—recommends that the project management team understand and use expertise in the following areas:

- The Project Management Body of Knowledge
- Application area knowledge, standards, and regulations
- Project environment knowledge
- General management knowledge and skills
- Soft skills or human relations skills

This chapter introduced the nine project management knowledge areas, as well as some general tools and techniques project managers use. The following section focuses on the IT application area, including skills required in the project environment, general management, and soft skills. Note that the *PMBOK® Guide, Fourth Edition* describes three dimensions of project management competency: project management knowledge and performance competency (knowing about project management and being able to apply that knowledge) as well as personal competency (attitudes and personality characteristics). Consult PMI's Web site at *www.pmi.org* for further information on skills for project managers and PMI's Career Framework for Practitioners.

The project environment differs from organization to organization and project to project, but some skills will help in almost all project environments. These skills include understanding change, and understanding how organizations work within their social, political, and physical environments. Project managers must be comfortable leading and handling change, since most projects introduce changes in organizations and involve changes within the projects themselves. Project managers need to understand the organization in which they work and how that organization develops products and provides services. The skills and behavior needed to manage a project for a Fortune 100 company in the United States may differ greatly from those needed to manage a government project in Poland. Chapter 2, The Project Management and Information Technology Context, provides detailed information on these topics.

Project managers should also possess general management knowledge and skills. They should understand important topics related to financial management, accounting, procurement, sales, marketing, contracts, manufacturing, distribution, logistics, the supply chain, strategic planning, tactical planning, operations management, organizational structures and behavior, personnel administration, compensation, benefits, career paths, and health and safety practices. On some projects, it will be critical for the project manager to have a lot of experience in one or several of these general management areas. On other projects, the project manager can delegate detailed responsibility for some of these areas to a team member, support staff, or even a supplier. Even so, the project manager must be intelligent and experienced enough to know which of these areas are most important and who is qualified to do the work. He or she must also make and/or take responsibility for all key project decisions.

Achieving high performance on projects requires soft skills, otherwise called human relations skills. Some of these soft skills include effective communication, influencing the organization to get things done, leadership, motivation, negotiation, conflict management, and problem solving. Why do project managers need good soft skills? One reason is that to understand, navigate, and meet stakeholders' needs and expectations, project managers need to lead, communicate, negotiate, solve problems, and influence the organization at large. They need to be able to listen actively to what others are saying, help develop new approaches for solving problems, and then persuade others to work toward achieving project goals. Project managers must lead their project teams by providing vision, delegating work, creating an energetic and positive environment, and setting an example of appropriate and effective behavior. Project managers must focus on teamwork skills to employ people effectively. They need to be able to motivate different types of people and develop *esprit de corps* within the project team and with other project stakeholders. Since most projects involve changes and trade-offs between competing goals, it is important for project managers to have strong coping skills as well. It helps project managers maintain their sanity and reduce their stress levels if they cope with criticism and constant change. Project managers must be flexible, creative, and sometimes patient in working toward project goals; they must also be persistent in making project needs known.

Lastly, project managers, especially those managing IT projects, must be able to make effective use of technology as it relates to the specific project. Making effective use of technology often includes special product knowledge or experience with a particular industry.

Project managers must make many decisions and deal with people in a wide variety of disciplines, so it helps tremendously to have a project manager who is confident in using the special tools or technologies that are the most effective in particular settings. Project managers do not normally have to be experts on any specific technology, but they have to know enough to build a strong team and ask the right questions to keep things on track. For example, project managers for large information technology projects do not have to be experts in the field of information technology, but they must have working knowledge of various technologies and understand how the project would enhance the business. Many companies have found that a good business manager can be a very good information technology project manager because they focus on meeting business needs and rely on key project members to handle the technical details.

All project managers should continue to develop their knowledge and experience in project management, general management, soft skills, and the industries they support. Non-IT business people are now very savvy with information technology, but few information technology professionals have spent the time developing their business savvy.[24] IT project managers must be willing to develop more than their technical skills to be productive team members and successful project managers. Everyone, no matter how technical they are, should develop business and soft skills.

## Importance of People and Leadership Skills

In a recent study, project management experts from various industries were asked to identify the ten most important skills and competencies for effective project managers. Table 1-3 shows the results.

Respondents were also asked what skills and competencies were most important in various project situations:

- *Large projects:* Leadership, relevant prior experience, planning, people skills, verbal communication, and team-building skills were most important.
- *High uncertainty projects:* Risk management, expectation management, leadership, people skills, and planning skills were most important.
- *Very novel projects:* Leadership, people skills, having vision and goals, self confidence, expectations management, and listening skills were most important.[25]

Notice that a few additional skills and competencies not cited in the top 10 list were mentioned when people thought about the context of a project. To be the most effective, project managers require a changing mix of skills and competencies depending on the project being delivered.

Also notice the general emphasis on people and leadership skills. As mentioned earlier, all project managers, especially those working on technical projects, need to demonstrate leadership and management skills. *Leadership* and *management* are terms often used interchangeably, although there are differences. Generally, a **leader** focuses on long-term goals and big-picture objectives, while inspiring people to reach those goals. A **manager** often deals with the day-to-day details of meeting specific goals. Some people say that, "Managers do things right, and leaders do the right things." "Leaders determine the vision, and managers achieve the vision." "You lead people and manage things."

**TABLE 1-3** Ten most important skills and competencies for project managers

| |
|---|
| 1. People skills |
| 2. Leadership |
| 3. Listening |
| 4. Integrity, ethical behavior, consistent |
| 5. Strong at building trust |
| 6. Verbal communication |
| 7. Strong at building teams |
| 8. Conflict resolution, conflict management |
| 9. Critical thinking, problem solving |
| 10. Understands, balances priorities |

Jennifer Krahn, "Effective Project Leadership: A Combination of Project Manager Skills and Competencies in Context," *PMI Research Conference Proceedings* ( July 2006).

However, project managers often take on the role of both leader and manager. Good project managers know that people make or break projects, so they must set a good example to lead their team to success. They are aware of the greater needs of their stakeholders and organizations, so they are visionary in guiding their current projects and in suggesting future ones. As mentioned earlier, companies that excel in project management grow project "leaders," emphasizing development of business and communication skills. Yet good project managers must also focus on getting the job done by paying attention to the details and daily operations of each task. Instead of thinking of leaders and managers as specific people, it is better to think of people as having leadership skills, such as being visionary and inspiring, and management skills, such as being organized and effective. Therefore, the best project managers have leadership and management characteristics; they are visionary yet focused on the bottom line. Above all else, good project managers focus on achieving positive results!

## Careers for Information Technology Project Managers

A recent article suggests that, "The most sought-after corporate IT workers in 2010 may be those with no deep-seated technical skills at all. The nuts-and-bolts programming and easy-to-document support jobs will have all gone to third-party providers in the U.S. or abroad. Instead, IT departments will be populated with 'versatilists'—those with a technology background who also know the business sector inside and out, can architect and carry out IT plans that will add business value, and can cultivate relationships both inside and outside the company."[26]

A recent survey by CIO.com supports this career projection. IT executives listed the skills they predicted would be the most in demand in the next two to five years. Project/

program management came in first place, followed by business process management, business analysis, and application development. Table 1-4 shows these results, as well as the percentage of respondents who listed the skill as most in demand. Even if you choose to stay in a technical role, you still need project management knowledge and skills to help your team and your organization succeed.

**TABLE 1-4** Top information technology skills

| Skill | Percentage of Respondents |
| --- | --- |
| Project/program management | 60% |
| Business process management | 55% |
| Business analysis | 53% |
| Application development | 52% |
| Database management | 49% |
| Security | 42% |
| Enterprise architect | 41% |
| Strategist/internal consultant | 40% |
| Systems analyst | 39% |
| Relationship management | 39% |
| Web services | 33% |
| Help desk/user support | 32% |
| Networking | 32% |
| Web site development | 30% |
| QA/testing | 28% |
| IT finance | 28% |
| Vendor management/ procurement | 27% |
| IT HR | 21% |
| Other | 3% |

Carolyn Johnson, "2006 Midyear Staffing Updates," *CIO Research Reports*, October 2, 2006.

## THE PROJECT MANAGEMENT PROFESSION

The profession of project management is growing at a very rapid pace. To understand this line of work, it is helpful to briefly review the history of project management, introduce you to the Project Management Institute (PMI) and some of its services (such as certification), and discuss the growth in project management software.

### History of Project Management

Although people have worked on projects for centuries, most agree that the modern concept of project management began with the Manhattan Project, which the U.S. military led to develop the atomic bomb in World War II. The Manhattan Project involved many people with different skills at several different locations. It also clearly separated the overall management of the project's mission, schedule, and budget under General Leslie R. Groves and the technical management of the project under the lead scientist, Dr. Robert Oppenheimer. The Manhattan Project lasted about three years and cost almost $2 billion in 1946.

In developing the project, the military realized that scientists and other technical specialists often did not have the desire or the necessary skills to manage large projects. For example, after being asked several times for each team member's responsibilities at the new Los Alamos laboratory in 1943, Dr. Oppenheimer tossed a piece of paper with an organization chart on it at his director and said, "Here's your damn organization chart."[27] Project management was recognized as a distinct discipline requiring people with special skills and, more importantly, the desire to lead project teams.

In 1917, Henry Gantt developed the famous Gantt chart for scheduling work in factories. A **Gantt chart** is a standard format for displaying project schedule information by listing project activities and their corresponding start and finish dates in a calendar format. Initially, managers drew Gantt charts by hand to show project tasks and schedule information, and this tool provided a standard format for planning and reviewing all the work on early military projects.

Today's project managers still use the Gantt chart as the primary tool to communicate project schedule information, but with the aid of computers, it is no longer necessary to draw the charts by hand and they can be more easily shared and disseminated to project stakeholders. Figure 1-6 displays a Gantt chart created with Microsoft Project, the most widely used project management software today. You will learn more about using Project 2007 in Appendix A.

During the Cold War years of the 1950s and '60s, the military continued to be key in refining several project management techniques. Members of the U.S. Navy Polaris missile/submarine project first used network diagrams in 1958. These diagrams helped managers model the relationships among project tasks, which allowed them to create schedules that were more realistic. Figure 1-7 displays a network diagram created using Microsoft Project. Note that the diagram includes arrows that show which tasks are related and the sequence in which team members must perform the tasks. The concept of determining relationships among tasks is essential in helping to improve project scheduling. This concept allows you to find and monitor the **critical path**—the longest path through a network diagram that determines the earliest completion of a project. You will learn more about Gantt charts, network diagrams, critical path analysis, and other time management concepts in Chapter 6, Project Time Management.

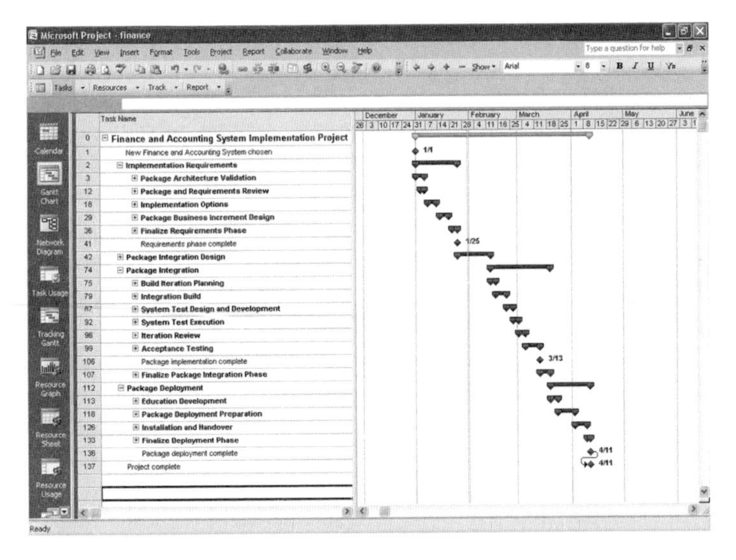


**FIGURE 1-6** Sample Gantt chart created with Project 2007

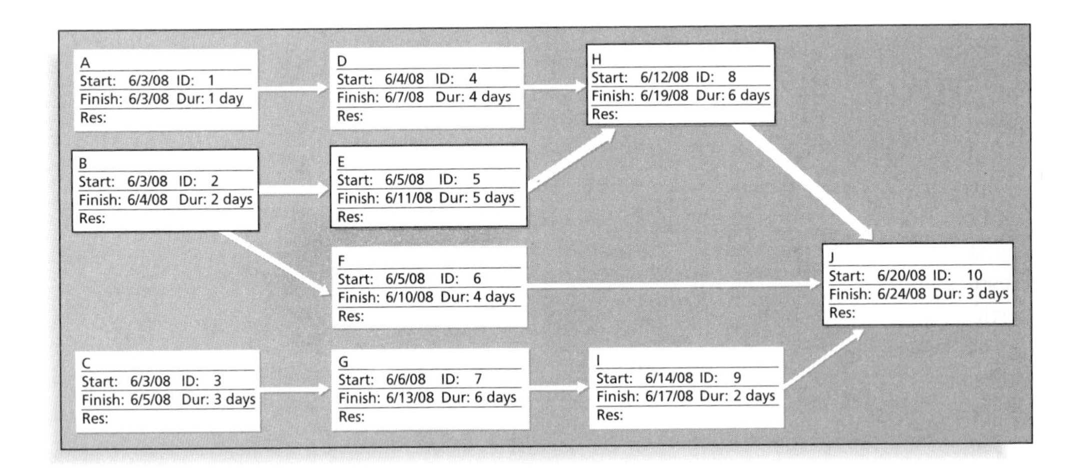


**FIGURE 1-7** Sample network diagram in Microsoft Project

By the 1970s, the U.S. military and its civilian suppliers developed software to assist in managing large projects. Early project management software was very expensive to purchase and it ran exclusively on mainframe computers. For example, Artemis was an early project management software product that helped managers analyze complex schedules for designing aircraft. A full-time employee was often required to run the complicated software, and expensive pen plotters were used to draw network diagrams and Gantt charts.

As computer hardware became smaller and more affordable and software included graphical, easy-to-use interfaces, project management software became less expensive and more widely used. This made it possible—and affordable—for many industries worldwide to use project management software on all types and sizes of projects. New software makes basic tools, such as Gantt charts and network diagrams, inexpensive, easy to create, and available for anyone to update. See the section in this chapter on project management software for more information.

In the 1990s, many companies began creating project management offices to help them handle the increasing number and complexity of projects. A **Project Management Office (PMO)** is an organizational group responsible for coordinating the project management function throughout an organization. There are different ways to structure a PMO, and they can have various roles and responsibilities. Below are possible goals of a PMO:

- Collect, organize, and integrate project data for the entire organization.
- Develop and maintain templates for project documents.
- Develop or coordinate training in various project management topics.
- Develop and provide a formal career path for project managers.
- Provide project management consulting services.
- Provide a structure to house project managers while they are acting in those roles or are between projects.

By the end of the twentieth century, people in virtually every industry around the globe began to investigate and apply different aspects of project management to their projects. The sophistication and effectiveness with which project management tools are being applied and used today is influencing the way companies do business, use resources, and respond to market requirements with speed and accuracy. As mentioned earlier in this chapter, many organizations are now using enterprise or project portfolio management software to help manage portfolios of projects.

Many colleges, universities, and companies around the world now offer courses related to various aspects of project management. You can even earn bachelor's, master's, and doctoral degrees in project management. PMI reported in 2008 that of the 280 institutions it has identified that offer degrees in project management, 103 are in mainland China. "When Western companies come into China they are more likely to hire individuals who have PMP certification as an additional verification of their skills. In our salary survey, the salary difference in IT, for example, was dramatic. A person with certification could make five to six times as much salary, so there is terrific incentive to get certified and work for these Western companies."[28]

The problems in managing projects, the publicity about project management, and the belief that it really can make a difference continue to contribute to the growth of this field.

## The Project Management Institute

Although many professional societies suffer from declining membership, the **Project Management Institute (PMI)**, an international professional society for project managers founded in 1969, has continued to attract and retain members, reporting 277,221 members worldwide by August 31, 2008. A large percentage of PMI members work in the information technology field and more than 13,000 pay additional dues to join the Information Systems Specific Interest Group. Because there are so many people working on projects in various industries, PMI has created specific interest groups (SIGs) that enable members to share ideas about project management in their particular application areas, such as information systems. PMI also has SIGs for aerospace/defense, financial services, healthcare, hospitality management, manufacturing, new product development, retail, and urban development, to name a few. Note that there are also other project management professional societies. See the companion Web site for more information.

## PMI STUDENT MEMBERSHIP

As a student, you can join PMI for a reduced fee. Consult PMI's Web site (*www.pmi.org*) for more information. You can also network with other students studying project management by joining the Students of Project Management Specific Interest Group (SIG) at *www.studentsofpm.org*. Note that PMI is changing the SIGs into Virtual Communities, so you may see that term used.

### Project Management Certification

Professional certification is an important factor in recognizing and ensuring quality in a profession. PMI provides certification as a **Project Management Professional (PMP)**—someone who has documented sufficient project experience and education, agreed to follow the PMI code of professional conduct, and demonstrated knowledge of the field of project management by passing a comprehensive examination. Appendix B provides more information on PMP certification as well as other certification programs, such as CompTIA's Project+ certification.

The number of people earning PMP certification continues to increase. In 1993, there were about 1,000 certified project management professionals. By December 31, 2008, there were 318,289 active PMPs.[29] Figure 1-8 shows the rapid growth in the number of people earning project management professional certification from 1993 to 2008.

Several studies show that organizations supporting technical certification programs tend to operate in more complex information technology environments and are more

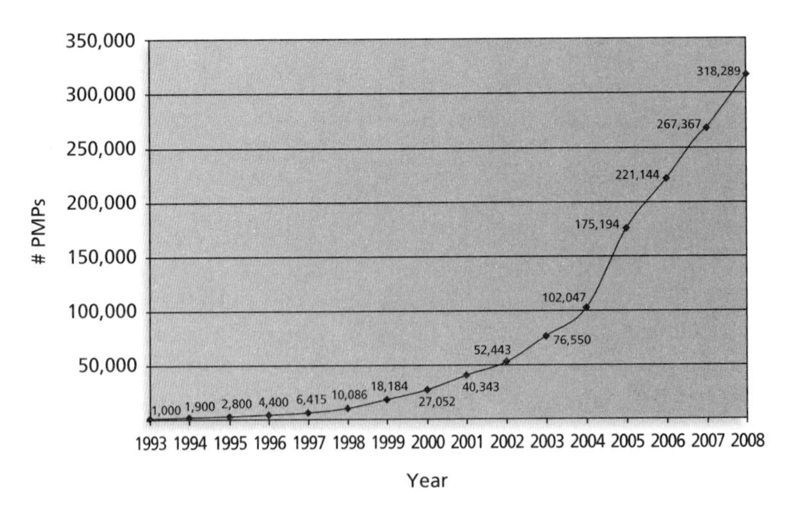


**FIGURE 1-8** Growth in PMP Certification, 1993–2008

efficient than companies that do not support certification. Likewise, organizations that support PMP certification see the value of investing in programs to improve their employees' knowledge in project management. Many employers today require specific certifications to ensure their workers have current skills, and job seekers find that they often have an advantage when they earn and maintain marketable certifications. A 2006 *Certification Magazine* survey of over 35,000 IT workers from 197 countries found that average salaries for workers in project management were among the highest for all IT specialties. IT workers with a PMP certification earned among the highest salaries for all IT workers who hold professional certifications.[30]

As information technology projects become more complex and global in nature, the need for people with demonstrated knowledge and skills in project management will continue. Just as passing the CPA exam is a standard for accountants, passing the PMP exam is becoming a standard for project managers. Some companies require that all project managers be PMP certified. Project management certification is also enabling professionals in the field to share a common base of knowledge. For example, any person with PMP certification can list, describe, and use the nine project management knowledge areas. Sharing a common base of knowledge is important because it helps advance the theory and practice of project management. PMI also offers additional certifications, including new ones in scheduling, risk, and program management. See Appendix B of this text for detailed information on certification.

## Ethics in Project Management

**Ethics**, loosely defined, is a set of principles that guide our decision making based on personal values of what is "right" and "wrong." Making ethical decisions is an important part of our personal and professional lives because it generates trust and respect with other people. Project managers often face ethical dilemmas. For example, several projects involve different payment methods. If a project manager can make more money by doing a job poorly, should he or she do the job poorly? No! If a project manager is personally opposed to the development of nuclear weapons, should he or she refuse to manage a project that helps produce them? Yes! Ethics guide us in making these types of decisions.

PMI approved a new Code of Ethics and Professional Conduct effective January 1, 2007. This new code applies not only to PMPs, but to all PMI members and individuals who hold a PMI certification, apply for a PMI certification, or serve PMI in a volunteer capacity. It is vital for project management practitioners to conduct their work in an ethical manner. Even if you are not affiliated with PMI, these guidelines can help you conduct your work in an ethical manner, which helps the profession earn the confidence of the public, employers, employees, and all project stakeholders. The PMI Code of Ethics and Professional Conduct includes short chapters addressing vision and applicability, responsibility, respect, fairness, and honestly. A few excerpts from this document include the following:

"As practitioners in the global project management community:

2.2.1 We make decisions and take actions based on the best interests of society, public safety, and the environment.

2.2.2 We accept only those assignment that are consistent with our background, experience, skills, and qualifications.

2.2.3 We fulfill the commitments that we undertake—we do what we say we will do.

3.2.1 We inform ourselves about the norms and customs of others and avoid engaging in behaviors they might consider disrespectful.

3.2.2 We listen to others' points of view, seeking to understand them.

3.2.3 We approach directly those persons with whom we have a conflict or disagreement.

4.2.1 We demonstrate transparency in our decision-making process.

4.2.2 We constantly reexamine our impartiality and objectivity, taking corrective action as appropriate.

4.3.1 We proactively and fully disclose any real or potential conflicts of interest to appropriate stakeholders.

5.2.1 We earnestly seek to understand the truth.

5.2.2 We are truthful in our communications and in our conduct."[31]

In addition, PMI added a new series of questions to the PMP certification exam in March 2002 to emphasize the importance of ethics and professional responsibility. See Appendix B for information on the PMP exam.

## Project Management Software

Unlike the cobbler neglecting to make shoes for his own children, the project management and software development communities have definitely responded to the need to provide more software to assist in managing projects. The Project Management Center, a Web site for people involved in project management, provides an alphabetical directory of more than 300 project management software solutions (*www.infogoal.com/pmc*). This site and others demonstrate the growth in available project management software products, especially Web-based tools. Deciding which project management software to use has become a project in itself. This section provides a summary of the basic types of project management software available and references for finding more information. In Appendix A, you will learn how to use Microsoft Project 2007, the most widely used project management software tool today.

### MICROSOFT PROJECT 2007

Appendix A includes a *Guide to Using Microsoft Project 2007,* which will help you develop hands-on skills using this most popular project management software tool. You can also access a trial version of VPMi Express—a Web-based product from VCS (*www.vcsonline.com*)—by following the information provided on the resources page in the front of this text or by going directly to the VCS Web site.

Many people still use basic productivity software such as Microsoft Word and Excel to perform many project management functions, including determining project scope, time, and cost, assigning resources, preparing project documentation, and so on. People often use productivity software instead of specialized project management software because they already have it and know how to use it. However, there are hundreds of project management software tools that provide specific functionality for managing projects. These project management software tools can be divided into three general categories based on functionality and price:

- *Low-end tools:* These tools provide basic project management features and generally cost less than $200 per user. They are often recommended for small projects and single users. Most of these tools allow users to create Gantt charts, which cannot be done easily using current productivity software. Top Ten Reviews listed MinuteMan ($49.95) and Project Kickstart ($199.95) in their list of top 10 project management software tools for 2008.[32] Basecamp (*www.basecamphq.com*) is another popular tool with low-end through high-end versions ranging in price from $24 to $149 per month. Several companies provide add-in features to Excel (see *www.business-spreadsheets.com*) to provide basic project management functions using a familiar software product.
- *Midrange tools:* A step up from low-end tools, midrange tools are designed to handle larger projects, multiple users, and multiple projects. All of these tools can produce Gantt charts and network diagrams, and can assist in critical path analysis, resource allocation, project tracking, status reporting, and so on. Prices range from about $200 to $600 per user, and several tools require additional server software for using workgroup features. Microsoft Project is still the most widely used project management software today in this category, and there is also an enterprise version, as described briefly below and in Appendix A. In the summer of 2008, Top Ten Reviews listed Microsoft Project as the number one choice ($599), along with Milestones ($249). A product called Copper also made the top ten list, with a price of $999 for up to 50 users. As noted earlier, this text includes a trial version of Project 2007 as well as one of VPMi Express, a totally Web-based tool. Note that students and educators can purchase software like Microsoft Project 2007 at reduced prices from sites like *www.journeyed.com* ($59.98 for Project 2007 Standard in October 2008), and anyone can download a trial version from Microsoft's Web site. Many other suppliers also provide trial versions of their products.
- *High-end tools:* Another category of project management software is high-end tools, sometimes referred to as enterprise project management software. These tools provide robust capabilities to handle very large projects, dispersed workgroups, and enterprise and portfolio management functions that summarize and combine individual project information to provide an enterprise view of all projects. These products are generally licensed on a per-user basis, integrate with enterprise database management software, and are accessible via the Internet. In mid-2002, Microsoft introduced the first version of their Enterprise Project Management software, and in 2003, they introduced the Microsoft Enterprise Project Management solution, which was updated in 2007 to include Microsoft Office Project Server 2007 and Microsoft Office Project Portfolio Server 2007A. Several inexpensive, Web-based products that provide enterprise and portfolio management capabilities are also on the market. For example, VPMi Enterprise Online (*www.vcsonline.com*) is available for a low monthly fee per user (see the front cover of this text for free trial information). See the Project Management Center Web site (*www.infogoal.com/pmc*) or Top Ten Reviews

(*http://project-management-software-review.toptenreviews.com*) for links to many companies that provide project management software.

There are also several free or open-source tools available. For example, Open Workbench (*www.openworkbench.org*), dotProject (*www.dotproject.net*), and TaskJuggler (*www.taskjuggler.org*) are all free online project management tools. Remember, however, that these tools are developed, managed, and maintained by volunteers. They also often run on limited platforms and may not be well supported.

As mentioned earlier, there are many reasons to study project management, particularly as it relates to information technology projects. The number of information technology projects continues to grow, the complexity of these projects continues to increase, and the profession of project management continues to expand and mature. As more people study and work in this important field, the success rate of information technology projects should improve.

## CASE WRAP-UP

Anne Roberts worked with the VPs and the CEO to form teams to help identify potential IT projects that would support their business strategies. They formed a project team to implement a project portfolio management software tool across the organization. They formed another team to develop project-based reward systems for all employees. They also authorized funds for a project to educate all employees in project management, to help people earn PMP and related certifications, and to develop a mentoring program. Anne had successfully convinced everyone that effectively managing projects was crucial to their company's future.

## Chapter Summary

There is a new or renewed interest in project management today as the number of projects continues to grow and their complexity continues to increase. The success rate of information technology projects has more than doubled since 1995, but still only about a third are successful in meeting scope, time, and cost goals. Using a more disciplined approach to managing projects can help projects and organizations succeed.

A project is a temporary endeavor undertaken to create a unique product, service, or result. An information technology project involves the use of hardware, software, and/or networks. Projects are unique, temporary, and developed incrementally; they require resources, have a sponsor, and involve uncertainty. The triple constraint of project management refers to managing the scope, time, and cost dimensions of a project.

Project management is the application of knowledge, skills, tools, and techniques to project activities to meet project requirements. Stakeholders are the people involved in or affected by project activities. A framework for project management includes the project stakeholders, project management knowledge areas, and project management tools and techniques. The nine knowledge areas are project integration management, scope, time, cost, quality, human resource, communications, risk, and procurement management. There are many tools and techniques in each knowledge area. There are different ways to define project success, and project managers must understand the success criteria for their unique projects.

A program is a group of related projects managed in a coordinated way to obtain benefits and control not available from managing them individually. Project portfolio management involves organizing and managing projects and programs as a portfolio of investments that contribute to theentire enterprise's success. Portfolio management emphasizes meeting strategic goals while project management focuses on tactical goals. Studies show that executive support is crucial to project success, as are other factors like user involvement, an experienced project manager, and clear business objectives.

Project managers play a key role in helping projects and organizations succeed. They must perform various job duties, possess many skills, and continue to develop skills in project management, general management, and their application area, such as information technology. Soft skills, especially leadership, are particularly important for project managers.

The profession of project management continues to grow and mature. In the U.S., the military took the lead in project management and developed many tools such as Gantt charts and network diagrams, but today people use project management in virtually every industry around the globe. The Project Management Institute (PMI) is an international professional society that provides certification as a Project Management Professional (PMP) and upholds a code of ethics. Today, hundreds of project management software products are available to assist people in managing projects.

## Quick Quiz

1. Approximately what percentage of the world's gross domestic product is spent on projects?
   a. 10 percent
   b. 25 percent
   c. 50 percent
   d. 75 percent

2. Which of the following is a not a potential advantage of using good project management?
    a. Shorter development times
    b. Higher worker morale
    c. Lower cost of capital
    d. Higher profit margins
3. A ______ is a temporary endeavor undertaken to create a unique product, service, or result.
    a. program
    b. process
    c. project
    d. portfolio
4. Which of the following is not an attribute of a project?
    a. projects are unique
    b. projects are developed using progressive elaboration
    c. projects have a primary customer or sponsor
    d. projects involve little uncertainty
5. Which of the following is not part of the triple constraint of project management?
    a. meeting scope goals
    b. meeting time goals
    c. meeting communications goals
    d. meeting cost goals
6. ______ is the application of knowledge, skills, tools and techniques to project activities to meet project requirements.
    a. Project management
    b. Program management
    c. Project portfolio management
    d. Requirements management
7. Project portfolio management addresses ______ goals of an organization, while project management addresses ______ goals.
    a. strategic, tactical
    b. tactical, strategic
    c. internal, external
    d. external, internal

8. Several application development projects done for the same functional group might best be managed as part of a ______.
   a. portfolio
   b. program
   c. investment
   d. collaborative
9. Which of the following is not one of the top ten skills or competencies of an effective project manager?
   a. people skills
   b. leadership
   c. integrity
   d. technical skills
10. What is the certification program called that the Project Management Institute provides?
   a. Certified Project Manager (CPM)
   b. Project Management Professional (PMP)
   c. Project Management Expert (PME)
   d. Project Management Mentor (PMM)

### Quick Quiz Answers

1. b; 2. c; 3. c; 4. d; 5. c; 6. a; 7. a; 8. b; 9. d; 10. b

## Discussion Questions

1. Why is there a new or renewed interest in the field of project management?
2. What is a project, and what are its main attributes? How is a project different from what most people do in their day-to-day jobs? What is the triple constraint?
3. What is project management? Briefly describe the project management framework, providing examples of stakeholders, knowledge areas, tools and techniques, and project success factors.
4. What is a program? What is a project portfolio? Discuss the relationship between projects, programs, and portfolio management and the contributions they each make to enterprise success.
5. What is the role of the project manager? What are suggested skills for all project managers and for information technology project managers? Why is leadership so important for project managers? How is the job market for information technology project managers?
6. Briefly describe some key events in the history of project management. What role does the Project Management Institute and other professional societies play in helping the profession?
7. What functions can you perform with project management software? What are some popular names of low-end, midrange, and high-end project management tools?

## Exercises

1. Visit the Standish Group's Web site at *www.standishgroup.com*. Read one of the CHAOS articles, and also read at least one report or article that questions the findings of the CHAOS studies. See the Suggested Readings by Robert L. Glass on the companion Web site for references. Write a two-page summary of the reports, key conclusions, and your opinion of them.
2. Find someone who works as a project manager or someone who works on projects, such as a worker in your school's IT department or the president of a social club. Prepare several interview questions to learn more about projects and project management, and then ask them your questions in person, through e-mail, or over the phone. Write a two-page summary of your findings. Guidelines for your interview and sample questions are available on the companion Web site.
3. Search the Internet for the terms *project management, project management careers, project portfolio management,* and *information technology project management.* Write down the number of hits that you received for each of these phrases. Find at least three Web sites that provide interesting information on one of the topics. Write a two-page paper summarizing key information about these three Web sites as well as the Project Management Institute's Web site (*www.pmi.org*).
4. Find any example of a real project with a real project manager. Feel free to use projects in the media (the Olympics, television shows, movies, etc.) or a project from your work, if applicable. Write a two-page paper describing the project in terms of its scope, time, and cost goals. Discuss what went right and wrong on the project and the role of the project manager and sponsor. Also describe if the project was a success or not and why. Include at least one reference and cite it on the last page.
5. Skim through Appendix A on Microsoft Project 2007. Review information about Project 2007 from Microsoft's Web site (*www.microsoft.com/project*) and information about VPMi Express from *www.vcsonline.com*. Also, visit The Project Management Center (*www.infogoal.com/pmc*) and Top Ten Reviews (*http://project-management-software-review.toptenreviews.com*). Research two project management software tools besides Project 2007. Write a two-page paper answering the following questions:
   a. What functions does project management software provide that you cannot do easily using other tools such as a spreadsheet or database?
   b. How do the different tools you reviewed compare, based on cost of the tool, key features, and other relevant criteria?
   c. How can organizations justify investing in enterprise or portfolio project management software?
6. Research information about PMP and related certifications. Skim through Appendix B for information and find at least two articles on this topic. What are benefits of certification in general? Do you think it is worthwhile for most project managers to get certified? Is it something you would consider? Write a two-page paper summarizing your findings and opinions.

## Companion Web Site



Visit the companion Web site for this text at *www.cengage.com/mis/schwalbe* to access:

- References cited in the text and additional suggested readings for each chapter
- Template files
- Lecture notes
- Interactive quizzes
- Podcasts
- Links to general project management Web sites
- And more

See the Preface of this text for additional information on accessing the companion Web site.

## Key Terms

**best practice** — An optimal way recognized by industry to achieve a stated goal or objective

**critical path** — The longest path through a network diagram that determines the earliest completion of a project

**enterprise project management software** — Software that integrates information from multiple projects to show the status of active, approved, and future projects across an entire organization; also called portfolio project management software

**ethics** — A set of principles that guide our decision making based on personal values of what is "right" and "wrong"

**Gantt chart** — A standard format for displaying project schedule information by listing project activities and their corresponding start and finish dates in a calendar format

**green IT** or **green computing** — Developing and using computer resources in an efficient way to improve economic viability, social responsibility, and environmental impact

**leader** — A person who focuses on long-term goals and big-picture objectives, while inspiring people to reach those goals

**manager** — A person who deals with the day-to-day details of meeting specific goals

**portfolio project management software** — Software that integrates information from multiple projects to show the status of active, approved, and future projects across an entire organization; also called enterprise project management software

**program** — A group of projects managed in a coordinated way to obtain benefits and control not available from managing them individually

**program manager** — A person who provides leadership and direction for the project managers heading the projects within a program

**project** — A temporary endeavor undertaken to create a unique product, service, or result

**project management** — The application of knowledge, skills, tools, and techniques to project activities to meet project requirements

**Project Management Institute (PMI)** — An international professional society for project managers

**project management knowledge areas** — Project integration management, scope, time, cost, quality, human resource, communications, risk, and procurement management

**Project Management Office (PMO)** — An organizational group responsible for coordinating the project management functions throughout an organization

**Project Management Professional (PMP)** — Certification provided by PMI that requires documenting project experience and education, agreeing to follow the PMI code of ethics, and passing a comprehensive exam

**project management tools and techniques** — Methods available to assist project managers and their teams; some popular tools in the time management knowledge area include Gantt charts, network diagrams, and critical path analysis

**project manager** — The person responsible for working with the project sponsor, the project team, and the other people involved in a project to meet project goals

**project portfolio management** or **portfolio management** — When organizations group and manage projects as a portfolio of investments that contribute to the entire enterprise's success

**project sponsor** — The person who provides the direction and funding for a project

**stakeholders** — People involved in or affected by project activities

**triple constraint** — Balancing scope, time, and cost goals

**virtualization** — Hiding the physical characteristics of computing resources from their users, such as making a single server, operating system, application, or storage device appear to function as multiple virtual resources

## End Notes

[1] Andrew H. Bartels, "Teleconference: Global IT 2008 Market Forecast," *Forrester Research* (February 11, 2008).

[2] Eric Chabrow, "Computer Jobs Hit Record High," *CIO Insight* (July 7, 2008).

[3] Project Management Institute, *Project Management Salary Survey*, Fifth Edition, 2007.

[4] Project Management Institute, *PMI Today*, October 2006 and June 2008.

[5] Standish Group, "The CHAOS Report" (*www.standishgroup.com*) (1995). Another reference is Jim Johnson, "CHAOS: The Dollar Drain of IT Project Failures," *Application Development Trends* (January 1995).

[6] PricewaterhouseCoopers, "Boosting Business Performance through Programme and Project Management" (June 2004).

[7] Project Management Institute, *A Guide to the Project Management Body of Knowledge (PMBOK® Guide)*, Fourth Edition (2008).

[8] Gartner, Inc. "Gartner Identifies the Top 10 Strategic Technologies for 2008," *Gartner Symposium/ITxpo* (October 9, 2007).

[9] Nicholas G. Carr, *Does IT Matter?* (Harvard Business School Press, 2004).

[10] "Where I.T. Matters: How 10 Technologies Transformed 10 Industries," *Baseline* (October 2, 2006).

[11] Project Management Institute, Inc., *A Guide to the Project Management Body of Knowledge* (*PMBOK® Guide*), Fourth Edition (2008).

[12] Claude Besner and Brian Hobbs, "The Perceived Value and Potential Contribution of Project Management Practices to Project Success," *PMI Research Conference Proceedings* (July 2006).

[13] Jim Johnson, "CHAOS 2006 Research Project," *CHAOS Activity News*, 2:no. 1 (2007).

[14] Standish Group, "CHAOS 2001: A Recipe for Success" (2001).

[15] Chang Dong, K.B. Chuah, and Li Zhai, "A Study of Critical Success Factor of Information System Projects in China," *Proceedings of PMI Research Conference* (2004).

[16] Dragan Milosevic and And Ozbay, "Delivering Projects: What the Winners Do," *Proceedings of the Project Management Institute Annual Seminars & Symposium* (November 2001).

[17] Project Management Institute, *A Guide to the Project Management Body of Knowledge (PMBOK® Guide)*, Fourth Edition (2008).

[18] Eric Burke, "Project Portfolio Management," *PMI Houston Chapter Meeting* (July 10, 2002).

[19] Project Management Institute, *Organizational Project Management Maturity Model (OPM3) Knowledge Foundation* (2003).

[20] Ibid., p. 13.

[21] Ultimate Business Library, *Best Practice: Ideas and Insights from the World's Foremost Business Thinkers* (New York: Perseus 2003), p. 1.

[22] Ibid., p. 8.

[23] Mary Brandel, "The Perfect Project Manager," *ComputerWorld* (August 6, 2001).

[24] Lauren Thomsen-Moore, "No 'soft skills' for us, we're techies," *ComputerWorld* (December 16, 2002).

[25] Jennifer Krahn, "Effective Project Leadership: A Combination of Project Manager Skills and Competencies in Context," *PMI Research Conference Proceedings* (July 2006).

[26] Stacy Collett, "Hot Skills: Cold Skills," *ComputerWorld* (July 17, 2006).

[27] Regents of the University of California, Manhattan Project History, "Here's Your Damned Organization Chart," (1998–2001).

[28] Venessa Wong, "PMI On Specialization and Globalization," *Projects@Work* (June 23, 2008)

[29] Project Management Institute, "PMI Today," (February 2009).

[30] Project Management Institute, "PMI Community Post," (February 9, 2007).

[31] Project Management Institute, "PMI Today," (December 2006), p. 12–13.

[32] Top Ten Reviews, "2006 Project Management Report," *http://project-management-software-review.toptenreviews.com*, (accessed September 10, 2008).

CHAPTER 2

# THE PROJECT MANAGEMENT AND INFORMATION TECHNOLOGY CONTEXT

## LEARNING OBJECTIVES

**After reading this chapter, you will be able to:**

- Describe the systems view of project management and how it applies to information technology projects
- Understand organizations, including the four frames, organizational structures, and organizational culture
- Explain why stakeholder management and top management commitment are critical for a project's success
- Understand the concept of a project phase and the project life cycle and distinguish between project development and product development
- Discuss the unique attributes and diverse nature of information technology projects
- Describe recent trends affecting IT project management, including globalization, outsourcing, and virtual teams

## OPENING CASE

Tom Walters recently accepted a new position at his college as the Director of Information Technology. Tom had been a respected faculty member at the college for the past 15 years. The college—a small, private institution in the Southwest—offered a variety of programs in the liberal arts and professional areas. Enrollment included 1,500 full-time traditional students and about 1,000 working-adult students attending evening programs. Many instructors supplemented their courses with information on the Internet and course Web sites, but they did not offer any distance-learning programs. The college's niche was serving students in that region who liked the setting of a small liberal arts college.

Like other institutions of higher learning, the use of information technology at the college had grown tremendously in the past 10 years. There were a few classrooms on campus with computers for the instructors and students, and a few more with just instructor stations and projection systems. Tom knew that several colleges throughout the country required that all students lease laptops and that these colleges incorporated technology components into most courses. This idea fascinated him. He and two other members of the Information Technology department visited a local college that had required all students to lease laptops for the past three years, and they were very impressed with what they saw and heard. Tom and his staff developed plans to start requiring students to lease laptops at their college the next year.

Tom sent an e-mail to all faculty and staff in September, which briefly described this and other plans. He did not get much response, however, until the February faculty meeting when, as he described some of the details of his plan, the chairs of the History, English, Philosophy, and Economics departments all voiced their opposition to the idea. They eloquently stated that the college was not a technical training school, and they thought the idea was ludicrous. Members of the Computer Science department voiced their concern that almost all of their students already had state-of-the art laptops and would not want to pay a mandatory fee to lease less-powerful ones. The director of the adult education program expressed her concern that many adult-education students would balk at an increase in fees. Tom was in shock to hear his colleagues' responses, especially after he and his staff had spent a lot of time planning details of how to implement laptops at their campus. Now what should he do?

Many of the theories and concepts of project management are not difficult to understand. What *is* difficult is implementing them in various environments. Project managers must consider many different issues when managing projects. Just as each project is unique, so is its environment. This chapter discusses some of the components involved in understanding the project environment, such as using a systems approach, understanding organizations, managing stakeholders, matching product life cycles to the project environment, understanding the context of information technology projects, and reviewing recent trends affecting IT project management.

## A SYSTEMS VIEW OF PROJECT MANAGEMENT

Even though projects are temporary and intended to provide a unique product or service, you cannot run projects in isolation. If project managers lead projects in isolation, it is unlikely that those projects will ever truly serve the needs of the organization. Therefore, projects must operate in a broad organizational environment, and project managers need to consider projects within the greater organizational context. To handle complex situations effectively, project managers need to take a holistic view of a project and understand how it relates to the larger organization. **Systems thinking** describes this holistic view of carrying out projects within the context of the organization.



## What Is a Systems Approach?

The term **systems approach** emerged in the 1950s to describe a holistic and analytical approach to solving complex problems that includes using a systems philosophy, systems analysis, and systems management. A **systems philosophy** is an overall model for thinking about things as systems. **Systems** are sets of interacting components working within an environment to fulfill some purpose. For example, the human body is a system composed of many subsystems—the nervous system, the skeletal system, the circulatory system, the digestive system, and so on. **Systems analysis** is a problem-solving approach that requires defining the scope of the system, dividing it into its components, and then identifying and evaluating its problems, opportunities, constraints, and needs. Once this is completed, the systems analyst then examines alternative solutions for improving the current situation, identifies an optimum, or at least satisfactory, solution or action plan, and examines that plan against the entire system. **Systems management** addresses the business, technological, and organizational issues associated with creating, maintaining, and making changes to a system.

Using a systems approach is critical to successful project management. Top management and project managers must follow a systems philosophy to understand how projects relate to the whole organization. They must use systems analysis to address needs with a problem-solving approach. They must use systems management to identify key business, technological, and organizational issues related to each project in order to identify and satisfy key stakeholders and do what is best for the entire organization.

In the opening case, when Tom Walters planned the laptop project, he did not use a systems approach. Members of his IT department did all of the planning. Even though Tom sent an e-mail describing the laptop project to all faculty and staff, he did not address many of the organizational issues involved in such a complex project. Most faculty and staff are very busy at the beginning of fall term and many may not have read the entire message. Others may have been too busy to communicate their concerns to the Information Technology department. Tom was unaware of the effects the laptop project would have on other parts of the college. He did not clearly define the business, technological, and organizational issues associated with the project. Tom and the Information Technology department began work on the laptop project in isolation. If they had taken a systems approach, considering other dimensions of the project, and involving key stakeholders, they could have identified and addressed many of the issues raised at the February faculty meeting *before* the meeting.

## The Three-Sphere Model for Systems Management

Many business and information technology students understand the concepts of systems and performing a systems analysis. However, they often gloss over the topic of systems management. The simple idea of addressing the three spheres of systems management—business, organization, and technology—can have a huge impact on selecting and managing projects successfully.

Figure 2-1 provides a sample of some of the business, organizational, and technological issues that could be factors in the laptop project. In this case, technological issues, though

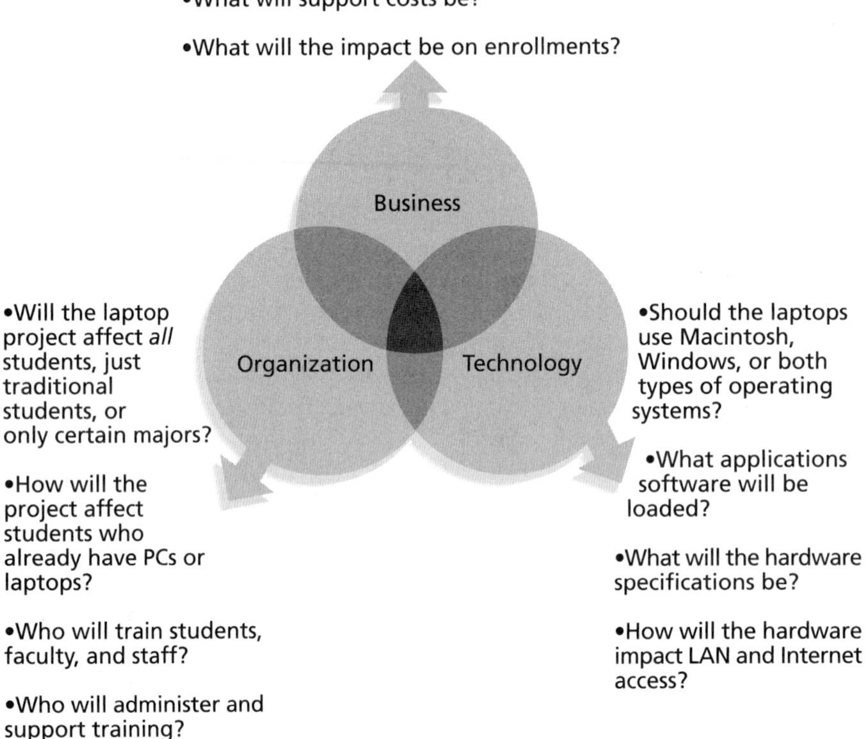


**FIGURE 2-1** Three-sphere model for systems management

not simple by any means, are probably the least difficult to identify and resolve. However, projects must address issues in all three spheres of the systems management model. Although it is easier to focus on the immediate and sometimes narrow concerns of a particular project, project managers and other staff must keep in mind the effects of any project on the interests and needs of the entire system or organization.

Many information technology professionals become captivated with the technology and day-to-day problem solving involved in working with information systems. They tend to become frustrated with many of the "people problems" or politics involved in most organizations. In addition, many information technology professionals ignore important business issues—such as, "Does it make financial sense to pursue this new technology?" or, "Should the company develop this software in-house or purchase it off-the-shelf?" Using a more holistic approach helps project managers integrate business and organizational issues into their planning. It also helps them look at projects as a series of interrelated phases. When you integrate business and organizational issues into project management planning and look at projects as a series of interrelated phases, you do a better job of ensuring project success.

## UNDERSTANDING ORGANIZATIONS

The systems approach requires that project managers always view their projects in the context of the larger organization. Organizational issues are often the most difficult part of working on and managing projects. For example, many people believe that most projects fail because of company politics. Project managers often do not spend enough time identifying all the stakeholders involved in projects, especially the people opposed to the projects. In fact, the latest edition of the *PMBOK® Guide* added a new initiating process under project communications management called "identify stakeholders." (See Chapter 10 for more information.) Project managers also often do not spend enough time considering the political context of a project or the culture of the organization. To improve the success rate of information technology projects, it is important for project managers to develop a better understanding of people as well as organizations.

### The Four Frames of Organizations

Organizations can be viewed as having four different frames: structural, human resources, political, and symbolic:[1]

- The **structural frame** deals with how the organization is structured (usually depicted in an organizational chart) and focuses on different groups' roles and responsibilities in order to meet the goals and policies set by top management. This frame is very rational and focuses on coordination and control. For example, within the structural frame, a key information technology issue is whether a company should centralize the information technology personnel in one department or decentralize across several departments. You will learn more about organizational structures in the next section.
- The **human resources (HR) frame** focuses on producing harmony between the needs of the organization and the needs of the people. It recognizes that there are often mismatches between the needs of the organization and the needs of individuals and groups and works to resolve any potential problems. For example, many projects might be more efficient for the organization if personnel worked 80 or more hours a week for several months. This work schedule would probably conflict with the personal lives of those people. Important issues in information technology related to the human resources frame are the shortage of skilled information technology workers within the organization and unrealistic schedules imposed on many projects.
- The **political frame** addresses organizational and personal politics. **Politics** in organizations take the form of competition among groups or individuals for power and leadership. The political frame assumes that organizations are coalitions composed of varied individuals and interest groups. Often, important decisions need to be made based on the allocation of scarce resources. Competition for scarce resources makes conflict a central issue in organizations, and power improves the ability to obtain scarce resources. Project managers must pay attention to politics and power if they are to be effective. It is important to know who opposes your projects as well as who supports them. Important issues in information technology related to the political frame are the power shifts from central functions to operating units or from functional managers to project managers.

- The **symbolic frame** focuses on symbols and meanings. What is most important about any event in an organization is not what actually happened, but what it means. Was it a good sign that the CEO came to a kickoff meeting for a project, or was it a threat? The symbolic frame also relates to the company's culture. How do people dress? How many hours do they work? How do they run meetings? Many information technology projects are international and include stakeholders from various cultures. Understanding those cultures is also a crucial part of the symbolic frame.

WHAT WENT WRONG?

Several large organizations have installed or tried to install enterprise resource planning (ERP) systems to integrate business functions such as ordering, inventory, delivery, accounting, and human resource management. They understand the potential benefits of an ERP system and can analyze its various technical issues, but many companies do not realize how important the organizational issues are to ERP implementations.

For example, in early 2001, Sobey's, Canada's second largest grocery store chain with 1,400 stores, abandoned its two-year, $90 million investment in an ERP system. The system was developed by SAP, the largest enterprise software company and the third-largest software supplier. Unfortunately, the system did not work properly due to several organizational challenges. People in different parts of the company had different terms for various items, and it was difficult to make the necessary decisions for the ERP system. Also, no one wanted to take the time required to get the new system to work because they had their daily work to do. Every department has to work together to implement an ERP system, and it is often difficult to get departments to communicate their needs. As Dalhousie University Associate Professor Sunny Marche states, "The problem of building an integrated system that can accommodate different people is a very serious challenge. You can't divorce technology from the sociocultural issues. They have an equal role." Sobey's ERP system shut down for five days and employees were scrambling to stock potentially empty shelves in several stores for weeks. The system failure cost Sobey's more than $90 million and caused shareholders to take an 82-cent after-tax hit per share.[2]

Project managers must learn to work within all four organizational frames to function well in organizations. Chapter 9, Project Human Resource Management, and Chapter 10, Project Communications Management, further develop some of the organizational issues. The following sections on organizational structures, organizational culture, stakeholder management, and the need for top management commitment provide additional information related to the structural and political frames.

## Organizational Structures

Many discussions of organizations focus on organizational structure. Three general classifications of organizational structures are functional, project, and matrix. Most companies today involve all three structures somewhere in the organization, but one is usually most common. Figure 2-2 portrays these three organizational structures. A **functional organizational structure** is the hierarchy most people think of when picturing an organizational

Functional

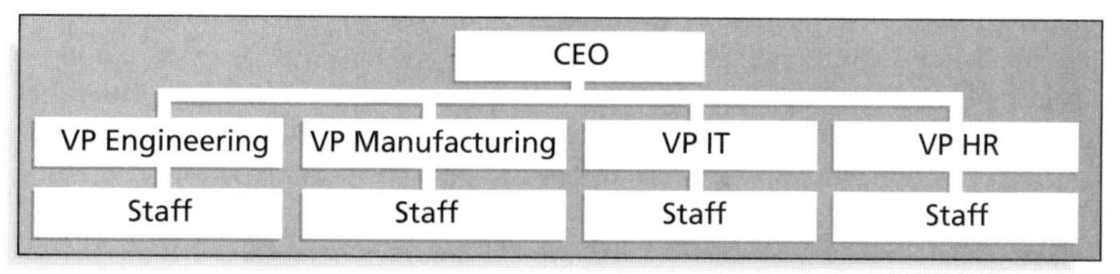


Project

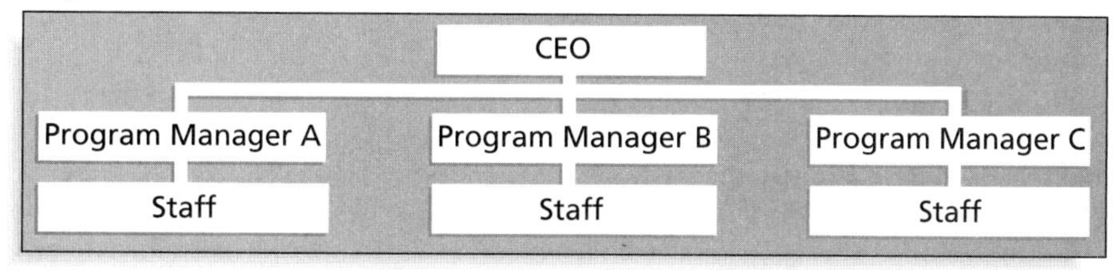


Matrix

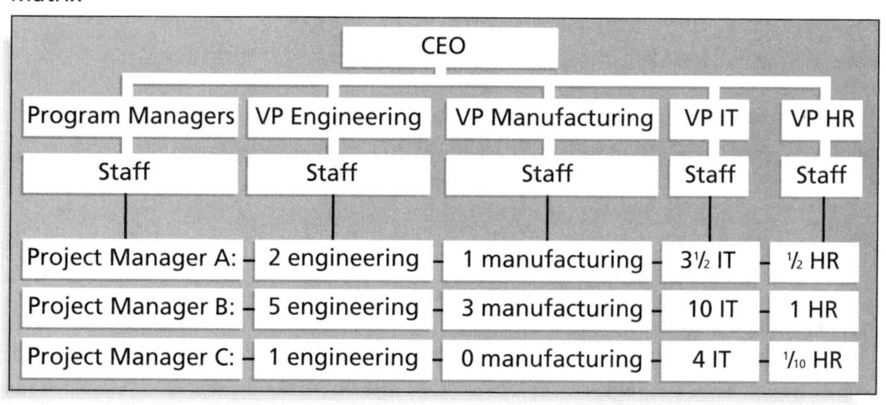


**FIGURE 2-2** Functional, project, and matrix organizational structures

chart. Functional managers or vice presidents in specialties such as engineering, manufacturing, information technology, and human resources report to the chief executive officer (CEO). Their staffs have specialized skills in their respective disciplines. For example, most colleges and universities have very strong functional organizations. Only faculty in the business department teach business courses; faculty in the history department teach history; faculty in the art department teach art, and so on.

A **project organizational structure** also has a hierarchical structure, but instead of functional managers or vice presidents reporting to the CEO, program managers report to the CEO. Their staffs have a variety of skills needed to complete the projects within their programs. An organization that uses this structure earns their revenue primarily from performing projects for other groups under contract. For example, many defense, architectural, engineering, and consulting companies use a project organizational structure. These companies often hire people specifically to work on particular projects.

A **matrix organizational structure** represents the middle ground between functional and project structures. Personnel often report to both a functional manager and one or more project managers. For example, information technology personnel at many companies often split their time between two or more projects, but they report to their manager in the information technology department. Project managers in matrix organizations have staff from various functional areas working on their projects, as shown in Figure 2-2. Matrix organizational structures can be strong, weak, or balanced, based on the amount of control exerted by the project managers.

Table 2-1 summarizes how organizational structures influence projects and project managers, based on information from several versions of the *PMBOK® Guide*. Project managers have the most authority in a pure project organizational structure and the least amount of authority in a pure functional organizational structure. It is important that project managers understand the current organizational structure under which they are working. For example, if someone in a functional organization is asked to lead a project that requires strong support from several different functional areas, he or she should ask for top management sponsorship. This sponsor should solicit support from all relevant functional managers to ensure that they cooperate on the project and that qualified people are

**TABLE 2-1** Organizational structure influences on projects

| Project Characteristics | Organizational Structure Type | | | | |
|---|---|---|---|---|---|
| | Functional | Matrix | | | Project |
| | | **Weak Matrix** | **Balanced Matrix** | **Strong Matrix** | |
| Project manager's authority | Little or none | Limited | Low to Moderate | Moderate to High | High to almost total |
| Percent of organization's personnel assigned full-time to project work | Virtually none | 0–25% | 15–60% | 50–95% | 85–100% |
| Who controls the project budget | Functional manager | Functional manager | Mixed | Project manager | Project manager |
| Project manager's role | Part-time | Part-time | Full-time | Full-time | Full-time |
| Common title for project manager's role | Project Coordinator/ Project Leader | Project Coordinator/ Project Leader | Project Manager/ Project Officer | Project Manager / Program Manager | Project Manager/ Program Manager |
| Project management administrative staff | Part-time | Part-time | Part-time | Full-time | Full-time |

available to work as needed. The project manager might also ask for a separate budget to pay for project-related trips, meetings, and training or to provide financial incentives to the people supporting the project.



Even though project managers have the most authority in the project organizational structure, this type of organization is often inefficient for the company as a whole. Assigning staff full-time to the project often creates underutilization and/or misallocation of staff resources. For example, if a technical writer is assigned full-time to a project, but there is no work for him or her on a particular day, the organization is wasting money by paying that person a full-time wage. Project organizations may also miss economies of scale available through the pooling of requests for materials with other projects.

Disadvantages such as these illustrate the benefit of using a systems approach to managing projects. For example, the project manager might suggest hiring an independent contractor to do the technical writing work instead of using a full-time employee. This approach would save the organization money while still meeting the needs of the project. When project managers use a systems approach, they are better able to make decisions that address the needs of the entire organization.

## Organizational Culture

Just as an organization's structure affects its ability to manage projects, so does an organization's culture. **Organizational culture** is a set of shared assumptions, values, and behaviors that characterize the functioning of an organization. It often includes elements of all four frames described previously. Organizational culture is very powerful, and many people believe the underlying causes of many companies' problems are not in the organizational structure or staff; they are in the culture. It is also important to note that the same organization can have different subcultures. The information technology department may have a different organizational culture than the finance department, for example. Some organizational cultures make it easier to manage projects.

According to Stephen P. Robbins and Timothy Judge, authors of a popular textbook on organizational behavior, there are ten characteristics of organizational culture:

1. *Member identity*: The degree to which employees identify with the organization as a whole rather than with their type of job or profession. For example, a project manager or team member might feel more dedicated to his or her company or project team than to their job or profession, or they might not have any loyalty to a particular company or team. As you can guess, an organizational culture where employees identify more with the whole organization are more conducive to a good project culture.
2. *Group emphasis*: The degree to which work activities are organized around groups or teams, rather than individuals. An organizational culture that emphasizes group work is best for managing projects.
3. *People focus*: The degree to which management's decisions take into account the effect of outcomes on people within the organization. A project manager might assign tasks to certain people without considering their individual needs, or the project manager might know each person very well and focus on individual needs when assigning work or making other decisions. Good project managers often balance the needs of individuals and the organization.

4. *Unit integration*: The degree to which units or departments within an organization are encouraged to coordinate with each other. Most project managers strive for strong unit integration to deliver a successful product, service, or result. An organizational culture with strong unit integration makes the project manager's job easier.
5. *Control*: The degree to which rules, policies, and direct supervision are used to oversee and control employee behavior. Experienced project managers know it is often best to balance the degree of control to get good project results.
6. *Risk tolerance*: The degree to which employees are encouraged to be aggressive, innovative, and risk seeking. An organizational culture with a higher risk tolerance is often best for project management since projects often involve new technologies, ideas, and processes.
7. *Reward criteria*: The degree to which rewards, such as promotions and salary increases, are allocated according to employee performance rather than seniority, favoritism, or other nonperformance factors. Project managers and their teams often perform best when rewards are based mostly on performance.
8. *Conflict tolerance*: The degree to which employees are encouraged to air conflicts and criticism openly. It is very important for all project stakeholders to have good communications, so it is best to work in an organization where people feel comfortable discussing conflict openly.
9. *Means-ends orientation*: The degree to which management focuses on outcomes rather than on techniques and processes used to achieve results. An organization with a balanced approach in this area is often best for project work.
10. *Open-systems focus*: The degree to which the organization monitors and responds to changes in the external environment. As discussed earlier in this chapter, projects are part of a larger organizational environment, so it is best to have a strong open-systems focus.[3]

As you can see, there is a definite relationship between organizational culture and successful project management. Project work is most successful in an organizational culture where employees identify more with the organization, where work activities emphasize groups, and where there is strong unit integration, high risk tolerance, performance-based rewards, high conflict tolerance, an open-systems focus, and a balanced focus on people, control, and means orientation.

## STAKEHOLDER MANAGEMENT

Recall from Chapter 1 that project stakeholders are the people involved in or affected by project activities. Stakeholders can be internal to the organization, external to the organization, directly involved in the project, or simply affected by the project. Internal project stakeholders generally include the project sponsor, project team, support staff, and internal customers for the project. Other internal stakeholders include top management, other functional managers, and other project managers. Since organizations have limited resources, projects affect top management, other functional managers, and other project managers by using some of the organization's limited resources. Thus, while additional internal stakeholders may not be directly involved in the project, they are still stakeholders because the project affects them in some way. External project stakeholders include the project's customers (if they are external to the organization), competitors, suppliers, and other external groups potentially involved in or affected by the project, such as government officials or concerned citizens. Since the purpose of project management is to meet project requirements and satisfy stakeholders, it is critical that project managers take adequate time to identify, understand, and manage relationships with all project stakeholders. Using the four frames of organizations to think about project stakeholders can help you meet their expectations.

Consider again the laptop project from the opening case. Tom Walters seemed to focus on just a few internal project stakeholders. He viewed only part of the structural frame of the college. Since his department would do most of the work in administering the laptop project, he concentrated on those stakeholders. Tom did not even involve the main customers for this project—the students at the college. Even though Tom sent an e-mail to faculty and staff, he did not hold meetings with senior administration or faculty at the college. Tom's view of who the stakeholders were for the laptop project was very limited.

During the faculty meeting, it became evident that the laptop project had many stakeholders in addition to the Information Technology department and students. If Tom had expanded his view of the structural frame of his organization by reviewing an organizational chart for the entire college, he could have identified other key stakeholders. He would have been able to see that the laptop project would affect academic department heads and members of different administrative areas. If Tom had focused on the human resources frame, he would have been able to tap his knowledge of the college and identify individuals who would most support or oppose requiring laptops. By using the political frame, Tom could have considered the main interest groups that would be most affected by this project's outcome. Had he used the symbolic frame, Tom could have tried to address what moving to a laptop environment would really mean for the college. He then could have anticipated some of the opposition from people who were not in favor of increasing the use of technology on campus. He also could have solicited a strong endorsement from the college president or dean before talking at the faculty meeting.

Tom Walters, like many new project managers, learned the hard way that his technical and analytical skills were not enough to guarantee success in project management. To be more effective, he had to identify and address the needs of different stakeholders and understand how his project related to the entire organization.

## MEDIA SNAPSHOT

The *New York Times* reported that the project to rebuild Ground Zero in New York City is having severe problems. Imagine all of the stakeholders involved in this huge, highly emotional project. A 34-page report (see the article reference for further information) describes the many challenges faced in the reconstruction of the former World Trade Center site nearly seven years after the terrorist attack of September 11, 2001. The report listed at least 15 fundamental unresolved issues, including the lack of final designs for the proposed World Trade Center Transportation Hub; the unfinished decontamination and

*continued*

dismantling of the former Deutsche Bank tower; and the resolution of a land-rights issue with the St. Nicholas Greek Orthodox Church.

"Perhaps most pressingly, the report identified a need for 'a more efficient, centralized decision-making structure—a steering committee—with authority to make final decisions on matters which fundamentally drive schedule and cost.'"[4] The "What Went Right?" example later in this chapter describes the benefits of having an executive steering committee to help projects succeed, especially when there are many stakeholders and challenges involved.

## The Importance of Top Management Commitment

People in top management positions, of course, are key stakeholders in projects. A very important factor in helping project managers successfully lead projects is the level of commitment and support they receive from top management. Without top management commitment, many projects will fail. Some projects have a senior manager called a **champion** who acts as a key proponent for a project. The sponsor can serve as the champion, but often another manager can more successfully take on this role. As described earlier, projects are part of the larger organizational environment, and many factors that might affect a project are out of the project manager's control. Several studies cite executive support as one of the key factors associated with virtually all project success.

Top management commitment is crucial to project managers for the following reasons:

- Project managers need adequate resources. The best way to kill a project is to withhold the required money, human resources, and visibility for the project. If project managers have top management commitment, they will also have adequate resources and not be distracted by events that do not affect their specific projects.
- Project managers often require approval for unique project needs in a timely manner. For example, on large information technology projects, top management must understand that unexpected problems may result from the nature of the products being produced and the specific skills of the people on the project team. For example, the team might need additional hardware and software halfway through the project for proper testing, or the project manager might need to offer special pay and benefits to attract and retain key project personnel. With top management commitment, project managers can meet these specific needs in a timely manner.
- Project managers must have cooperation from people in other parts of the organization. Since most information technology projects cut across functional areas, top management must help project managers deal with the political issues that often arise in these types of situations. If certain functional managers are not responding to project managers' requests for necessary information, top management must step in to encourage functional managers to cooperate.
- Project managers often need someone to mentor and coach them on leadership issues. Many information technology project managers come from technical positions and are inexperienced as managers. Senior managers should take the time to pass on advice on how to be good leaders. They should encourage new

project managers to take classes to develop leadership skills and allocate the time and funds for them to do so.



Information technology project managers work best in an environment in which top management values information technology. Working in an organization that values good project management and sets standards for its use also helps project managers succeed.

## BEST PRACTICE

A major element of good practice concerns **IT governance**, which addresses the authority and control for key IT activities in organizations, including IT infrastructure, IT use, and project management. (The term *project governance* can also used to describe a uniform method of controlling all types of projects.) The IT Governance Institute (ITGI) was established in 1998 to advance international thinking and standards in directing and controlling an organization's use of technology. Effective IT governance helps ensure that IT supports business goals, maximizes investment in IT, and addresses IT-related risks and opportunities. A 2004 book by Peter Weill and Jeanne Ross called *IT Governance: How Top Performers Manage IT Decision Rights for Superior Results*[5] includes research stating that firms with superior IT governance systems have 20 percent higher profits than firms with poor governance. (See the ITGI's Web site *www.itgi.org* for more information, including many case studies and best practices in this area.)

A lack of IT governance can be dangerous, as evidenced by three well-publicized IT project failures in Australia—Sydney Water's customer relationship management system, the Royal Melbourne Institute of Technology's academic management system, and One.Tel's billing system. Researchers explained how these projects were catastrophic for their organizations, primarily due to a severe lack of IT governance, which the authors dubbed *managerial IT unconsciousness*, the title of their article.

> "All three projects suffered from poor IT governance. Senior management in all three organizations had not ensured that prudent checks and balances were in place to enable them to monitor either the progress of the projects or the alignment and impact of the new systems on their business. Proper governance, particularly with respect to financial matters, auditing, and contract management, was not evident. Also, project-level planning and control were notably absent or inadequate—with the result that project status reports to management were unrealistic, inaccurate, and misleading."[6]

### The Need for Organizational Commitment to Information Technology

Another factor affecting the success of information technology projects is the organization's commitment to information technology in general. It is very difficult for a large information technology project (or a small one, for that matter) to be successful if the organization itself does not value information technology. Many companies have realized that information technology is integral to their business and have created a vice president or equivalent-level position for the head of information technology, often called the Chief Information Officer (CIO). Some companies assign people from non-information technology areas to work on large projects full-time to increase involvement from end users of the systems. Some CEOs even take a strong leadership role in promoting the use of information technology in their organizations.

Gartner, Inc., a well-respected information technology consulting firm, provides awards to organizations for excellence in applying various technologies. For example, in 2006, Gartner announced the winners of its eighth annual Customer Relationship Management (CRM) Excellence Awards. BNSF Railway received the award in the "Excellence in Enterprise CRM" category, and UnitedHealth Group received the award in the "Excellence in Sales, Marketing or Customer Service" category. (Electronic Arts, an independent producer of electronic games, won the award in 2007.) The 2006 award winners had the following to say:

- *Elizabeth Obermiller, director of ERM systems for BNSF Railway*: "Our success was driven by the ongoing executive commitment and passionate and talented teams, who were able to implement a planned and phased approach with advanced application of analytics to monitor, measure and drive success."
- *John Reinke, a senior vice president of Uniprise, a UnitedHealth Group*: "We are excited to receive this award for our partnership with eLoyalty to implement a new, cutting-edge call center technology application called Behavioral Analytics™. This technology allows us to engage in deeper, more personally relevant phone conversations with each consumer who speaks with a customer care professional. Health care consumers often face complex and emotional issues, and this is a great example of how technology can help improve their experience."[7]

## The Need for Organizational Standards

Another problem in most organizations is not having standards or guidelines to follow that could help in performing project management. These standards or guidelines might be as simple as providing standard forms or templates for common project documents, examples of good project management plans, or guidelines on how the project manager should provide status information to top management. The content of a project management plan and how to provide status information might seem like common sense to senior managers, but many new information technology project managers have never created plans or given a non-technical status report. Top management must support the development of these standards and guidelines and encourage or even enforce their use. For example, an organization might require all potential project information in a standard format to make project portfolio management decisions. If a project manager does not submit a potential project in the proper format, it could be rejected.

As described in Chapter 1, some organizations invest heavily in project management by creating a project management office or center of excellence, an organizational entity created to assist project managers in achieving project goals and maintaining project governance. Rachel Hollstadt, founder and CEO of a project management consulting firm, suggests that organizations consider adding a new position, a Chief Project Officer (CPO). Some organizations develop career paths for project managers. Some require that all project managers have Project Management Professional (PMP) certification and that all employees have some type of project management training. The implementation of all of these standards demonstrates an organization's commitment to project management.



## PROJECT PHASES AND THE PROJECT LIFE CYCLE

Since projects operate as part of a system and involve uncertainty, it is good practice to divide projects into several phases. A **project life cycle** is a collection of project phases. Some organizations specify a set of life cycles for use on all of their projects, while others follow common industry practices based on the types of projects involved. In general, project life cycles define what work will be performed in each phase, what deliverables will be produced and when, who is involved in each phase, and how management will control and approve work produced in each phase. A **deliverable** is a product or service, such as a technical report, a training session, a piece of hardware, or a segment of software code, produced or provided as part of a project. (See Chapter 5, Project Scope Management, for detailed information on deliverables.)

In early phases of a project life cycle, resource needs are usually lowest and the level of uncertainty is highest. Project stakeholders have the greatest opportunity to influence the final characteristics of the project's products, services, or results during the early phases of a project life cycle. It is much more expensive to make major changes to a project during latter phases. During the middle phases of a project life cycle, the certainty of completing a project improves as a project continues, more information is known about the project requirements and objectives, and more resources are usually needed than during the initial or final phase. The final phase of a project focuses on ensuring that project requirements were met and that the project sponsor approves completion of the project.

Project phases vary by project or industry, but some general phases in traditional project management are often called the concept, development, implementation, and close-out phases. The *PMBOK Guide®, Fourth Edition* calls these phases starting the project, organizing and preparing, carrying out the project work, and finishing the project. These phases should not be confused with the project management process groups of initiating, planning, executing, monitoring and controlling, and closing, as described in Chapter 3. The first two traditional project phases (concept and development) focus on planning and are often referred to as **project feasibility**. The last two phases (implementation and close-out) focus on delivering the actual work and are often referred to as **project acquisition**. A project should successfully complete each phase before moving on to the next. This project life cycle approach provides better management control and appropriate links to the ongoing operations of the organization.

Figure 2-3 provides a summary framework for the general phases of the traditional project life cycle. In the concept phase of a project, managers usually develop some type of business case, which describes the need for the project and basic underlying concepts. A preliminary or rough cost estimate is developed in this first phase, and an overview of the work involved is created. A work breakdown structure (WBS) outlines project work by decomposing the work activities into different levels of tasks. The WBS is a deliverable-oriented document that defines the total scope of the project. (You will learn more about the work breakdown structure in Chapter 5, Project Scope Management.) For example, if Tom

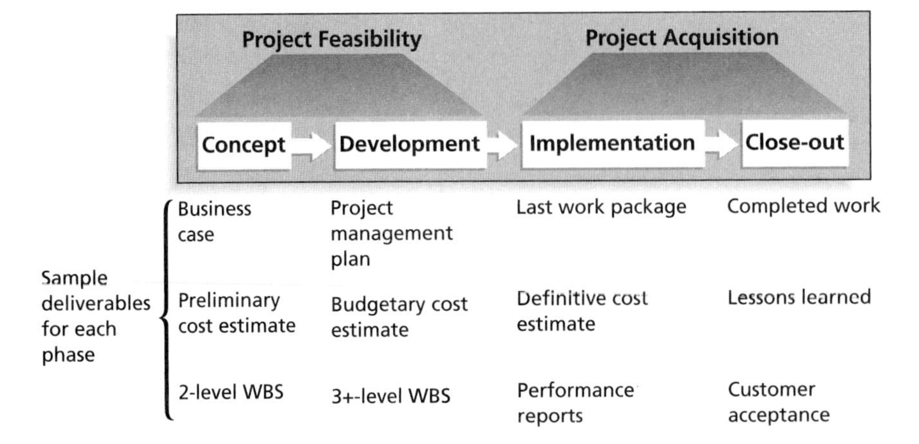


**FIGURE 2-3** Phases of the traditional project life cycle

Walters (from the opening case) had followed the project life cycle instead of moving full-steam ahead with the laptop project, he could have created a committee of faculty and staff to study the concept of increasing the use of technology on campus. This committee might have developed a business case and plan that included an initial, smaller project to investigate alternative ways of increasing the use of technology. They might have estimated that it would take six months and $20,000 to conduct a detailed technology study. The WBS at this phase of the study might have three levels and partition the work to include a competitive analysis of what five similar campuses were doing, a survey of local students, staff, and faculty, and a rough assessment of how using more technology would affect costs and enrollments. At the end of the concept phase, the committee would be able to deliver a report and presentation on its findings. The report and presentation would be an example of a deliverable.

After the concept phase is completed, the next project phase—development—begins. In the development phase, the project team creates more detailed project management plans, a more accurate cost estimate, and a more thorough WBS. In the example under discussion, suppose the concept phase report suggested that requiring students to have laptops was one means of increasing the use of technology on campus. The project team could then further expand this idea in the development phase. They would have to decide if students would purchase or lease the laptops, what type of hardware and software the laptops would require, how much to charge students, how to handle training and maintenance, how to integrate the use of the new technology with current courses, and so on. If, however, the concept phase report showed that the laptop idea was not a good idea for the college, then the project team would no longer consider increasing the use of technology by requiring laptops in the development phase and would cancel the project before development. This phased approach minimizes the time and money spent developing inappropriate projects. A project idea must pass the concept phase before evolving into the development phase.

The third phase of the traditional project life cycle is implementation. In this phase, the project team creates a definitive or very accurate cost estimate, delivers the required work, and provides performance reports to stakeholders. Suppose Tom Walters' college took the idea of requiring students to have laptops through the development phase. During the implementation phase, the project team would need to obtain the required hardware and software, install the necessary network equipment, deliver the laptops to the students, create a process for collecting fees, provide training to students, faculty, and staff, and so on. Other people on campus would also be involved in the implementation phase. Faculty would need to consider how best to take advantage of the new technology. The recruiting staff would have to update their materials to reflect this new feature of the college. Security would need to address new problems that might result from having students carry around expensive equipment. The project team usually spends the bulk of their efforts and money during the implementation phase of projects.



The last phase of the traditional project life cycle is close-out. In the close-out phase, all of the work is completed, and there should be some sort of customer acceptance of the entire project. The project team should document its experiences on the project in a lessons-learned report. If the laptop idea made it all the way through the implementation phase and all students received laptops, the project team would then complete the project by closing out any related activities. Team members might administer a survey to students, faculty, and staff in order to gather opinions on how the project fared. They would ensure that any contracts with suppliers were completed and appropriate payments made. They would transition future work related to the laptop project to other parts of the organization. The project team could also share its lessons-learned report with other college campuses that are considering implementing a similar program.

Many projects, however, do not follow this traditional project life cycle. They still have general phases with some similar characteristics as the traditional project life cycle, but they are much more flexible. For example, there may be just three phases, the initial, intermediate, and final phase. Or there may be multiple intermediate phases. There might be a separate project just to complete a feasibility study. Regardless of the project life cycle's specific phases, it is good practice to think of projects as having phases that connect the beginning and the end of the project, so that people can measure progress toward achieving project goals during each phase.

Just as a *project* has a life cycle, so does a *product*. Information technology projects help produce products and services such as new software, hardware, networks, research reports, and training on new systems. Understanding the product life cycle is just as important to good project management as understanding the phases of the traditional project life cycle.

## Product Life Cycles

Recall from Chapter 1 that a project is defined as "a temporary endeavor undertaken to create a unique product, service, or result," and a program is defined as "a group of projects managed in a coordinated way." A program often refers to the creation of a product, like an automobile or a new operating system. Therefore, developing a product often involves many projects.

All products follow some type of life cycle—cars, buildings, even amusement parks. The Walt Disney Company, for example, follows a rigorous process to design, build, and test new products. They assign project managers to oversee the development of all new products, such as rides, parks, and cruise lines. Likewise, major automotive companies follow product life cycles to produce new cars, trucks, and other products. Most information technology professionals are familiar with the concept of a product life cycle, especially for developing software.



Software development projects are one subset of information technology projects. Many information technology projects involve researching, analyzing, and then purchasing and installing new hardware and software with little or no actual software development required. However, some projects involve minor software modifications to enhance existing software or to integrate one application with another. Other projects involve a major amount of software development. Many argue that developing software requires project managers to modify traditional project management methods, depending on a particular product's life cycle.

A **systems development life cycle (SDLC)** is a framework for describing the phases involved in developing information systems. Some popular models of an SDLC include the waterfall model, the spiral model, the incremental build model, the prototyping model, and the Rapid Application Development (RAD) model. These life cycle models are examples of a **predictive life cycle**, meaning that the scope of the project can be clearly articulated and the schedule and cost can be accurately predicted. The project team spends a large portion of the project effort attempting to clarify the requirements of the entire system and then producing a design. Users are often unable to see any tangible results in terms of working software for an extended period. Below are brief descriptions of several predictive SDLC models.[8]

- The waterfall life cycle model has well-defined, linear stages of systems analysis, design, construction, testing, and support. This life cycle model assumes that requirements will remain stable after they are defined.
- The spiral life cycle model was developed based on experience with various refinements of the waterfall model as applied to large government software projects. It recognizes the fact that most software is developed using an iterative or spiral approach rather than a linear approach.
- The incremental build life cycle model provides for progressive development of operational software, with each release providing added capabilities.
- The prototyping life cycle model is used for developing software prototypes to clarify user requirements for operational software. It requires heavy user involvement, and developers use a model to generate functional requirements and physical design specifications simultaneously. Developers can throw away or keep prototypes, depending on the project.
- The RAD life cycle model uses an approach in which developers work with an evolving prototype. This life cycle model also requires heavy user involvement and helps produce systems quickly without sacrificing quality. Developers use RAD tools such as CASE (computer-aided software engineering), JRP (joint requirements planning), and JAD (joint application design) to facilitate rapid prototyping and code generation.

In contrast to the predictive life cycle models, the **Adaptive Software Development (ASD)** life cycle model assumes that software development follows an adaptive approach because the requirements cannot be clearly expressed early in the life cycle. An adaptive approach is also used to provide more freedom than the prescriptive approaches. It allows the development to proceed by creating components that provide the functionality specified by the business group as these needs are discovered in a more free-form approach.

Important attributes of this approach are that the projects are mission driven and component based, using time-based cycles to meet target dates. Requirements are developed using an iterative approach, and development is risk driven and change tolerant to address and incorporate rather than mitigate risks. More recently, the term **agile software development** has become popular to describe new approaches that focus on close collaboration between programming teams and business experts. (See the companion Web site for the Suggested Readings related to agile and other software development methodologies.)

These life cycle models are all examples of SDLCs. Many Web sites and introductory management information systems texts describe each of them in detail. The type of software and complexity of the information system in development determines which life cycle model to use. It is important to understand the product life cycle to meet the needs of the project environment.

Most large information technology products are developed as a series of projects. For example, the systems planning phase for a new information system can include a project to hire an outside consulting firm to help identify and evaluate potential strategies for developing a particular business application, such as a new order processing system or general ledger system. It can also include a project to develop, administer, and evaluate a survey of users to get their opinions on the current information systems used for performing that business function in the organization. The systems analysis phase might include a project to create process models for certain business functions in the organization. It can also include a project to create data models of existing databases in the company related to the business function and application. The implementation phase might include a project to hire contract programmers to code a part of the system. The close-out phase might include a project to develop and run several training sessions for users of the new application. All of these examples show that large information technology projects are usually composed of several smaller projects. It is often good practice to view large projects as a series of smaller, more manageable ones, especially when there is a lot of uncertainty involved. Successfully completing one small project at a time will help the project team succeed in completing the larger project.

Because some aspects of project management need to occur during each phase of the product life cycle, it is critical for information technology professionals to understand and practice good project management throughout the product life cycle.

## The Importance of Project Phases and Management Reviews

Due to the complexity and importance of many information technology projects and their resulting products, it is important to take time to review the status of a project at each phase. A project should successfully pass through each of the main project or product phases before continuing to the next. Since the organization usually commits more money as a project continues, a management review should occur after each phase to evaluate progress, potential success, and continued compatibility with organizational goals. These management reviews, called **phase exits** or **kill points**, are very important for keeping projects on track and determining if they should be continued, redirected, or terminated. Recall that projects are just one part of the entire system of an organization. Changes in other parts of the organization might affect a project's status, and a project's status might likewise affect what is happening in other parts of the organization. By breaking projects into phases, top management can make sure that the projects are still compatible with the needs of the rest of the organization.



Let's take another look at the opening case. Suppose Tom Walters' college did a study on increasing the use of technology that was sponsored by the college president. At the end of the concept phase, the project team could have presented information to the faculty, president, and other staff members that described different options for increasing the use of technology, an analysis of what competing colleges were doing, and results of a survey of local stakeholders' opinions on the subject. This presentation at the end of the concept phase represents one form of a management review. Suppose the study reported that 90 percent of students, faculty, and staff surveyed strongly opposed the idea of requiring all students to have laptops and that many adult students said they would attend other colleges if they were required to pay for the additional technology. The college would probably decide not to pursue this idea any further. Had Tom taken a phased approach, he and his staff would not have wasted the time and money it took to develop detailed plans.

In addition to formal management reviews, it is important to have top management involvement throughout the life cycle of most projects. It is unwise to wait for the end of project or product phases to have management inputs. Many projects are reviewed by management on a regular basis, such as weekly or even daily, to make sure they are progressing well. Everyone wants to be successful in accomplishing goals at work, and having management involvement ensures that they are on track in accomplishing both project and organizational goals.

## WHAT WENT RIGHT?

Having specific deliverables and kill points at the end of project or product phases helps managers make better decisions about whether to proceed, redefine, or kill a project. Improvement in information technology project success rates reported by the Standish Group has been due, in part, to an increased ability to know when to cancel failing projects. Standish Group Chairman Jim Johnson made the following observation: "The real improvement that I saw was in our ability to—in the words of Thomas Edison—know when to stop beating a dead horse.... Edison's key to success was that he failed fairly often; but as he said, he could recognize a dead horse before it started to smell.... In information technology we ride dead horses—failing projects—a long time before we give up. But what we are seeing now is that we are able to get off them; able to reduce cost overrun and time overrun. That's where the major impact came on the success rate."[9]

Another example of the power of management oversight comes from Huntington Bancshares, Inc. This company, like many others, had an **executive steering committee**, a group of senior executives from various parts of the organization who regularly reviewed important corporate projects and issues. This Ohio-based, $26 billion bank holding company completed a year-long Web site redesign effort using XML technology to give its online customers access to real-time account information as well as other banking services. The CIO, Joe Gottron, said there were "four or five very intense moments" when the whole project was almost stopped due to its complexity. An executive steering committee met

*continued*

weekly to review the project's progress and discuss work planned for the following week. Gottron said the meetings ensured that "if we were missing a beat on the project, no matter which company [was responsible], we were on top of it and adding additional resources to make up for it."[10]

Managers in the motorcycle industry now understand the importance of overseeing their IT projects. Harley-Davidson Motor Company used to focus only on producing and selling high-quality motorcycles. In 2003, however, management realized that it had to improve its IT operations and control to stay in business and adhere to new government laws such as the accounting reporting regulations of Sarbanes-Oxley. Harley-Davidson had no standardized processes for user access, change management, or backup and recovery at that time. "Although complying with Sarbanes-Oxley was going to be a challenge, the company took strong action, utilized COBIT (Control Objectives for Information and related Technology) and passed Sarbanes-Oxley year one compliance .... One of the major benefits of using COBIT as its overall internal control and compliance model was getting everyone—especially non-technical motorcycle experts—revved up about control activities and why controls are important.[11]



## THE CONTEXT OF INFORMATION TECHNOLOGY PROJECTS

As described earlier, software development projects can follow several different product life cycles based on the project context. There are several other issues related to managing information technology projects. This section highlights some of the issues unique to the information technology industry that affect project management, including the nature of projects, the characteristics of project team members, and the diverse nature of technologies involved.

### The Nature of Information Technology Projects

Unlike projects in many other industries, projects labeled as information technology projects can be very diverse. Some involve a small number of people installing off-the-shelf hardware and associated software. Others involve hundreds of people analyzing several organizations' business processes and then developing new software in a collaborative effort with users to meet business needs. Even for small hardware-oriented projects, there is a wide diversity in the types of hardware that could be involved—personal computers, mainframe computers, network equipment, kiosks, or small mobile devices, to name a few. The network equipment might be wireless, phone-based, cable-based, or require a satellite connection. The nature of software development projects is even more diverse than hardware-oriented projects. A software development project might include developing a simple, standalone Microsoft Excel or Access application or a sophisticated, global e-commerce system using state-of-the-art programming languages.

Information technology projects also support every possible industry and business function. Managing an information technology project for a film company's animation department would require different knowledge and skills of the project manager and team members than a project to improve a federal tax collection system or install a communication infrastructure in a third-world country. Because of the diversity of information technology projects and the newness of the field, it is important to develop and follow best practices in managing these varied projects. That way, information technology project managers will have a common starting point and method to follow with every project.



## Characteristics of Information Technology Project Team Members

Because of the nature of information technology projects, the people involved come from very diverse backgrounds and possess different skill sets. Most trade schools, colleges, and universities did not start offering degrees in computer technology, computer science, management information systems, or other information technology areas until the 1970s. Therefore, many people in the field do not have a common educational background. Many companies purposely hire graduates with degrees in other fields such as business, mathematics, or the liberal arts to provide different perspectives on information technology projects. Even with these different educational backgrounds, there are some common job titles for people working on most information technology projects such as business analyst, programmer, network specialist, database analyst, quality assurance expert, technical writer, security specialist, hardware engineer, software engineer, and system architect. Within the category of programmer, there are several other job titles used to describe the specific technologies the programmer uses, such as Java programmer, XML programmer, C/C++ programmer, and so on.

Some information technology projects require the skills of people in just a few of these job functions, but many require inputs from many or all of them. Occasionally, information technology professionals move around between these job functions, but more often people become technical experts in one area or they decide to move into a management position. It is also rare for technical specialists or project managers to remain with the same company for a long time, and in fact, many information technology projects include a large number of contract workers. Working with this "army of free agents," as Rob Thomsett, author and consultant for the Cutter Consortium, calls them, creates special challenges. (See the companion Web site for an article on this topic by Thomsett and other suggested readings.)

## Diverse Technologies

Many of the job titles for IT professionals reflect the different technologies required to hold that position. Unfortunately, hardware specialists might not understand the language of database analysts, and vice versa. Security specialists may have a hard time communicating with business analysts. It is also unfortunate that people within the same information technology job function often do not understand each other because each uses different technology. For example, someone with the title of programmer can often use several different programming languages. A COBOL programmer, however, cannot be of much help on a Java project. These highly specialized positions also make it difficult for project managers to form and lead project teams.

Another problem with diverse technologies is that many of them change rapidly. A project team might be close to finishing a project when it discovers a new technology that can greatly enhance the project and better meet long-term business needs. New technologies have also shortened the time frame many businesses have to develop, produce, and distribute new products and services. This fast-paced environment requires equally fast-paced processes to manage and produce information technology projects and products.

# RECENT TRENDS AFFECTING INFORMATION TECHNOLOGY PROJECT MANAGEMENT

Additional challenges and opportunities face IT project managers and their teams in the form of the recent trends of increased globalization, outsourcing, and virtual teams. Each of these trends and suggestions for addressing them are provided in this section.

## Globalization

In his popular book, *The World Is Flat*, Thomas L. Friedman describes the effects of globalization, which has created a "flat" world where everyone is connected and the "playing field" is level for many more participants.[12] Lower trade and political barriers and the digital revolution have made it possible to interact almost instantaneously with billions of other people across the planet, and for individuals and small companies to compete with large corporations. Friedman also discusses the increase in "uploading," where people share information through blogging, podcasts, and open-source software.

Information technology is a key enabler of globalization, and globalization has significantly affected the field of IT. Even though major IT companies such as Microsoft and IBM started in the United States, much of their business is global—indeed, companies and individuals throughout the world contribute to the growth of information technologies and work and collaborate on various IT projects. As mentioned in Chapter 1, the total global spending on technology goods, services, and staff was projected to reach $2.4 trillion in 2008, and the main engines of growth were Asia Pacific and the oil-exporting areas of Eastern Europe, the Middle East, and Africa.

It is important for project managers to address several issues when working on global projects. Several key issues include the following:

- *Communications*: Since people will be working in different time zones, speak different languages, have different cultural backgrounds, celebrate different holidays, etc., it is important to address how people will communicate in an efficient and timely manner. A communications management plan (like the one described in Chapter 10, Project Communications Management) is vital.
- *Trust*: Trust is an important issue for all teams, especially when they are global teams. It is important to start building trust immediately by recognizing and respecting others' differences and the value they add to the project.
- *Common work practices*: It is important to align work processes to come up with an agreed-upon modus operandi with which everyone is comfortable. Project managers must allow time for the team to develop these common work practices. Using special tools, as described in the next section, can facilitate this process.
- *Tools*: Information technology plays a vital role in globalization, especially in enhancing communications and work practices. For example, Timothy Porter, a project manager for Hundsun Technologies, a Chinese domestic software company building a global services business, describes several tools they use as follows:
    - XPlanner is used for project planning and project monitoring. This tool is suitable for agile software development and is Web-based for ease of distributed geographic access.

- TRAC is an enhanced issue-tracking system for software development projects. TRAC includes features such as defect management, source code control, project roadmap management, and an integrated wiki—a collaborative, Web-based feedback system—for project documentation that is very easy for stakeholders to review.
- CruiseControl is a framework for a continuous build process. It includes plug-ins for e-mail notification, source control tools, and so on. A Web interface is provided to view the details of the current and previous builds.
- WebEx, a Web-based conferencing tool, is used to record each development cycle's demo, which is stored on our wikis. These demos provide stakeholders visibility into our progress and can also be used as training materials for new staff members or the test team.
- E-mail, telephone, SKYPE (software that allows users to make telephone calls over the Internet), and instant messaging (IM) are used for routine daily communication among team members.[13]

After researching over 600 global organizations, KPMG International summarized several suggestions for managing global project teams:

- Employ greater project discipline for global projects, otherwise weaknesses within the traditional project disciplines may be amplified by the geographical differences.
- Think global, but act local to align and integrate stakeholders at all project levels.
- Consider collaboration over standardization to help balance the goals and project approach.
- Keep project momentum going for projects, which will typically have a long duration.
- Consider the use of newer, perhaps more innovative, tools and technology.[14]

## Outsourcing

As described in detail in Chapter 12, Project Procurement Management, **outsourcing** is when an organization acquires goods and/or sources from an outside source. The term **offshoring** is sometimes used to describe outsourcing from another country. Offshoring is a natural outgrowth of globalization. IT projects continue to rely more and more on outsourcing, both within and outside of their country boundaries.

Organizations remain competitive by using outsourcing to their advantage. For example, many organizations have found ways to reduce costs by outsourcing. Their next challenge is to make strategic IT investments with outsourcing by improving their enterprise architecture to ensure that IT infrastructure and business processes are integrated and standardized. (See the Suggested Readings on the companion Web site for this chapter by Ross and Beath and KPMG International. Chapter 12, Project Procurement Management, also features more information.)

Because of the increased use of outsourcing for IT projects, project managers should become more familiar with negotiating contracts and many other issues, including working on and managing virtual teams.

## Virtual Teams



Increased globalization and outsourcing have increased the need for virtual teams. A **virtual team** is a group of individuals who work across time and space using communication technologies. Team members might all work for the same company in the same country, or they might include employees as well as independent consultants, suppliers, or even volunteers providing their expertise from around the globe.

The main advantages of virtual teams include:

- Increasing competiveness and responsiveness by having a team of workers available 24/7.
- Lowering costs because many virtual workers do not require office space or support beyond their home offices.
- Providing more expertise and flexibility by having team members from across the globe working any time of day or night.
- Increasing the work/life balance for team members by eliminating fixed office hours and the need to travel to work.

Disadvantages of virtual teams include:

- Isolating team members who may not adjust well to working in a virtual environment.
- Increasing the potential for communications problems since team members cannot use body language or other communications to understand each other and build relationships and trust.
- Reducing the ability for team members to network and transfer information informally.
- Increasing the dependence on technology to accomplish work.

Like any team, a virtual team should focus on achieving a common goal.

Research on virtual teams reveals a growing list of factors that influence their success including:

- *Team processes*: It is important to define how the virtual team will operate. For examples, teams must agree on how and when work will be done, what technologies will be used, how decisions will be made, and other important process issues.
- *Leadership style*: The project manager's leadership style affects all teams, especially virtual ones.
- *Trust and relationships*: Many virtual teams fail because of a lack of trust. It is difficult to build relationships and trust from a distance. Some project managers like to have a face-to-face meeting so team members can get to know each other and build trust. If that is not possible, phone or video conferences can help.
- *Team member selection and role preferences*: Dr. Meredith Belbin defined a team role as "a tendency to behave, contribute and interrelate with others in a particular way."[15] It is important to select team members carefully and to form a team where all roles are covered. Each virtual team member must also understand his or her role(s) on the team. (Visit *www.belbin.com* for more information on this topic.)

- *Task-technology fit*: IT is more likely to have a positive impact on individual performance if the capabilities of the technologies match the tasks that the user must perform.
- *Cultural differences*: It is important to address cultural differences, including the dimensions of directness, hierarchy, consensus, and individualism. These dimensions will affect many aspects of the team such as communications and decision making.
- *Computer-mediated communication*: It is crucial to provide reliable and appropriate computer-mediated communication to virtual team members, including e-mail, instant messaging, text messaging, chat rooms, and so on.
- *Team life cycles*: Just as projects and products have life cycles, so do teams. Project managers must address the team life cycle especially in assigning team members and determining deliverable schedules.
- *Incentives*: Virtual teams may require different types of incentives in order to accomplish quality work on time. They do not have the benefit of physical contact with their project managers or other team members, so it important to provide frequent positive incentives like a thank you via e-mail or phone, or even a bonus on occasion. Negative incentives, such as payment withholding or fines, can also be effective if virtual team members are not being productive.
- *Conflict management*: Even though they may never physically meet, virtual teams will still have conflict. It is important to address conflict management, as described in more detail in Chapter 10, Project Communications Management.

Several studies have been done to try to determine factors that are positively correlated to the effectiveness of virtual teams. Research suggests that team processes, trust/relationships, leadership style, and team member selection provide the strongest relationships to team performance and team member satisfaction.[16] See the companion Web site for suggested readings on virtual teams and other topics discussed in this chapter.

As you can see, working as an information technology project manager or team member is an exciting and challenging job. It's important to focus on successfully completing projects that will have a positive impact on the organization as a whole.

## CASE WRAP-UP

After several people voiced concerns about the laptop idea at the faculty meeting, the president of the college directed that a committee be formed to formally review the concept of requiring students to have laptops in the near future. Because the college was dealing with several other important enrollment-related issues, the president named the vice president of enrollment to head the committee. Other people soon volunteered or were assigned to the committee, including Tom Walters as head of Information Technology, the director of the adult education program, the chair of the Computer Science department, and the chair of the History department. The president also insisted that the committee include at least two members of the student body. The president knew everyone was busy, and he questioned whether the laptop idea was a high-priority issue for the college. He directed the committee to present a proposal at the next month's faculty meeting, either to recommend the creation of a formal project team (of which these committee members would commit to be a part) to fully investigate requiring laptops, or to recommend terminating the concept. At the next faculty meeting, few people were surprised to hear the recommendation to terminate the concept. Tom Walters learned that he had to pay much more attention to the needs of the entire college before proceeding with detailed information technology plans.

## Chapter Summary



Projects operate in an environment broader than the project itself. Project managers need to take a systems approach when working on projects; they need to consider projects within the greater organizational context.

Organizations have four different frames: structural, human resources, political, and symbolic. Project managers need to understand all of these aspects of organizations to be successful. The structural frame focuses on different groups' roles and responsibilities to meet the goals and policies set by top management. The human resources frame focuses on producing harmony between the needs of the organization and the needs of people. The political frame addresses organizational and personal politics. The symbolic frame focuses on symbols and meanings.

The structure of an organization has strong implications for project managers, especially in terms of the amount of authority the project manager has. The three basic organizational structures include functional, matrix, and project. Project managers have the most authority in a pure project organization, an intermediate amount of authority in a matrix organization, and the least amount of authority in a pure functional organization.

Organizational culture also affects project management. A culture where employees have a strong identity with the organization, where work activities emphasize groups, where there is strong unit integration, high risk tolerance, performance-based rewards, high conflict tolerance, an open-systems focus, and a balance on the dimensions of people focus, control, and means-orientation is more conducive to project work.

Project stakeholders are individuals and organizations who are actively involved in the project or whose interests may be positively or negatively affected because of project execution or successful project completion. Project managers must identify and understand the different needs of all stakeholders on their projects.

Top management commitment is crucial for project success. Since projects often affect many areas in an organization, top management must assist project managers if they are to do a good job of project integration. Organizational commitment to information technology is also important to the success of information technology projects. Development standards and guidelines assist most organizations in managing projects.

A project life cycle is a collection of project phases. Traditional project phases include concept, development, implementation, and close-out phases. Projects often produce products, which follow product life cycles. Examples of product life cycles for software development include the waterfall, spiral, incremental build, prototyping, RAD, and the adaptive software development models. Project managers must understand the specific life cycle of the products they are producing as well as the general project life cycle model.

A project should successfully pass through each of the project phases in order to continue to the next phase. A management review should occur at the end of each project phase, and more frequent management inputs are often needed. These management reviews and inputs are important for keeping projects on track and determining if projects should be continued, redirected, or terminated.

Project managers need to consider several factors due to the unique context of information technology projects. The diverse nature of these projects and the wide range of business areas and technologies involved make information technology projects especially challenging to manage. Leading project team members with a wide variety of specialized skills and understanding rapidly changing technologies are also important considerations.

Several recent trends have affected information technology project management. Increased globalization, outsourcing, and virtual teams have changed the way many IT projects are staffed and managed. Project managers must stay abreast of these and other trends and discover ways to use them to their advantage.



## Quick Quiz

1. Which of the following is not part of the three-sphere model for systems management?
   a. business
   b. information
   c. technology
   d. organization
2. Which of the four frames of organizations addresses how meetings are run, employee dress codes, and expected work hours?
   a. structural
   b. human resources
   c. political
   d. symbolic
3. Personnel in a ______ organizational structure often report to two or more bosses.
   a. functional
   b. project
   c. matrix
   d. hybrid
4. Project work is most successful in an organizational culture where all of the following characteristics are high except ______.
   a. member identity
   b. group emphasis
   c. risk tolerance
   d. control
5. A ______ is a product or service, such as a technical report, a training session, or hardware, produced or provided as part of a project.
   a. deliverable
   b. product
   c. work package
   d. tangible goal

6. Which of the following is not a phase of the traditional project life cycle?
   a. systems analysis
   b. concept
   c. development
   d. implementation
7. What is the term used to describe a framework of the phases involved in developing information systems?
   a. systems development life cycle
   b. rapid application development
   c. predictive life cycle
   d. extreme programming
8. Another name for a phase exit is a _____ point.
   a. review
   b. stage
   c. meeting
   d. kill
9. The nature of information technology projects is different from projects in many other industries because they are very ______.
   a. expensive
   b. technical
   c. diverse
   d. challenging
10. What term is used to describe when an organization acquires goods and/or sources from an outside source in another country?
    a. globalization
    b. offshoring
    c. exporting
    d. global sourcing

### Quick Quiz Answers

1. b; 2. d; 3. c; 4. d; 5. a; 6. a; 7. a; 8. d; 9. c; 10. b

## Discussion Questions

1. What does it mean to take a systems view of a project? How does taking a systems view of a project apply to project management?
2. Explain the four frames of organizations. How can they help project managers understand the organizational context for their projects?

3. Briefly explain the differences between functional, matrix, and project organizations. Describe how each structure affects the management of the project.
4. Describe how organizational culture is related to project management. What type of culture promotes a strong project environment?
5. Discuss the importance of top management commitment and the development of standards for successful project management. Provide examples to illustrate the importance of these items based on your experience on any type of project.
6. What are the phases in a traditional project life cycle? How does a project life cycle differ from a product life cycle? Why does a project manager need to understand both?
7. What makes information technology projects different from other types of projects? How should project managers adjust to these differences?
8. Define globalization, outsourcing, and virtual teams and describe how these trends are changing IT project management.



## Exercises

1. Summarize the three-sphere model of systems management in your own words. Then use your own experience or interview someone who recently completed an information technology project and list several business, technology, and organizational issues addressed during the project. Which issues were most important to the project and why? Summarize your answers in a two-page paper.
2. Apply the four frames of organizations to an information technology project with which you are familiar. If you cannot think of a good information technology project, use your personal experience in deciding where to attend college to apply this framework. Write a two-page paper describing key issues related to the structural, human resources, political, and symbolic frames. Which frame seemed to be the most important and why? For example, did you decide where to attend college primarily because of the curriculum and structure of the program? Did you follow your friends? Did your parents have a lot of influence in your decision? Did you like the culture of the campus?
3. Search the Internet for two interesting articles about software development life cycles, including agile software development. Also review the Web site *www.agilealliance.org*. What do these sources say about project management? Write a two-page summary of your findings, citing your references.

   *Note*: For this exercise and others, remember that you can find references cited in this text, suggested readings, and links to general project management Web sites on the companion Web site.
4. Search the Internet and scan information technology industry magazines or Web sites to find an example of an information technology project that had problems due to organizational issues. Write a two-page paper summarizing who the key stakeholders were for the project and how they influenced the outcome.
5. Write a two-page summary of an article about the importance of top management support for successful information technology projects and your opinion on this topic.

6. Research the trend of using virtual teams. Review the information on team role theory from *www.belbin.com* and other related sources. Write a two-page summary of your findings, citing at least three references. Also include your personal experience and/or opinion on the topic. For example, what role(s) would you prefer to play on a team? Do you like working on virtual teams? If you have not yet worked on one, how do you think it would be different from working on a face-to-face team?

## Companion Web Site

Visit the companion Web site for this text at *www.cengage.com/mis/schwalbe* to access:

- References cited in the text and additional suggested readings for each chapter
- Template files
- Lecture notes
- Interactive quizzes
- Podcasts
- Links to general project management Web sites
- And more

See the Preface of this text for additional information on accessing the companion Web site.

## Key Terms

**adaptive software development (ASD)** — A software development approach used when requirements cannot be clearly expressed early in the life cycle

**agile software development** — A method for software development that uses new approaches, focusing on close collaboration between programming teams and business experts

**champion** — A senior manager who acts as a key proponent for a project

**deliverable** — A product or service, such as a technical report, a training session, a piece of hardware, or a segment of software code, produced or provided as part of a project

**executive steering committee** — A group of senior executives from various parts of the organization who regularly review important corporate projects and issues

**functional organizational structure** — An organizational structure that groups people by functional areas such as information technology, manufacturing, engineering, and human resources

**human resources frame** — Focuses on producing harmony between the needs of the organization and the needs of people

**IT governance** — Addresses the authority and control for key IT activities in organizations, including IT infrastructure, IT use, and project management

**kill point** — Management review that should occur after each project phase to determine if projects should be continued, redirected, or terminated; also called a phase exit

**matrix organizational structure** — An organizational structure in which employees are assigned to both functional and project managers

**offshoring** — Outsourcing from another country

**organizational culture** — A set of shared assumptions, values, and behaviors that characterize the functioning of an organization

**outsourcing** — When an organization acquires goods and/or sources from an outside source

**phase exit** — Management review that should occur after each project phase to determine if projects should be continued, redirected, or terminated; also called a kill point

**political frame** — Addresses organizational and personal politics

**politics** — Competition between groups or individuals for power and leadership

**predictive life cycle** — A software development approach used when the scope of the project can be clearly articulated and the schedule and cost can be accurately predicted

**project acquisition** — The last two phases in a project (implementation and close-out) that focus on delivering the actual work

**project feasibility** — The first two phases in a project (concept and development) that focus on planning

**project life cycle** — A collection of project phases, such as concept, development, implementation, and close-out

**project organizational structure** — An organizational structure that groups people by major projects, such as specific aircraft programs

**structural frame** — Deals with how the organization is structured (usually depicted in an organizational chart) and focuses on different groups' roles and responsibilities to meet the goals and policies set by top management

**symbolic frame** — Focuses on the symbols, meanings, and culture of an organization

**systems** — Sets of interacting components working within an environment to fulfill some purpose

**systems analysis** — A problem-solving approach that requires defining the scope of the system to be studied, and then dividing it into its component parts for identifying and evaluating its problems, opportunities, constraints, and needs

**systems approach** — A holistic and analytical approach to solving complex problems that includes using a systems philosophy, systems analysis, and systems management

**systems development life cycle (SDLC)** — A framework for describing the phases involved in developing and maintaining information systems

**systems management** — Addressing the business, technological, and organizational issues associated with creating, maintaining, and making changes to a system

**systems philosophy** — An overall model for thinking about things as systems

**systems thinking** — Taking a holistic view of an organization to effectively handle complex situations

**virtual team** — A group of individuals who work across time and space using communication technologies

## End Notes

[1] Lee G. Bolman and Terrence E. Deal, *Reframing Organizations* (San Francisco: Jossey-Bass, 1991).

[2] Eva Hoare, "Software hardships," *Herald* (Halifax: Nova Scotia) (February 4, 2001).

[3] Stephen P. Robbins and Timothy A. Judge, *Organizational Behavior, 13th Edition* (Prentice Hall, 2008).

[4] Charles V. Bagli, "Higher Costs and Delays Expected at Ground Zero," *New York Times* (June 30, 2008).

[5] Peter Weill and Jeanne Ross, *IT Governance: How Top Performers Manage IT Decision Rights for Superior Results* (Harvard Business School Press, 2004).

[6] David Avison, Shirely Gregor, and David Wilson, "Managerial IT Unconsciousness," *Communications of the ACM* 49, no. 7 (July 2006), p. 92.

[7] Gartner Inc., "BNSF and UnitedHealth Group Win 2006 Gartner CRM Excellence Awards," press release (September 25, 2006).

[8] Douglas H. Desaulniers and Robert J. Anderson, "Matching Software Development Life Cycles to the Project Environment," Proceedings of the Project Management Institute Annual Seminars & Symposium, (November 1–10, 2001).

[9] Jeannette Cabanis, "A Major Import: The Standish Group's Jim Johnson on Project Management and IT Project Success," *PM Network* (PMI), (September 1998), p. 7.

[10] Lucas Mearian, "Bank Hones Project Management Skills with Redesign," *ComputerWorld* (April 29, 2002).

[11] IT Governance Institute, "COBIT and IT Governance Case Study: Harley-Davidson," *www.itgi.org* (September 2006).

[12] Thomas L. Friedman, *The World Is Flat: A Brief History of the Twenty-First Century* (Farrar, Straus, and Giroux, 2005).

[13] Timothy Porter, "Tools for Facilitating Project Communication in an Onshore-Offshore Engagement Model," *PMI-ISSIG.org* (June 30, 2008).

[14] KPMG International, "Managing Global Projects: Observations from the front-line," *www.kpmg.com* (2007).

[15] Belbin® Team Role Theory, *Belbin.com* (accessed July 1, 2008).

[16] Jeremy S. Lurey and Mahesh S. Raisinghani, "An Empirical Study of Best Practices in Virtual Teams," *Information & Management* 38, no. 8 (2001).

CHAPTER 3

# THE PROJECT MANAGEMENT PROCESS GROUPS: A CASE STUDY

## LEARNING OBJECTIVES

**After reading this chapter, you will be able to:**

- Describe the five project management process groups, the typical level of activity for each, and the interactions among them
- Understand how the project management process groups relate to the project management knowledge areas
- Discuss how organizations develop information technology project management methodologies to meet their needs
- Review a case study of an organization applying the project management process groups to manage an information technology project, describe outputs of each process group, and understand the contribution that effective initiating, planning, executing, monitoring and controlling, and closing make to project success

## OPENING CASE

Erica Bell was in charge of the Project Management Office (PMO) for her consulting firm. The firm, JWD—for Job Well Done—Consulting, had grown to include more than 200 full-time consultants and even more part-time consultants. JWD Consulting provides a variety of consulting services to assist organizations in selecting and managing information technology projects. The firm focuses on finding and managing high-payoff projects and developing strong metrics to measure project performance and benefits to the organization after the project is implemented. The firm's emphasis on metrics and working collaboratively with its customers gives it an edge over many competitors.

Joe Fleming, the CEO, wanted his company to continue to grow and become a world-class consulting organization. Since the core of the business was helping other organizations with project management, he felt it was crucial for JWD Consulting to have an exemplary process for managing its own projects. He asked Erica to work with her team and other consultants in the firm to develop several intranet site applications that would allow them to share their project management knowledge. He also thought it would make sense to make some of the information available to the firm's clients. For example, the firm could provide project management templates, tools, articles, links to other sites, and an "Ask the Expert" feature to help build relationships with current and future clients. Since JWD Consulting emphasizes the importance of high-payoff projects, Joe also wanted to see a business case for this project before proceeding.

Recall from Chapter 1 that project management consists of nine knowledge areas: integration, scope, time, cost, quality, human resources, communications, risk, and procurement. Another important concept to understand is that projects involve five project management process groups: initiating, planning, executing, monitoring and controlling, and closing. Tailoring these process groups to meet individual project needs increases the chance of success in managing projects. This chapter describes each project management process group in detail through a simulated case study based on JWD Consulting. It also includes samples of typical project documents applied to this case. You can download templates for these and other project documents from the companion Web site for this text. Although you will learn more about each knowledge area in Chapters 4 though 12, it is important first to learn how they fit into the big picture of managing a project. Understanding how the knowledge areas and project management process groups function together will lend context to the remaining chapters.

## PROJECT MANAGEMENT PROCESS GROUPS

Project management is an integrative endeavor; decisions and actions taken in one knowledge area at a certain time usually affect other knowledge areas. Managing these interactions often requires making trade-offs among the project's scope, time, and cost—the triple constraint of project management described in Chapter 1. A project manager may also need to make trade-offs between other knowledge areas, such as between risk and human resources. Consequently, you can view project management as a number of related processes.

- A **process** is a series of actions directed toward a particular result. **Project management process groups** progress from initiating activities to planning activities, executing activities, monitoring and controlling activities, and closing activities. **Initiating processes** include defining and authorizing a project or project phase. Initiating processes take place during *each* phase of a project. Therefore, you cannot equate process groups with project phases. Recall that there can be different project phases, but all projects will include all five process groups. For example, project managers and teams should reexamine the business need for the project during every phase of the project life cycle to determine if the project is worth continuing. Initiating processes are also required to end a project. Someone must initiate activities to ensure that the project team completes all the work, documents lessons learned, assigns project resources, and that the customer accepts the work.
- **Planning processes** include devising and maintaining a workable scheme to ensure that the project addresses the organization's needs. There are several plans for projects, such as the scope management plan, schedule management plan, cost management plan, procurement management plan, and so on, defining each knowledge area as it relates to the project at that point in time. For example, a project team must develop a plan to define the work that needs to be done for the project, to schedule activities related to that work, to estimate costs for performing the work, to decide what resources to procure to accomplish the work, and so on. To account for changing conditions on the project and in the organization, project teams often revise plans during each phase of the project life cycle. The project management plan, described in Chapter 4, coordinates and encompasses information from all other plans.
- **Executing processes** include coordinating people and other resources to carry out the various plans and produce the products, services, or results of the project or phase. Examples of executing processes include acquiring and developing the project team, performing quality assurance, distributing information, managing stakeholder expectations, and conducting procurements.
- **Monitoring and controlling processes** include regularly measuring and monitoring progress to ensure that the project team meets the project objectives. The project manager and staff monitor and measure progress against the plans and take corrective action when necessary. A common monitoring and controlling process is reporting performance, where project stakeholders can identify any necessary changes that may be required to keep the project on track.
- **Closing processes** include formalizing acceptance of the project or project phase and ending it efficiently. Administrative activities are often involved in this process group, such as archiving project files, closing out contracts, documenting lessons learned, and receiving formal acceptance of the delivered work as part of the phase or project.

The process groups are not mutually exclusive. For example, project managers must perform monitoring and controlling processes throughout the project's life span. The level of activity and length of each process group varies for every project. Normally, executing tasks requires the most resources and time, followed by planning tasks. Initiating and closing tasks are usually the shortest (at the beginning and end of a project or phase, respectively), and they require the least amount of resources and time. However, every project is unique, so there can be exceptions. You can apply the process groups for each major phase of a project, or you can apply the process groups to an entire project, as the JWD Consulting case study does in this chapter.

Many people ask for guidelines on how much time to spend in each process group. In his book, *Alpha Project Managers: What the Top 2% Know That Everyone Else Does Not*, Andy Crowe collected data from 860 project managers in various companies and industries in the United States. He found that the best—the "alpha"—project managers spent more time on every process group than their counterparts except for executing, as shown in Figure 3-1. This breakdown suggests that the most time should be spent on executing, followed by planning. Spending a fair amount of time on planning should lead to less time spent on execution. Notice that the alpha project managers spent almost twice as much time on planning (21 percent versus 11 percent) as other project managers.[1]

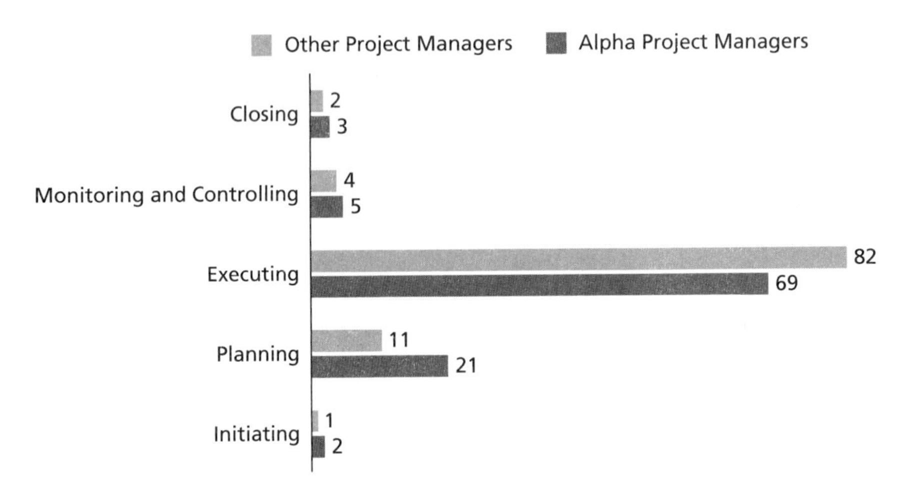


**FIGURE 3-1** Percentage of time spent on each process group

Many readers of *CIO Magazine* commented on its cover story about problems with information systems at the U.S. Internal Revenue Service (IRS). The article described serious problems the IRS has had in managing information technology projects. Philip A. Pell, PMP,

*continued*

believes that having a good project manager and following a good project management process would help the IRS and many organizations tremendously. Mr. Pell provided the following feedback:

> Pure and simple, good, methodology-centric, predictable, and repeatable project management is the SINGLE greatest factor in the success (or in this case failure) of any project. When a key stakeholder says, 'I didn't know how bad things were,' it is a direct indictment of the project manager's communications management plan. When a critical deliverable like the middleware infrastructure that makes the whole thing work is left without assigned resources and progress tracking, the project manager has failed in his duty to the stakeholders. When key stakeholders (people and organizations that will be affected by the project, not just people who are directly working on the project) are not informed and their feedback incorporated into the project plan, disaster is sure to ensue. The project manager is ultimately responsible for the success or failure of the project.[2]

The IRS continues to have problems managing IT projects. A 2008 U.S. Government Accountability Office (GAO) report stated that IRS had fixed just 29 of 98 information security weaknesses identified the previous year. The report stated that the IRS has "persistent information security weaknesses that place [it] at risk of disruption, fraud or inappropriate disclosure of sensitive information."[3]

Each of the five project management process groups is characterized by the completion of certain tasks. During initiating processes for a new project, the organization recognizes that a new project exists, and completes a project charter as part of this recognition (see Chapter 4 for more information on project charters). Tables are provided later in this chapter with detailed lists of possible outputs for each process group by knowledge area. For example, Tables 3-3 through 3-7 list potential outputs for the initiating and planning process groups. Samples of some outputs are provided for each process group in a case study of JWD Consulting's Project Management Intranet Site project. Project managers and their teams must decide which outputs are required for their particular projects.

Outputs of the planning process group include completing the project scope statement, the work breakdown structure, the project schedule, and many other items. Planning processes are especially important for information technology projects. Everyone who has ever worked on a large information technology project that involves new technology knows the saying, "A dollar spent up front in planning is worth one hundred dollars spent after the system is implemented." Planning is crucial in information technology projects because once a project team implements a new system, it takes a considerable amount of effort to change the system. Research suggests that companies working to implement best practices should spend at least 20 percent of project time in initiating and planning.[4] This percentage is backed up by evidence from Alpha Project Managers, as described earlier.

The executing process group takes the actions necessary to complete the work described in the planning activities. The main outcome of this process group is delivering the actual work of the project. For example, if an information technology project involves providing new hardware, software, and training, the executing processes would include leading the project team and other stakeholders to purchase the hardware, develop and test the software, and deliver and participate in the training. The executing process group should overlap the other process groups and generally requires the most resources.

Monitoring and controlling processes measure progress toward the project objectives, monitor deviation from the plan, and take corrective action to match progress with the plan. Performance reports are common outputs of monitoring and controlling. The project manager should be monitoring progress closely to ensure that deliverables are being completed and objectives are being met. The project manager must work closely with the project team and other stakeholders and take appropriate actions to keep the project running smoothly. The ideal outcome of the monitoring and controlling process group is to complete a project successfully by delivering the agreed-upon project scope within time, cost, and quality constraints. If changes to project objectives or plans are required, monitoring and controlling processes ensure that these changes are made efficiently and effectively to meet stakeholder needs and expectations. Monitoring and controlling processes overlap all of the other project management process groups because changes can occur at any time.

During the closing processes, the project team works to gain acceptance of the end products, services, or results and bring the phase or project to an orderly end. Key outcomes of this process group are formal acceptance of the work and creation of closing documents, such as a final project report and lessons-learned report.

## MEDIA SNAPSHOT

Just as information technology projects need to follow the project management process groups, so do other projects, such as the production of a movie. Processes involved in making movies might include screenwriting (initiating), producing (planning), acting and directing (executing), editing (monitoring and controlling), and releasing the movie to theaters (closing). Many people enjoy watching the extra features on a DVD that describe how these processes lead to the creation of a movie. For example, the DVD for *Lord of the Rings: The Two Towers Extended Edition* includes detailed descriptions of how the script was created, how huge structures were built, how special effects were made, and how talented professionals overcame numerous obstacles to complete the project. This acted "not as promotional filler but as a serious and meticulously detailed examination of the entire filmmaking process."[5] New Line Cinema made history by shooting all three Lord of the Rings films consecutively during one massive production. It took three years of preparation to build the sets, find the locations, write the scripts, and cast the actors. Director Peter Jackson said that the amount of early planning they did made it easier than he imagined to produce the films. Project managers in any field know how important it is to have good plans and to follow a good process.

## MAPPING THE PROCESS GROUPS TO THE KNOWLEDGE AREAS

You can map the main activities of each project management process group into the nine project management knowledge areas. Table 3-1 provides a big-picture view of the relationships among the 42 project management activities, the process groups in which they are typically completed, and the knowledge areas into which they fit. The activities listed in the table are the main processes for each knowledge area listed in the *PMBOK® Guide, Fourth Edition*. This text also includes additional activities not listed in the *PMBOK® Guide*, such as creating a business case and team contract, which can also assist in managing projects.

Several organizations use PMI's *PMBOK® Guide* information as a foundation for developing their own project management methodologies, as described in the next section. Notice in Table 3-1 that many of the project management processes occur as part of the planning process group. Since each project is unique, project teams are always trying to do something that has not been done before. To succeed at unique and new activities, project teams must do a fair amount of planning. Recall, however, that the most time and money is normally spent on executing. It is good practice for organizations to determine how project management will work best in their own organizations.

**TABLE 3-1** Project management process groups and knowledge area mapping

| Knowledge Area | Project Management Process Groups | | | | |
|---|---|---|---|---|---|
| | Initiating | Planning | Executing | Monitoring and Controlling | Closing |
| ***Project Integration Management*** | Develop project charter | Develop project management plan | Direct and manage project execution | Monitor and control project work, Perform integrated change control | Close project or phase |
| ***Project Scope Management*** | | Collect requirements, Define scope, Create WBS | | Verify scope, Control scope | |
| ***Project Time Management*** | | Define activities, Sequence activities, | | Control schedule | |

*(continued)*

**TABLE 3-1** Project management process groups and knowledge area mapping (*continued*)

| Knowledge Area | Project Management Process Groups | | | | |
|---|---|---|---|---|---|
| | Initiating | Planning | Executing | Monitoring and Controlling | Closing |
| ***Project Time Management (continued)*** | | Estimate activity resources, Estimate activity durations, Develop schedule | | | |
| ***Project Cost Management*** | | Estimate costs, Determine budget | | Control costs | |
| ***Project Quality Management*** | | Plan quality | Perform quality assurance | Perform quality control | |
| ***Project Human Resource Management*** | | Develop human resource plan | Acquire project team, Develop project team, Manage project team | | |
| ***Project Communi-cations Management*** | Identify stake-holders | Plan communi-cations | Distribute information, Manage stakeholders expectations | Report performance | |
| ***Project Risk Management*** | | Plan risk man-agement, Identi-fy risks, Perform qualitative risk analysis, Perform quantitative risk analysis, Plan risk responses | | Monitor and control risks | |
| ***Project Procurement Management*** | | Plan procurements | Conduct procurements | Administer procurements | Close procurements |

*Source: PMBOK® Guide, Fourth Edition*, 2008.

## DEVELOPING AN INFORMATION TECHNOLOGY PROJECT MANAGEMENT METHODOLOGY

Some organizations spend a great deal of time and money on training efforts for general project management skills, but after the training, project managers may still not know how to tailor their project management skills to the organization's particular needs. Because of this problem, some organizations develop their own internal information technology project management methodologies. The *PMBOK® Guide* is a **standard** that describes best practices for *what* should be done to manage a project. A **methodology** describes *how* things should be done, and different organizations often have different ways of doing things.

In addition to using the *PMBOK® Guide* as a basis for project management methodology, many organizations use others, such as the following:

- **PRojects IN Controlled Environments (PRINCE2):** Originally developed for information technology projects, PRINCE2 was released in 1996 as a generic project management methodology by the U.K. Office of Government Commerce (OCG). It is the de facto standard in the United Kingdom and is used in over 50 countries. (See *www.prince2.com* for more information.) PRINCE2 defines 45 separate subprocesses and organizes these into eight process groups as follows:
    1. Starting Up a Project
    2. Planning
    3. Initiating a Project
    4. Directing a Project
    5. Controlling a Stage
    6. Managing Product Delivery
    7. Managing Stage Boundaries
    8. Closing a Project
- **Agile methodologies:** As described in Chapter 2, agile software development is a form of adaptive software development. All agile methodologies include an iterative workflow and incremental delivery of software in short iterations. Several popular agile methodologies include extreme programming, scrum, feature driven development, lean software development, Agile Unified Process (AUP), Crystal, and Dynamic Systems Development Method (DSDM). (See Web sites like *www.agilealliance.org* and the Suggested Readings on the companion Web site for this text for more information.)
- **Rational Unified Process (RUP) framework:** RUP is an iterative software development process that focuses on team productivity and delivers software best practices to all team members. According to RUP expert Bill Cottrell, "RUP embodies industry-standard management and technical methods and techniques to provide a software engineering process particularly suited to creating and maintaining component-based software system solutions."[6] Cottrell explains that you can tailor RUP to include the PMBOK process groups, since several customers asked for that capability. There are several other project management methodologies specifically for software development projects such as Joint Application Development (JAD) and Rapid Application Development

(RAD). (See Web sites such as *www.ibm.com/software/awdtools/rup* for more information.)

- **Six Sigma methodologies:** Many organizations have projects underway that use Six Sigma methodologies. The work of many project quality experts contributed to the development of today's Six Sigma principles. Two main methodologies are used on Six Sigma projects: DMAIC (Define, Measure, Analyze, Improve, and Control) is used to improve an existing business process, and DMADV (Define, Measure, Analyze, Design, and Verify) is used to create new product or process designs to achieve predictable, defect-free performance. (See Chapter 8, Project Quality Management, for more information on Six Sigma.)

Many organizations tailor a standard or methodology to meet their unique needs. For example, if organizations use the *PMBOK® Guide* as the basis for their project management methodology, they still have to do a fair amount of work to adapt it to their work environment. See the suggested reading on the companion Web site by William Munroe for an example of how Blue Cross Blue Shield of Michigan developed its IT project management methodology.

### WHAT WENT RIGHT?

AgênciaClick, an interactive advertising and online communications company based in São Paulo, Brazil, made PMI's list of outstanding organizations in project management in 2007. Since 2002, the company saw revenues jump 132 percent, primarily due to their five-year emphasis on practicing good project management across the entire company. AgênciaClick launched a PMO in 2002 and used the *PMBOK® Guide* as the basis for developing their methodology and project management training program. The company also developed a custom project tracking system to help calculate physical work progress each day and alert managers of any schedule or cost issues. PMO Director Fabiano D'Agostinho said, "We realized the only way to manage multiple dynamic projects and deliver great products is to focus on project management ... By monitoring and controlling projects and programs more efficiently, senior managers can focus on issues within the portfolio that need more attention."[7]

The following section describes an example of applying the project management process groups to a project at JWD Consulting. It uses some of the ideas from the *PMBOK® Guide, Fourth Edition*, some ideas from other methodologies, and some new ideas to meet unique project needs.

## CASE STUDY: JWD CONSULTING'S PROJECT MANAGEMENT INTRANET SITE PROJECT

The following fictitious case provides an example of the elements involved in managing a project from start to finish. This example also uses Microsoft Project to demonstrate how project management software can assist in several aspects of managing a project. Several templates illustrate how project teams prepare various project management documents. Files for these and other templates are available on the companion Web site for this text. Details on creating many of the documents shown are provided in later chapters, so do not worry if you do not understand everything right now. You might want to read this section again to enhance your learning.

## Project Pre-Initiation and Initiation

In project management, initiating includes recognizing and starting a new project. An organization should put considerable thought into project selection to ensure that it initiates the right kinds of projects for the right reasons. *It is better to have a moderate or even small amount of success on an important project than huge success on one that is unimportant.* The selection of projects for initiation, therefore, is crucial, as is the selection of project managers. Ideally, the project manager would be involved in initiating a project, but often the project manager is selected after many initiation decisions have already been made. You will learn more about project selection in Chapter 4, Project Integration Management. Organizations must also understand and plan for the ongoing support that is often required after implementing a new system or other product or service resulting from a project.

It is important to remember that strategic planning should serve as the foundation for deciding which projects to pursue. The organization's strategic plan expresses the vision, mission, goals, objectives, and strategies of the organization. It also provides the basis for information technology project planning. Information technology is usually a support function in an organization, so it is critical that the people initiating information technology projects understand how those projects relate to current and future needs of the organization. For example, JWD Consulting's main business is providing consulting services to other organizations, not developing its own intranet site applications. Information systems, therefore, must support the firm's business goals, such as providing consulting services more effectively and efficiently.

An organization may initiate information technology projects for several reasons, but the most important reason is to support business objectives. Providing a good return on investment at a reasonable level of risk is also important, especially in tough economic times. As mentioned in the opening case, JWD Consulting wants to follow an exemplary process for managing its projects since its core business is helping other organizations manage projects. Developing an intranet to share its project management knowledge could help JWD Consulting reduce internal costs by working more effectively, and by allowing existing and potential customers to access some of the firm's information. JWD Consulting could also increase revenues by bringing in more business. Therefore, they will use these metrics—reducing internal costs and increasing revenues—to measure their own performance on this project.

### Pre-Initiation Tasks

It is good practice to lay the groundwork for a project *before* it officially starts. Senior managers often perform several tasks, sometimes called pre-initiation tasks, including the following:

- Determine the scope, time, and cost constraints for the project
- Identify the project sponsor

- Select the project manager
- Develop a business case for a project
- Meet with the project manager to review the process and expectations for managing the project
- Determine if the project should be divided into two or more smaller projects



As described in the opening case, the CEO of JWD Consulting, Joe Fleming, defined the high-level scope of the project, and he wanted to sponsor it himself since it was his idea and it was of strategic importance to the business. He wanted Erica Bell, the PMO Director, to manage the project after proving there was a strong business case for it. If there was a strong business case for pursuing the project, then Joe and Erica would meet to review the process and expectations for managing the project. If there was not a strong business case, the project would not continue. As for the necessity of the last pre-initiation task, many people know from experience that it is easier to successfully complete a small project than a large one, especially for IT projects. It often makes sense to break large projects down into two or more smaller ones to help increase the odds of success. In this case, however, Joe and Erica decided that the work could be done in one project that would last about six months. To justify investing in this project, Erica drafted a business case for the project, getting input and feedback from Joe, from one of her senior staff members in the PMO, and from a member of the Finance department. She also used a corporate template and sample business cases from past projects as a guide. Table 3-2 provides the business case. (Note that this example and others are abbreviated examples. See the companion Web site for additional examples of project documents and to download a business case template and other templates.) Notice that the following information is included in this business case:

- Introduction/background
- Business objective
- Current situation and problem/opportunity statement
- Critical assumptions and constraints
- Analysis of options and recommendation
- Preliminary project requirements
- Budget estimate and financial analysis
- Schedule estimate
- Potential risks
- Exhibits

Since this project is relatively small and is for an internal sponsor, the business case is not as long as many other business cases. Erica reviewed the business case with Joe, and he agreed that the project was definitely worth pursuing. He was quite pleased to see that payback was estimated within a year, and the return on investment was projected to be 112 percent. He told Erica to proceed with the formal initiation tasks for this project, as described in the next section.

**TABLE 3-2** JWD Consulting's business case



**1.0 Introduction/Background**

JWD Consulting's core business goal is to provide world-class project management consulting services to various organizations. The CEO, Joe Fleming, believes the firm can streamline operations and increase business by providing information related to project management on its intranet site, making some information and services accessible to current and potential clients.

**2.0 Business Objective**

JWD Consulting's strategic goals include continuing growth and profitability. The Project Management Intranet Site Project will support these goals by increasing visibility of the firm's expertise to current and potential clients by allowing client and public access to some sections of the intranet. It will also improve profitability by reducing internal costs by providing standard tools, techniques, templates, and project management knowledge to all internal consultants. Since JWD Consulting focuses on identifying profitable projects and measuring their value after completion, this project must meet those criteria.

**3.0 Current Situation and Problem/Opportunity Statement**

JWD Consulting has a corporate Web site as well as an intranet. The firm currently uses the Web site for marketing information. The primary use of the intranet is for human resource information, such as where consultants enter their hours on various projects, change and view their benefits information, access an online directory and Web-based e-mail system, and so on. The firm also uses an enterprise-wide project management system to track all project information, focusing on the status of deliverables and meeting scope, time, and cost goals. There is an opportunity to provide a new section on the intranet dedicated to sharing consultants' project management knowledge across the organization. JWD Consulting only hires experienced consultants and gives them freedom to manage projects as they see fit. However, as the business grows and projects become more complex, even experienced project managers are looking for suggestions on how to work more effectively.

**4.0 Critical Assumption and Constraints**

The proposed intranet site must be a valuable asset for JWD Consulting. Current consultants and clients must actively support the project, and it must pay for itself within one year by reducing internal operating costs and generating new business. The Project Management Office manager must lead the effort, and the project team must include participants from several parts of the company, as well as current client organizations. The new system must run on existing hardware and software, and it should require minimal technical support. It must be easily accessible by consultants and clients and be secure from unauthorized users.

**5.0 Analysis of Options and Recommendation**

There are three options for addressing this opportunity:

1. Do nothing. The business is doing well, and we can continue to operate without this new project.
2. Purchase access to specialized software to support this new capability with little in-house development.

*(continued)*

**TABLE 3-2** JWD Consulting's business case (*continued*)



3. Design and implement the new intranet capabilities in-house using mostly existing hardware and software.

Based on discussions with stakeholders, we believe that option 3 is the best option.

**6.0 Preliminary Project Requirements**

The main features of the project management intranet site include the following:

1. Access to several project management templates and tools. Users must be able to search for templates and tools, read instructions on using these templates and tools, and see examples of how to apply them to real projects. Users must also be able to submit new templates and tools, which should be first screened or edited by the Project Management Office.
2. Access to relevant project management articles. Many consultants and clients feel as though there is an information overload when they research project management information. They often waste time they should be spending with their clients. The new intranet should include access to several important articles on various project management topics, which are searchable by topic, and allow users to request the Project Management Office staff to find additional articles to meet their needs.
3. Links to other, up-to-date Web sites, with brief descriptions of the main features of the external site.
4. An "Ask the Expert" feature to help build relationships with current and future clients and share knowledge with internal consultants.
5. Appropriate security to make the entire intranet site accessible to internal consultants and certain sections accessible to others.
6. The ability to charge money for access to some information. Some of the information and features of the intranet site should prompt external users to pay for the information or service. Payment options should include a credit card option or similar online payment transactions. After the system verifies payment, the user should be able to access or download the desired information.
7. Other features suggested by users, if they add value to the business.

**7.0 Budget Estimate and Financial Analysis**

A preliminary estimate of costs for the entire project is $140,000. This estimate is based on the project manager working about 20 hours per week for six months and other internal staff working a total of about 60 hours per week for six months. The customer representatives would not be paid for their assistance. A staff project manager would earn $50 per hour. The hourly rate for the other project team members would be $70 per hour, since some hours normally billed to clients may be needed for this project. The initial cost estimate also includes $10,000 for purchasing software and services from suppliers. After the project is completed, maintenance costs of $40,000 are included for each year, primarily to update the information and coordinate the "Ask the Expert" feature and online articles.

Projected benefits are based on a reduction in hours consultants spend researching project management information, appropriate tools and templates, and so on. Projected benefits are also based on a small increase in profits due to new business generated by this project. If each of more than 400 consultants saved just 40 hours each year (less than one hour per week) and could bill that time to other projects that generate a conservative estimate of $10 per hour in *profits*, then the projected benefit would be $160,000 per year. If the new intranet increased business by just 1 percent, using past profit information,

**TABLE 3-2** JWD Consulting's business case (*continued*)



increased profits due to new business would be at least $40,000 each year. Total projected benefits, therefore, are about $200,000 per year.

Exhibit A summarizes the projected costs and benefits and shows the estimated net present value (NPV), return on investment (ROI), and year in which payback occurs. It also lists assumptions made in performing this preliminary financial analysis. All of the financial estimates are very encouraging. The estimated payback is within one year, as requested by the sponsor. The NPV is $272,800, and the discounted ROI based on a three-year system life is excellent at 112 percent.

### 8.0 Schedule Estimate

The sponsor would like to see the project completed within six months, but there is some flexibility in the schedule. We also assume that the new system will have a useful life of at least three years.

### 9.0 Potential Risks

There are several risks involved with this project. The foremost risk is a lack of interest in the new system by our internal consultants and external clients. User inputs are crucial for populating information into this system and realizing the potential benefits from using the system. There are some technical risks in choosing the type of software used to search the system, check security, process payments, and so on, but the features of this system all use proven technologies. The main business risk is investing the time and money into this project and not realizing the projected benefits.

### 10.0 Exhibits

Exhibit A: Financial Analysis for Project Management Intranet Site Project

| Discount rate | 8% | | | | |
|---|---|---|---|---|---|
| Assume the project is done in about 6 months | Year | | | | |
| | 0 | 1 | 2 | 3 | Total |
| Costs | 140,000 | 40,000 | 40,000 | 40,000 | |
| Discount factor | 1 | 0.93 | 0.86 | 0.79 | |
| Discounted costs | 140,000 | 37,037 | 34,294 | 31,753 | 243,084 |
| | | | | | |
| Benefits | 0 | 200,000 | 200,000 | 200,000 | |
| Discount factor | 1 | 0.93 | 0.86 | 0.79 | |
| Discounted benefits | 0 | 186,185 | 171,468 | 158,766 | 515,419 |
| | | | | | |
| Discounted benefits - costs | (140,000) | 148,148 | 137,174 | 127,013 | |
| Cumulative benefits - costs | (140,000) | 8,148 | 145,322 | 272,336 | ← NPV |
| | Payback in Year 1 | | | | |
| **Discounted life cycle ROI------------>** | **112%** | | | | |
| | | | | | |
| **Assumptions** | | | | | |
| Costs | # hours | | | | |
| PM (500 hours, $50/hour) | 25,000 | | | | |
| Staff (1500 hours, $70/hour) | 105,000 | | | | |
| Outsourced software and services | 10,000 | | | | |
| Total project costs (all applied in year 0) | 140,000 | | | | |
| Benefits | | | | | |
| # consultants | 400 | | | | |
| Hours saved | 40 | | | | |
| $/hour profit | 10 | | | | |
| Benefits from saving time | 160,000 | | | | |
| Benefits from 1% increase in profits | 40,000 | | | | |
| Total annual projected benefits | 200,000 | | | | |

### Initiating



To officially initiate the Project Management Intranet Site project, Erica knew that main tasks were to identify all of the project stakeholders and to develop the project charter. Table 3-3 shows these processes and their outputs, based on the *PMBOK® Guide Fourth Edition*. The main outputs are a project charter, stakeholder register, and stakeholder management strategy. Another output that Erica found very useful for initiating projects was a formal project kick-off meeting. Descriptions of how these outputs were created and sample documents related to each of them are provided for this particular project. Recall that every project and every organization is unique, so not all project charters, stakeholder registers, etc. will look the same. You will see examples of several of these documents in later chapters.

**TABLE 3-3** Project initiation knowledge areas, processes, and outputs

| Knowledge Area | Initiating Process | Outputs |
|---|---|---|
| ***Project Integration Management*** | Develop project charter | Project charter |
| ***Project Communications Management*** | Identify stakeholders | Stakeholder register<br>Stakeholder management strategy |

*Identifying Project Stakeholders*

Erica met with Joe Fleming, the project's sponsor, to help identify key stakeholders for this project. Recall from Chapter 1 that stakeholders are people involved in or affected by project activities and include the project sponsor, project team, support staff, customers, users, suppliers, and even opponents to the project. Joe, the project sponsor, knew it would be important to assemble a strong project team, and he was very confident in Erica's ability to lead that team. They decided that key team members should include one of their full-time consultants with an outstanding record, Michael Chen, one part-time consultant, Jessie Faue, who was new to the company and supported the Project Management Office, and two members of the Information Technology (IT) department who supported the current intranet, Kevin Dodge and Cindy Dawson. They also knew that client inputs would be important for this project, so Joe agreed to call the CEOs of two of the firm's largest clients to see if they would be willing to provide representatives to work on this project at their own expense. All of the internal staff Joe and Erica recommended agreed to work on the project, and the two client representatives would be Kim Phuong and Page Miller. Since many other people would be affected by this project as future users of the new intranet, they also identified other key stakeholders including their directors of IT, Human Resources (HR), and Public Relations (PR), as well as Erica's administrative assistant.

After Joe and Erica made the preliminary contacts, Erica documented the stakeholders' roles, names, organizations, and contact information in a **stakeholder register**, a document that includes details related to the identified project stakeholders. Table 3-4 provides an example of part of the initial stakeholder register. Since this document would be public, Erica was careful not to include information that might be sensitive, such as how strongly the stakeholder supported the project, potential influence on the project, requirements and

**TABLE 3-4** Stakeholder register

| Name | Position | Internal/ External | Project Role | Contact Information |
|---|---|---|---|---|
| Joe Fleming | CEO | Internal | Sponsor | joe_fleming@jwdconsulting.com |
| Erica Bell | PMO Director | Internal | Project manager | erica_bell@jwdconsulting.com |
| Michael Chen | Team member | Internal | Team member | michael_chen@jwdconsulting.com |
| Kim Phuong | Business analyst | External | Advisor | kim_phuong@client1.com |
| Louise Mills | PR Director | Internal | Advisor | louise_mills@jwdconsulting.com |

expectations, etc. She would keep these issues in mind discretely and use them in developing the stakeholder management strategy.

A stakeholder management strategy is an approach to help increase the support of stakeholders throughout the project. It includes basic information such as stakeholder names, level of interest in the project, level of influence on the project, and potential management strategies for gaining support or reducing obstacles from that particular stakeholder. Since much of this information can be sensitive, it should be considered confidential. Some project managers do not even write down this information, but they do consider it since stakeholder management is a crucial part of their jobs. Table 3-5 provides an example of part of Erica's stakeholder management strategy for the Project Management Intranet Site project. You will see other examples of documenting stakeholder information in later chapters.

**TABLE 3-5** Stakeholder management strategy

| Name | Level of Interest | Level of Influence | Potential Management Strategies |
|---|---|---|---|
| Joe Fleming | High | High | Joe likes to stay on top of key projects and make money. Have a lot of short, face-to-face meetings and focus on achieving the financial benefits of the project. |
| Louise Mills | Low | High | Louise has a lot of things on her plate, and she does not seem excited about this project. She may be looking at other job opportunities. Show her how this project will help the company and her resume. |

*Drafting the Project Charter*



Erica drafted a project charter and had the project team members review it before showing it to Joe. Joe made a few minor changes, which Erica incorporated. Table 3-6 shows the final project charter (see Chapter 4 for more information on project charters). Note the items included on the project charter and its short length. JWD Consulting believes that project charters should preferably be one or two pages long, and they may refer to other documents, such as a business case, as needed. Erica felt the most important parts of the project charter were the signatures of key stakeholders (not included for brevity) and their individual comments. It is hard to get stakeholders to agree on even a one-page project charter, so everyone has a chance to make their concerns known in the comments section. Note that Michael Chen, the senior consultant asked to work on the project, was concerned about working on this project when he felt that his other assignments with external clients might have a higher priority. He offered to have an assistant help as needed. The information technology staff members mentioned their concerns about testing and security issues. Erica knew that she would have to consider these concerns when managing the project.

**TABLE 3-6** Project charter

**Project Title:** Project Management Intranet Site Project
**Project Start Date:** May 2 **Projected Finish Date:** November 4

**Budget Information:** The firm has allocated $140,000 for this project. The majority of costs for this project will be internal labor. An initial estimate provides a total of 80 hours per week.

**Project Manager:** Erica Bell, (310) 555-5896, erica_bell@jwdconsulting.com

**Project Objectives:** Develop a new capability accessible on JWD Consulting's intranet site to help internal consultants and external customers manage projects more effectively. The intranet site will include several templates and tools that users can download, examples of completed templates and related project management documents used on real projects, important articles related to recent project management topics, an article retrieval service, links to other sites with useful information, and an "Ask the Expert" feature, where users can post questions they have about their projects and receive advice from experts in the field. Some parts of the intranet site will be accessible free to the public, other parts will only be accessible to current customers and/or internal consultants, and other parts of the intranet site will be accessible for a fee.

**Main Project Success Criteria:** The project should pay for itself within one year of completion.

**Approach:**

- Develop a survey to determine critical features of the new intranet site and solicit input from consultants and customers.
- Review internal and external templates and examples of project management documents.
- Research software to provide security, manage user inputs, and facilitate the article retrieval and "Ask the Expert" features.
- Develop the intranet site using an iterative approach, soliciting a great deal of user feedback.

TABLE 3-6 Project charter (*continued*)

- Determine a way to measure the value of the intranet site in terms of reduced costs and new revenues, both during the project and one year after project completion.

ROLES AND RESPONSIBILITIES

| Name | Role | Position | Contact Information |
|---|---|---|---|
| Joe Fleming | Sponsor | JWD Consulting, CEO | joe_fleming@jwdconsulting.com |
| Erica Bell | Project Manager | JWD Consulting, manager | erica_bell@jwdconsulting.com |
| Michael Chen | Team Member | JWD Consulting, senior consultant | michael_chen@jwdconsulting.com |
| Jessie Faue | Team Member | JWD Consulting, consultant | jessie_faue@jwdconsulting.com |
| Kevin Dodge | Team Member | JWD Consulting, IT department | kevin_dodge@jwdconsulting.com |
| Cindy Dawson | Team Member | JWD Consulting, IT department | cindy_dawson@jwdconsulting.com |
| Kim Phuong | Advisor | Client representative | kim_phuong@client1.com |
| Page Miller | Advisor | Client representative | page_miller@client2.com |

**Sign-Off:** (Signatures of all the above stakeholders)
**Comments:** (Handwritten or typed comments from above stakeholders, if applicable)
*"I will support this project as time allows, but I believe my client projects take priority. I will have one of my assistants support the project as needed." —Michael Chen*
"We need to be extremely careful testing this new system, especially the security in giving access to parts of the intranet site to the public and clients." —Kevin Dodge and Cindy Dawson



*Holding a Project Kick-off Meeting*

Experienced project managers like Erica know that it is crucial to get projects off to a great start. Holding a good kick-off meeting is an excellent way to do this. A **kick-off meeting** is a meeting held at the beginning of a project so that stakeholders can meet each other, review the goals of the project, and discuss future plans. The kick-off meeting is often held after the business case and project charter are completed, but it could be held sooner, as needed. Even if some or even all project stakeholders must meet virtually, it is still important to have a kick-off meeting.

Erica also knows that all project meetings with major stakeholders should include an agenda. Figure 3-2 shows the agenda that Erica provided for the Project Management Intranet Site project kick-off meeting. Notice the main topics in an agenda:

- Meeting objective
- Agenda (lists in order the topics to be discussed)

- A section for documenting action items, who they are assigned to, and when each person will complete the action
- A section to document the date and time of the next meeting



**Kick-Off Meeting**
**[Date of Meeting]**

**Project Name:** Project Management Intranet Site Project

**Meeting Objective:** Get the project off to an effective start by introducing key stakeholders, reviewing project goals, and discussing future plans

**Agenda:**

- Introductions of attendees
- Review of the project background
- Review of project-related documents (i.e., business case, project charter)
- Discussion of project organizational structure
- Discussion of project scope, time, and cost goals
- Discussion of other important topics
- List of action items from meeting

| Action Item | Assigned To | Due Date |
|---|---|---|
| | | |
| | | |
| | | |

**Date and time of next meeting:**

**FIGURE 3-2** Kick-off meeting agenda

It is good practice to focus on results of meetings, and having sections for documenting action items and deciding on the next meeting date and time on the agenda helps to do so. It is also good practice to document meeting minutes, focusing on key decisions and action items. Erica planned to send the meeting minutes to all meeting participants and other appropriate stakeholders within a day or two of the meeting.

## Project Planning

Planning is often the most difficult and unappreciated process in project management. Because planning is not always used to facilitate action, many people view planning negatively. The main purpose of project plans, however, is *to guide project execution*. To guide execution, plans must be realistic and useful, so a fair amount of time and effort must go into the planning process; people knowledgeable with the work need to plan the work.

Chapter 4, Project Integration Management, provides detailed information on preparing a project management plan, and Chapters 5 through 12 describe planning processes for each of the other knowledge areas.



Table 3-7 lists the project management knowledge areas, processes, and outputs of project planning according to the *PMBOK® Guide, Fourth Edition*. There are many potential outputs from the planning process group, and every knowledge area is included. Just a few planning documents from JWD Consulting's Project Management Intranet Site Project are provided in this chapter as examples, and later chapters include many more examples. Recall that the *PMBOK® Guide* is only a guide, so many organizations may have different planning outputs based on their particular needs, as is the case in this example. There are many templates related to planning as well, with several listed in the last section of this chapter.

**TABLE 3-7** Planning processes and outputs

| Knowledge Area | Planning Process | Outputs |
|---|---|---|
| ***Project Integration Management*** | Develop project management plan | Project management plan |
| ***Project Scope Management*** | Collect requirements | Requirements documents<br>Requirements management plan<br>Requirements traceability matrix |
| | Define scope | Project scope statement<br>Project document updates |
| | Create WBS | WBS<br>WBS dictionary<br>Scope baseline<br>Project document updates |
| ***Project Time Management*** | Define activities | Activity list<br>Activity attributes<br>Milestone list |
| | Sequence activities | Project schedule network diagrams<br>Project document updates |
| | Estimate activity resources | Activity resource requirements<br>Resource breakdown structure<br>Project document updates |
| | Estimate activity durations | Activity duration estimates<br>Project document updates |
| | Develop schedule | Project schedule<br>Schedule baseline |

*(continued)*

TABLE 3-7 Planning processes and outputs (*continued*)

| Knowledge Area | Planning Process | Outputs |
|---|---|---|
| | | Schedule data<br>Project document updates |
| ***Project Cost Management*** | Estimate costs | Activity cost estimates<br>Basis of estimates<br>Project document updates |
| | Determine budget | Cost performance baseline<br>Project funding requirements<br>Project document updates |
| ***Project Quality Management*** | Plan quality | Quality management plan<br>Quality metrics<br>Quality checklists<br>Process improvement plan<br>Project document updates |
| ***Project Human Resource Management*** | Develop human resource plan | Human resource plan |
| ***Project Communications Management*** | Plan communications | Communications management plan<br>Project document updates |
| ***Project Risk Management*** | Plan risk management | Risk management plan |
| | Identify risks | Risk register |
| | Perform qualitative risk analysis | Risk register updates |
| | Perform quantitative risk analysis | Risk register updates |
| | Plan risk responses | Risk register updates<br>Project management plan updates<br>Risk related contract decisions<br>Project document updates |
| ***Project Procurement Management*** | Plan procurements | Procurement management plan<br>Procurement statement of work<br>Make-or-buy decisions<br>Procurement documents<br>Source selection criteria<br>Change requests |

Since the Project Management Intranet Site project is relatively small, Erica believes some of the most important planning documents to focus on are the following:

- A team contract (not listed in Table 3-7, which is based only on the *PMBOK® Guide*)
- A project scope statement
- A work breakdown structure (WBS)
- A project schedule, in the form of a Gantt chart with all dependencies and resources entered
- A list of prioritized risks (part of a risk register)

All of these documents, as well as other project-related information, will be available to all team members on a project Web site. JWD Consulting has used project Web sites for several years, and has found that they really help facilitate communications and document project information. For larger projects, JWD Consulting also creates many of the other outputs listed in Table 3-7. (You will learn more about these documents by knowledge area in the following chapters.)

Soon after the project team signed the project charter, Erica organized a team-building meeting for the Project Management Intranet Site Project. An important part of the meeting was helping the project team get to know each other. Erica had met and talked to each member separately, but this was the first time the project team would spend much time together. Jessie Faue worked in the Project Management Office with Erica, so they knew each other well, but Jessie was new to the company and did not know any of the other team members. Michael Chen was a senior consultant and often worked on the highest priority projects for external clients. He attended the meeting with his assistant, Jill Anderson, who would also support the project when Michael was too busy. Everyone valued Michael's expertise, and he was extremely straightforward in dealing with people. He also knew both of the client representatives from past projects. Kevin Dodge was JWD Consulting's intranet guru who tended to focus on technical details. Cindy Dawson was also from the Information Technology department and had experience working as a business consultant and negotiating with outside suppliers. Kim Phuong and Page Miller, the two client representatives, were excited about the project, but they were wary of sharing sensitive information about their company.

Erica had everyone introduce him or herself, and then she facilitated an icebreaker activity so everyone would be more relaxed. She asked everyone to describe his or her dream vacation, assuming cost was no issue. This activity helped everyone get to know each other and show different aspects of their personalities. Erica knew that it was important to build a strong team and have everyone work well together.

Erica then explained the importance of the project, again reviewing the signed project charter. She explained that an important tool to help a project team work together was to have members develop a team contract that everyone felt comfortable signing. JWD Consulting believed in using team contracts for all projects to help promote teamwork and clarify team communications. She explained the main topics covered in a team contract and showed them a team contract template. She then had the team members form two smaller groups, with one consultant, one Information Technology department member, and one client representative in each group. These smaller groups made it easier for everyone to contribute ideas. Each group shared their ideas for what should go into the contract, and then they worked together to form one project team contract. Table 3-8 shows the resulting team contract, which took about 90 minutes to create. Erica could see that

**TABLE 3-8** Team contract

| |
|---|
| **Code of Conduct:** As a project team, we will:<br>• Work proactively, anticipating potential problems and working to prevent them.<br>• Keep other team members informed of information related to the project.<br>• Focus on what is best for the entire project team. |
| **Participation:** We will:<br>• Be honest and open during all project activities.<br>• Encourage diversity in team work.<br>• Provide the opportunity for equal participation.<br>• Be open to new approaches and consider new ideas.<br>• Have one discussion at a time.<br>• Let the project manager know well in advance if a team member has to miss a meeting or may have trouble meeting a deadline for a given task. |
| **Communication:** We will:<br>• Decide as a team on the best way to communicate. Since a few team members cannot meet often for face-to-face meetings, we will use e-mail, a project Web site, and other technology to assist in communicating.<br>• Have the project manager facilitate all meetings and arrange for phone and video conferences, as needed.<br>• Work together to create the project schedule and enter actuals into our enterprise-wide project management system by 4 p.m. every Friday.<br>• Present ideas clearly and concisely.<br>• Keep discussions on track. |
| **Problem Solving:** We will:<br>• Encourage everyone to participate in solving problems.<br>• Only use constructive criticism and focus on solving problems, not blaming people.<br>• Strive to build on each other's ideas. |
| **Meeting Guidelines:** We will:<br>• Plan to have a face-to-face meeting the first and third Tuesday morning of every month.<br>• Meet more frequently the first month.<br>• Arrange for telephone or videoconferencing for participants as needed.<br>• Hold other meetings as needed.<br>• Record meeting minutes and send them out via e-mail within 24 hours of all project meetings, focusing on decisions made and action items from each meeting. |

there were different personalities on this team, but she felt they all could work together well.

Erica wanted to keep their meeting to its two-hour time limit. Their next task would be to clarify the scope of the project by developing a project scope statement and WBS. She knew it took time to develop these documents, but she wanted to get a feel for what everyone thought were the main deliverables for this project, their roles in producing those deliverables, and what areas of the project scope needed clarification. She reminded everyone what their budget and schedule goals were so they would keep that in mind as they discussed the scope of the project. She also asked each person to provide the number of hours he or she would be available to work on this project each month for the next six months. She then had each person write down his or her answers to the following questions:

1. List one item that is most unclear to you about the scope of this project.
2. What other questions do you have or issues do you foresee about the scope of the project?
3. List what you believe to be the main deliverables for this project.
4. Which deliverables do you think you will help create or review?

Erica collected everyone's inputs. She explained that she would take this information and work with Jessie to develop the first draft of the scope statement that she would e-mail to everyone by the end of the week. She also suggested that they all meet again in one week to develop the scope statement further and to start creating the WBS for the project.

Erica and Jessie reviewed all the information and created the first draft of the scope statement. At their next team meeting, they discussed the scope statement and got a good start on the WBS. Table 3-9 shows a portion of the scope statement that Erica created after a few more e-mails and another team meeting. Note that the scope statement lists the product characteristics and requirements, summarizes the deliverables, and describes project success criteria in detail.

**TABLE 3-9** Scope statement (draft version)

**Project Title: Project Management Intranet Site Project**
**Date:** May 18 **Prepared by:** Erica Bell, Project Manager, erica_bell@jwdconsulting.com

**Project Summary and Justification:** Joe Fleming, CEO of JWD Consulting, requested this project to assist the company in meeting its strategic goals. The new intranet site will increase visibility of the company's expertise to current and potential clients. It will also help reduce internal costs and improve profitability by providing standard tools, techniques, templates, and project management knowledge to all internal consultants. The budget for the project is $140,000. An additional $40,000 per year will be required for operational expenses after the project is completed. Estimated benefits are $200,000 each year. It is important to focus on the system paying for itself within one year of its completion.

**Product Characteristics and Requirements:**

1. Templates and tools: The intranet site will allow authorized users to download files they can use to create project management documents and to help them use project

*(continued)*

management tools. These files will be in Microsoft Word, Excel, Access, Project, or in HTML or PDF format, as appropriate.

2. User submissions: Users will be encouraged to e-mail files with sample templates and tools to the Webmaster. The Webmaster will forward the files to the appropriate person for review and then post the files to the intranet site, if desired.
3. Articles: Articles posted on the intranet site will have appropriate copyright permission. The preferred format for articles will be PDF. The project manager may approve other formats.
4. Requests for articles: The intranet site will include a section for users to request someone from the Project Management Office (PMO) at JWD Consulting to research appropriate articles for them. The PMO manager must first approve the request and negotiate payments, if appropriate.
5. Links: All links to external sites will be tested on a weekly basis. Broken links will be fixed or removed within five working days of discovery.
6. The "Ask the Expert" feature must be user-friendly and capable of soliciting questions and immediately acknowledging that the question has been received in the proper format. The feature must also be capable of forwarding the question to the appropriate expert (as maintained in the system's expert database) and capable of providing the status of questions that are answered. The system must also allow for payment for advice, if appropriate.
7. Security: The intranet site must provide several levels of security. All internal employees will have access to the entire intranet site when they enter their security information to access the main, corporate intranet. Part of the intranet will be available to the public from the corporate Web site. Other portions of the intranet will be available to current clients based on verification with the current client database. Other portions of the intranet will be available after negotiating a fee or entering a fixed payment using pre-authorized payment methods.
8. Search feature: The intranet site must include a search feature for users to search by topic, key words, etc.
9. The intranet site must be accessible using a standard Internet browser. Users must have appropriate application software to open several of the templates and tools.
10. The intranet site must be available 24 hours a day, 7 days a week, with one hour per week for system maintenance and other periodic maintenance, as appropriate.

**Summary of Project Deliverables**

**Project management-related deliverables:** Business case, charter, team contract, scope statement, WBS, schedule, cost baseline, progress reports, final project presentation, final project report, lessons-learned report, and any other documents required to manage the project.

**Product-related deliverables:**

1. Survey: Survey current consultants and clients to help determine desired content and features for the intranet site.
2. Files for templates: The intranet site will include templates for at least 20 documents when the system is first implemented, and it will have the capacity to store up to 100 documents. The project team will decide on the initial 20 templates based on survey results.
3. Examples of completed templates: The intranet site will include examples of projects that have used the templates available on the intranet site. For example, if there is a

**TABLE 3-9** Scope statement (draft version) (*continued*)

template for a business case, there will also be an example of a real business case that uses the template.

4. Instructions for using project management tools: The intranet site will include information on how to use several project management tools, including the following as a minimum: work breakdown structures, Gantt charts, network diagrams, cost estimates, and earned value management. Where appropriate, sample files will be provided in the application software appropriate for the tool. For example, Microsoft Project files will be available to show sample work breakdown structures, Gantt charts, network diagrams, cost estimates, and applications of earned value management. Excel files will be available for sample cost estimates and earned value management charts.
5. Example applications of tools: The intranet site will include examples of real projects that have applied the tools listed in number 4 above.
6. Articles: The intranet site will include at least 10 useful articles about relevant topics in project management. The intranet site will have the capacity to store at least 1,000 articles in PDF format with an average length of 10 pages each.
7. Links: The intranet site will include links with brief descriptions for at least 20 useful sites. The links will be categorized into meaningful groups.
8. Expert database: In order to deliver an "Ask the Expert" feature, the system must include and access a database of approved experts and their contact information. Users will be able to search for experts by pre-defined topics.
9. User Requests feature: The intranet site will include an application to solicit and process requests from users.
10. Intranet site design: An initial design of the new intranet site will include a site map, suggested formats, appropriate graphics, etc. The final design will incorporate comments from users on the initial design.
11. Intranet site content: The intranet site will include content for the templates and tools section, articles section, article retrieval section, links section, "Ask the Expert" section, User Requests feature, security, and payment features.
12. Test plan: The test plan will document how the intranet site will be tested, who will do the testing, and how bugs will be reported.
13. Promotion: A plan for promoting the intranet site will describe various approaches for soliciting inputs during design. The promotion plan will also announce the availability of the new intranet site.
14. Project benefit measurement plan: A project benefit plan will measure the financial value of the intranet site.

**Project Success Criteria:** Our goal is to complete this project within six months for no more than $140,000. The project sponsor, Joe Fleming, has emphasized the importance of the project paying for itself within one year after the intranet site is complete. To meet this financial goal, the intranet site must have strong user inputs. We must also develop a method for capturing the benefits while the intranet site is being developed and tested, and after it is rolled out. If the project takes a little longer to complete or costs a little more than planned, the firm will still view it as a success if it has a good payback and helps promote the firm's image as an excellent consulting organization.

As the project team worked on the scope statement, they also developed the work breakdown structure (WBS) for the project. The WBS is a very important tool in project management because it provides the basis for deciding how to do the work. The WBS also provides a basis for creating the project schedule and performing earned value management for measuring and forecasting project performance. Erica and her team decided to use the project management process groups as the main categories for the WBS, as shown in Figure 3-3. They included completed work from the initiating process to provide a complete picture of the project's scope. The group also wanted to list several milestones on their schedule, such as the completion of key deliverables, so they prepared a separate list of milestones that they would include on the Gantt chart. You will learn more about creating a WBS in Chapter 5, Project Scope Management.

After preparing the WBS, the project team held another face-to-face meeting to develop the project schedule, following the steps outlined in section 2.5 of the WBS. Several of the project schedule tasks are dependent on one another. For example, the intranet site testing was dependent on the construction and completion of the content tasks. Everyone participated in the development of the schedule, especially the tasks on which each would be working. Some of the tasks were broken down further so the team members had a better understanding of what they had to do and when. They also kept their workloads and cost constraints in mind when developing the duration estimates. For example, Erica was scheduled to work 20 hours per week on this project, and the other project team members combined should not spend more than 60 hours per week on average for the project. As team members provided duration estimates, they also estimated how many work hours they would spend on each task.

After the meeting, Erica worked with Jessie to enter all of the information into Microsoft Project. Erica was using the intranet site project to train Jessie in applying several project management tools and templates. They entered all of the tasks, duration estimates, and dependencies to develop the Gantt chart. Erica decided to enter the resource and cost information after reviewing the schedule. Their initial inputs resulted in a completion date a few weeks later than planned. Erica and Jessie reviewed the critical path for the project, and Erica had to shorten the duration estimates for a few critical tasks in order to meet their schedule goal of completing the project within six months. She talked to the team members working on those tasks, and they agreed that they could plan to work more hours each week on those tasks, if required, in order to complete them on time. Figure 3-4 shows the resulting Gantt chart created in Microsoft Project. Only the executing tasks are expanded to show the subtasks under that category. (You will learn how to use Project 2007 in Appendix A. Chapter 6, Project Time Management, explains Gantt charts and other time management tools.) The baseline schedule projects a completion date of November 1. The project charter had a planned completion date of November 4. Erica wanted to complete the project on time, and although three extra days was not much of a buffer, she felt the baseline schedule was very realistic. She would do her best to help everyone meet their deadlines.

The majority of the costs for this project were internal labor, and the team kept their labor hour constraints in mind when developing task duration estimates. Erica and Jessie entered each project team member's name and labor rate in the resource sheet for their Microsoft Project file. The client representatives were not being paid for their time, so she left their labor rates at the default value of zero. Erica had also included $10,000 for procurement in the financial analysis she prepared for the business case, and she showed Jessie how to enter that amount as a fixed cost split equally between the "Ask the Expert" and User Requests features,

- 1.0 Initiating
  - 1.1 Identify key stakeholders
  - 1.2 Prepare project charter
  - 1.3 Hold project kick-off meeting
- 2.0 Planning
  - 2.1 Hold team planning meeting
  - 2.2 Prepare team contract
  - 2.3 Prepare scope statement
  - 2.4 Prepare WBS
  - 2.5 Prepare schedule and cost baseline
    - 2.5.1 Determine task resources
    - 2.5.2 Determine task durations
    - 2.5.3 Determine task dependencies
    - 2.5.4 Create draft Gantt chart
    - 2.5.5 Review and finalize Gantt chart
  - 2.6 Identify, discuss, and prioritize risks
- 3.0 Executing
  - 3.1 Survey
  - 3.2 User inputs
  - 3.3 Intranet site content
    - 3.3.1 Templates and tools
    - 3.3.2 Articles
    - 3.3.3 Links
    - 3.3.4 Ask the Expert
    - 3.3.5 User requests feature
  - 3.4 Intranet site design
  - 3.5 Intranet site construction
  - 3.6 Intranet site testing
  - 3.7 Intranet site promotion
  - 3.8 Intranet site roll-out
  - 3.9 Project benefits measurement
- 4.0 Monitoring and Controlling
  - 4.1 Progress reports
- 5.0 Closing
  - 5.1 Prepare final project report
  - 5.2 Prepare final project presentation
  - 5.3 Lessons learned

**FIGURE 3-3** JWD Consulting intranet project work breakdown structure (WBS)

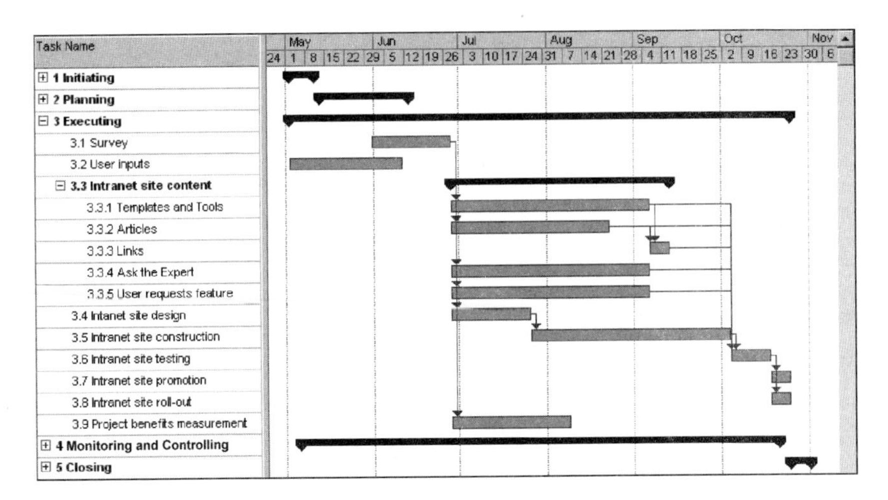


**FIGURE 3-4** JWD Consulting intranet site project baseline Gantt chart

where she thought they would have to purchase some external software and/or services. Erica then helped Jessie assign resources to tasks, entering the projected number of hours everyone planned to work each week on each task. They then ran several cost reports and made a few minor adjustments to resource assignments to make their planned total cost meet their budget constraints. Their cost baseline was very close to their planned budget of $140,000.

The last deliverable her team needed to create within the planning process group was a list of prioritized risks. This information will be updated and expanded as the project progresses in a risk register, which also includes information on root causes of the risks, warning signs that potential risks might occur, and response strategies for the risks. (See Chapter 12, Project Risk Management, for more information on risk registers.) Erica reviewed the risks she had mentioned in the business case as well as the comments team members made on the project charter and in their team meetings. She held a special meeting for everyone to brainstorm and discuss potential risks. They posted all of the risks they identified on a probability/impact matrix, and then they grouped some of the ideas. There was only one risk in the high probability and high impact category, and several with medium impact in one or both categories. They chose not to list the low probability and low impact risks. After some discussion, the team developed the list of prioritized risks shown in Table 3-10.

## Project Execution

Executing the project involves taking the actions necessary to ensure that activities in the project plan are completed. It also includes work required to introduce any new hardware, software, and procedures into normal operations. The products of the project are produced during project execution, and it usually takes the most resources to accomplish this process. Table 3-11 lists the knowledge areas, executing processes, and outputs of project execution listed in the *PMBOK® Guide, Fourth Edition*. Many project sponsors and customers focus on deliverables related to providing the products, services, or results desired from the project. It is also important to document change requests and prepare updates to planning documents as part of execution. Templates related to this process group are also listed later in this chapter.

**TABLE 3-10** List of prioritized risks

| Ranking | Potential Risk |
|---|---|
| 1 | Lack of inputs from internal consultants |
| 2 | Lack of inputs from client representatives |
| 3 | Security of new system |
| 4 | Outsourcing/purchasing for the article retrieval and "Ask the Expert" features |
| 5 | Outsourcing/purchasing for processing online payment transactions |
| 6 | Organizing the templates and examples in a useful fashion |
| 7 | Providing an efficient search feature |
| 8 | Getting good feedback from Michael Chen and other senior consultants |
| 9 | Effectively promoting the new system |
| 10 | Realizing the benefits of the new system within one year |

For this relatively small project, Erica would work closely with all the team members to make sure they were producing the desired work results. She also used her networking skills to get input from other people in the firm and from external sources at no additional cost to the project. She made sure that everyone who would use the resulting intranet application also understood what they were producing as part of the project and how it would help them in the future. She knew that providing strong leadership and using good communication skills were crucial to good project execution. The firm did have a formal change request form, but primarily used it for external projects. The firm also had contract specialists and templates for several procurement documents that the project team would use for the portions of the project they planned to outsource.

As mentioned earlier, Erica knew that Joe, the CEO and project sponsor, liked to see progress on projects through milestone reports. He also wanted Erica to alert him to any potential issues or problems. Table 3-12 shows a sample of a milestone report for the Project Management Intranet Site Project that Erica reviewed with Joe in mid-June. Erica met with most of her project team members often, and she talked to Joe about once a week to review progress on completing milestones and to discuss any other project issues. Although Erica could have used project management software to create milestone reports, she used word processing software instead because this project was small and she could more easily manipulate the report format.

Human resource issues often occur during project execution, especially conflicts. At several of the team meetings, Erica could see that Michael seemed to be bored and often left the room to make phone calls to clients. She talked to Michael about the situation, and she discovered that Michael was supportive of the project, but he knew he could only spend a minimal amount of time on it. He was much more productive outside of meetings, so Erica agreed to have Michael attend a minimal amount of project team meetings. She could see

**TABLE 3-11** Executing processes and outputs

| Knowledge Area | Executing Process | Outputs |
| --- | --- | --- |
| ***Project Integration Management*** | Direct and manage project execution | Deliverables<br>Work performance information<br>Change requests<br>Project management plan updates<br>Project document updates |
| ***Project Quality Management*** | Perform quality assurance | Organizational process asset updates<br>Change requests<br>Project management plan updates<br>Project document updates |
| ***Project Human Resource Management*** | Acquire project team<br>Develop project team<br>Manage project team | Project staff assignments<br>Resource calendars<br>Project management plan updates<br>Team performance assessment<br>Enterprise environmental factor updates<br>Enterprise environmental factors updates<br>Organizational process assets updates<br>Project management plan updates<br>Change requests |
| ***Project Communications Management*** | Distribute information<br>Manage stake-holders expectations | Organizational process assets updates<br>Organizational process assets updates<br>Change requests<br>Project management plan updates<br>Project document updates |
| ***Project Procurement Management*** | Conduct procurements | Selected sellers<br>Procurement contract award<br>Resource calendars<br>Change requests<br>Project management plan updates<br>Project documents updates |

that Michael was contributing to the team by the feedback he provided and his leadership on the "Ask the Expert" feature for the intranet site. Erica adjusted her communication style to meet his specific needs.

Another problem occurred when Cindy was contacting potential suppliers for software to help with the "Ask the Expert" and User Requests features. Kevin wanted to write all of the software for the project himself, but Cindy knew it made better business sense to

**TABLE 3-12** Milestone Report as of June 17



| Milestone | Date | Status | Responsible | Issues/ Comments |
|---|---|---|---|---|
| ***Initiating*** | | | | |
| Stakeholders identified | May 2 | Completed | Erica and Joe | |
| Project charter signed | May 10 | Completed | Erica | |
| Project kick-off meeting held | May 13 | Completed | Erica | Went very well |
| ***Planning*** | | | | |
| Team contract signed | May 13 | Completed | Erica | |
| Scope statement completed | May 27 | Completed | Erica | |
| WBS completed | May 31 | Completed | Erica | |
| List of prioritized risks completed | June 3 | Completed | Erica | Reviewed with sponsor and team |
| Schedule and cost baseline completed | June 13 | Completed | Erica | |
| ***Executing*** | | | | |
| Survey completed | June 28 | | Erica | Poor response so far! |
| Intranet site design completed | July 26 | | Kevin | |
| Project benefits measurement completed | August 9 | | Erica | |
| User inputs collected | August 9 | | Jessie | |
| Articles completed | August 23 | | Jessie | |
| Templates and tools completed | September 6 | | Erica | |
| Ask the Expert completed | September 6 | | Michael | |
| User Requests feature completed | September 6 | | Cindy | |
| Links completed | September 13 | | Kevin | |

*(continued)*

TABLE 3-12 Milestone Report as of June 17 (*continued*)



| Milestone | Date | Status | Responsible | Issues/ Comments |
|---|---|---|---|---|
| Intranet site construction completed | October 4 | | Kevin | |
| Intranet site testing completed | October 18 | | Cindy | |
| Intranet site promotion completed | October 25 | | Erica | |
| Intranet site roll-out completed | October 25 | | Kevin | |
| ***Monitoring and Controlling*** Progress reports | Every Friday | | All | |
| ***Closing*** Final project presentation completed | October 27 | | Erica | |
| Sponsor sign-off on project completed | October 27 | | Joe | |
| Final project report completed | October 28 | | Erica | |
| Lessons-learned reports submitted | November 1 | | All | |

purchase these new software capabilities from a reliable source. Cindy had to convince Kevin that it was worth buying some software from other sources.

Cindy also discovered that their estimate of $10,000 was only about half the amount they needed. She discussed the problem with Erica, explaining the need for some custom development no matter which supplier they chose. Erica agreed that they should go with an outside source, and she asked their sponsor to approve the additional funds. Joe agreed, but he stressed the importance of still having the system pay for itself within a year.

Erica also had to ask Joe for help when the project team received a low response rate to their survey and requests for user inputs. Joe sent out an e-mail to all of JWD Consulting's consultants describing the importance of the project. He also offered five extra vacation days to the person who provided the best examples of how they used tools and templates to manage their projects. Erica then received informative input from the consultants. Having effective communication skills and strong top management support are essential to good project execution.

One way to learn about best practices in project management is by studying recipients of PMI's Project of the Year award. The Quartier International de Montréal (QIM), Montreal's international district, was a 66-acre urban revitalization project in the heart of downtown Montreal. This $90 million, five-year project turned a once unpopular area into a thriving section of the city with a booming real estate market and has generated $770 million in related construction. Clement Demers, PMP, was the director general for the QIM project. He said the team "took a unique project execution approach by dividing work into packages that allowed for smaller-scale testing of management techniques and contract awards. Benefiting from experience gained in each stage, managers could then adjust future work segments and management styles accordingly."[8]

Other strategies that helped the team succeed included the following:

- The team identified champions in each stakeholder group to help inspire others to achieve project goals.
- The team's communications plan included a Web site dedicated to public concerns.
- There were two-day reviews at the beginning of each project phase to discuss problems and develop solutions to prevent conflict.
- Financial investors were asked for input to increase their stake in the project.
- The team recognized the cost value of hiring high-quality experts, such as architects, engineers, lawyers, and urban planners. They paid all professionals a fixed price for their services and paid their fees quickly.

## Project Monitoring and Controlling

Monitoring and controlling is the process of measuring progress toward project objectives, monitoring deviation from the plan, and taking corrective action to match progress with the plan. Monitoring and controlling is done throughout the life of a project. It also involves eight of the nine project management knowledge areas. Table 3-13 lists the knowledge areas, monitoring and controlling processes, and outputs, according to the *PMBOK® Guide, Fourth Edition*. Templates related to this process group are listed later in this chapter.

**TABLE 3-13** Monitoring and controlling processes and outputs

| Knowledge Area | Monitoring and Controlling Process | Outputs |
|---|---|---|
| ***Project Integration Management*** | Monitor and control project work | Change requests<br>Project management plan updates<br>Project document updates |
| | Perform integrated change control | Change request status updates<br>Project management plan updates<br>Project document updates |
| ***Project Scope Management*** | Verify scope | Accepted deliverables<br>Change requests<br>Project document updates |

**TABLE 3-13** Monitoring and controlling processes and outputs (*continued*)

| Knowledge Area | Monitoring and Controlling Process | Outputs |
|---|---|---|
| | Control scope | Work performance measurements<br>Organizational process assets updates<br>Change requests<br>Project management plan updates<br>Project document updates |
| ***Project Time Management*** | Control schedule | Work performance measurements<br>Organizational process assets updates<br>Change requests<br>Project management plan updates<br>Project document updates |
| ***Project Cost Management*** | Control cost | Work performance measurements<br>Budget forecasts<br>Organizational process assets updates<br>Change requests<br>Project management plan updates<br>Project document updates |
| ***Project Quality Management*** | Perform quality control | Quality control measurements<br>Validated deliverables<br>Organizational process assets updates<br>Change requests<br>Project management plan updates<br>Project document updates |
| ***Project Communications Management*** | Report performance | Performance reports<br>Organizational process assets updates<br>Change requests |
| ***Project Risk Management*** | Monitor and control risks | Risk register updates<br>Organizational process assets updates<br>Change requests<br>Project management plan updates<br>Project document updates |
| ***Project Procurement Management*** | Administer procurements | Procurement documentation<br>Organizational process assets updates<br>Change requests<br>Project management plan updates |

On the Project Management Intranet Site Project, there were several updates to the project management plan to reflect changes made to the project scope, schedule, and budget. Erica and other project team members took corrective action when necessary. For example, when they were not getting many responses to their survey, Erica asked Joe for help. When Cindy had trouble negotiating with a supplier, she got help from another senior consultant who had worked with that supplier in the past. Erica also had to request more funds for that part of the project.



Project team members submitted a brief progress report every Friday. They were originally using a company template for progress reports, but Erica found that by modifying the old template, she received better information to help her team work more effectively. She wanted team members not only to report what they did but also to focus on what was going well or not going well and why. This extra information helped team members reflect on the project's progress and identify areas in need of improvement. Table 3-14 is an example of one of Cindy's progress reports.

**TABLE 3-14** Sample weekly progress report

| |
|---|
| **Project Name:** Project Management Intranet Project<br>**Team Member Name:** Cindy Dawson, cindy_dawson@jwdconsulting.com<br>**Date:** August 5 |
| **Work completed this week:**<br>–Worked with Kevin to start the intranet site construction<br>–Organized all the content files<br>–Started developing a file naming scheme for content files<br>–Continued work on "Ask the Expert" and User Requests features<br>–Met with preferred supplier<br>–Verified that their software would meet our needs<br>–Discovered the need for some customization |
| **Work to complete next week:**<br>–Continue work on intranet site construction<br>–Prepare draft contract for preferred supplier<br>–Develop new cost estimate for outsourced work |
| **What's going well and why:**<br>The intranet site construction started well. The design was very clear and easy to follow. Kevin really knows what he's doing. |
| **What's not going well and why:**<br>It is difficult to decide how to organize the templates and examples. Need more input from senior consultants and clients. |
| **Suggestions/Issues:**<br>–Hold a special meeting to decide how to organize the templates and examples on the intranet site.<br>–Get some sample contracts and help in negotiating with the preferred supplier. |

*(continued)*

TABLE 3-14 Sample weekly progress report (*continued*)



**Project changes:**
I think we can stay on schedule, but it looks like we'll need about $10,000 more for outsourcing. That's doubling our budget in that area.

In addition to progress reports, an important tool for monitoring and controlling the project was using project management software. Each team member submitted his or her actual hours worked on tasks each Friday afternoon by 4 p.m. via the firm's enterprise-wide project management software. They were using the enterprise version of Microsoft Project 2007, so they could easily update their task information via the Web. Erica worked with Jessie to analyze the information, paying special attention to the critical path and earned value data. (See Chapter 6 on Project Time Management for more information on critical path analysis; Chapter 7 on Project Cost Management for a description of earned value management; and Appendix A for more information on using Project 2007 to help control projects.) Erica wanted to finish the project on time, even if it meant spending more money. Joe agreed with that approach, and approved the additional funding Erica projected they would need based on the earned value projections and the need to make up a little time on critical tasks.

Joe again emphasized the importance of the new system paying for itself within a year. Erica was confident that they could exceed the projected financial benefits, and she decided to begin capturing benefits as soon as the project team began testing the system. When she was not working on this project, Erica was managing JWD Consulting's Project Management Office (PMO), and she could already see how the intranet site would help her staff save time and make their consultants more productive. One of her staff members wanted to move into the consulting group, and she believed the PMO could continue to provide its current services with one less person due to this new system—a benefit she had not considered before. Several of the firm's client contracts were based on performance and not hours billed, so she was excited to start measuring the value of the new intranet site to their consultants as well.

## Project Closing

The closing process involves gaining stakeholder and customer acceptance of the final products and services and bringing the project, or project phase, to an orderly end. It includes verifying that all of the deliverables are complete, and it often includes a final project report and presentation. Even though many information technology projects are canceled before completion, it is still important to formally close any project and reflect on what can be learned to improve future projects. As philosopher George Santayana said, "Those who cannot remember the past are condemned to repeat it."

It is also important to plan for and execute a smooth transition of the project into the normal operations of the company. Most projects produce results that are integrated into the existing organizational structure. For example, JWD Consulting's Project Management Intranet Site Project will require staff to support the intranet site after it is operational. Erica included support costs of $40,000 per year for the projected three-year life of the new system. She also created a transition plan as part of the final report to provide for a smooth transition of the system into the firm's operations. The plan included a list of issues that had to be resolved before the firm could put the new intranet site into production. For example,

Michael Chen would not be available to work on the intranet site after the six-month project was complete, so they had to know who would support the "Ask the Expert" feature and plan some time for Michael to work with him or her.



Table 3-15 lists the knowledge areas, processes, and outputs of project closing based on the *PMBOK® Guide, Fourth Edition*. During the closing processes of any project, project team members must deliver the final product, service, or result of the project, and update organizational process assets, such as project files and a lessons-learned report. If the project team procured items during the project, they must formally complete or close out all contracts. Templates related to project closing are listed later in this chapter.

**TABLE 3-15** Closing processes and output

| Knowledge Area | Closing Process | Outputs |
| --- | --- | --- |
| ***Project Integration Management*** | Close project or phase | Final product, service, or result transition<br>Organizational process assets updates |
| ***Project Procurement Management*** | Close procurements | Closed procurements<br>Organizational process assets updates |

Erica and her team prepared a final report, final presentation, contract files, and lessons-learned report in closing the project. Erica reviewed the confidential, individual lessons-learned report from each team member and wrote one summary lessons-learned report to include in the final documentation, part of which is provided in Table 3-16. Notice the bulleted items in the fourth question, such as the importance of having a good kick-off meeting, working together to develop a team contract, using project management software, and communicating well with the project team and sponsor.

**TABLE 3-16** Lessons-learned report (abbreviated)

| | |
| --- | --- |
| **Project Name:** | JWD Consulting Project Management Intranet Site Project |
| **Project Sponsor:** | Joe Fleming |
| **Project Manager:** | Erica Bell |
| **Project Dates:** | May 2 – November 4 |
| **Final Budget:** | $150,000 |

1. Did the project meet scope, time, and cost goals?

   *We did meet scope and time goals, but we had to request an additional $10,000, which the sponsor did approve.*

*(continued)*

**TABLE 3-16** Lessons-learned report (abbreviated) (*continued*)



2. What were the success criteria listed in the project scope statement?

   *Below is what we put in our project scope statement under project success criteria:*

   *"Our goal is to complete this project within six months for no more than $140,000. The project sponsor, Joe Fleming, has emphasized the importance of the project paying for itself within one year after the intranet site is complete. To meet this financial goal, the intranet site must have strong user input. We must also develop a method for capturing the benefits while the intranet site is being developed and tested, and after it is rolled out. If the project takes a little longer to complete or costs a little more than planned, the firm will still view it as a success if it has a good payback and helps promote the firm's image as an excellent consulting organization."*

3. Reflect on whether or not you met the project success criteria.

   *As stated above, the sponsor was not too concerned about going over budget as long as the system would have a good payback period and help promote our firm's image. We have already documented some financial and image benefits of the new intranet site. For example, we have decided that we can staff the PMO with one less person, resulting in substantial cost savings. We have also received excellent feedback from several of our clients about the new intranet site.*

4. In terms of managing the project, what were the main lessons your team learned from this project?

   *The main lessons we learned include the following:*

   - *Having a good project sponsor was instrumental to project success. We ran into a couple of difficult situations, and Joe was very creative in helping us solve problems.*
   - *Teamwork was essential. It really helped to take time for everyone to get to know each other at the kick-off meeting. It was also helpful to develop and follow a team contract.*
   - *Good planning paid off in execution. We spent a fair amount of time developing a good project charter, scope statement, WBS, schedules, and so on. Everyone worked together to develop these planning documents, and there was strong buy-in.*
   - *Project management software was very helpful throughout the project.*

5. Describe one example of what went right on this project.

6. Describe one example of what went wrong on this project.

7. What will you do differently on the next project based on your experience working on this project?

Erica also had Joe sign a client acceptance form, one of the sample templates on the new intranet site that the project team suggested all consultants use when closing their projects. (You can find this and other templates on the companion Web site for this text.)



Table 3-17 provides the table of contents for the final project report. The cover page included the project title, date, and team member names. Notice the inclusion of a

**TABLE 3-17** Final project report table of contents

1. Project Objectives
2. Summary of Project Results
3. Original and Actual Start and End Dates
4. Original and Actual Budget
5. Project Assessment (Why did you do this project? What did you produce? Was the project a success? What went right and wrong on the project?)
6. Transition Plan
7. Annual Project Benefits Measurement Approach

*Attachments*:

A. Project Management Documentation

- Business case
- Project charter
- Team contract
- Scope statement
- WBS and WBS dictionary
- Baseline and actual Gantt chart
- List of prioritized risks
- Milestone reports
- Progress reports
- Contract files
- Lessons-learned reports
- Final presentation
- Client acceptance form

B. Product-Related Documentation

- Survey and results
- Summary of user inputs
- Intranet site content
- Intranet site design documents
- Test plans and reports
- Intranet site promotion information
- Intranet site roll-out information
- Project benefits measurement information

transition plan and a plan to analyze the benefits of the system each year in the final report. Also, notice that the final report includes attachments for all the project management and product-related documents. Erica knew how important it was to provide good final documentation on projects. The project team produced a hard copy of the final documentation and an electronic copy to store on the new intranet site for other consultants to use as desired.



Erica also organized a project closure luncheon for the project team right after their final project presentation. She used the luncheon to share lessons learned and celebrate a job well done!

As you can see, there are many documents that project teams prepare throughout the life of a project. Many people use templates as a standard format for preparing those documents. Table 3-18 lists templates used in this text for preparing the documents

**TABLE 3-18** Templates by process group

| Template Name | Process Group | Chapter(s) Where Used | Application Software | File Name |
|---|---|---|---|---|
| Business Case | Pre-initiating | 3 | Word | business_case.doc |
| Business Case Financial Analysis | Pre-initiating | 3, 4 | Excel | business_case_ financials.xls |
| Stakeholder Register | Initiating | 3, 10 | Word | stakeholder_register.doc |
| Stakeholder Management Strategy | Initiating | 3, 10 | Word | stakeholder_strategy.doc |
| Kick-off Meeting | Initiating | 3 | Word | kick-off_meeting.doc |
| Payback Chart | Initiating | 4 | Excel | payback.xls |
| Weighted Decision Matrix | Initiating | 4, 12 | Excel | wtd_decision_matrix.xls |
| Project Charter | Initiating | 3, 4, 5 | Word | charter.doc |
| Team Contract | Planning | 3 | Word | team_contract.doc |
| Requirements Traceability Matrix | Planning | 5 | Word | reqs_matrix.xls |
| Scope Statement | Planning | 3, 4, 5 | Word | scope_ statement.doc |
| Statement of Work | Planning | 12 | Word | statement_of_work.doc |

**TABLE 3-18** Templates by process group (*continued*)

| Template Name | Process Group | Chapter(s) Where Used | Application Software | File Name |
|---|---|---|---|---|
| Request for Proposal | Planning | 12 | Word | rfp_outline.doc |
| Software Project Management Plan | Planning | 4 | Word | sw_project_mgt_plan.doc |
| Work Breakdown Structure | Planning | 3, 5, 6 | Word | wbs.doc |
| Gantt Chart | Planning, Executing | 3, 5, 6 | Project | Gantt_chart.mpp |
| Network Diagram | Planning, Executing | 3, 6 | Project | network_ diagram.mpp |
| Project Cost Estimate | Planning | 7 | Excel | cost_estimate.xls |
| Earned Value Data and Chart | Monitoring and Controlling | 7 | Excel | earned_value.xls |
| Quality Assurance Plan | Executing | 8 | Word | quality_assurance_ plan.doc |
| Pareto Chart | Monitoring and Controlling | 8 | Excel | pareto_chart.xls |
| Project Organizational Chart | Planning, Executing | 9 | PowerPoint | project_org_chart.ppt |
| Responsibility Assignment Matrix | Planning, Executing | 9 | Excel | ram.xls |
| Resource Histogram | Planning, Executing | 9 | Excel | resource_histogram.xls |
| Communications Management Plan | Planning | 10 | Word | comm_plan.doc |

(*continued*)

**TABLE 3-18** Templates by process group (*continued*)



| Template Name | Process Group | Chapter(s) Where Used | Application Software | File Name |
|---|---|---|---|---|
| Project Description (text) | Planning | 10 | Word | project_desc_text.doc |
| Project Description (Gantt chart) | Planning | 10 | Project | project_desc_Gantt.mpp |
| Milestone Report | Executing | 3, 6 | Word | milestone_report.doc |
| Change Request Form | Planning, Monitoring and Controlling | 4 | Word | change_request.doc |
| Progress Report | Monitoring and Controlling | 3, 10 | Word | progress_report.doc |
| Expectations Management Matrix | Monitoring and Controlling | 10 | Word | expectations.doc |
| Issue Log | Monitoring and Controlling | 10 | Word | issue_log.doc |
| Probability/ Impact Matrix | Planning, Executing, Monitoring and Controlling | 11 | PowerPoint | prob_impact_matrix.ppt |
| List of Prioritized Risks | Planning, Executing, Monitoring and Controlling | 3, 11 | Word | list_of_risks.doc |
| Risk Register | Planning, Monitoring and Controlling | 11 | Excel | risk_register.xls |

TABLE 3-18 Templates by process group (*continued*)

| Template Name | Process Group | Chapter(s) Where Used | Application Software | File Name |
|---|---|---|---|---|
| Top 10 Risk Item Tracking | Planning, Monitoring and Controlling | 11 | Excel | top_10.xls |
| Breakeven/ Sensitivity Analysis | Planning | 11 | Excel | breakeven.xls |
| Client Acceptance Form | Closing | 3, 10 | Word | client_acceptance.doc |
| Lessons-Learned Report | Closing | 3, 10 | Word | lessons_learned_ report.doc |
| Final Project Documentation | Closing | 3, 10 | Word | final_documentation.doc |



shown in this chapter and in later chapters. It lists the template name, chapter number, process group(s) where you normally use the template, application software used to create it, and the file name for the template. You can download all of these files in one compressed file from the companion Web site for this text or from the author's Web site at *www.kathyschwalbe.com*. Note that the templates were saved in Office 2003 and 2007 format to allow for easier compatibility. Feel free to modify the templates to meet your needs.

The project management process groups—initiating, planning, executing, monitoring and controlling, and closing—provide a useful framework for understanding project management. They apply to most projects (information technology and non-information technology) and, along with the project management knowledge areas, help project managers see the big picture of managing a project in their particular organization.

## CASE WRAP-UP

Erica Bell and her team finished the Project Management Intranet Site Project on November 4, as planned in their project charter. They did go over budget, however, but Joe had approved Erica's request for additional funds, primarily for purchasing external software and customization. Like any project, they had a few challenges, but they worked together as a team and used good project management to meet their sponsor's and users' needs. They received positive initial feedback from internal consultants and some of their clients on the new intranet site. People were asking for templates, examples, and expert advice even before the system was ready. About a year after the project was completed, Erica worked with a member of the Finance department to review the benefits of the new system. The Project Management Office did lose one of its staff members, but it did not request a replacement since the new system helped reduce the PMO's workload. This saved the firm about $70,000 a year for the salary and benefits of that staff position. They also had data to show that the firm saved more than $180,000 on contracts with clients due to the new system, while they had projected just $160,000. The firm was breaking even with the "Ask the Expert" feature the first year, and Erica estimated that the system provided $30,000 in additional profits the first year by generating new business, not the $40,000 they had projected. However, savings from the PMO staff position salary and the extra savings on contracts more than made up for the $10,000 difference. Joe was proud of the project team and the system they produced to help make JWD Consulting a world-class organization.

## Chapter Summary



Project management involves a number of interlinked processes. The five project management process groups are initiating, planning, executing, monitoring and controlling, and closing. These processes occur at varying levels of intensity throughout each phase of a project, and specific outcomes are produced as a result of each process. Normally the executing processes require the most resources and time, followed by the planning processes.

Mapping the main activities of each project management process group into the nine project management knowledge areas provides a big picture of what activities are involved in project management.

Some organizations develop their own information technology project management methodologies, often using the standards found in the *PMBOK® Guide* as a foundation. It is important to tailor project management methodologies to meet the organization's particular needs. Popular methodologies like PRINCE2, agile methodologies, RUP, and Six Sigma include project management processes.

The JWD Consulting case study demonstrates how one organization managed an information technology project from its initiation through its closure. The case study provides several samples of outputs produced for initiating (including pre-initiating), planning, executing, monitoring and controlling, and closing as follows:

- Business case
- Stakeholder register
- Stakeholder management strategy
- Project charter
- Kick-off meeting agenda
- Team contract
- Work breakdown structure
- Gantt chart
- List of prioritized risks
- Milestone report
- Progress report
- Lessons-learned report
- Final project report

Later chapters in this text provide detailed information on creating these and other project management documents and using several of the tools and techniques described in this case study.

## Quick Quiz

1. A ______ is a series of actions directed toward a particular result.
   a. goal
   b. process
   c. plan
   d. project

2. ______ processes include coordinating people and other resources to carry out the project plans and produce the products, services, or results of the project or phase.
   a. Initiating
   b. Planning
   c. Executing
   d. Monitoring and controlling
   e. Closing



3. Which process group normally requires the most resources and time?
   a. Initiating
   b. Planning
   c. Executing
   d. Monitoring and controlling
   e. Closing
4. What methodology was developed in the U.K., defines 45 separate subprocesses, and organizes these into eight process groups?
   a. Six Sigma
   b. RUP
   c. *PMBOK® Guide*
   d. PRINCE2
5. Which of the following outputs is often completed before initiating a project?
   a. stakeholder register
   b. business case
   c. project charter
   d. kick-off meeting
6. A work breakdown structure, project schedule, and cost estimates are outputs of the ______ process.
   a. initiating
   b. planning
   c. executing
   d. monitoring and controlling
   e. closing
7. Initiating involves developing a project charter, which is part of the project ______ management knowledge area.
   a. integration
   b. scope
   c. communications
   d. risk

8. ______ involves measuring progress toward project objectives and taking corrective actions.
   a. Initiating
   b. Planning
   c. Executing
   d. Monitoring and controlling
   e. Closing
9. What type of report do project teams create to reflect on what went right and what went wrong with the project?
   a. lessons-learned report
   b. progress report
   c. final project report
   d. business case
10. Many people use ______ to have a standard format for preparing various project management documents.
    a. methodologies
    b. templates
    c. project management software
    d. standards

### Quick Quiz Answers

1. b; 2. c; 3. c; 4. d; 5. b; 6. b; 7. a; 8. d; 9. a; 10. b

## Discussion Questions

1. Briefly describe what happens in each of the five project management process groups (initiating, planning, executing, monitoring and controlling, and closing). What types of activities are done before initiating a project?
2. Approximately how much time do good project managers spend on each process group and why?
3. Why do organizations need to tailor project management concepts, such as those found in the *PMBOK® Guide*, to create their own methodologies?
4. What are some of the key outputs of each process group?
5. What are some of the typical challenges project teams face during each of the five process groups?

## Exercises

1. Study the WBS and Gantt charts provided in Figures 3-3 and 3-4. Enter the WBS into Project 2007, indenting tasks as shown to create the WBS hierarchy. Do not enter durations or dependencies. Print the resulting Gantt chart. See the scope management section of Appendix A for help using Project 2007.

2. Read the article by William Munroe regarding BlueCross BlueShield of Michigan's information technology project management methodology (available on the companion Web site for this text under Chapter 3). Or, research another methodology, such as PRINCE2, an agile methodology, RUP, or Six Sigma, and how organizations use it, citing at least two references. Why do you think organizations spend time and money tailoring a methodology to their environment? Write a two-page summary of your findings and your opinion on the topic.
3. Read the "ResNet Case Study" (available from the companion Web site for this text under Chapter 3). This real case study about Northwest Airlines' reservation system illustrates another application of the project management process groups. Write a three-page paper summarizing the main outputs produced during each project process group in this case. Also, include your opinion of whether or not Peeter Kivestu was an effective project manager. If you prefer, find another well-documented project and summarize it instead.
4. JWD Consulting wrote a business case before officially initiating the Project Management Intranet Site project. Review the contents of this document (Table 3-2) and find two articles describing the need to justify investing in IT projects. In addition, describe whether you think most projects should include a business case before the project sponsors officially approve the project. Write a two-page paper summarizing your findings and opinions.
5. Read an article about a recipient of PMI's Project of the Year award. Past winners include Fluor Corporation's Fernald Closure Project, Kaiser-Hill's Rocky Flats Nuclear Plant Closing, the Quartier International de Montréal district revitalization project, Saudi Aramco Haradh Gas Plant, and the Winter Olympics Salt Lake Organizing Committee. Write a one-page paper summarizing the project, focusing on how the project manager and team used good project management practices.
6. Download the template files used in this text from the companion Web site or from *www.kathyschwalbe.com*. Review several of them, and look at examples of how they are used in this text. Also search the Internet for other template files. Summarize what you think about using templates and how you think they can help project managers and their teams in a two-page paper. Also discuss potential problems with using templates.

## Companion Web Site

Visit the companion Web site for this text at *www.cengage.com/mis/schwalbe* to access:

- References cited in the text and additional suggested readings for each chapter
- Template files
- Lecture notes
- Interactive quizzes
- Podcasts
- Links to general project management Web sites
- And more

See the Preface of this text for more information on accessing the companion Web site.

## Key Terms



**closing processes** — formalizing acceptance of the project or project phase and ending it efficiently

**executing processes** — coordinating people and other resources to carry out the project plans and produce the products, services, or results of the project or project phase

**initiating processes** — defining and authorizing a project or project phase

**kick-off meeting** — a meeting held at the beginning of a project so that stakeholders can meet each other, review the goals of the project, and discuss future plans

**methodology** — describes *how* things should be done

**monitoring and controlling processes** — regularly measuring and monitoring progress to ensure that the project team meets the project objectives

**planning processes** — devising and maintaining a workable scheme to ensure that the project addresses the organization's needs

**process** — a series of actions directed toward a particular result

**project management process groups** — the progression of project activities from initiation to planning, executing, monitoring and controlling, and closing

**PRojects IN Controlled Environments (PRINCE2)** — a project management methodology developed in the U.K. that defines 45 separate sub-processes and organizes these into eight process groups

**Rational Unified Process (RUP)** — an iterative software development process that focuses on team productivity and delivers software best practices to all team members

**Six Sigma methodologies** — DMAIC (Define, Measure, Analyze, Improve, and Control) is used to improve an existing business process and DMADV (Define, Measure, Analyze, Design, and Verify) is used to create new product or process designs

**stakeholder register** — a document that includes details related to the identified project stakeholders

**standard** — describes best practices for *what* should be done

## End Notes


[1] Andy Crowe, Alpha Project Managers: What the Top 2% Know That Everyone Else Does Not, Velociteach Press, Atlanta, GA, (2006).

[2] Phillip A. Pell, Comments posted at Elaine Varron, "No Easy IT Fix for IRS [formerly "For the IRS, There's No EZ Fix"], *CIO.com* (April 1, 2004).

[3] Grant Gross, "Report: IRS information security still poor," *InfoWorld* (January 8, 2008).

[4] PCI Group, "PM Best Practices Report," (October 2001).

[5] Brian Jacks, "Lord of the Rings: The Two Towers Extended Edition (New Line)," *Underground Online (UGO.com)* (accessed August 4, 2004).

[6] Bill Cottrell, "Standards, compliance, and Rational Unified Process, Part I: Integrating RUP and the PMBOK," IBM Developerworks (May 10, 2004).

[7] Sarah Fister Gale, "Outstanding Organizations 2007," *PM Network*, (October 2007), p. 5.

[8] Libby Elis, "Urban Inspiration," *PM Network,* (January 2006), p. 30.

CHAPTER 4

# PROJECT INTEGRATION MANAGEMENT

## LEARNING OBJECTIVES

**After reading this chapter, you will be able to:**

- Describe an overall framework for project integration management as it relates to the other project management knowledge areas and the project life cycle
- Explain the strategic planning process and apply different project selection methods
- Explain the importance of creating a project charter to formally initiate projects
- Describe project management plan development, understand the content of these plans, and review approaches for creating them
- Explain project execution, its relationship to project planning, the factors related to successful results, and tools and techniques to assist in managing project execution
- Describe the process of monitoring and controlling a project
- Understand the integrated change control process, planning for and managing changes on information technology projects, and developing and using a change control system
- Explain the importance of developing and following good procedures for closing projects
- Describe how software can assist in project integration management

## OPENING CASE

Nick Carson recently became project manager of a critical biotech enterprise at his Silicon Valley company. This project involved creating the hardware and software for a DNA-sequencing instrument used in assembling and analyzing the human genome. The biotech project was the company's largest endeavor, and it had tremendous potential for future growth and revenues. Unfortunately, there were problems managing this large project. It had been underway for three years and had already gone through three different project managers. Nick had been the lead software developer on the project before top management made him the project manager. The CEO told him to do whatever it took to deliver the first version of the software for the DNA-sequencing instrument in four months and a production version in nine months. Negotiations for a potential corporate buyout with a larger company influenced top management's sense of urgency to complete the project.

Highly energetic and intelligent, Nick had the technical background to make the project a success. He delved into the technical problems and found some critical flaws that kept the DNA-sequencing instrument from working. Nevertheless, he was having difficulty in his new role as project manager. Although Nick and his team got the product out on time, top management was upset because Nick did not focus on managing all aspects of the project. He never provided them with accurate schedules or detailed plans of what was happening on the project. Instead of performing the work of project manager, Nick had taken on the role of software integrator and troubleshooter. Nick, however, did not understand top management's problem—he delivered the product, didn't he? Didn't they realize how valuable he was?

## WHAT IS PROJECT INTEGRATION MANAGEMENT?

**Project integration management** involves coordinating all of the other project management knowledge areas throughout a project's life cycle. This integration ensures that all the elements of a project come together at the right times to complete a project successfully. According to the *PMBOK® Guide, Fourth Edition* there are six main processes involved in project integration management:

1. *Developing the project charter* involves working with stakeholders to create the document that formally authorizes a project—the charter.
2. *Developing the project management plan* involves coordinating all planning efforts to create a consistent, coherent document—the project management plan.
3. *Directing and managing project execution* involves carrying out the project management plan by performing the activities included in it. The outputs of this process are deliverables, work performance information, change requests, project management plan updates, and project document updates.
4. *Monitoring and controlling project work* involves overseeing activities to meet the performance objectives of the project. The outputs of this process are

change requests, project management plan updates, and project document updates.

5. *Performing integrated change control* involves identifying, evaluating, and managing changes throughout the project life cycle. The outputs of this process include change request status updates, project management plan updates, and project document updates.
6. *Closing the project or phase* involves finalizing all activities to formally close the project or phrase. Outputs of this process include final product, service, or result transition and organizational process assets updates. Figure 4-1 summarizes these processes and outputs, showing when they occur in a typical project.



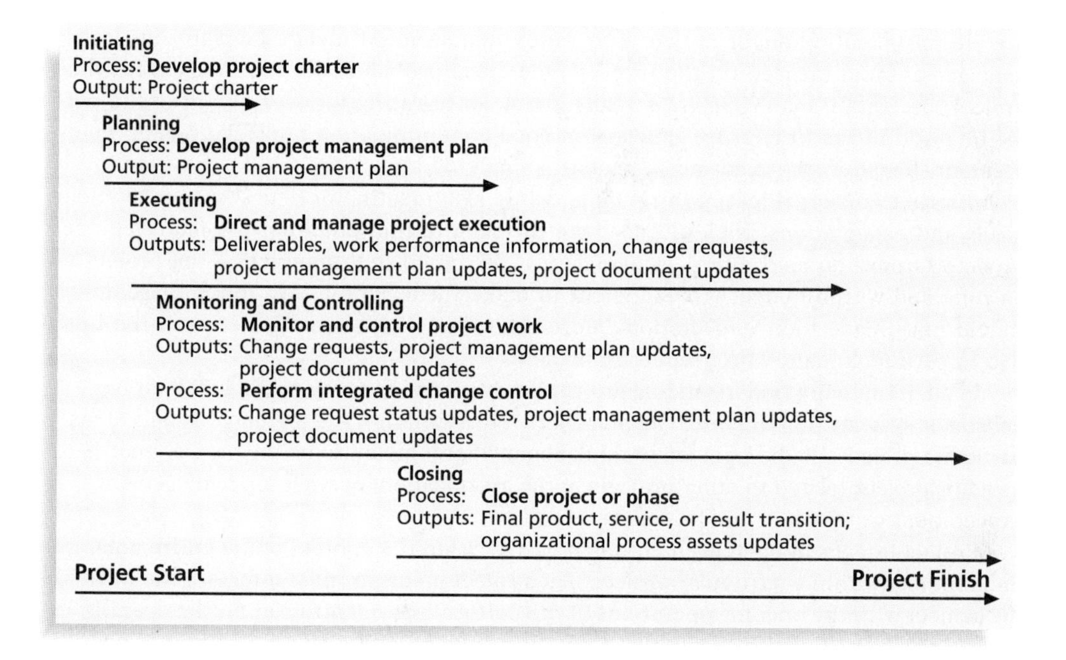


**FIGURE 4-1** Project integration management summary

Many people consider project integration management the key to overall project success. Someone must take responsibility for coordinating all of the people, plans, and work required to complete a project. Someone must focus on the big picture of the project and steer the project team toward successful completion. Someone must make the final decisions when there are conflicts among project goals or people. Someone must communicate key project information to top management. This someone is the project manager, and the project manager's chief means for accomplishing all these tasks is project integration management.

Good project integration management is critical to providing stakeholder satisfaction. Project integration management includes interface management. **Interface management** involves identifying and managing the points of interaction between various elements of the project. The number of interfaces can increase exponentially as the number of people involved in a project increases. Thus, one of the most important jobs of a project manager is to establish and maintain good communication and relationships across organizational interfaces. The project manager must communicate well with all project stakeholders, including customers, the project team, top management, other project managers, and opponents of the project.



What happens when a project manager does not communicate well with all stakeholders? In the opening case, Nick Carson seemed to ignore a key stakeholder for the DNA-sequencing instrument project—his top management. Nick was comfortable working with other members of the project team, but he was not familiar with his new job as project manager or the needs of the company's top management. Nick continued to do his old job of software developer and took on the added role of software integrator. He mistakenly thought project integration management meant software integration management and focused on the project's technical problems. He totally ignored what project integration management is really about—integrating the work of all of the people involved in the project by focusing on good communication and relationship management. Recall that project management is applying knowledge, skills, tools, and techniques to meet project requirements, while also meeting or exceeding stakeholder needs and expectations. Nick did not take the time to find out what top management expected from him as the project manager; he assumed that completing the project on time and within budget was sufficient to make them happy. Yes, top management should have made their expectations more clear, but Nick should have taken the initiative to get the guidance he needed from them.

In addition to not understanding project integration management, Nick did not use holistic or systems thinking (see Chapter 2). He burrowed into the technical details of his particular project. He did not stop to think about what it meant to be the project manager, how this project related to other projects in the organization, or what top management's expectations were of him and his team.

Project integration management must occur within the context of the entire organization, not just within a particular project. The project manager must integrate the work of the project with the ongoing operations of the performing organization. In the opening case, Nick's company was negotiating a potential buyout with a larger company. Consequently, top management needed to know when the DNA-sequencing instrument would be ready, how big the market was for the product, and if they had enough in-house staff to continue to manage projects like this one in the future. They wanted to see a project management plan and a schedule to help them monitor the project's progress and show their potential buyer what was happening. When top managers tried to talk to Nick about these issues, Nick soon returned to discussing the technical details of the project. Even though Nick was very bright, he had no experience or real interest in many of the business aspects of how the company operated. Project managers must always view their projects in the context of the changing needs of their organizations and respond to requests from top management. Likewise, top management must keep project managers informed of major issues that could affect their projects and strive to make processes consistent throughout their organization.

**WHAT WENT WRONG?**

The Airbus A380 megajet project was two years behind schedule in October 2006, causing Airbus' parent company to face an expected loss of $6.1 billion over the next four years. Why? The project suffered from severe integration management problems, or "*integration disintegration*. … [W]hen pre-assembled bundles containing hundreds of miles of cabin wiring were delivered from a German factory to the assembly line in France, workers discovered that the bundles, called harnesses, didn't fit properly into the plane. Assembly slowed to a near-standstill, as workers tried to pull the bundles apart and re-thread them through the fuselage. Now Airbus will have to go back to the drawing board and redesign the wiring system."[1]

How did this lack of integration occur? At the end of 2000, just as Airbus was giving the go-ahead to the A380 project, the company announced that it was completing the process of transforming itself into an integrated corporation. Since its founding in 1970, Airbus had operated as a loose consortium of aerospace companies in several countries, including France, Germany, Britain, and Spain. The company wanted to integrate all of its operations into one cohesive business. Unfortunately, that integration was much easier said than done and caused major problems on the A380 project. For example, the Toulouse assembly plant used the latest version of a sophisticated design software tool called CATIA, but the design center at the Hamburg factory used an earlier version—a completely different system dating from the 1980s. As a result, design specs could not flow easily back and forth between the two systems. Airbus's top managers should have made it a priority to have all sites use the latest software, but they didn't, resulting in this project disaster.

Following a standard process for managing projects can help prevent some of the typical problems new and experienced project managers face, including communicating with and managing stakeholders. Before organizations begin projects, however, they should go through a formal process to decide what projects to pursue.

# STRATEGIC PLANNING AND PROJECT SELECTION

Successful leaders look at the big picture or strategic plan of the organization to determine what types of projects will provide the most value. Some may argue that project managers should not be involved in strategic planning and project selection because top management is usually responsible for these types of business decisions. But successful organizations know that project managers can provide valuable insight into the project selection process.

## Strategic Planning

**Strategic planning** involves determining long-term objectives by analyzing the strengths and weaknesses of an organization, studying opportunities and threats in the business environment, predicting future trends, and projecting the need for new products and services. Strategic planning provides important information to help organizations identify and then select potential projects.

Many people are familiar with **SWOT analysis**—analyzing Strengths, Weaknesses, Opportunities, and Threats—which is used to aid in strategic planning. For example, a group of four people who want to start a new business in the film industry could perform a SWOT analysis to help identify potential projects. They might determine the following based on a SWOT analysis:



Strengths:

- As experienced professionals, we have numerous contacts in the film industry.
- Two of us have strong sales and interpersonal skills.
- Two of us have strong technical skills and are familiar with several filmmaking software tools.
- We all have impressive samples of completed projects.

Weaknesses:

- None of us have accounting/financial experience.
- We have no clear marketing strategy for products and services.
- We have little money to invest in new projects.
- We have no company Web site and limited use of technology to run the business.

Opportunities:

- A current client has mentioned a large project she would like us to bid on.
- The film industry continues to grow.
- There are two major conferences this year where we could promote our company.

Threats:

- Other individuals or companies can provide the services we can.
- Customers might prefer working with more established individuals/organizations.
- There is high risk in the film business.

Based on their SWOT analysis, the four entrepreneurs outline potential projects as follows:

- Find an external accountant or firm to help run the business.
- Hire someone to develop a company Web site, focusing on our experience and past projects.
- Develop a marketing plan.
- Develop a strong proposal to get the large project the current client mentioned.
- Plan to promote the company at two major conferences this year.

Some people like to perform a SWOT analysis by using **mind mapping**, which is a technique that uses branches radiating out from a core idea to structure thoughts and ideas. The human brain does not work in a linear fashion. People come up with many unrelated ideas. By putting those ideas down in a visual mind map format, you can often generate more ideas than by just creating lists. You can create mind maps by hand, using sticky notes, using presentation software such as Microsoft PowerPoint, or by using mind mapping software.

Figure 4-2 shows a sample mind map for the SWOT analysis presented earlier. This diagram was created using MindManager software by Mindjet. (You can download a free trial of this software from *www.mindjet.com* or use a similar free tool called FreeMind available at *www.freemind.sourceforge.net*.) Notice that this map has four main branches representing strengths, weaknesses, opportunities, and threats. Icons are added to each of those main branches to more visually identify them, such as the thumbs up for strengths and thumbs down for weaknesses. Ideas in each category are added to the appropriate branch. You could also add sub-branches to show ideas under those categories. For example, under the first branch for strengths, you could start adding sub-branches to list the most important contacts you have. This mind map includes branches for project ideas related to different categories, with text markers used to identify the project names. From this visual example, you can see that there are no project ideas identified to address strengths or threats, so these areas should be discussed further.



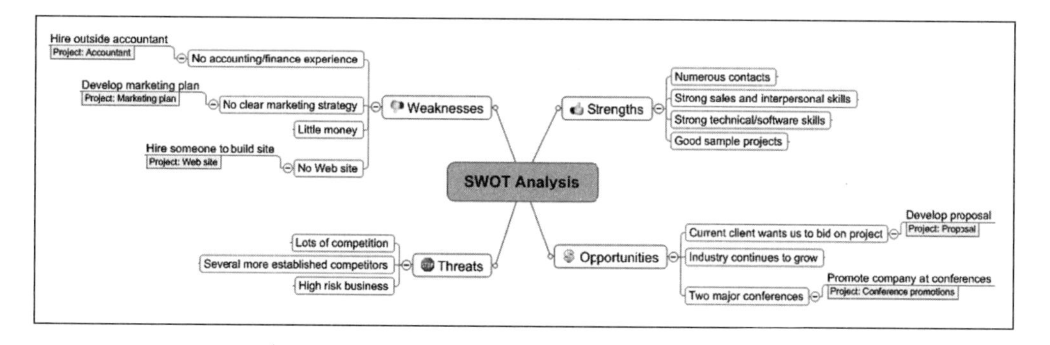


**FIGURE 4-2** Mind map of a SWOT analysis to help identify potential projects

## Identifying Potential Projects

The first step in project management is deciding what projects to do in the first place. Therefore, project initiation starts with identifying potential projects, using realistic methods to select which projects to work on, and then formalizing their initiation by issuing some sort of project charter.

In addition to using a SWOT analysis, organizations often follow a detailed process for project selection. Figure 4-3 shows a four-stage planning process for selecting information technology projects. Note the hierarchical structure of this model and the results produced from each stage. The first step in this process, starting at the top of the hierarchy, is to tie the information technology strategic plan to the organization's overall strategic plan. It is very important to have managers from outside the information technology department assist in the information technology planning process, as they can help information technology personnel understand organizational strategies and identify the business areas that support them.

After identifying strategic goals, the next step in the planning process for selecting information technology projects is to perform a business area analysis. This analysis outlines

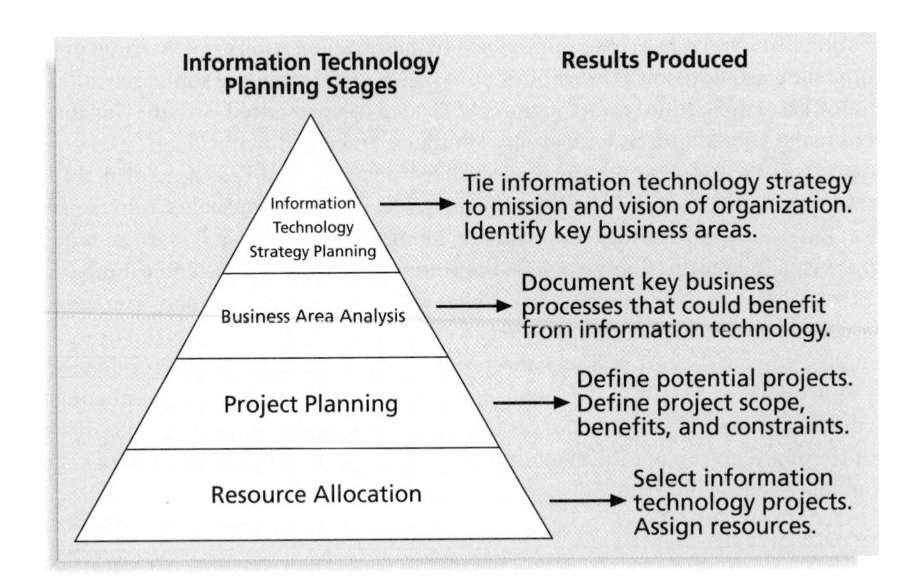


**FIGURE 4-3** Planning process for selecting information technology projects

business processes that are central to achieving strategic goals and helps determine which ones could most benefit from information technology. The next step is to start defining potential information technology projects, their scope, benefits, and constraints. The last step in the planning process for selecting information technology projects is choosing which projects to do and assigning resources for working on them.

## Aligning Information Technology with Business Strategy

Aligning IT projects with business strategy is consistently the top concern for CIOs. It is often difficult to educate line managers on technology's possibilities and limitations and keep IT professionals in tune with changing business needs. Most organizations face thousands of problems and opportunities for improvement. Therefore, an organization's strategic plan should guide the information technology project selection process. Recall from Chapter 2's Best Practice feature that IT governance is also important in ensuring that IT supports business goals. IT governance helps organizations maximize their investments in IT and address IT-related risks and opportunities.

An organization must develop a strategy for using information technology to define how it will support the organization's objectives. This information technology strategy must align with the organization's strategic plans and strategy. In fact, research shows that supporting explicit business objectives is the number one reason cited for why organizations invest in information technology projects. Other top criteria for investing in information technology projects include supporting implicit business objectives and providing financial incentives, such as a good internal rate of return (IRR) or net present value (NPV).[2] You will learn more about these financial criteria later in this section.

Information systems can be and often are central to business strategy. Author Michael Porter, who developed the concept of the strategic value of competitive advantage, has written several books and articles on strategic planning and competition. He and many other experts have emphasized the importance of using information technology to support strategic plans and provide a competitive advantage. Many information systems are classified as "strategic" because they directly support key business strategies. For example, information systems can help an organization support a strategy of being a low-cost producer. As one of the largest retailers in the United States, Wal-Mart's inventory control system is a classic example of such a system. Information systems can support a strategy of providing specialized products or services that set a company apart from others in the industry. Consider the classic example of Federal Express's introduction of online package tracking systems. They were the first company to provide this type of service, which gave them a competitive advantage until others developed similar systems. Information systems can also support a strategy of selling to a particular market or occupying a specific product niche. Owens-Corning developed a strategic information system that boosted the sales of its home-insulation products by providing its customers with a system for evaluating the energy efficiency of building designs.

## BEST PRACTICE

Many organizations rely on effective new product development (NPD) to increase growth and profitability, yet according to Robert Cooper, of McMaster University and New Product Development Institute in Ontario, Canada, only one in seven product concepts comes to fruition. Why is it that some companies such as Procter & Gamble, Johnson & Johnson, Hewlett-Packard, and Sony are consistently successful in NPD? Because they use a disciplined, systematic approach to NPD projects based on best practices. Four important forces behind NPD success include the following:

1. A product innovation and technology strategy for the business
2. Resource commitment and focusing on the right projects, or solid portfolio management
3. An effective, flexible and streamlined idea-to-launch process
4. The right climate and culture for innovation, true cross-functional teams, and senior management commitment to NPD

Cooper's study compared companies that were the best at performing NPD with those that were the worst. For example, 65.5 percent of companies performing the best at NPD align projects with business strategy. However, within the group of companies performing the worst at NPD, only 46 percent align projects with business strategy. Even more telling is that 65.5 percent of best performing NPD companies have their resource breakdown aligned to business strategy while only 8 percent of worst performing companies do. It's easy for a company to say that its projects are aligned with business strategy, but assigning its resources based on that strategy is a measurable action that produces results. Best performing NPD companies are also more customer-focused in identifying new product ideas. Sixty-nine percent of them identify customer needs and problems based on customer input, while only 15 percent of worst performing companies do. Also, 80 percent of best performing companies have an identifiable NPD project manager compared to only 50 percent of worst performing companies.[3] These best practices apply to all projects: align projects *and* resources with business strategy, focus on customer needs when identifying potential projects, and assign project managers to lead the projects.

## Methods for Selecting Projects

Organizations identify many potential projects as part of their strategic planning processes, and they often rely on experienced project managers to help them make project selection decisions. However, organizations need to narrow down the list of potential projects to those projects that will be of most benefit. Selecting projects is not an exact science, but it is a necessary part of project management. Many methods exist for selecting from among possible projects. Five common techniques are:



- Focusing on broad organizational needs
- Categorizing information technology projects
- Performing net present value or other financial analyses
- Using a weighted scoring model
- Implementing a balanced scorecard

In practice, organizations usually use a combination of these approaches to select projects. Each approach has advantages and disadvantages, and it is up to management to decide the best approach for selecting projects based on their particular organization.

### Focusing on Broad Organizational Needs

Top managers must focus on meeting their organization's many needs when deciding what projects to undertake, when to undertake them, and to what level. Projects that address broad organizational needs are much more likely to be successful because they will be important to the organization. For example, a broad organizational need might be to improve safety, increase morale, provide better communications, or improve customer service. However, it is often difficult to provide a strong justification for many information technology projects related to these broad organizational needs. For example, it is often impossible to estimate the financial value of such projects, but everyone agrees that they do have a high value. As the old proverb says, "It is better to measure gold roughly than to count pennies precisely."

One method for selecting projects based on broad organizational needs is to determine whether they first meet three important criteria: *need*, *funding*, and *will*. Do people in the organization agree that the project needs to be done? Does the organization have the desire and the capacity to provide adequate funds to perform the project? Is there a strong will to make the project succeed? For example, many visionary CEOs can describe a broad need to improve certain aspects of their organizations, such as communications. Although they cannot specifically describe how to improve communications, they might allocate funds to projects that address this need. As projects progress, the organization must reevaluate the need, funding, and will for each project to determine if the project should be continued, redefined, or terminated.

### Categorizing Information Technology Projects

Another method for selecting projects is based on various categorizations, such as the impetus for the project, the time window for the project, and the general priority for the project. The impetus for a project is often to respond to a problem, an opportunity, or a directive.

- **Problems** are undesirable situations that prevent an organization from achieving its goals. These problems can be current or anticipated. For example, users of an information system may be having trouble logging onto the system or getting information in a timely manner because the system has reached its capacity. In response, the company could initiate a project to enhance the current system by adding more access lines or upgrading the hardware with a faster processor, more memory, or more storage space.

- **Opportunities** are chances to improve the organization. For example, the project described in the opening case involves creating a new product that can make or break the entire company.
- **Directives** are new requirements imposed by management, government, or some external influence. For example, many projects involving medical technologies must meet rigorous government requirements.



Organizations select projects for any of these reasons. It is often easier to get approval and funding for projects that address problems or directives because the organization must respond to these categories of projects to avoid hurting their business. Many problems and directives must be resolved quickly, but managers must also apply systems thinking and seek opportunities for improving the organization through information technology projects.

Another categorization for information technology projects is based on the time it will take to complete a project or the date by which it must be done. For example, some potential projects must be finished within a specific time window. If they cannot be finished by this set date, they are no longer valid projects. Some projects can be completed very quickly—within a few weeks, days, or even minutes. Many organizations have an end user support function to handle very small projects that can be completed quickly. Even though many information technology projects can be completed quickly, it is still important to prioritize them.

Organizations can also prioritize information technology projects as being high-, medium-, or low-priority based on the current business environment. For example, if it is crucial to cut operating costs quickly, projects that have the most potential to do so would be given a high priority. The organization should always complete high-priority projects first, even if a low- or medium-priority project could be finished in less time. Usually there are many more potential information technology projects than an organization can undertake at any one time, so it is very important to work on the most important ones first.

### Performing Net Present Value Analysis, Return on Investment, and Payback Analysis

Financial considerations are often an important aspect of the project selection process, especially during tough economic times. As authors Dennis Cohen and Robert Graham put it, "Projects are never ends in themselves. Financially they are always a means to an end, cash."[4] Many organizations require an approved business case before pursuing projects, and financial projections are a critical component of the business case. (See Chapter 3 for a sample business case.) Three primary methods for determining the projected financial value of projects include net present value analysis, return on investment, and payback analysis. Because project managers often deal with business executives, they must understand how to speak their language, which often boils down to these important financial concepts.

### Net Present Value Analysis

Everyone knows that a dollar earned today is worth more than a dollar earned five years from now. **Net present value (NPV) analysis** is a method of calculating the expected net monetary gain or loss from a project by discounting all expected future cash inflows and outflows to the present point in time. An organization should consider only projects with a positive NPV if financial value is a key criterion for project selection. This is because a positive NPV means the return from a project exceeds the **cost of capital**—the return available by investing the capital elsewhere. Projects with higher NPVs are preferred to projects with lower NPVs, if all other factors are equal.



Figure 4-4 illustrates this concept for two different projects. Note that this example starts discounting right away in Year 1 and uses a 10 percent discount rate. You can use the NPV function in Microsoft Excel to calculate the NPV quickly. Detailed steps on performing this calculation manually are provided later in this section. Note that Figure 4-4 lists the projected benefits first, followed by the costs, and then the calculated cash flow amount. Note that the sum of the **cash flow**—benefits minus costs or income minus expenses—is the same for both projects at $5,000. The net present values are different, however, because they account for the time value of money. Project 1 has a negative cash flow of $5,000 in the first year, while Project 2 has a negative cash flow of only $1,000 in the first year. Although both projects have the same total cash flows without discounting, these cash flows are not of comparable financial value. Project 2's NPV of $3,201 is better than Project 1's NPV of $2,316. NPV analysis, therefore, is a method for making equal comparisons between cash flows for multi-year projects.

There are some items to consider when calculating NPV. Some organizations refer to the investment years for project costs as Year 0 instead of Year 1 and do not discount costs in Year 0. Other organizations start discounting immediately based on their financial procedures; it's simply a matter of preference for the organization. The discount rate can also vary, often based on the prime rate and other economic considerations. Financial experts in your organization can tell you what discount rate to use. Some people consider it to be the rate at which you could borrow money for the project. You can enter costs as negative numbers

| | A | B | C | D | E | F | G |
|---|---|---|---|---|---|---|---|
| 1 | Discount rate | 10% | | | | | |
| 2 | | | | | | | |
| 3 | PROJECT 1 | YEAR 1 | YEAR 2 | YEAR 3 | YEAR 4 | YEAR 5 | TOTAL |
| 4 | Benefits | $0 | $2,000 | $3,000 | $4,000 | $5,000 | $14,000 |
| 5 | Costs | $5,000 | $1,000 | $1,000 | $1,000 | $1,000 | $9,000 |
| 6 | Cash flow | ($5,000) | $1,000 | $2,000 | $3,000 | $4,000 | $5,000 |
| 7 | NPV → | $2,316 | | | | | |
| 8 | | Formula =npv(b1,b6:f6) | | | | | |
| 9 | | | | | | | |
| 10 | PROJECT 2 | YEAR 1 | YEAR 2 | YEAR 3 | YEAR 4 | YEAR 5 | TOTAL |
| 11 | Benefits | $1,000 | $2,000 | $4,000 | $4,000 | $4,000 | $15,000 |
| 12 | Costs | $2,000 | $2,000 | $2,000 | $2,000 | $2,000 | $10,000 |
| 13 | Cash flow | ($1,000) | $0 | $2,000 | $2,000 | $2,000 | $5,000 |
| 14 | NPV → | $3,201 | | | | | |
| 15 | | Formula =npv(b1,b13:f13) | | | | | |
| 16 | | | | | | | |
| 17 | | | | | | | |



**FIGURE 4-4** Net present value example

instead of positive numbers, and you can list costs first and then benefits. For example, Figure 4-5 shows the financial calculations JWD Consulting provided in the business case for the Project Management Intranet Site Project described in Chapter 3. Note that the discount rate is 8 percent, costs are not discounted right away (note the Year 0), the discount factors are rounded to two decimal places, costs are listed first, and costs are entered as positive numbers. The NPV and other calculations are the same; only the format is different. A project manager must be sure to check with his or her organization to find out its guidelines for when discounting starts, what discount rate to use, and what format the organization prefers.



To determine NPV, follow these steps:

1. Determine the estimated costs and benefits for the life of the project and the products it produces. For example, JWD Consulting assumed the project would produce a system in about six months that would be used for three years, so costs are included in Year 0, when the system is developed, and ongoing system costs and projected benefits are included for Years 1, 2, and 3.
2. Determine the discount rate. A **discount rate** is the rate used in discounting future cash flow. It is also called the **capitalization rate** or **opportunity cost of capital**. In Figure 4-4, the discount rate is 10 percent per year, and in Figure 4-5, the discount rate is 8 percent per year.
3. Calculate the net present value. There are several ways to calculate NPV. Most spreadsheet software has a built-in function to calculate NPV. For example, Figure 4-4 shows the formula that Microsoft Excel uses: =npv(discount rate, range of cash flows), where the discount rate is in cell B1 and the range of cash flows for Project 1 are in cells B6 through F6. (See Chapter 7, Project Cost Management, for more information on cash flow and other cost-related terms.) To use the NPV function, there must be a row in the spreadsheet (or column,

| **Discount rate** | 8% | | | | | |
|---|---|---|---|---|---|---|
| Assume the project is completed in Year 0 | | | Year | | | |
| | 0 | 1 | 2 | 3 | **Total** | |
| Costs | 140,000 | 40,000 | 40,000 | 40,000 | | |
| Discount factor | 1 | 0.93 | 0.86 | 0.79 | | |
| **Discounted costs** | **140,000** | **37,200** | **34,400** | **31,600** | **243,200** | |
| | | | | | | |
| Benefits | 0 | 200,000 | 200,000 | 200,000 | | |
| Discount factor | 1 | 0.93 | 0.86 | 0.79 | | |
| **Discounted benefits** | **0** | **186,000** | **172,000** | **158,000** | **516,000** | |
| | | | | | | |
| Discounted benefits - costs | (140,000) | 148,800 | 137,600 | 126,400 | **272,800** | ←**NPV** |
| Cumulative benefits - costs | (140,000) | 8,800 | 146,400 | 272,800 | | |
| | | ↑ | | | | |
| **ROI** ⟶ | **112%** | | | | | |
| | **Payback In Year 1** | | | | | |

**FIGURE 4-5** JWD Consulting net present value example

depending how it is organized) for the cash flow each year, which is the benefit amount for that year minus the cost amount. The result of the formula yields an NPV of $2316 for Project 1 and $3201 for Project 2. Since both projects have positive NPVs, they are both good candidates for selection. However, since Project 2 has a higher NPV than Project 1 (38 percent higher), it would be the better choice. If the two numbers are close, then other methods should be used to help decide which project to select.



The mathematical formula for calculating NPV is:

$$NPV = \sum_{t=0...n} A_t/(1+r)^t$$

where $t$ equals the year of the cash flow, $n$ is the last year of the cash flow, $A$ is the amount of cash flow each year, and $r$ is the discount rate. If you cannot enter the data into spreadsheet software, you can perform the calculations by hand or with a simple calculator. First, determine the annual **discount factor**—a multiplier for each year based on the discount rate and year—and then apply it to the costs and benefits for each year. The formula for the discount factor is $1/(1 + r)t$ where $r$ is the discount rate, such as 8 percent, and $t$ is the year. For example, the discount factors used in Figure 4-5 are calculated as follows:

$$\text{Year 0 : discount factor} = 1/(1+0.08)^0 = 1$$

$$\text{Year 1 : discount factor} = 1/(1+0.08)^1 = .93$$

$$\text{Year 2 : discount factor} = 1/(1+0.08)^2 = .86$$

$$\text{Year 3 : discount factor} = 1/(1+0.08)^3 = .79$$

After determining the discount factor each year, multiply the costs and benefits each year by the appropriate discount factor. For example, in Figure 4-5, the discounted cost for Year 1 is $40,000 * .93 = $37,200. Next, sum all of the discounted costs and benefits each year to get a total. For example, the total discounted costs in Figure 4-5 are $243,200. To calculate the NPV, take the total discounted benefits and subtract the total discounted costs. In this example, the NPV is $516,000 – $243,200 = $272,800.

### Return on Investment

Another important financial consideration is return on investment. **Return on investment (ROI)** is the result of subtracting the project costs from the benefits and then dividing by the costs. For example, if you invest $100 today and next year it is worth $110, your ROI is ($110 – 100)/100 or 0.10 (10 percent). Note that the ROI is always a percentage. It can be positive or negative. It is best to consider discounted costs and benefits for multi-year projects when calculating ROI. Figure 4-5 shows an ROI of 112 percent. You calculate this number as follows:

$$\text{ROI} = (\text{total discounted benefits} - \text{total discounted costs})/\text{discounted costs}$$

$$\text{ROI} = (516,000 - 243,200)/243,200 = 112\%$$

The higher the ROI, the better. An ROI of 112 percent is outstanding. Many organizations have a required rate of return for projects. The **required rate of return** is the minimum acceptable rate of return on an investment. For example, an organization might have a required rate of return of at least 10 percent for projects. The organization bases the required rate of return on what it could expect to receive elsewhere for an investment of comparable risk. You can also determine a project's **internal rate of return (IRR)** by finding what discount rate results in an NPV of zero for the project. You can use the Goal Seek function in Excel (use Excel's Help function for more information on Goal Seek) to determine the IRR quickly. Simply set the cell containing the NPV calculation to zero while changing the cell containing the discount rate. For example, in Figure 4-4, you could set cell b7 to zero while changing cell b1 to find that the IRR for Project 1 is 27 percent.



Many organizations use ROI in the project selection process. In a recent *Information Week* study, more than 82 percent of IT decisions required an ROI analysis.[5]

## Payback Analysis

Payback analysis is another important financial tool to use when selecting projects. **Payback period** is the amount of time it will take to recoup, in the form of net cash inflows, the total dollars invested in a project. In other words, payback analysis determines how much time will lapse before accrued benefits overtake accrued and continuing costs. Payback occurs when the net cumulative benefits equals the net cumulative costs, or when the net cumulative benefits minus costs equals zero. Figure 4-6 shows how to find the payback period. The cumulative benefits minus costs for Year 0 are ($140,000). Adding that number

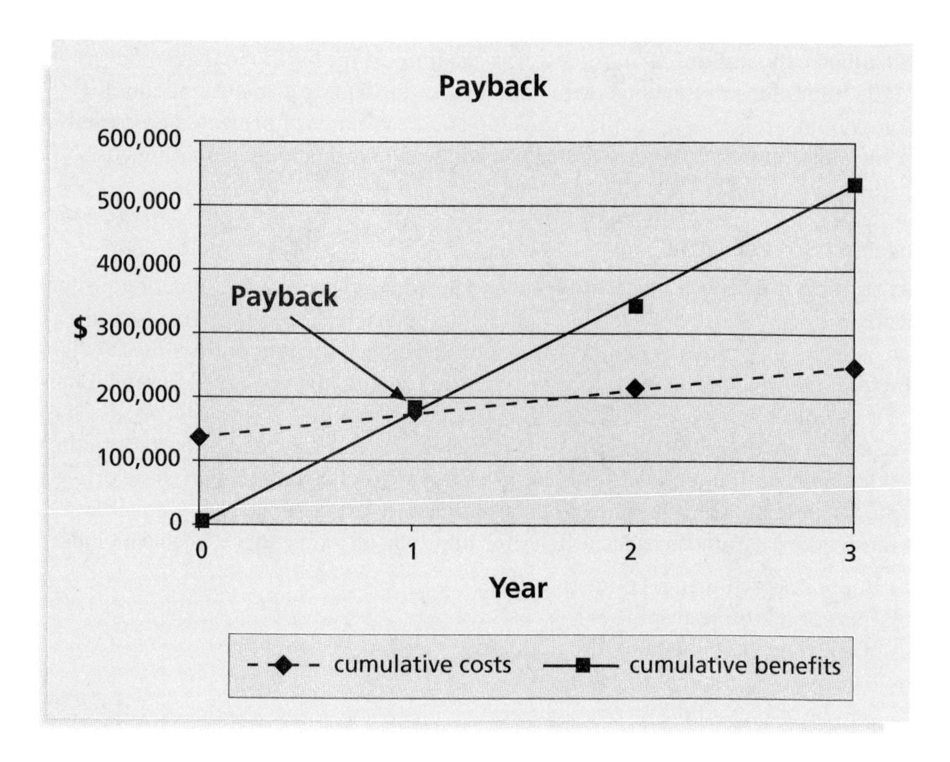


**FIGURE 4-6** Charting the payback period

to the discounted benefits minus costs for Year 1 results in $8,800. Since that number is positive, the payback occurs in Year 1.

Creating a chart helps illustrate more precisely when the payback period occurs. Figure 4-6 charts the cumulative discounted costs and cumulative discounted benefits each year using the numbers from Figure 4-5. Note that the lines cross right around Year 1. This is the point where the cumulative discounted benefits equal the cumulative discounted costs, so that the cumulative discounted benefits minus costs are zero. Beyond this point, discounted benefits exceed discounted costs and the project is showing a profit. Since this project started in Year 0, a payback in Year 1 actually means the project reached payback in its second year. The cumulative discounted benefits and costs are equal to zero where the lines cross. An early payback period, such as in the first or second year, is normally considered very good.



Many organizations have certain recommendations for the length of the payback period of an investment. They might require all information technology projects to have a payback period of less than two years or even one year, regardless of the estimated NPV or ROI. Dan Hoover, vice president and area director at Ciber Inc., an international systems integration consultancy, suggests that organizations, especially small firms, should focus on payback period when making IT investment decisions. "If your costs are recovered in the first year," Hoover says, "the project is worthy of serious consideration, especially if the benefits are high. If the payback period is more than a year, it may be best to look elsewhere."[6] However, organizations must also consider long-range goals when making technology investments. Many crucial projects cannot achieve a payback so quickly or be completed in such a short time period.

To aid in project selection, it is important for project managers to understand the organization's financial expectations for projects. It is also important for top management to understand the limitations of financial estimates, particularly for information technology projects. For example, it is very difficult to develop good estimates of projected costs and benefits for information technology projects. You will learn more about estimating costs and benefits in Chapter 7, Project Cost Management.

### Using a Weighted Scoring Model

A **weighted scoring model** is a tool that provides a systematic process for selecting projects based on many criteria. These criteria can include factors such as meeting broad organizational needs; addressing problems, opportunities, or directives; the amount of time it will take to complete the project; the overall priority of the project; and projected financial performance of the project.

The first step in creating a weighted scoring model is to identify criteria important to the project selection process. It often takes time to develop and reach agreement on these criteria. Holding facilitated brainstorming sessions or using groupware to exchange ideas can aid in developing these criteria. Some possible criteria for information technology projects include:

- Supports key business objectives
- Has strong internal sponsor
- Has strong customer support
- Uses realistic level of technology
- Can be implemented in one year or less
- Provides positive NPV
- Has low risk in meeting scope, time, and cost goals

Next, you assign a weight to each criterion. Once again, determining weights requires consultation and final agreement. These weights indicate how much you value each criterion or how important each criterion is. You can assign weights based on percentages, and the sum of all of the criteria's weights must total 100 percent. You then assign numerical scores to each criterion (e.g. 0 to 100) for each project. The scores indicate how much each project meets each criterion. At this point, you can use a spreadsheet application to create a matrix of projects, criteria, weights, and scores. Figure 4-7 provides an example of a weighted scoring model to evaluate four different projects. After assigning weights for the criteria and scores for each project, you calculate a weighted score for each project by multiplying the weight for each criterion by its score and adding the resulting values.

For example, you calculate the weighted score for Project 1 in Figure 4-7 as:

$$25\% * 90 + 15\% * 70 + 15\% * 50 + 10\% * 25 + 5\% * 20 + 20\% * 50 + 10\% * 20 = 56$$

Note that in this example, Project 2 would be the obvious choice for selection because it has the highest weighted score. Creating a bar chart to graph the weighted

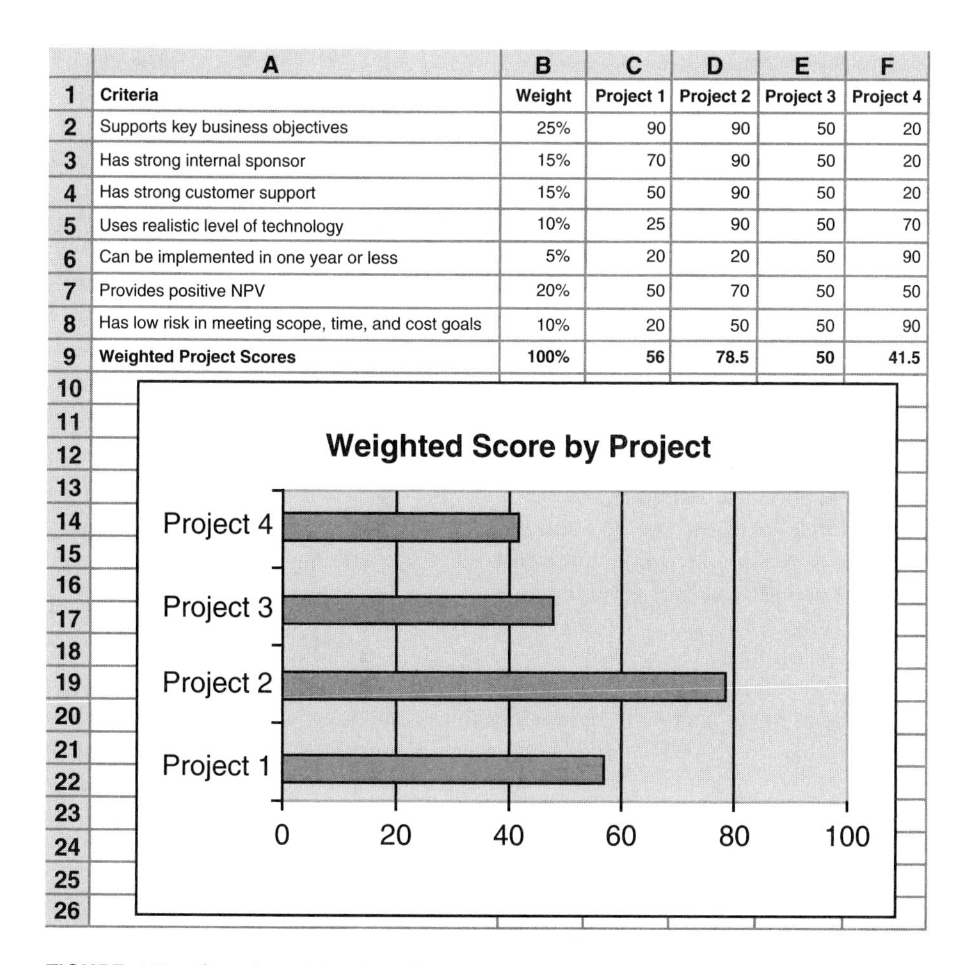

| | A | B | C | D | E | F |
|---|---|---|---|---|---|---|
| 1 | **Criteria** | **Weight** | **Project 1** | **Project 2** | **Project 3** | **Project 4** |
| 2 | Supports key business objectives | 25% | 90 | 90 | 50 | 20 |
| 3 | Has strong internal sponsor | 15% | 70 | 90 | 50 | 20 |
| 4 | Has strong customer support | 15% | 50 | 90 | 50 | 20 |
| 5 | Uses realistic level of technology | 10% | 25 | 90 | 50 | 70 |
| 6 | Can be implemented in one year or less | 5% | 20 | 20 | 50 | 90 |
| 7 | Provides positive NPV | 20% | 50 | 70 | 50 | 50 |
| 8 | Has low risk in meeting scope, time, and cost goals | 10% | 20 | 50 | 50 | 90 |
| 9 | **Weighted Project Scores** | **100%** | **56** | **78.5** | **50** | **41.5** |



**FIGURE 4-7** Sample weighted scoring model for project selection

scores for each project allows you to see the results at a glance. If you create the weighted scoring model in a spreadsheet, you can enter the data, create and copy formulas, and perform a "what-if" analysis. For example, suppose you change the weights for the criteria. By having the weighted scoring model in a spreadsheet, you can easily change the weights to update the weighted scores and charts automatically. This capability allows you to investigate various options for different stakeholders quickly. Ideally, the result should be reflective of the group's consensus, and any major disagreements should be documented.



Teachers often use a weighted scoring model to determine grades. Suppose grades for a class are based on two homework assignments and two exams. To calculate final grades, the teacher would assign a weight to each of these items. Suppose Homework One is worth 10 percent of the grade, Homework Two is worth 20 percent of the grade, Test One is worth 20 percent of the grade, and Test Two is worth 50 percent of the grade. Students would want to do well on each of these items, but they should focus on performing well on Test Two since it is 50 percent of the grade.

You can also establish weights by assigning points. For example, a project might receive 10 points if it definitely supports key business objectives, 5 points if it somewhat supports them, and 0 points if it is totally unrelated to key business objectives. With a point model, you can simply add all the points to determine the best projects for selection, without having to multiply weights and scores and sum the results.

You can also determine minimum scores or thresholds for specific criteria in a weighted scoring model. For example, suppose an organization really should not consider a project if it does not score at least 50 out of 100 on every criterion. You can build this type of threshold into the weighted scoring model to reject projects that do not meet these minimum standards. As you can see, weighted scoring models can aid in project selection decisions.

### Implementing a Balanced Scorecard

Drs. Robert Kaplan and David Norton developed another approach to help select and manage projects that align with business strategy. A **balanced scorecard** is a methodology that converts an organization's value drivers, such as customer service, innovation, operational efficiency, and financial performance, to a series of defined metrics. Organizations record and analyze these metrics to determine how well projects help them achieve strategic goals. Using a balanced scorecard involves several detailed steps. You can learn more about how balanced scorecards work from the Balanced Scorecard Institute (*www.balancedscorecard.org*) or other sources. Although this concept can work within an information technology department specifically, it is best to implement a balanced scorecard throughout an organization because it helps foster alignment between business and information technology.[7] The Balanced Scorecard Institute's Web site includes several examples of how organizations use this methodology, For example, the U.S. Defense Finance and Accounting Services (DFAS) organization uses a balanced scorecard to measure performance and track progress in achieving its strategic goals. Its strategy focuses on four perspectives: customer, financial, internal, and growth and learning. Figure 4-8 shows how the balanced scorecard approach ties together the organization's mission, vision, and goals based on these four perspectives. The DFAS continuously monitors this corporate scorecard and revises it based on identified priorities.

**FIGURE 4-8** Balanced scorecard example

Defense Finance and Accounting Service, "DFAS Strategic Plan," Nov 2001 *(http:/balancedscorecard.org/Portals/0/PDF/DFAS-strategic-plan.pdf )*, p. 13.

As you can see, organizations can use many approaches to select projects. Many project managers have some say in which projects their organization select for implementation. Even if they do not, they need to understand the motives and overall business strategies for the projects they are managing. Project managers and team members are often called upon to explain the importance of their projects, and understanding many of these project selection methods can help them represent the project effectively.

## Developing a Project Charter

After top management decides on which projects to pursue, it is important to let the rest of the organization know about these projects. Management needs to create and distribute documentation to authorize project initiation. This documentation can take many different forms, but one common form is a project charter. A **project charter** is a document that formally recognizes the existence of a project and provides direction on the project's objectives and management. It authorizes the project manager to use organizational resources to complete the project. Ideally, the project manager will provide a major role in developing the project charter. Instead of project charters, some organizations initiate projects using a simple letter of agreement, while others use much longer documents or formal contracts.

Key project stakeholders should sign a project charter to acknowledge agreement on the need for and intent of the project. A project charter is a key output of the initiation process, as described in Chapter 3.

The *PMBOK® Guide, Fourth Edition* lists inputs, tools and techniques, and outputs of the seven project integration management processes. For example, inputs that are helpful in developing a project charter include the following:



- *A project statement of work*: A statement of work is a document that describes the products or services to be created by the project team. It usually includes a description of the business need for the project, a summary of the requirements and characteristics of the products or services, and organizational information, such as appropriate parts of the strategic plan, showing the alignment of the project with strategic goals.
- *A business case*: As explained in Chapter 3, many projects require a business case to justify their investment. Information in the business case, such as the project objective, high-level requirements, and time and cost goals are included in the project charter.
- *A contract*: If you are working on a project under contract for an external customer, the contract should include much of the information needed for creating a good project charter. Some people might use a contract in place of a charter; however, many contracts are difficult to read and can often change, so it is still a good idea to create a project charter.
- *Enterprise environmental factors*: These factors include relevant government or industry standards, the organization's infrastructure, and marketplace conditions. Managers should review these factors when developing a project charter.
- *Organizational process assets*: **Organizational process assets** include formal and informal plans, policies, procedures, guidelines, information systems, financial systems, management systems, lessons learned, and historical information that can be used to influence a project's success.

The main tool and technique for developing a project charter is expert judgment. Experts from within as well as outside the organization should be consulted when creating a project charter to make sure it is useful and realistic.

The only output of the process to develop a project charter is a project charter. Although the format of project charters can vary tremendously, they should include at least the following basic information:

- The project's title and date of authorization
- The project manager's name and contact information
- A summary schedule, including the planned start and finish dates; if a summary milestone schedule is available, it should also be included or referenced
- A summary of the project's budget or reference to budgetary documents
- A brief description of the project objectives, including the business need or other justification for authorizing the project
- Project success criteria, including project approval requirements and who signs off on the project
- A summary of the planned approach for managing the project, which should describe stakeholder needs and expectations, important assumptions, and

constraints, and refer to related documents, such as a communications management plan, as available
- A roles and responsibilities matrix
- A sign-off section for signatures of key project stakeholders
- A comments section in which stakeholders can provide important comments related to the project



Unfortunately, many internal projects, like the one in the opening case of this chapter, do not have project charters. They often have a budget and general guidelines, but no formal, signed documentation. If Nick had a project charter to refer to—especially if it included information on the approach for managing the project—top management would have received the business information they needed, and managing the project might have been easier. Project charters are usually not difficult to write. What is difficult is getting people with the proper knowledge and authority to write and sign the project charters. Top management should have reviewed the charter with Nick, since he was the project manager. In their initial meeting, they should have discussed roles and responsibilities, as well as their expectations of how Nick should work with them. If there is no project charter, the project manager should work with key stakeholders, including top management, to create one. Table 4-1 shows a possible charter that Nick could have created for completing the DNA-sequencing instrument project.

Many projects fail because of unclear requirements and expectations, so starting with a project charter makes a lot of sense. If project managers are having difficulty obtaining support from project stakeholders, for example, they can refer to what everyone agreed to in the project charter. Note that the sample project charter in Table 4-1 includes several items under the Approach section to help Nick in managing the project and the sponsor in overseeing it. To help Nick transition to the role of project manager, the charter said that they would hire a technical replacement and part-time assistant for Nick as soon as possible. To help Ahmed, the project sponsor, feel more comfortable with how the project was being managed, there were items included to ensure proper planning and communications. Recall from Chapter 2 that executive support contributes the most to successful information technology projects. Since Nick was the fourth project manager on this project, top management at his company obviously had some problems choosing and working with project managers.

**TABLE 4-1** Project charter for the DNA-sequencing instrument completion project

**Project Title:** DNA-Sequencing Instrument Completion Project
**Date of Authorization:** February 1
**Project Start Date:** February 1 **Projected Finish Date:** November 1

**Key Schedule Milestones:**

- Complete first version of the software by June 1
- Complete production version of the software by November 1

**Budget Information:** The firm has allocated $1.5 million for this project, and more funds are available if needed. The majority of costs for this project will be internal labor. All hardware will be outsourced.

**TABLE 4-1** Project charter for the DNA-sequencing instrument completion project (cont)

**Project Manager:** Nick Carson, (650) 949-0707, ncarson@dnaconsulting.com

**Project Objectives:** The DNA-sequencing instrument project has been underway for three years. It is a crucial project for our company. This is the first charter for the project, and the objective is to complete the first version of the software for the instrument in four months and a production version in nine months.

**Main Project Success Criteria:** The software must meet all written specifications, be thoroughly tested, and be completed on time. The CEO will formally approve the project with advice from other key stakeholders.

**Approach:**

- Hire a technical replacement for Nick Carson and a part-time assistant as soon as possible.
- Within one month, develop a clear work breakdown structure, scope statement, and Gantt chart detailing the work required to complete the DNA sequencing instrument.
- Purchase all required hardware upgrades within two months.
- Hold weekly progress review meetings with the core project team and the sponsor.
- Conduct thorough software testing per the approved test plans.

ROLES AND RESPONSIBILITIES

| Name | Role | Position | Contact Information |
|---|---|---|---|
| Ahmed Abrams | Sponsor | CEO | aabrams@dnaconsulting.com |
| Nick Carson | Project Manager | Manager | ncarson@dnaconsulting.com |
| Susan Johnson | Team Member | DNA expert | sjohnson@dnaconsulting.com |
| Renyong Chi | Team Member | Testing expert | rchi@dnaconsulting.com |
| Erik Haus | Team Member | Programmer | ehaus@dnaconsulting.com |
| Bill Strom | Team Member | Programmer | bstrom@dnaconsulting.com |
| Maggie Elliot | Team Member | Programmer | melliot@dnaconsulting.com |

Sign-off: (Signatures of all the above stakeholders)

*Ahmed Abrams* *Nick Carson*
*Susan Johnson* *Renyong Chi*
*Erik Haus* *Bill Strom*
*Maggie Elliot*

**Comments:** (Handwritten or typed comments from above stakeholders, if applicable)

*"I want to be heavily involved in this project. It is crucial to our company's success, and I expect everyone to help make it succeed." —Ahmed Abrams*

"The software test plans are complete and well documented. If anyone has questions, do not hesitate to contact me." —Renyong Chi

Taking the time to discuss, develop, and sign off on a simple project charter could have prevented several problems in this case.

After creating a project charter, the next step in project integration management is preparing a project management plan.



## DEVELOPING A PROJECT MANAGEMENT PLAN

To coordinate and integrate information across project management knowledge areas and across the organization, there must be a good project management plan. A **project management plan** is a document used to coordinate all project planning documents and help guide a project's execution and control. Plans created in the other knowledge areas are considered subsidiary parts of the overall project management plan. Project management plans also document project planning assumptions and decisions regarding choices, facilitate communication among stakeholders, define the content, extent, and timing of key management reviews, and provide a baseline for progress measurement and project control. Project management plans should be dynamic, flexible, and subject to change when the environment or project changes. These plans should greatly assist the project manager in leading the project team and assessing project status.

To create and assemble a good project management plan, the project manager must practice the art of project integration management, since information is required from all of the project management knowledge areas. Working with the project team and other stakeholders to create a project management plan will help the project manager guide the project's execution and understand the overall project. The main inputs for developing a project management plan include the project charter, outputs from planning processes, enterprise environment factors, and organizational process assets. The main tool and technique is expert judgment, and the output is a project management plan.

### Project Management Plan Contents

Just as projects are unique, so are project management plans. A small project involving a few people over a couple of months might have a project management plan consisting of only a project charter, scope statement, and Gantt chart. A large project involving a hundred people over three years would have a much more detailed project management plan. It is important to tailor project management plans to fit the needs of specific projects. The project management plans should guide the work, so they should be only as detailed as needed for each project.

There are, however, common elements to most project management plans. Parts of a project management plan include an introduction or overview of the project, a description of how the project is organized, the management and technical processes used on the project, and sections describing the work to be performed, the schedule, and the budget.

The introduction or overview of the project should include, as a minimum, the following information:

- *The project name*: Every project should have a unique name. Unique names help distinguish each project and avoid confusion among related projects.

- *A brief description of the project and the need it addresses*: This description should clearly outline the goals of the project and reason for the project. It should be written in layperson's terms, avoid technical jargon, and include a rough time and cost estimate.
- *The sponsor's name*: Every project needs a sponsor. Include the name, title, and contact information of the sponsor in the introduction.
- *The names of the project manager and key team members*: The project manager should always be the contact for project information. Depending on the size and nature of the project, names of key team members may also be included.
- *Deliverables of the project*: This section should briefly list and describe the products that will be produced as part of the project. Software packages, pieces of hardware, technical reports, and training materials are examples of deliverables.
- *A list of important reference materials*: Many projects have a history preceding them. Listing important documents or meetings related to a project helps project stakeholders understand that history. This section should reference the plans produced for other knowledge areas. (Recall from Chapter 3 that every single knowledge area includes some planning processes.) Therefore, the project management plan should reference and summarize important parts of the scope management, schedule management, cost management, quality management, human resource management, communications management, risk management, and procurement management plans.
- *A list of definitions and acronyms, if appropriate*: Many projects, especially information technology projects, involve terminology unique to a particular industry or technology. Providing a list of definitions and acronyms will help avoid confusion.

The description of how the project is organized should include the following information:

- *Organizational charts*: In addition to an organizational chart for the company sponsoring the project and for the customer's company (if it is an external customer), there should be a project organizational chart to show the lines of authority, responsibilities, and communication for the project. For example, the Manhattan Project introduced in Chapter 1 had a very detailed organizational chart to show all the people working on the project.
- *Project responsibilities*: This section of the project plan should describe the major project functions and activities and identify those individuals who are responsible for them. A responsibility assignment matrix (described in Chapter 9) is a tool often used for displaying this information.
- *Other organizational or process-related information*: Depending on the nature of the project, there may be a need to document major processes followed on the project. For example, if the project involves releasing a major software upgrade, it might help everyone involved in the project to see a diagram or timeline of the major steps involved in this process.

The section of the project management plan describing management and technical approaches should include the following information:

- *Management objectives*: It is important to understand top management's view of the project, what the priorities are for the project, and any major assumptions or constraints.
- *Project controls*: This section describes how to monitor project progress and handle changes. Will there be monthly status reviews and quarterly progress reviews? Will there be specific forms or charts to monitor progress? Will the project use earned value management (described in Chapter 7) to assess and track performance? What is the process for change control? What level of management is required to approve different types of changes? (You will learn more about change control later in this chapter.)
- *Risk management*: This section briefly addresses how the project team will identify, manage, and control risks. It should refer to the risk management plan, if one is required for the project.
- *Project staffing*: This section describes the number and types of people required for the project. It should refer to the human resource plan, if one is required for the project.
- *Technical processes*: This section describes specific methodologies a project might use and explains how to document information. For example, many information technology projects follow specific software development methodologies or use particular Computer Aided Software Engineering (CASE) tools. Many companies or customers also have specific formats for technical documentation. It is important to clarify these technical processes in the project management plan.

The next section of the project management plan should describe the work to perform and reference the scope management plan. It should summarize the following:

- *Major work packages*: A project manager usually organizes the project work into several work packages using a work breakdown structure (WBS), and produces a scope statement to describe the work in more detail. This section should briefly summarize the main work packages for the project and refer to appropriate sections of the scope management plan.
- *Key deliverables*: This section lists and describes the key products produced as part of the project. It should also describe the quality expectations for the product deliverables.
- *Other work-related information*: This section highlights key information related to the work performed on the project. For example, it might list specific hardware or software to use on the project or certain specifications to follow. It should document major assumptions made in defining the project work.

The project schedule information section should include the following:

- *Summary schedule*: It is helpful to see a one-page summary of the overall project schedule. Depending on the size and complexity of the project, the summary schedule might list only key deliverables and their planned completion dates. For smaller projects, it might include all of the work and associated dates for the entire project in a Gantt chart. For example, the Gantt chart and milestone schedule provided in Chapter 3 for JWD Consulting were fairly short and simple.



- *Detailed schedule*: This section provides information on the project schedule that is more detailed. It should reference the schedule management plan and discuss dependencies among project activities that could affect the project schedule. For example, it might explain that a major part of the work cannot start until an external agency provides funding. A network diagram can show these dependencies (see Chapter 6, Project Time Management).
- *Other schedule-related information*: Many assumptions are often made in preparing project schedules. This section should document major assumptions and highlight other important information related to the project schedule.

The budget section of the project management plan should include the following:

- *Summary budget*: The summary budget includes the total estimate of the overall project's budget. It could also include the budget estimate for each month or year by certain budget categories. It is important to provide some explanation of what these numbers mean. For example, is the total budget estimate a firm number that cannot change, or is it a rough estimate based on projected costs over the next three years?
- *Detailed budget*: This section summarizes what is in the cost management plan and includes more detailed budget information. For example, what are the fixed and recurring cost estimates for the project each year? What are the projected financial benefits of the project? What types of people are needed to do the work, and how are the labor costs calculated? (See Chapter 7, Project Cost Management, for more information on creating cost estimates and budgets.)
- *Other budget-related information*: This section documents major assumptions and highlights other important information related to financial aspects of the project.

## Using Guidelines to Create Project Management Plans

Many organizations use guidelines to create project management plans. Microsoft Project 2007 and other project management software packages come with several template files to use as guidelines. However, do not confuse a project management plan with a Gantt chart. The project management plan is much more than a Gantt chart, as described earlier.

Many government agencies also provide guidelines for creating project management plans. For example, the U.S. Department of Defense (DOD) Standard 2167, Software Development Plan, describes the format for contractors to use in creating a plan for software development for DOD projects. The Institute of Electrical and Electronics Engineers (IEEE) Standard 1058–1998 describes the contents of a Software Project Management Plan (SPMP). Table 4-2 provides some of the categories for the IEEE SPMP. Companies working on software development projects for the Department of Defense must follow this or a similar standard.

In many private organizations, specific documentation standards are not as rigorous; however, there are usually guidelines for developing project management plans. It is good

**TABLE 4-2** Sample contents for a software project management plan (SPMP)

| Major section headings | Section topics |
|---|---|
| **Overview** | Purpose, scope, and objectives; assumptions and constraints; project deliverables; schedule and budget summary; evolution of the plan |
| **Project Organization** | External interfaces; internal structure; roles and responsibilities |
| **Managerial Process Plan** | Start-up plans (estimation, staffing, resource acquisition, and project staff training plans); work plan (work activities, schedule, resource, and budget allocation); control plan; risk management plan; closeout plan |
| **Technical Process Plans** | Process model; methods, tools, and techniques; infrastructure plan; product acceptance plan |
| **Supporting Process Plans** | Configuration management plan; verification and validation plan; documentation plan; quality assurance plan; reviews and audits; problem resolution plan; subcontractor management plan; process improvement plan |

*Source:* IEEE Standard 1058–1998.



practice to follow standards or guidelines for developing project management plans in an organization to facilitate the development and execution of those plans. The organization can work more efficiently if all project management plans follow a similar format. Recall from Chapter 1 that companies that excel in project management develop and deploy standardized project delivery systems.

> The winners clearly spell out what needs to be done in a project, by whom, when, and how. For this they use an integrated toolbox, including PM tools, methods, and techniques … If a scheduling template is developed and used over and over, it becomes a repeatable action that leads to higher productivity and lower uncertainty. Sure, using scheduling templates is neither a breakthrough nor a feat. But laggards exhibited almost no use of the templates. Rather, in constructing schedules their project managers started with a clean sheet, a clear waste of time.[8]

For example, in the opening case, Nick Carson's top managers were disappointed because he did not provide them with the project planning information they needed to make important business decisions. They wanted to see detailed project management plans, including schedules and a means for tracking progress. Nick had never created a project management plan or even a simple progress report before, and the organization did not provide templates or examples to follow. If it had, Nick might have been able to deliver the information top management was expecting.

# DIRECTING AND MANAGING PROJECT EXECUTION



Directing and managing project execution involves managing and performing the work described in the project management plan, one of the main inputs for this process. Other inputs include approved change requests, enterprise environmental factors, and organizational process assets. The majority of time on a project is usually spent on execution, as is most of the project's budget. The application area of the project directly affects project execution because the products of the project are produced during project execution. For example, the DNA-sequencing instrument project from the opening case and all associated software and documentation would be produced during project execution. The project team would need to use their expertise in biology, hardware and software development, and testing to produce the product successfully.

The project manager would also need to focus on leading the project team and managing stakeholder relationships to execute the project management plan successfully. Project human resource management and project communications management are crucial to a project's success. See Chapters 9 and 10 respectively for more information on those knowledge areas. If the project involves a significant amount of risk or outside resources, the project manager also needs to be well versed in project risk management and project procurement management. See Chapters 11 and 12 for details on those knowledge areas. Many unique situations occur during project execution, so project managers must be flexible and creative in dealing with them. Review the situation Erica Bell faced during project execution in Chapter 3. Also review the ResNet case study (available on the companion Web site for this text) to understand the execution challenges project manager Peeter Kivestu and his project team faced.

## Coordinating Planning and Execution

In project integration management, project planning and execution are intertwined and inseparable activities. The main function of creating a project management plan is to guide project execution. A good plan should help produce good products or work results. Plans should document what good work results consist of. Updates to plans should reflect knowledge gained from completing work earlier in the project. Anyone who has tried to write a computer program from poor specifications appreciates the importance of a good plan. Anyone who has had to document a poorly programmed system appreciates the importance of good execution.

A commonsense approach to improving the coordination between project plan development and execution is to follow this simple rule: Those who will do the work should plan the work. All project personnel need to develop both planning and executing skills and need experience in these areas. In information technology projects, programmers who have had to write detailed specifications and then create the code from their own specifications become better at writing specifications. Likewise, most systems analysts begin their careers as programmers, so they understand what type of analysis and documentation they need to write good code. Although project managers are responsible for developing the overall project management plan, they must solicit input from the project team members who are developing plans in each knowledge area.

## Providing Strong Leadership and a Supportive Culture

Strong leadership and a supportive organizational culture are crucial during project execution. Project managers must lead by example to demonstrate the importance of creating good project plans and then following them in project execution. Project managers often create plans for things they need to do themselves. If project managers follow through on their own plans, their team members are more likely to do the same.



Good project execution also requires a supportive organizational culture. For example, organizational procedures can help or hinder project execution. If an organization has useful guidelines and templates for project management that everyone in the organization follows, it will be easier for project managers and their teams to plan and do their work. If the organization uses the project plans as the basis for performing and monitoring progress during execution, the culture will promote the relationship between good planning and execution. On the other hand, if organizations have confusing or bureaucratic project management guidelines that hinder getting work done or measuring progress against plans, project managers and their teams will be frustrated.

Even with a supportive organizational culture, project managers may sometimes find it necessary to break the rules to produce project results in a timely manner. When project managers break the rules, politics will play a role in the results. For example, if a particular project requires use of nonstandard software, the project manager must use his or her political skills to convince concerned stakeholders of the need to break the rules on using only standard software. Breaking organizational rules—and getting away with it—requires excellent leadership, communication, and political skills.

## Capitalizing on Product, Business, and Application Area Knowledge

In addition to possessing strong leadership, communication, and political skills, project managers also need to possess product, business, and application area knowledge to execute projects successfully. It is often helpful for information technology project managers to have prior technical experience or at least a working knowledge of information technology products. For example, if the project manager were leading a Joint Application Design (JAD) team to help define user requirements, it would be helpful for him or her to understand the language of the business and technical experts on the team. See Chapter 5, Project Scope Management, for more information on JAD and other methods for collecting requirements.

Many information technology projects are small, so project managers may be required to perform some technical work or mentor team members to complete the project. For example, a three-month project to develop a Web-based application with only three team members would benefit most from a project manager who can complete some of the technical work. On larger projects, however, the project manager's primary responsibility is to lead the team and communicate with key project stakeholders. He or she would not have time to do any of the technical work. In this case, it is usually best that the project manager understand the business and application area of the project more than the technology involved.

However, it is very important on large projects for the project manager to understand the business and application area of his or her project. For example, Northwest Airlines completed a series of projects in the last several years to develop and upgrade its reservation systems. The company spent millions of dollars and had more than 70 full-time people working on the projects at peak periods. The project manager, Peeter Kivestu, had never worked in an information technology department, but he had extensive knowledge of the airline industry and the reservations process. He carefully picked his team leaders, making sure they had the required technical and product knowledge. ResNet was the first large information technology project at Northwest Airlines led by a business manager instead of a technical expert, and it was a roaring success. Many organizations have found that large information technology projects require experienced general managers who understand the business and application area of the technology, not the technology itself. (You can find the entire ResNet case study on the companion Web site for this text.)



## Project Execution Tools and Techniques

Directing and managing project execution requires specialized tools and techniques, some of which are unique to project management. Project managers can use specific tools and techniques to perform activities that are part of execution processes. These include:

- *Expert judgment*: Anyone who has worked on a large, complex project appreciates the importance of expert judgment in making good decisions. Project managers should not hesitate to consult experts on different topics, such as what methodology to follow, what programming language to use, what training approach to follow, and so on.
- *Project management information systems*: As described in Chapter 1, there are hundreds of project management software products on the market today. Many large organizations use powerful enterprise project management systems that are accessible via the Internet and tie into other systems, such as financial systems. Even in smaller organizations, project managers or other team members can create Gantt charts that include links to other planning documents on an internal network. For example, Nick or his assistant could have created a detailed Gantt chart for their project in Project 2007 and created a link to other key planning documents created in Word, Excel, or PowerPoint. Nick could have shown the summary tasks during the progress review meetings, and if top management had questions, Nick could have shown them supporting details. Nick's team could also have set baselines for completing the project and tracked their progress toward achieving those goals. See Appendix A for details on using Project 2007 to perform these functions.

Although project management information systems can aid in project execution, project managers must remember that positive leadership and strong teamwork are critical to successful project management. Project managers should delegate the detailed work involved in using these tools to other team members and focus on providing leadership for the whole project to ensure project success. Stakeholders often focus on what to them is the most important output of execution: the deliverables. For example, a production version of the DNA-sequencing instrument was the main deliverable for the project in the opening case. Of course there were many other deliverables created along the way, such as software modules, tests, reports, and so on. Other outputs of project execution include work performance information, change requests, and updates to the project management plan and project documents.

## WHAT WENT RIGHT?

Malaysia's capital, Kuala Lumpur, has become one of Asia's busiest, most exciting cities. With growth, however, came traffic. To help alleviate this problem, the city hired a local firm in mid-2003 to manage a MYR $400 million (U.S. $105 million) project to develop a state-of-the-art Integrated Transport Information System (ITIS). The Deputy Project Director, Lawrence Liew, explained that they broke the project into four key phases and focused on several key milestones. They deliberately kept the work loosely structured to allow the team to be more flexible and creative in handling uncertainties. They based the entire project team within a single project office to streamline communications and facilitate quick problem solving through ad hoc working groups. They also used a dedicated project intranet to exchange information between the project team and sub-contractors. The project was completed in 2005, and ITIS continues to improve traffic flow into Kuala Lumpur.[9]

Project managers and their teams are most often remembered for how well they executed a project and handled difficult situations. Likewise, sports teams around the world know that, the key to winning is good execution. Team coaches can be viewed as project managers, with each game a separate project. Coaches are often judged primarily based on their win-loss record, not on how well they planned for each game. On a humorous note, when one *losing* coach was asked what he thought about his team's execution, he responded, "I'm all for it!"

# MONITORING AND CONTROLLING PROJECT WORK

On large projects, many project managers say that 90 percent of the job is communicating and managing changes. Changes are inevitable on most projects, so it's important to develop and follow a process to monitor and control changes.

Monitoring project work includes collecting, measuring, and disseminating performance information. It also involves assessing measurements and analyzing trends to determine what process improvements can be made. The project team should continuously monitor project performance to assess the overall health of the project and identify areas that require special attention.

The project management plan, performance reports, enterprise environmental factors, and organizational process assets are all important inputs for monitoring and controlling project work.

The project management plan provides the baseline for identifying and controlling project changes. A **baseline** is the approved project management plan plus approved changes. For example, the project management plan includes a section describing the work to perform on a project. This section of the plan describes the key deliverables for the project, the products of the project, and quality requirements. The schedule section of the project management plan lists the planned dates for completing key deliverables, and the budget section of the project management plan provides the planned cost for these deliverables. The project team must focus on delivering the work as planned. If the project team or someone else causes changes during project execution, they must revise the project management plan and have it approved by the project sponsor. Many people refer to different types of baselines, such as a cost baseline or schedule baseline, to describe different project goals more clearly and performance toward meeting them.



Performance reports use this data to provide information on how project execution is going. The main purpose of these reports is to alert the project manager and project team of issues that are causing problems or might cause problems in the future. The project manager and project team must continuously monitor and control project work to decide if corrective or preventive actions are needed, what the best course of action is, and when to act.

## MEDIA SNAPSHOT

Few events get more media attention than the Olympic Games. Imagine all the work involved in planning and executing an event that involves thousands of athletes from around the world with millions of spectators. The 2002 Olympic Winter Games and Paralympics took five years to plan and cost more than $1.9 billion. PMI awarded the Salt Lake Organizing Committee (SLOC) the Project of the Year award for delivering world-class games that, according to the International Olympic Committee, "made a profound impact upon the people of the world."[10]

Four years before the Games began, the SLOC used a Primavera software-based system with a cascading color-coded WBS to integrate planning. A year before the Games, they added a Venue Integrated Planning Schedule to help the team integrate resource needs, budgets, and plans. For example, this software helped the team coordinate different areas involved in controlling access into and around a venue, such as roads, pedestrian pathways, seating and safety provisions, and hospitality areas, saving nearly $10 million.

When the team experienced a budget deficit three years before the games, they separated "must-have" from "nice-to-have" items and implemented a rigorous expense approval process. According to Matthew Lehman, SLOC managing director, using classic project management tools turned a $400 million deficit into a $100 million surplus.

The SLOC also used an Executive Roadmap, a one-page list of the top 100 Games-wide activities, to keep executives apprised of progress. Activities were tied to detailed project information within each department's schedule. A 90-day highlighter showed which managers were accountable for each integrated activity. Fraser Bullock, SLOC Chief Operating Officer and Chief, said, "We knew when we were on and off schedule and where we had to apply additional resources. The interrelation of the functions meant they could not run in isolation—it was a smoothly running machine."[11]

An important output of monitoring and controlling project work is a change request, which includes recommended corrective and preventive actions and defect repairs. Corrective actions should result in improvements in project performance. Preventive actions reduce the probability of negative consequences associated with project risks. Defect repairs involve bringing defective deliverables into conformance with requirements. For example, if project team members have not been reporting hours that they worked, a corrective action would be to show them how to enter the information and let them know that they need to do it. A preventive action might be modifying a time-tracking system screen to avoid common errors people made in the past. A defect repair might be having someone redo an entry that was incorrect. Many organizations use a formal change request process and forms to keep track of project changes, as described in the next section.



## PERFORMING INTEGRATED CHANGE CONTROL

**Integrated change control** involves identifying, evaluating, and managing changes throughout the project life cycle. The three main objectives of integrated change control are:

- *Influencing the factors that create changes to ensure that changes are beneficial*: To ensure that changes are beneficial and that a project is successful, project managers and their teams must make trade-offs among key project dimensions, such as scope, time, cost, and quality.
- *Determining that a change has occurred*: To determine that a change has occurred, the project manager must know the status of key project areas at all times. In addition, the project manager must communicate significant changes to top management and key stakeholders. Top management and other key stakeholders do not like surprises, especially ones that mean the project might produce less, take longer to complete, cost more than planned, or be of lower quality than desired.
- *Managing actual changes as they occur*: Managing change is a key role of project managers and their teams. It is important that project managers exercise discipline in managing the project to help minimize the number of changes that occur.

Important inputs to the integrated change control process include the project management plan, work performance information, change requests, enterprise environmental factors, and organizational process assets. Important outputs include updates to change request status, the project management plan, and project documents.

Change requests are common on projects and occur in many different forms. They can be oral or written, formal or informal. For example, a project team member responsible for installing a server needed to support the project might ask the project manager at a progress review meeting if it is all right to order a server with a faster processor than planned, from the same manufacturer for the same approximate cost. Since this change is positive and should have no negative effects on the project, the project manager might give a verbal approval at the progress review meeting. Nevertheless, it is still important that the project manager document this change to avoid any potential problems. The appropriate team member should update the section of the scope statement with the new specifications for the server. Still, keep in mind that many change requests can have a major impact on a project. For example, customers changing their minds about the number of pieces of hardware they want as part of a project will have a definite impact on the scope and cost of the project. Such a change might also affect the project's schedule. The project team must present such significant changes in written form, and there should be a formal review process for analyzing and deciding whether to approve these changes.

Change is unavoidable and often expected on most information technology projects. Technologies change, personnel change, organizational priorities change, and so on. Careful change control on information technology projects is a critical success factor. A good change control system is also important for project success.



## Change Control on Information Technology Projects

From the 1950s to the 1980s, a widely held view of information technology (then often referred to as data automation or data processing) project management was that the project team should strive to do exactly what they planned on time and within budget. The problem with this view was that project teams could rarely meet original project goals, especially on projects involving new technologies. Stakeholders rarely agreed up front on what the scope of the project really was or what the finished product should really look like. Time and cost estimates created early in a project were rarely accurate.

Beginning in the 1990s, most project managers and top management realized that project management is a process of constant communication and negotiation about project objectives and stakeholder expectations. This view assumes that changes happen throughout the project life cycle and recognizes that changes are often beneficial to some projects. For example, if a project team member discovers a new hardware or software technology that could satisfy the customers' needs for less time and money, the project team and key stakeholders should be open to making major changes in the project.

All projects will have some changes, and managing them is a key issue in project management, especially for information technology projects. Many information technology projects involve the use of hardware and software that is updated frequently. For example, the initial plan for the specifications of the server described earlier may have been cutting-edge technology at that time. If the actual ordering of the server occurred six months later, it is quite possible that a more powerful server could be ordered at the same cost. This example illustrates a positive change. On the other hand, the manufacturer of the server specified in the project plan could go out of business, which would result in a negative change. Information technology project managers should be accustomed to changes such as these and build some flexibility into their project plans and execution. Customers for information technology projects should also be open to meeting project objectives in different ways.

Even if project managers, project teams, and customers are flexible, it is important that projects have a formal change control system. This formal change control system is necessary to plan for managing change.

## Change Control System

A **change control system** is a formal, documented process that describes when and how official project documents may be changed. It also describes the people authorized to make changes, the paperwork required for this change, and any automated or manual tracking systems the project will use. A change control system often includes a change control board, configuration management, and a process for communicating changes.

A **change control board (CCB)** is a formal group of people responsible for approving or rejecting changes to a project. The primary functions of a change control board are to provide guidelines for preparing change requests, evaluating change requests, and managing the implementation of approved changes. An organization could have key stakeholders for the entire organization on this board, and a few members could rotate based on the unique needs of each project. By creating a formal board and process for managing changes, better overall change control should result.



However, CCBs can have some drawbacks. One drawback is the time it takes to make decisions on proposed changes. CCBs often meet only once a week or once a month and may not make decisions in one meeting. Some organizations have streamlined processes for making quick decisions on smaller project changes. One company created a "48-hour policy," in which task leaders on a large information technology project would reach agreements on key decisions or changes within their expertise and authority. The person in the area most affected by this decision or change then had 48 hours to go to his or her top management to seek approval. If for some reason the project team's decision could not be implemented, the top manager consulted would have 48 hours to reverse a decision; otherwise, the project team's decision was approved. This type of process is a great way to deal with the many time-sensitive decisions or changes that project teams must make on many information technology projects.

Configuration management is another important part of integrated change control. **Configuration management** ensures that the descriptions of the project's products are correct and complete. It involves identifying and controlling the functional and physical design characteristics of products and their support documentation. Members of the project team, frequently called configuration management specialists, are often assigned to perform configuration management for large projects. Their job is to identify and document the functional and physical characteristics of the project's products, control any changes to such characteristics, record and report the changes, and audit the products to verify conformance to requirements. Visit the Institute of Configuration Management's Web site (*www.icmhq.com*) for more information on this topic.

Another critical factor in change control is communication. Project managers should use written and oral performance reports to help identify and manage project changes. For example, on software development projects, most programmers must make their edits to one master file in a database that requires the programmers to "check out" the file to edit it. If two programmers check out the same file, they must coordinate their work before they can check the file back in to the database. In addition to written or formal communication methods, oral and informal communications are also important. Some project managers have stand-up meetings once a week or even every morning, depending on the nature of the project. The goal of a stand-up meeting is to communicate what is most important on the project quickly. For example, the project manager might have an early morning stand-up meeting every day with all of his or her team leaders. There might be a weekly stand-up meeting every Monday morning with all interested stakeholders. Requiring participants to stand keeps meetings short and forces everyone to focus on the most important project events.

Why is good communication so critical to success? One of the most frustrating aspects of project change is not having everyone coordinated and informed about the latest project information. Again, it is the project manager's responsibility to integrate all project changes so that the project stays on track. The project manager and his or her staff must develop a system for notifying everyone affected by a change in a timely manner. E-mail, real-time databases, cell phones, and the Web make it easier to disseminate the most current project information. You will learn more about good communication in Chapter 10, Project Communications Management.

Table 4-3 lists suggestions for performing integrated change control. As described earlier, project management is a process of constant communication and negotiation. Project managers should plan for changes and use appropriate tools and techniques such as a change control board, configuration management, and good communication. It is helpful to define procedures for making timely decisions on small changes, use written and oral performance reports to help identify and manage changes, and use software to assist in planning, updating, and controlling projects.



**TABLE 4-3** Suggestions for performing integrated change control

| Suggestions for performing integrated change control |
|---|
| View project management as a process of constant communication and negotiation |
| Plan for change |
| Establish a formal change control system, including a change control board (CCB) |
| Use effective configuration management |
| Define procedures for making timely decisions on smaller changes |
| Use written and oral performance reports to help identify and manage change |
| Use project management and other software to help manage and communicate changes |
| Focus on leading the project team and meeting overall project goals and expectations |

Project managers must also provide strong leadership to steer the project to successful completion. They must not get too involved in managing project changes. Project managers should delegate much of the detailed work to project team members and focus on providing overall leadership for the project in general. Remember, project managers must focus on the big picture and perform project integration management well to lead their team and organization to success.

## CLOSING PROJECTS OR PHASES

The last process in project integration management is closing the project or phase. In order to close a project or phase, you must finalize all activities and transfer the completed or cancelled work to the appropriate people. The main inputs to this process are the project management plan, accepted deliverables, and organizational process assets. The main tool and technique is again expert judgment. The outputs of closing projects are:

- *Final product, service, or result transition*: Project sponsors are usually most interested in making sure they receive delivery of the final products, services, or results they expected when they authorized the project. For items produced under contract, formal acceptance or handover includes a written statement

that the terms of the contract were met. Internal projects can also include some type of project completion form.
- *Organizational process asset updates*: The project team should provide a list of project documentation, project closure documents, and historical information produced by the project in a useful format. This information is considered a process asset. Project teams normally produce a final project report, which often includes a transition plan describing work to be done as part of operations after the project is completed. They also often write a lessons-learned report at the end of a project, and this information can be a tremendous asset for future projects. (See Chapter 10, Project Communications Management for more information on creating project final reports, lessons-learned reports, and other project communications.) Several organizations also conduct a post-implementation review to analyze whether or not the project achieved what it set out to do. Information from this type of review also becomes an organizational process asset for future projects.



## USING SOFTWARE TO ASSIST IN PROJECT INTEGRATION MANAGEMENT

As described throughout this chapter, project teams can use various types of software to assist in project integration management. Project teams can create documents with word processing software, give presentations with presentation software, track information with spreadsheets, databases, or customized software, and transmit information using various types of communication software.

Project management software is also an important tool for developing and integrating project planning documents, executing the project management plan and related project plans, monitoring and controlling project activities, and performing integrated change control. Small project teams can use low-end or midrange project management software products to coordinate their work. For large projects, however, such as managing the Olympic Games described in the Media Snapshot, organizations may benefit most from high-end tools that provide enterprise project management capabilities and integrate all aspects of project management. All projects can benefit from using some type of project management information system to coordinate and communicate project information.

Another category of software that can help align projects with business strategy, as described in this chapter, is called **business service management (BSM) tools**. "BSM tools track the execution of business process flows and expose how the state of supporting IT systems and resources is impacting end-to-end business process performance in real time. Consider, for example, the difference between IT working to increase network capacity and having IT able to demonstrate that because of their efforts, they increased the ability to process new customer orders by 15 percent."[12] BSM tools can help improve alignment between information technology projects, such as upgrading network capacity, and business goals, such as reducing costs by processing customer orders more quickly. In addition, BSM tools can help validate the projects' contributions to the success of the business. However, recent studies suggest that successfully implementing

BSM tools, like many other new tools, is far from easy. "Like those work-from-home, get-rich-quick schemes, you will hear from vendors claiming fast and easy business services management. Smaller upstart vendors will often oversimplify problems and oversell their products' capabilities, while most larger vendors still have work to do integrating disparate product portfolios. Both tend to understate the effort and cost to deploy and configure."[13]



As you can see, there is a lot of work involved in project integration management. Project managers and their teams must focus on pulling all the elements of a project together to successfully complete projects.

## CASE WRAP-UP

Without consulting Nick Carson or his team, Nick's CEO hired a new person, Jim, to act as a middle manager between himself and the people in Nick's department. The CEO and other top managers really liked Jim, the new middle manager. He met with them often, shared ideas, and had a great sense of humor. He started developing standards the company could use to help manage projects in the future. For example, he developed templates for creating plans and progress reports and put them on the company's intranet. However, Jim and Nick did not get along. Jim accidentally sent an e-mail to Nick that was supposed to go to the CEO. In this e-mail, Jim said that Nick was hard to work with and preoccupied with the birth of his son.

Nick was furious when he read the e-mail and stormed into the CEO's office. The CEO suggested that Nick move to another department, but Nick did not like that option. Without considering the repercussions, the CEO offered Nick a severance package to leave the company. Because of the planned corporate buyout, the CEO knew they might have to let some people go anyway. Nick talked the CEO into giving him a two-month sabbatical he had not yet taken plus a higher percentage on his stock options. After discussing the situation with his wife and realizing that he would get over $70,000 if he resigned, Nick took the severance package. He had such a bad experience as a project manager that he decided to stick with being a technical expert. Jim, however, thrived in his position and helped the company improve their project management practices and ensure success in a highly competitive market.

## Chapter Summary

Project integration management is usually the most important project management knowledge area, since it ties together all the other areas of project management. A project manager's primary focus should be on project integration management.



Before selecting projects to pursue, it is important for organizations to follow a strategic planning process. Many organizations perform a SWOT analysis to help identify potential projects based on their strengths, weaknesses, opportunities, and threats. Information technology projects should support the organization's overall business strategy. Common techniques for selecting projects include focusing on broad organizational needs, categorizing projects, performing financial analyses, developing weighted scoring models, and using balanced scorecards.

Project integration management includes the following processes:

- Developing the project charter involves working with stakeholders to create the document that formally authorizes a project. Project charters can have different formats, but they should include basic project information and signatures of key stakeholders.
- Developing the project management plan involves coordinating all planning efforts to create a consistent, coherent document—the project management plan. The main purpose of project plans is to facilitate action.
- Directing and managing project execution involves carrying out the project plans by performing the activities included in it. Project plan execution should require the majority of a project's budget.
- Monitoring and controlling project work is needed to meet the performance objectives of the project. The project team should continuously monitor project performance to assess the overall health of the project.
- Performing integrated change control involves identifying, evaluating, and managing changes throughout the project life cycle. A change control system often includes a change control board (CCB), configuration management, and a process for communicating changes.
- Closing the project or phase involves finalizing all project activities. It is important to follow good procedures to ensure that all project activities are completed and that the project sponsor accepts delivery of the final products, services, or results of the project.

There are several types of software products available to assist in project integration management. There are also several tools to assist in project selection and to ensure that projects align with business strategy.

## Quick Quiz



1. Which of the following processes is not part of project integration management?
   a. develop the project business case
   b. develop the project charter
   c. develop the project management plan
   d. close the project or phase
2. What is the last step in the four-stage planning process for selecting information technology projects?
   a. information technology strategy planning
   b. business area analysis
   c. mind mapping
   d. resource allocation
3. Which of the following is not a best practice for new product development projects?
   a. align projects and resources with business strategy
   b. select projects that will take less than two years to provide payback
   c. focus on customer needs in identifying projects
   d. assign project managers to lead projects
4. A new government law requires an organization to report data in a new way. Under which category would a new information system project to provide this data fall?
   a. problem
   b. opportunity
   c. directive
   d. regulation
5. If estimates for total discounted benefits for a project are $120,000 and total discounted costs are $100,000, what is the estimated return on investment (ROI)?
   a. $20,000
   b. $120,000
   c. 20 percent
   d. 120 percent
6. A ____________ is a document that formally recognizes the existence of a project and provides direction on the project's objectives and management.
   a. project charter
   b. contract
   c. business case
   d. project management plan

7. Which of the following items is not normally included in a project charter?
   a. the name of the project manager
   b. budget information
   c. stakeholder signatures
   d. a Gantt chart



8. __________ ensures that the descriptions of the project's products are correct and complete.
   a. Configuration management
   b. Integrated change control
   c. Integration management
   d. A change control board
9. Which of the following is not a suggestion for performing integrated change control?
   a. use good configuration management
   b. minimize change
   c. establish a formal change control system
   d. view project management as a process of constant communication and negotiation
10. What tool and technique is used for all of the other project integration management processes?
    a. project management software
    b. templates
    c. expert judgment
    d. all of the above

### Quick Quiz Answers

1. a; 2. d; 3. b; 4. c; 5. c; 6. a; 7. d; 8. a; 9. b; 10. c

## Discussion Questions

1. Describe project integration management. How does project integration management relate to the project life cycle, stakeholders, and the other project management knowledge areas?
2. Briefly describe the strategic planning process, including a SWOT analysis. Which project selection method(s) do you think organizations use most often for justifying information technology projects?
3. Summarize key work involved in each of the six processes for project integration management.
4. Either from your own experience or by searching the Internet, describe a well-planned and executed project. Describe a disastrous project. What were some of the main differences between these projects?
5. Discuss the importance of following a well-integrated change control process on information technology projects. What do you think of the suggestions made in this chapter? Think of three additional suggestions for integrated change control on information technology projects.

## Exercises



1. Write a two-page paper based on the opening case. Answer the following questions:
   a. What do you think the real problem was in this case?
   b. Does the case present a realistic scenario? Why or why not?
   c. Was Nick Carson a good project manager? Why or why not?
   d. What should top management have done to help Nick?
   e. What could Nick have done to be a better project manager?
2. Download a free trial of mind mapping software and create a mind map of a SWOT analysis for your organization or your personal life. Include at least two strengths, weaknesses, opportunities, and threats and ideas for at least three potential projects.
3. Use spreadsheet software to create Figure 4-4 through Figure 4-7 in this text. Make sure your formulas work correctly.
4. Perform a financial analysis for a project using the format provided in Figure 4-5. Assume the projected costs and benefits for this project are spread over four years as follows: Estimated costs are $200,000 in Year 1 and $30,000 each year in Years 2, 3, and 4. Estimated benefits are $0 in Year 1 and $100,000 each year in Years 2, 3, and 4. Use a 9 percent discount rate, and round the discount factors to two decimal places. Create a spreadsheet (or use the business case financials template provided on the companion Web site) to calculate and clearly display the NPV, ROI, and year in which payback occurs. In addition, write a paragraph explaining whether you would recommend investing in this project, based on your financial analysis.
5. Create a weighted scoring model to determine grades for a course. Final grades are based on three exams worth 20 percent, 15 percent, and 25 percent, respectively; homework is worth 15 percent; and a group project is worth 25 percent. Enter scores for three students. Assume Student 1 earns 100 percent (or 100) on every item. Assume Student 2 earns 70 percent on each of the exams, 80 percent on the homework, and 95 percent on the group project. Assume Student 3 earns 90 percent on Exam 1, 80 percent on Exam 2, 75 percent on Exam 3, 80 percent on the homework, and 70 percent on the group project. You can use the weighted scoring model template, create your own spreadsheet, or make the matrix by hand.
6. Develop an outline (major headings and subheadings only) for a project management plan to create a Web site for your class, and then fill in the details for the introduction or overview section. Assume that this Web site would include a home page with links to a syllabus for the class, lecture notes or other instructional information, links to the Web site for this textbook, links to other Web sites with project management information, and links to personal pages for each member of your class and future classes. Also, include a bulletin board and chat room feature where students and the instructor can exchange information. Assume your instructor is the project's sponsor, you are the project manager, your classmates are your project team, and you have three months to complete the project.
7. Research software mentioned in this chapter, such as software for assisting in project selection, enterprise project management software, BSM tools, etc. Find at least two references and summarize your findings in a two-page paper.

8. Read and critique two of the Suggested Readings provided on the companion Web site for this book, or find similar articles related to topics discussed in this chapter. Write a two-page paper summarizing your ideas.

## Running Case



*Note*: Additional running cases are provided in Appendix C and on the companion Web site. Template files are also available on the companion Web site.

Manage Your Health, Inc. (MYH) is a Fortune 500 company that provides a variety of health care services across the globe. MYH has more than 20,000 full-time employees and more than 5,000 part-time employees. MYH recently updated its strategic plan, and key goals include reducing internal costs, increasing cross-selling of products, and exploiting new Web-based technologies to help employees, customers, and suppliers work together to improve the development and delivery of its health care products and services. Below are some ideas the Information Technology department has developed for supporting these strategic goals:

1. *Recreation and Wellness Intranet Project*: Provide an application on the current intranet to help employees improve their health. A recent study found that MYH, Inc. pays 20 percent more than the industry average for employee health care premiums, primarily due to the poor health of its employees. You believe that this application will help improve employee health within one year of its rollout so that you can negotiate lower health insurance premiums, providing net savings of at least $30/employee/year for full-time employees over the next four years. This application would include the following capabilities:
   - Allow employees to register for company-sponsored recreational programs, such as soccer, softball, bowling, jogging, walking, and other sports
   - Allow employees to register for company-sponsored classes and programs to help them manage their weight, reduce stress, stop smoking, and manage other health-related issues.
   - Track data on employee involvement in these recreational and health-management programs
   - Offer incentives for people to join the programs and do well in them (e.g., incentives for achieving weight goals, winning sports team competitions, etc.).
2. *Health Coverage Costs Business Model*: Develop an application to track employee health care expenses and company health care costs. Health care premiums continue to increase, and the company has changed insurance carriers several times in the past ten years. This application should allow business modeling of various scenarios as well as tracking and analyzing current and past employee health care expenses and company health care costs. This application must be secure and run on the current intranet so several managers and analysts could access it and download selective data for further analysis. The new application must also import data from the current systems that track employee expenses submitted to the company and the company's costs to the insurance provider. You believe that having this data will help you revise policies concerning employee contributions to health care premiums and help you negotiate for lower premiums with insurance companies. You estimate that this application would save your company about $20/employee/year for full-time employees over the next four years and cost about $100,000 to develop

3. *Cross-Selling System*: Develop an application to improve cross-selling to current customers. The current sales management system has separate sections for major product/service categories and different sales reps based on those products and services. You see great opportunities to increase sales to current customers by providing them discounts when they purchase multiple products/services. You estimate that this system would increase profits by $1 million each year for the next three years and cost about $800,000 each year for development and maintenance.



4. *Web-Enhanced Communications System*: Develop a Web-based application to improve development and delivery of products and services. There are currently several incompatible systems related to the development and delivery of products and services to customers. This application would allow customers and suppliers to provide suggestions, enter orders, view the status and history of orders, and use electronic commerce capabilities to purchase and sell their products. You estimate that this system would save your company about $2 million each year for three years after implementation. You estimate it will take one year and $3 million to develop and require 20 percent of development costs each year to maintain.

## Tasks

1. Summarize each of the above-proposed projects in a simple table format suitable for presentation to top management. Include the name for each project, identify how each one supports business strategies, assess the potential financial benefits and other benefits of each project, and provide your initial assessment of the value of each project. Write your results in a one- to two-page memo to top management, including appropriate back-up information and calculations.
2. Prepare a weighted scoring model using the template provided on the companion Web site for this text to evaluate these four projects. Develop at least four criteria, assign weights to each criterion, assign scores, and then calculate the weighted scores. Print the spreadsheet and bar chart with the results. Also write a one-page paper describing this weighted scoring model and what the results show.
3. Prepare a business case for the Recreation and Wellness Intranet Project. Assume the project will take six months to complete and cost about $200,000. Use the business case template provided on the companion Web site for this text.
4. Prepare a project charter for the Recreation and Wellness Intranet Project. Assume the project will take six months to complete and cost about $200,000. Use the project charter template provided in this text and the sample project charter provided in Table 4-1 as a guide.
5. Prepare a change request for this project, using the template provided on the companion Web site for this text. Be creative in making up information.

## Companion Web Site

Visit the companion Web site for this text at *www.cengage.com/mis/schwalbe* to access:

- References cited in the text and additional suggested readings for each chapter
- Template files
- Lecture notes
- Interactive quizzes
- Podcasts
- Links to general project management Web sites
- And more



See the Preface of this text for more information on accessing the companion Web site.

## Key Terms

**balanced scorecard** — a methodology that converts an organization's value drivers to a series of defined metrics

**baseline** — the approved project management plan plus approved changes

**business service management (BSM) tools** — tools that help track the execution of business process flows and expose how the state of supporting IT systems and resources is impacting end-to-end business process performance in real time

**capitalization rate** — the rate used in discounting future cash flow; also called the discount rate or opportunity cost of capital

**cash flow** — benefits minus costs or income minus expenses

**change control board (CCB)** — a formal group of people responsible for approving or rejecting changes on a project

**change control system** — a formal, documented process that describes when and how official project documents may be changed

**configuration management** — a process that ensures that the descriptions of the project's products are correct and complete

**cost of capital** — the return available by investing the capital elsewhere

**directives** — new requirements imposed by management, government, or some external influence

**discount factor** — a multiplier for each year based on the discount rate and year

**discount rate** — the rate used in discounting future cash flow; also called the capitalization rate or opportunity cost of capital

**integrated change control** — identifying, evaluating, and managing changes throughout the project life cycle

**interface management** — identifying and managing the points of interaction between various elements of a project

**internal rate of return (IRR)** — the discount rate that results in an NPV of zero for a project

**mind mapping** — a technique that uses branches radiating out from a core idea to structure thoughts and ideas

**net present value (NPV) analysis** — a method of calculating the expected net monetary gain or loss from a project by discounting all expected future cash inflows and outflows to the present point in time

**opportunities** — chances to improve the organization

**opportunity cost of capital** — the rate used in discounting future cash flow; also called the capitalization rate or discount rate



**organizational process assets** — formal and informal plans, policies, procedures, guidelines, information systems, financial systems, management systems, lessons learned, and historical information that can be used to influence a project's success

**payback period** — the amount of time it will take to recoup, in the form of net cash inflows, the total dollars invested in a project

**problems** — undesirable situations that prevent the organization from achieving its goals

**project charter** — a document that formally recognizes the existence of a project and provides direction on the project's objectives and management

**project integration management** — processes that coordinate all project management knowledge areas throughout a project's life, including developing the project charter, developing the preliminary project scope statement, developing the project management plan, directing and managing the project, monitoring and controlling the project, providing integrated change control, and closing the project

**project management plan** — a document used to coordinate all project planning documents and guide project execution and control

**required rate of return** — the minimum acceptable rate of return on an investment

**return on investment (ROI)** — (benefits minus costs) divided by costs

**strategic planning** — determining long-term objectives by analyzing the strengths and weaknesses of an organization, studying opportunities and threats in the business environment, predicting future trends, and projecting the need for new products and services

**SWOT analysis** — analyzing **S**trengths, **W**eaknesses, **O**pportunities, and **T**hreats; used to aid in strategic planning

**weighted scoring model** — a technique that provides a systematic process for basing project selection on numerous criteria

## End Notes


[1] Carol Matlack, "First, Blame the Software," *BusinessWeek* Online (October 5, 2006).

[2] James Bacon, "The Use of Decision Criteria in Selecting Information Systems/Technology Investments," *MIS Quarterly*, Vol. 16, No. 3 (September 1992).

[3] Robert G. Cooper, "Winning at New Products: Pathways to Profitable Intervention," PMI Research Conference Proceedings (July 2006).

[4] Dennis J. Cohen and Robert J. Graham, *The Project Manager's MBA*, San Francisco, Jossey-Bass (2001), p. 31.

[5] CIO View Corp., "White Papers: Business Benefits of Utilizing ROI Analysis," *Information Week* (2007).

[6] Jake Widman, "Big IT to small biz: Listen up, little dudes!" *ComputerWorld* (January 24, 2008).

[7] Eric Berkman, "How to Use the Balanced Scorecard," *CIO Magazine* (May 15, 2002).

[8] Fragan Milosevic and A. Ozbay. "Delivering Projects: What the Winners Do." Proceedings of the Project Management Institute Annual Seminars & Symposium (November 2001).

[9] Sarah Parkes, "Crosstown Traffic," *PM Network* (August 2004).

[10] Ross Foti, "The Best Winter Olympics, Period," *PM Network* (January 2004) p. 23.

[11] Ibid, 23.

[12] Mary Johnson Turner, Beyond ITIL: Process-Aware BSM Connects IT to Business Priorities, Summit Strategies, (July 2005).

[13] Michael Biddick, "Uncertain Future," *Information Week Research & Reports* (May 12, 2008), p. 47.

CHAPTER 5

# PROJECT SCOPE MANAGEMENT

## LEARNING OBJECTIVES

**After reading this chapter, you will be able to:**

- Understand the importance of good project scope management
- Discuss methods for collecting and documenting requirements in order to meet stakeholder needs and expectations
- Explain the scope definition process and describe the contents of a project scope statement
- Discuss the process for creating a work breakdown structure using the analogy, top-down, bottom-up, and mind-mapping approaches
- Explain the importance of verifying scope and how it relates to defining and controlling scope
- Understand the importance of controlling scope and approaches for preventing scope-related problems on information technology projects
- Describe how software can assist in project scope management

## OPENING CASE

Kim Nguyen was leading a meeting to create the work breakdown structure (WBS) for the IT Upgrade Project. This project was necessary because of several high-priority, Internet-based applications the company was developing. The IT Upgrade Project involved creating and implementing a plan to get all employees' information technology assets to meet new corporate standards within nine months. These standards specified the minimum equipment required for each desktop or laptop computer, including the type of processor, amount of memory, hard disk size, type of network connection, and software. Kim knew that to perform the upgrades, they would first have to create a detailed inventory of all of the current hardware, networks, and software in the entire company of 2000 employees.

Kim had worked with other stakeholders to develop a project charter and initial scope statement. The project charter included rough cost and schedule estimates for the project and signatures of key stakeholders; the initial scope statement provided a start in defining the hardware, software, and network requirements as well as other information related to the project scope. Kim called a meeting with her project team and other stakeholders to further define the scope of the project. She wanted to get everyone's ideas on what the project involved, who would do what, and how they could avoid scope creep. The company's new CEO, Walter Schmidt, was known for keeping a close eye on major projects like this one. The company had started using a new project management information system that let everyone know the status of projects at a detailed and high level. Kim knew that a good WBS was the foundation for scope, time, and cost performance, but she had never led a team in creating one or allocating costs based on a WBS. Where should she begin?

## WHAT IS PROJECT SCOPE MANAGEMENT?

Recall from Chapter 1 that several factors are associated with project success. Many of these factors, such as user involvement, clear business objectives, a minimized or clearly defined scope, and firm basic requirements, are elements of project scope management.

One of the most important and most difficult aspects of project management is defining the scope of a project. **Scope** refers to *all* the work involved in creating the products of the project and the processes used to create them. Recall from Chapter 2 that the term **deliverable** describes a product produced as part of a project. Deliverables can be product-related, such as a piece of hardware or software, or process-related, such as a planning document or meeting minutes. Project stakeholders must agree on what the products of the project are and, to some extent, how they should produce them to define all of the deliverables.

**Project scope management** includes the processes involved in defining and controlling what work is or is not included in a project. It ensures that the project team and stakeholders have the same understanding of what products the project will produce and what processes the project team will use to produce them. There are five main processes involved in project scope management:

1. *Collecting requirements* involves defining and documenting the features and functions of the products produced during the project as well as the processes used for creating them. The project team creates stakeholder requirements

documentation, a requirements management plan, and a requirements traceability matrix as outputs of the requirements collection process.

2. *Defining scope* involves reviewing the project charter, requirements documents, and organizational process assets to create a scope statement, adding more information as requirements are developed and change requests are approved. The main outputs of scope definition are the project scope statement and updates to project documents.
3. *Creating the WBS* involves subdividing the major project deliverables into smaller, more manageable components. The main outputs include a work breakdown structure, a WBS dictionary, a scope baseline, and updates to project documents.
4. *Verifying scope* involves formalizing acceptance of the project deliverables. Key project stakeholders, such as the customer and sponsor for the project, inspect and then formally accept the deliverables during this process. If the deliverables are not acceptable, the customer or sponsor usually requests changes. The main outputs of this process, therefore, are accepted deliverables and change requests.
5. *Controlling scope* involves controlling changes to project scope throughout the life of the project—a challenge on many information technology projects. Scope changes often influence the team's ability to meet project time and cost goals, so project managers must carefully weigh the costs and benefits of scope changes. The main outputs of this process are change requests, work performance measurements, and updates to organizational process assets, the project management plan, and project documents.



Figure 5-1 summarizes these processes and outputs and shows when they occur in a typical project.

## COLLECTING REQUIREMENTS

The first step in project scope management is often the most difficult: collecting requirements. A major consequence of not defining requirements well is rework, which can consume up to half of project costs, especially for software development projects. As illustrated in Figure 5-2, it costs much more to correct a software defect that is found in later development phases than to fix it in the requirements phase.

Part of the difficulty is that people often don't have a consistent definition of what requirements are, how to collect them, and how to document them.

### What Are Requirements?

The 1990 IEEE Standard Glossary of Software Engineering Terminology defines a requirement as follows:

> "1. A condition or capability needed by a user to solve a problem or achieve an objective.
> 2. A condition or capability that must be met or possessed by a system or system component to satisfy a contract, standard, specification, or other formally imposed document.
> 3. A documented representation of a condition or capability as in 1 or 2."[1]

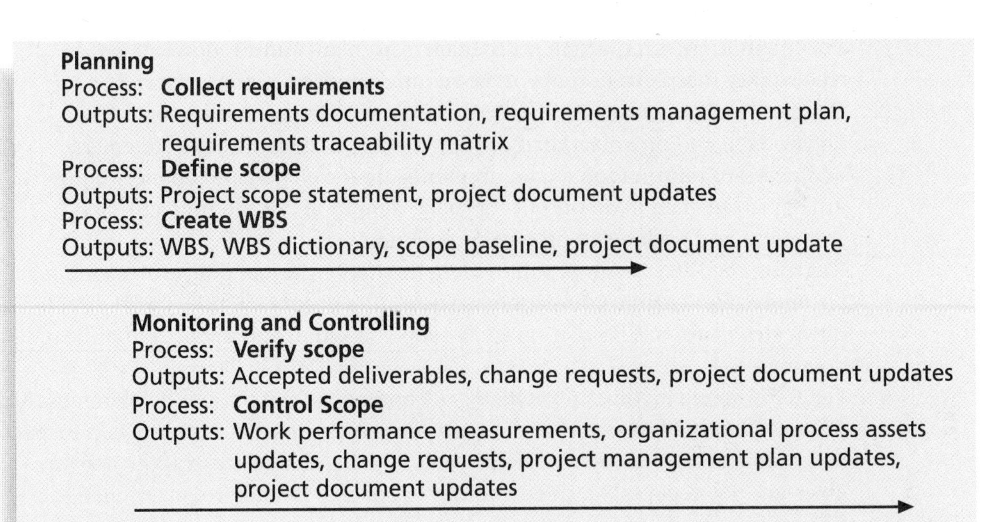


**FIGURE 5-1** Project scope management summary

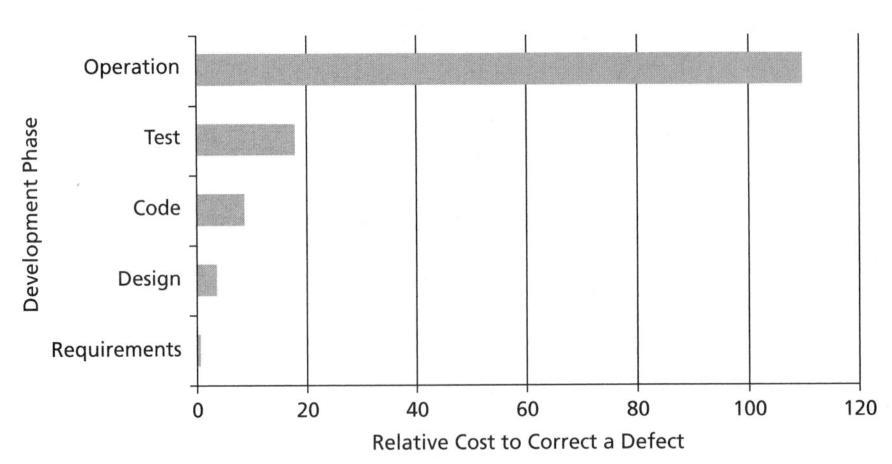


*Source:* Robert B. Grady, "An Economic Release Decision Model: Insights into Software Project Management." *Proceedings of the Applications of Software Measurement Conference* (Orange Park, FL: Software Quality Engineering, 1999), pp.227–239.

**FIGURE 5-2** Relative cost to correct a software requirement defect

The *PMBOK® Guide, Fourth Edition*, defines a requirement almost identical to item 2 above: It says that a **requirement** is "a condition or capability that must be met or possessed by a system, product, service, result, or component to satisfy a contract, standard, specification, or other formal document." It is important to document requirements in enough detail so that they can be measured during project execution. After all, meeting scope goals is often based on meeting documented requirements.



For example, the opening case describes a project for upgrading IT assets to meet corporate standards. It says that these standards specify the minimum equipment required for each desktop or laptop computer, such as the type of processor, amount of memory, and hard disk size. The documented requirements for this project might include, therefore, that all computers include an Intel processor, 4GB of memory, and a 160GB hard drive.

For some IT projects, it is helpful to divide requirements development into categories called *elicitation*, *analysis*, *specification*, and *validation*. These categories include all the activities involved in gathering, evaluating, and documenting requirements for a software or software-containing product. It is also important to use an iterative approach to defining requirements since requirements are often unclear early in a project. (See the Suggested Readings on the companion Web site for further information.)

## How Do You Collect Requirements?

There are several ways to collect requirements. Interviewing stakeholders one-on-one is often very effective, although it can be very expensive and time-consuming. Holding focus groups, facilitated workshops, and using group creativity and decision-making techniques to collect requirements are normally faster and less expensive than one-on-one interviews. Questionnaires and surveys can be very efficient ways to collect requirements as long as key stakeholders provide honest and thorough information. Observation can also be a good technique for collecting requirements, especially for projects that involve improving work processes and procedures. Prototyping is a commonly used technique for collecting requirements for software development projects. There are also several software tools available to assist in collecting and managing requirements, as described later in this chapter and in the following What Went Right? example.

### WHAT WENT RIGHT?

With over 4,000 customers and 1,500 employees worldwide, Genesys Telecommunications Laboratories has a reputation for pioneering telephony solutions by developing software to manage customer interactions over the phone, via the Web, and with e-mail. However, as the company grows, it has to make changes to meet the challenges of maintaining its competitive edge in new product development. For example, Genesys now uses Accept software, a product planning and innovation management application and winner of the Excellence in Product Management Award from 2006–2008. Before implementing Accept, Genesys' product planning process was time-consuming and difficult to replicate from release to release. Paul Lang, Vice President of Product Management and Strategy, says that

*continued*

Accept provides them a single system that contains all of their planning data across multiple products for a multi-year, multi-phase program. With Accept, Lang says they can instill a consistent, repeatable, and predictable process for new product definition and development. They can define what information comprises a requirement and enforce discipline around that process. "Among other things, each requirement must show who requested that enhancement and why, and these must be linked back to the corporate strategy theme, product theme, and the release theme," Lang says.[2]

The project's size, complexity, importance, and other factors will affect how much effort is spent on collecting requirements. For example, a team working on a project to upgrade the entire corporate accounting system for a multibillion dollar company with more than 50 geographic locations should spend a fair amount of time collecting requirements. A project to upgrade the hardware and software for a small accounting firm with only five employees, on the other hand, would need a much smaller effort. In any case, it is important for a project team to decide how they will collect and manage requirements. It is crucial to gather inputs from key stakeholders and align the scope, a key aspect of the entire project, with business strategy, as described in Chapter 4.

## How Do You Document Requirements?

Just as there are several ways to collect requirements, there are several ways to document them. Project teams should first review the project charter since it includes high-level requirements for the project and may refer to other documents that include requirements. They should also review the stakeholder register to ensure that all key stakeholders have a say in determining requirements. The format for documenting stakeholder requirements can range from a listing of all requirements on a single piece of paper to a room full of notebooks documenting requirements. People who have worked on complex projects, such as building a new airplane, know that the paper documenting requirements for a plane can weigh more than the plane itself! Requirements documents are often generated by software and include text, images, diagrams, videos, and other media. Requirements are also often broken down into different categories such as functional requirements, service requirements, performance requirements, quality requirements, training requirements, and so on.

In addition to preparing stakeholder requirements documentation as an output of the collecting requirements process, project teams often create a requirements management plan and a requirements traceability matrix. The **requirements management plan** describes how project requirements will be analyzed, documented, and managed. A **requirements traceability matrix (RTM)** is a table that lists requirements, various attributes of each requirement, and the status of the requirements to ensure that all requirements are addressed. Table 5-1 provides an example of an RTM entry for the IT Upgrade Project described in the opening case. There are many variations of what can be included in an RTM. For example, software requirements are often documented in an RTM that cross-references each requirement with related ones and lists specific tests to verify that they are met. Remember that the main purpose of an RTM is to maintain the linkage from the source of each requirement through its decomposition to implementation and verification.

TABLE 5-1 Sample requirements traceability matrix

| Requirement No. | Name | Category | Source | Status |
|---|---|---|---|---|
| R32 | Laptop memory | Hardware | Project charter and corporate laptop specifications | Complete. Laptops ordered meet requirement by having 4GB of memory. |



## DEFINING SCOPE

The next step in project scope management is to define in detail the scope or work required for the project. Good scope definition is very important to project success because it helps improve the accuracy of time, cost, and resource estimates, it defines a baseline for performance measurement and project control, and it aides in communicating clear work responsibilities. The main tools and techniques used in defining scope include expert judgment, product analysis, alternatives identification, and facilitated workshops. The main outputs of scope definition are the project scope statement and project document updates.

Key inputs for preparing the project scope statement include the project charter, requirements documentation, and organizational process assets such as policies and procedures related to scope statements as well as project files and lessons learned from previous, similar projects. Table 5-2 shows the project charter for the IT Upgrade Project described in the opening case. Notice how information from the project charter provides a basis for further defining the project scope. The charter describes the high-level scope, time, and cost goals for the project objectives and success criteria, a general approach to accomplishing the project's goals, and the main roles and responsibilities of important project stakeholders.

TABLE 5-2 Sample project charter

**Project Title:** Information Technology (IT) Upgrade Project

**Project Start Date:** March 4 **Projected Finish Date:** December 4

**Key Schedule Milestones:**

- Inventory update completed April 15
- Hardware and software acquired August 1
- Installation completed October 1
- Testing completed November 15

**Budget Information:** Budgeted $1,000,000 for hardware and software costs and $500,000 for labor costs.

**Project Manager:** Kim Nguyen, (310) 555–2784, knguyen@course.com

*(continued)*

**Project Objectives:** Upgrade hardware and software for all employees (approximately 2,000) within nine months based on new corporate standards. See attached sheet describing the new standards. Upgrades may affect servers, as well as associated network hardware and software.

**Main Project Success Criteria:** The hardware, software, and network upgrades must meet all written specifications, be thoroughly tested, and be completed in less than ten months. Employee work disruptions will be minimal.



**Approach:**

- Update the information technology inventory database to determine upgrade needs
- Develop detailed cost estimate for project and report to CIO
- Issue a request for quote to obtain hardware and software
- Use internal staff as much as possible for planning, analysis, and installation

ROLES AND RESPONSBILITES

| Name | Role | Responsibility |
|---|---|---|
| Walter Schmidt | CEO | Project sponsor, monitor project |
| Mike Zwack | CIO | Monitor project, provide staff |
| Kim Nguyen | Project Manager | Plan and execute project |
| Jeff Johnson | Director of Information Technology Operations | Mentor Kim |
| Nancy Reynolds | VP, Human Resources | Provide staff, issue memo to all employees about project |
| Steve McCann | Director of Purchasing | Assist in purchasing hardware and software |

Sign–off: (Signatures of all the above stakeholders)

**Comments:** (Handwritten or typed comments from above stakeholders, if applicable)

"This project must be done within ten months at the absolute latest." Mike Zwack, CIO

"We are assuming that adequate staff will be available and committed to supporting this project. Some work must be done after hours to avoid work disruptions, and overtime will be provided." Jeff Johnson and Kim Nguyen, Information Technology department.

Although contents vary, project scope statements should include, at a minimum, a product scope description, product user acceptance criteria, and detailed information on all project deliverables. It is also helpful to document other scope-related information, such as the project boundaries, constraints, and assumptions. The project scope statement should also reference supporting documents, such as product specifications that will affect what products are produced or purchased, or corporate policies, which might affect how products or services are produced. Many information technology projects also require detailed functional and design specifications for developing software, which also should be referenced in the detailed scope statement.



As time progresses, the scope of a project should become more clear and specific. For example, the project charter for the IT Upgrade Project shown in Table 5-2 includes a short statement about the servers and other computers and software that the IT Upgrade Project may affect. Table 5-3 provides an example of how the scope becomes progressively elaborated or more detailed in scope statements labeled Version 1 and Version 2.

**TABLE 5-3** Further defining project scope

| **Project Charter:** |
| --- |
| Upgrades may affect servers . . . (listed under Project Objectives) |
| **Project Scope Statement, Version 1:** |
| Servers: If additional servers are required to support this project, they must be compatible with existing servers. If it is more economical to enhance existing servers, a detailed description of enhancements must be submitted to the CIO for approval. See current server specifications provided in Attachment 6. The CEO must approve a detailed plan describing the servers and their location at least two weeks before installation. |
| **Project Scope Statement, Version 2:** |
| Servers: This project will require purchasing ten new servers to support Web, network, database, application, and printing functions. Virtualization will be used to maximize efficiency. Detailed descriptions of the servers are provided in a product brochure in Appendix 8 along with a plan describing where they will be located. |

Notice in Table 5-3 that the project scope statements often refer to related documents, which can be product specifications, product brochures, or other plans. As more information becomes available and decisions are made related to project scope, such as specific products that will be purchased or changes that have been approved, the project team should update the project scope statement. They might name different iterations of the scope statement Version 1, Version 2, and so on. These updates may also require changes to be made to other project documents. For example, if the company must purchase servers for the project from a supplier it has never worked with before, the procurement management plan should include information on working with that new supplier.

An up-to-date project scope statement is an important document for developing and confirming a common understanding of the project scope. It describes in detail the work to be accomplished on the project and is an important tool for ensuring customer satisfaction and preventing scope creep, as described later in this chapter.

Recall from Chapter 1 the importance of addressing the triple constraint of project management—meeting scope, time, and cost goals for a project. Time and cost goals are normally straightforward. For example, the time goal for the IT Upgrade Project is nine months, and the cost goal is $1.5 million. It is much more difficult to describe, agree upon, and meet the scope goal of many projects.

## MEDIA SNAPSHOT

Many people enjoy watching television shows like *Trading Spaces*, where participants have two days and $1,000 to update a room in their neighbor's house. Since the time and cost are set, it's the scope that has the most flexibility. Examples of some of the work completed include new flooring, light fixtures, paint, new shelves, or artwork to brighten up a dull room.

Designers on these shows often have to change initial scope goals due to budget or time constraints. For example, designers often go back to local stores to exchange items, such as lights, artwork, or fabric, for less expensive items to meet budget constraints. Or they might describe a new piece of furniture they'd like the carpenter to build, but the carpenter changes the design or materials to meet time constraints. Occasionally designers can buy more expensive items or have more elaborate furniture built because they underestimated costs and schedules.

Another important issue related to project scope management is meeting customer expectations. Who wouldn't be happy with a professionally designed room at no cost to them? Although most homeowners are very happy with work done on the show, some are obviously disappointed. Unlike most projects where the project team works closely with the customer, homeowners have little say in what gets done and cannot inspect the work along the way. They walk into their newly decorated room with their eyes closed. Modernizing a room can mean something totally different to a homeowner and the interior designer. For example, one woman was obviously shocked when she saw her bright orange kitchen with black appliances. Another couple couldn't believe there was moss on their bedroom walls. What happens when the homeowners don't like the work that's been done? Part of agreeing to be on the show includes signing a release statement acknowledging that you will accept whatever work has been done. Too bad you can't get sponsors for most projects to sign a similar release statement. It would make project scope management much easier!

# CREATING THE WORK BREAKDOWN STRUCTURE

After collecting requirements and defining scope, the next step in project scope management is to create a work breakdown structure. A **work breakdown structure (WBS)** is a deliverable-oriented grouping of the work involved in a project that defines the total scope of the project. Because most projects involve many people and many different deliverables, it is important to organize and divide the work into logical parts based on how the work will be performed. The WBS is a foundation document in project management because it provides the basis for planning and managing project schedules, costs, resources, and changes. Since the WBS defines the total scope of the project, some project management experts believe that work should not be done on a project if it is not included in the WBS. Therefore, it is crucial to develop a good WBS.

The project scope statement, stakeholder requirements documentation, and organizational process assets are the primary inputs for creating a WBS. The main tool or technique is **decomposition**, that is, subdividing project deliverables into smaller pieces. The outputs of the process of creating the WBS are the WBS itself, the WBS dictionary, a scope baseline, and project document updates.



What does a WBS look like? A WBS is often depicted as a task-oriented family tree of activities, similar to an organizational chart. A project team often organizes the WBS around project products, project phases, or using the project management process groups. Many people like to create a WBS in chart form first to help them visualize the whole project and all of its main parts. For example, Figure 5-3 shows a WBS for an intranet project. Notice that product areas provide the basis for its organization. In this case, there are main boxes or groupings on the WBS for developing the Web site design, the home page for the intranet, the marketing department's pages, and the sales department's pages.

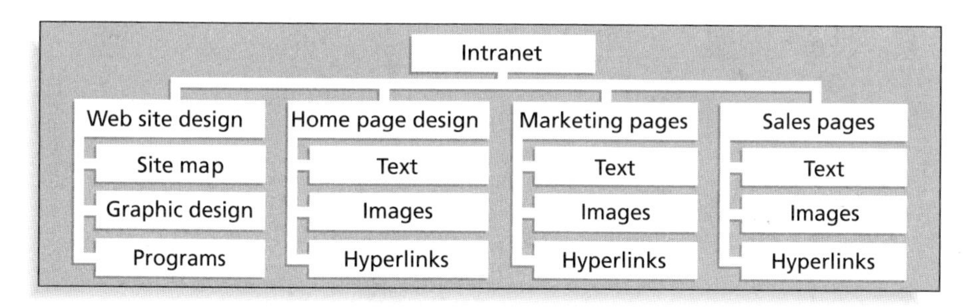


**FIGURE 5-3** Sample intranet WBS organized by product

In contrast, a WBS for the same intranet project can be organized around project phases, as shown in Figure 5-4.[3] Notice that project phases of concept, Web site design, Web site development, roll out, and support provide the basis for its organization.

Also note the levels in Figure 5-4. The name of the entire project is the top box, called Level 1, and the main groupings for the work are listed in the second tier of boxes, called Level 2. This level numbering is based on PMI's *Practice Standard for Work Breakdown Structures, Second Edition* (2006). Each of those boxes can be broken down into subsequent tiers of boxes to show the hierarchy of the work. PMI uses the term "task" to describe each level of work in the WBS. For example, in Figure 5-4, the following items can be referred to as tasks: the Level 2 item called Concept, the Level 3 item called Define requirements, and the Level 4 item called Define user requirements. Tasks that are decomposed into smaller tasks are called summary tasks. Figure 5-4 shows a sample WBS in both chart and tabular form. Notice that both of these formats show the same information. Many documents, such as contracts, use the tabular format. Project management software also uses this format. The WBS becomes the contents of the Task Name column in Microsoft Project, and the hierarchy or level of tasks is shown by indenting and numbering tasks within the software. The numbering shown in the tabular form on the left in Figure 5-4 coincides with numbering in Microsoft

Level 1 - Entire Project
Intranet project
Level 2
Concept
Web site design
Web site development
Roll out
Support
Chart form
Level 3
Evaluate current systems
Define requirements
Define specific functionality
Define risks & risk management approach
Develop project plan
Brief Web development team
Level 4
Define user requirements
Define content requirements
Define system requirements
Define server owner requirements

**Tabular form with Microsoft Project numbering**

1.0 Concept
- 1.1 Evaluate current systems
- 1.2 Define requirements
  - 1.2.1 Define user requirements
  - 1.2.2 Define content requirements
  - 1.2.3 Define system requirements
  - 1.2.4 Define server owner requirements
- 1.3 Define specific functionality
- 1.4 Define risks and risk management approach
- 1.5 Develop project plan
- 1.6 Brief Web development team

2.0 Web site design
3.0 Web site development
4.0 Roll out
5.0 Support

**Tabular form with PMI numbering**

1.1 Concept
- 1.1.1 Evaluate current systems
- 1.1.2 Define requirements
  - 1.1.2.1 Define user requirements
  - 1.1.2.2 Define content requirements
  - 1.1.2.3 Define system requirements
  - 1.1.2.4 Define server owner requirements
- 1.1.3 Define specific functionality
- 1.1.4 Define risks and risk management approach
- 1.1.5 Develop project plan
- 1.1.6 Brief Web development team

1.2 Web site design
1.3 Web site development
1.4 Roll out
1.5 Support

**FIGURE 5-4** Sample intranet WBS organized by phase in chart and tabular form

Project and other sources. The numbering shown in the tabular form on the right in Figure 5-4 is based on PMI's *Practice Standard for Work Breakdown Structures, Second Edition*. Be sure to check with your organization to see what numbering scheme they prefer to use for work breakdown structures.

In Figure 5-4, the lowest level of the WBS is Level 4. A **work package** is a task at the lowest level of the WBS. In Figure 5-4, tasks 1.2.1, 1.2.2, 1.2.3, and 1.2.4 (based on the numbering on the left) are work packages. The other tasks would probably be broken down further. However, some tasks can remain at Level 2 or 3 in the WBS. Some might be broken down to Level 5 or 6, depending on the complexity of the work. A work package also represents the level of work that the project manager monitors and controls. You can think of work packages in terms of accountability and reporting. If a project has a relatively short time frame and requires weekly progress reports, a work package might represent work completed in one week or less. If a project has a very long time frame and requires quarterly progress reports, a work package might represent work completed in one month or more. A work package might also be the procurement of a specific product or products, such as an item or items purchased from an outside source.

Another way to think of work packages relates to entering data into project management software. *You can only enter duration estimates for work packages*. The rest of the WBS items are just groupings or summary tasks for the work packages. The software automatically calculates duration estimates for various WBS levels based on data entered for each work package and the WBS hierarchy. See Appendix A for detailed information on using Project 2007.



Figure 5-5 shows the phase-oriented intranet WBS, using the Microsoft Project numbering scheme from Figure 5-4, in the form of a Gantt chart created in Project 2007. You can see from this figure that the WBS is the basis for project schedules. Notice that the WBS is in the left part of the figure under the Task Name column. The resulting schedule is in the right part of the figure. You will learn more about Gantt charts in Chapter 6, Project Time Management.

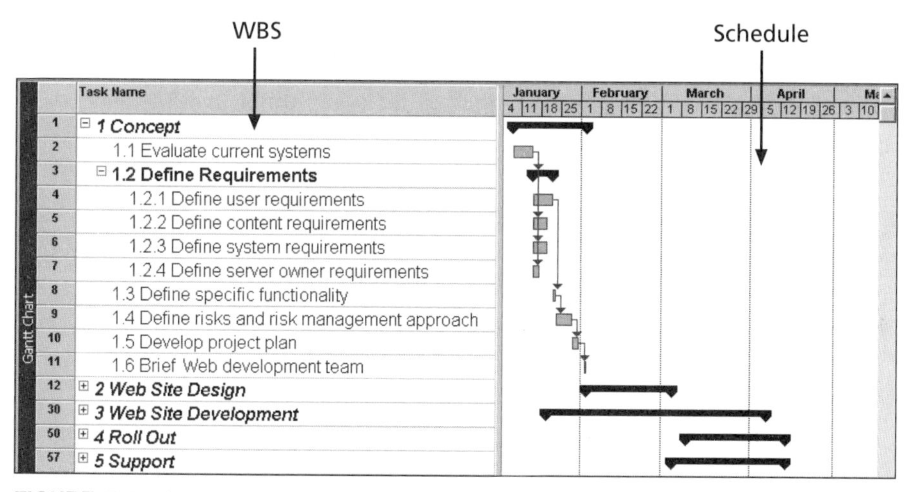


**FIGURE 5-5** Intranet Gantt chart in Microsoft Project

The sample WBSs shown here seem somewhat easy to construct and understand. *Nevertheless, it is very difficult to create a good WBS*. To create a good WBS, you must understand the project and its scope and incorporate the needs and knowledge of the stakeholders. The project manager and the project team must decide as a group how to organize the work and how many levels to include in the WBS. Many project managers have found that it is better to focus on getting the top levels done well before getting too bogged down in more detailed levels.

Many people confuse tasks on a WBS with specifications. Tasks on a WBS represent work that needs to be done to complete the project. For example, if you are creating a WBS to redesign a kitchen, you might have Level 2 categories called design, purchasing, flooring, walls, cabinets, and appliances. Under flooring, you might have tasks to remove the old flooring, install the new flooring, and install the trim. You would not have tasks like "12 ft. by 14 ft. of light oak" or "flooring must be durable."

Another concern when creating a WBS is how to organize it so that it provides the basis for the project schedule. You should focus on what work needs to be done and how it will be done, not when it will be done. In other words, the tasks do not have to be developed as a sequential list of steps. If you do want some time-based flow for the work, you can create a WBS using the project management process groups of initiating, planning, executing, monitoring and controlling, and closing as Level 2 in the WBS. By doing this, not only does the project team follow good project management practice, but the WBS tasks can also be mapped more easily against time. For example, Figure 5-6 shows a WBS and Gantt chart for the intranet project, organized by the five project management process groups. Tasks under initiating include selecting a project manager, forming the project team, and developing the project charter. Tasks under planning include developing a scope statement, creating a WBS, and developing and refining other plans, which would be broken down in more detail for a real project. The tasks of concept, Web site design, Web site development, and roll out, which were WBS Level 2 items in Figure 5-4, now become WBS Level 3 items under executing. The executing tasks vary the most from project to project, but many of the tasks under the other project management process groups would be similar for all projects. If you do not use the project management process groups in the WBS, you can have a Level 2 category called project management to make sure that tasks related to managing the project are accounted for. Remember that all work should be included in the WBS, including project management.

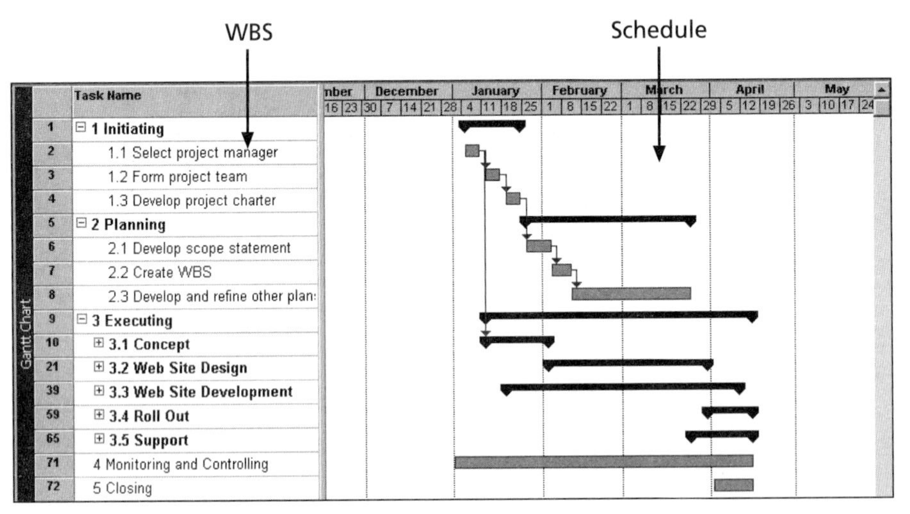


**FIGURE 5-6** Intranet project Gantt chart organized by project management process groups

JWD Consulting used the project management process groups for the Level 2 items in its WBS for the Project Management Intranet Site Project in Chapter 3. The project team focused on the product deliverables they had to produce for the project in breaking down the executing task. Table 5-4 shows the categories they used for that part of the WBS. Some project teams like to list every deliverable they need to produce and then use those as the basis for creating all or part of their WBS. Recall that the scope statement should list and describe all of the deliverables required for the project. It is very important to ensure consistency between the project charter, scope statement, WBS, and Gantt chart to define the scope of the project accurately.

TABLE 5-4 Executing tasks for JWD Consulting's WBS

3.0 Executing
- 3.1 Survey
- 3.2 User inputs
- 3.3 Intranet site content
  - 3.3.1 Templates and tools
  - 3.3.2 Articles
  - 3.3.3 Links
  - 3.3.4 Ask the Expert
  - 3.3.5 User requests
- 3.4 Intranet site design
- 3.5 Intranet site construction
- 3.6 Site testing
- 3.7 Site promotion
- 3.8 Site roll out
- 3.9 Project benefits measurement



It is also very important to involve the entire project team and the customer in creating and reviewing the WBS. *People who will do the work should help to plan the work* by creating the WBS. Having group meetings to develop a WBS helps everyone understand *what* work must be done for the entire project and *how* it should be done, given the people involved. It also helps to identify where coordination between different work packages will be required.

## Approaches to Developing Work Breakdown Structures

There are several approaches you can use to develop a work breakdown structure. These approaches include:

- Using guidelines
- The analogy approach
- The top-down approach
- The bottom-up approach
- The mind-mapping approach

### Using Guidelines

If guidelines for developing a WBS exist, it is very important to follow them. Some organizations—the U.S. Department of Defense (DOD) for example—prescribe the form and content for WBSs for particular projects. Many DOD projects require contractors to prepare their proposals based on the DOD-provided WBS. These proposals must include cost estimates for each task in the WBS at a detailed and summary level. The cost for the entire project must be calculated by summing the costs of all of the lower-level WBS tasks. When DOD personnel evaluate cost proposals, they must compare the contractors' costs with the DOD's estimates. A large variation in costs for a certain WBS task often indicates confusion as to what work must be done.

Consider a large automation project for the U.S. Air Force. In the mid-1980s, the Air Force developed a request for proposals for the Local On-Line Network System (LONS) to automate 15 Air Force Systems Command bases. This $250 million project involved providing the hardware and developing software for sharing documents such as contracts, specifications, requests for proposals, and so on. The Air Force proposal guidelines included a WBS that contractors were required to follow in preparing their cost proposals. Level 2 WBS items included hardware, software development, training, project management, and the like. The hardware item was composed of several Level 3 items, such as servers, workstations, printers, network hardware, and so on. Air Force personnel reviewed the contractors' cost proposals against their internal cost estimate, which was also based on this WBS. Having a prescribed WBS helped contractors to prepare their cost proposals and the Air Force to evaluate them.



Many organizations provide guidelines and templates for developing WBSs, as well as examples of WBSs from past projects. Microsoft Project 2007 comes with several templates, and more are available on Microsoft's Web site and other sites. At the request of many of its members, the Project Management Institute developed a WBS Practice Standard to provide guidance for developing and applying the WBS to project management (see the Suggested Readings for this chapter on the companion Web site). This document includes sample WBSs for a wide variety of projects in various industries, including projects for Web design, telecom, service industry outsourcing, and software implementation.

Project managers and their teams should review appropriate information to develop their unique project WBSs more efficiently. For example, Kim Nguyen and key team members from the opening case should review their company's WBS guidelines, templates, and other related information before and during the team meetings to create their WBS.

### The Analogy Approach

Another approach for constructing a WBS is the analogy approach. In the **analogy approach**, you use a similar project's WBS as a starting point. For example, Kim Nguyen from the opening case might learn that one of her organization's suppliers did a similar information technology upgrade project last year. She could ask them to share their WBS for that project to provide a starting point for her own project.

McDonnell Aircraft Company, now part of Boeing, provides an example of using an analogy approach when creating WBSs. McDonnell Aircraft Company designed and manufactured several different fighter aircraft. When creating a WBS for a new aircraft design, it started by using 74 predefined subsystems for building a fighter aircraft based on past experience. There was a Level 2 WBS item for the airframe that was composed of Level 3 items such as a forward fuselage, center fuselage, aft fuselage, and wings. This generic product-oriented WBS provided a starting point for defining the scope of new aircraft projects and developing cost estimates for new aircraft designs.

Some organizations keep a repository of WBSs and other project documentation on file to assist people working on projects. Project 2007 and many other software tools include sample files to assist users in creating a WBS and Gantt chart. Viewing examples of other similar projects' WBSs allows you to understand different ways to create a WBS.

### The Top-down and Bottom-up Approaches

Two other approaches for creating WBSs are the top-down and bottom-up approaches. Most project managers consider the top-down approach of WBS construction to be conventional.

To use the **top-down approach,** start with the largest items of the project and break them into their subordinate items. This process involves refining the work into greater and greater levels of detail. For example, Figure 5-4 shows how work was broken down to Level 4 for part of the intranet project. After finishing the process, all resources should be assigned at the work package level. The top-down approach is best suited to project managers who have vast technical insight and a big-picture perspective.



In the **bottom-up approach,** team members first identify as many specific tasks related to the project as possible. They then aggregate the specific tasks and organize them into summary activities, or higher levels in the WBS. For example, a group of people might be responsible for creating a WBS to create an e-commerce application. Instead of looking for guidelines on how to create a WBS or viewing similar projects' WBSs, they could begin by listing detailed tasks they think they would need to do in order to create the application. After listing these detailed tasks, they would group the tasks into categories. Then they would group these categories into higher-level categories. Some people have found that writing all possible tasks down on notes and then placing them on a wall helps them see all the work required for the project and develop logical groupings for performing the work. For example, a business analyst on the project team might know that they had to define user requirements and content requirements for the e-commerce application. These tasks might be part of the requirements documents they would have to create as one of the project deliverables. A hardware specialist might know they had to define system requirements and server requirements, which would also be part of a requirements document. As a group, they might decide to put all four of these tasks under a higher-level item called "define requirements" that would result in the delivery of a requirements document. Later, they might realize that defining requirements should fall under a broader category of concept design for the e-commerce application, along with other groups of tasks related to the concept design. The bottom-up approach can be very time-consuming, but it can also be a very effective way to create a WBS. Project managers often use the bottom-up approach for projects that represent entirely new systems or approaches to doing a job, or to help create buy-in and synergy with a project team.

### Mind Mapping

Some project managers like to use mind mapping to help develop WBSs. As described in Chapter 4 when showing an example of performing a SWOT analysis, mind mapping is a technique that uses branches radiating out from a core idea to structure thoughts and ideas. Instead of writing tasks down in a list or immediately trying to create a structure for tasks, mind mapping allows people to write and even draw pictures of ideas in a nonlinear format. This more visual, less-structured approach to defining and then grouping tasks can unlock creativity among individuals and increase participation and morale among teams.[4]

Figure 5-7 shows a diagram that uses mind mapping to create a WBS for the IT Upgrade Project from Chapter 3. The figure was created using MindManager software by Mindjet (*www.mindjet.com*). The circle in the center represents the entire project. Each of the four main branches radiating out from the center represents the main tasks or Level 2 items for the WBS. Different people at the meeting creating this mind map might have different roles in the project, which could help in deciding the tasks and WBS structure. For example, Kim would want to focus on all of the project management tasks, and she might also know that they will be tracked in a separate budget category. People who are familiar with acquiring or installing hardware and software might focus on that work, and so on. Branching off from the main task called "Update inventory" are two subtasks, "Perform physical inventory" and "Update database." Branching off from the "Perform physical inventory" subtask are three further subdivisions, labeled Building A, Building B, and Building C, and so on. The team would continue to add branches and items until they have exhausted ideas on what work needs to be performed.



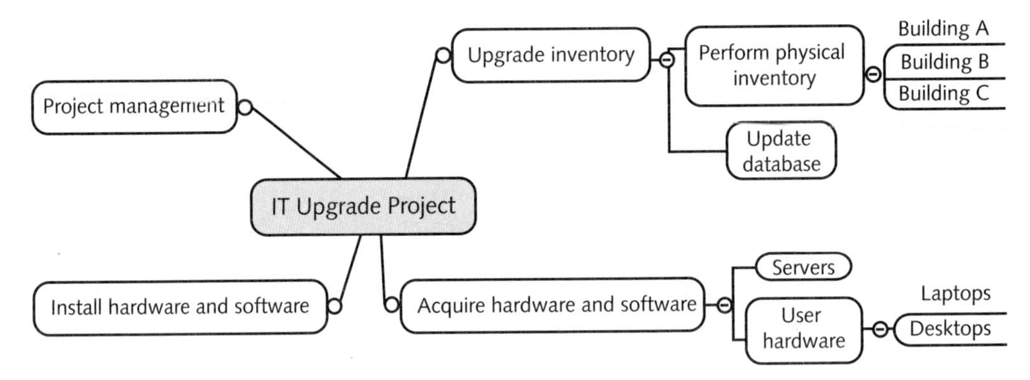


**FIGURE 5-7** Sample mind-mapping technique for creating a WBS

After discovering WBS items and their structure using the mind-mapping technique, you could then translate the information into chart or tabular form, as described earlier. A feature of MindManager software is that you can export your map into Microsoft Project. The WBS is entered in the Task List column, with the structure automatically created based on the mind map. Figure 5-8 shows the resulting Project 2007 file for the IT Upgrade Project.

Mind mapping can be used for developing WBSs using the top-down or bottom-up approach. For example, you could conduct mind mapping for an entire project by listing the whole project in the center of a document, adding the main categories on branches radiating out from the center, and then adding branches for appropriate subcategories. You could also develop a separate mind-mapping diagram for each deliverable and then merge them to create one large diagram for the entire project. You can also add items anywhere on a mind-mapping document without following a strict top-down or bottom-up approach. After the mind-mapping documents are complete, you can convert them into a chart or tabular WBS form.

## The WBS Dictionary and Scope Baseline

As you can see from these sample WBSs, many of the items listed on them are rather vague. What exactly does "Update database" mean, for example? The person responsible for this task might think that it does not need to be broken down any further, which could be fine. However, the task should be described in more detail so everyone has the same understanding of what it involves. What if someone else has to perform the task? What would you tell him/her to do? What will it cost to complete the task? Information that is more detailed is needed to answer these and other questions.

A **WBS dictionary** is a document that describes detailed information about each WBS item. The format of the WBS dictionary can vary based on project needs. It might be appropriate to have just a short paragraph describing each work package. For a more complex project, an entire page or more might be needed for the work package descriptions. Some

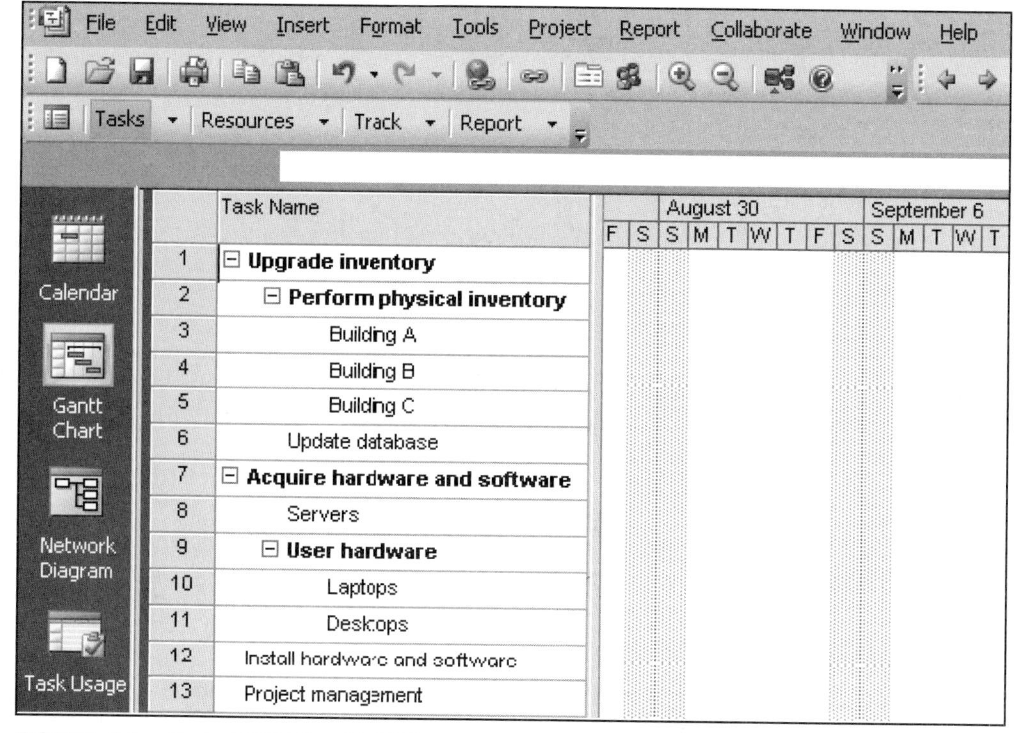


**FIGURE 5-8** Project 2007 file with WBS generated from a mind map

projects might require that each WBS item describe the responsible organization, resource requirements, estimated costs, and other information.

Kim should work with her team and sponsor to determine the level of detail needed in the WBS dictionary. They should also decide where this information will be entered and how it will be updated. Project teams often review WBS dictionary entries from similar tasks to get a better idea of how to create these entries. For the IT Upgrade Project, Kim and her team decided to enter all of the WBS dictionary information into their enterprise project management system, following departmental guidelines. Table 5-5 is an example of one entry.

**TABLE 5-5** Sample WBS dictionary entry

| WBS Dictionary Entry March 20 |
|---|
| **Project Title:** Information Technology (IT) Upgrade Project |
| **WBS Item Number:** 2.2 |
| **WBS Item Name:** Update Database |
| **Description:** The IT department maintains an online database of hardware and software on the corporate Intranet. However, we need to make sure that we know exactly what |

*(continued)*

hardware and software employees are currently using and if they have any unique needs before we decide what to order for the upgrade. This task will involve reviewing information from the current database, producing reports that list each department's employees and location, and updating the data after performing the physical inventory and receiving inputs from department managers. Our project sponsor will send out a notice to all department managers to communicate the importance of this project and this particular task. In addition to general hardware and software upgrades, the project sponsors will ask the department managers to provide information for any unique requirements they might have that could affect the upgrades. This task also includes updating the inventory data for network hardware and software. After updating the inventory database, we will send an e-mail to each department manager to verify the information and make changes online, as needed. Department managers will be responsible for ensuring that their people are available and cooperative during the physical inventory. Completing this task is dependent on WBS Item Number 2.1, Perform Physical Inventory and must precede WBS Item Number 3.0, Acquire Hardware and Software.



The approved project scope statement and its associated WBS and WBS dictionary form the **scope baseline**. Performance in meeting project scope goals is based on this scope baseline.

### Advice for Creating a WBS and WBS Dictionary

As stated previously, creating a good WBS is no easy task and usually requires several iterations. Often, it is best to use a combination of approaches to create a project's WBS. There are some basic principles, however, that apply to creating any good WBS and its WBS dictionary.

- A unit of work should appear at only one place in the WBS.
- The work content of a WBS item is the sum of the WBS items below it.
- A WBS item is the responsibility of only one individual, even though many people may be working on it.
- The WBS must be consistent with the way in which work is actually going to be performed; it should serve the project team first, and other purposes only if practical.
- Project team members should be involved in developing the WBS to ensure consistency and buy-in.
- Each WBS item must be documented in a WBS dictionary to ensure accurate understanding of the scope of work included and not included in that item.
- The WBS must be a flexible tool to accommodate inevitable changes while properly maintaining control of the work content in the project according to the scope statement.[5]

## VERIFYING SCOPE

It is difficult to create a good project scope statement and WBS for a project. It is even more difficult, especially on information technology projects, to verify the project scope and minimize scope changes. Some project teams know from the start that the scope is very unclear and that they must work closely with the project customer to design and produce various deliverables. In this case, the project team must develop a process for scope verification that meets unique project needs. Careful procedures must be developed to ensure the customer is getting what they want and the project team has enough time and money to produce the desired products and services.

Even when the project scope is fairly well defined, many information technology projects suffer from **scope creep**—the tendency for project scope to keep getting bigger and bigger. There are many horror stories about information technology projects failing due to scope problems such as scope creep, with a few classic examples in the following What Went Wrong? feature. For this reason, it is very important to verify the project scope with users throughout the life of the project and develop a process for controlling scope changes.

## WHAT WENT WRONG?

A project scope that is too broad and grandiose can cause severe problems. Scope creep and an overemphasis on technology for technology's sake resulted in the bankruptcy of a large pharmaceutical firm, Texas-based FoxMeyer Drug. In 1994, the CIO was pushing for a $65 million system to manage the company's critical operations. He did not believe in keeping things simple, however. The company spent nearly $10 million on state-of-the-art hardware and software and contracted the management of the project to a prestigious (and expensive) consulting firm. The project included building an $18 million robotic warehouse, which looked like something out of a science fiction movie, according to insiders. The scope of the project kept getting bigger and more impractical. The elaborate warehouse was not ready on time, and the new system generated erroneous orders that cost FoxMeyer Drug more than $15 million in unrecovered excess shipments. In July 1996, the company took a $34 million charge for its fourth fiscal quarter, and by August of that year, FoxMeyer Drug filed for bankruptcy.[6] Another example of scope creep comes from McDonald's Restaurants. In 2001, the fast-food chain initiated a project to create an intranet that would connect its headquarters with all of its restaurants to provide detailed operational information in real time. For example, headquarters would know if sales were slowing or if the grill temperature was correct in every single store—all 30,000 of them in more than 120 countries. McDonald's would not divulge detailed information, but they admitted that the project was too large in scale and scope. After spending $170 million on consultants and initial implementation planning, McDonald's realized that the project was too much to handle and terminated it.[7]

Another major scope problem on information technology projects is a lack of user involvement. For example, in the late 1980s at Northrop Grumman, which specializes in defense electronics, information technology, advanced aircraft, shipbuilding, and space technology, an information technology project team became convinced that it could, and should, automate the review and approval process of government proposals. The team implemented a powerful workflow system to manage the whole process. Unfortunately, the end users for the system were aerospace engineers who preferred to work in a more casual, ad hoc fashion. They dubbed the system "Naziware" and refused to use it. This example illustrates an information technology project that wasted millions of dollars developing a system that was not in touch with the way end users did their work.[8]

*continued*

Failing to follow good project management processes and use off-the-shelf software also results in scope problems. 21st Century Insurance Group in Woodland Hills, California, paid Computer Sciences Corporation $100 million on a project to develop a system for managing business applications, including managing insurance policies, billing, claims, and customer service. After five years, the system was still in development and used to support less than 2 percent of the company's business. Joshua Greenbaum, an analyst at Enterprise Applications Consulting, called the project a "huge disaster" and questioned the insurance company's ability "to manage a process that is pretty well known these days . . . . I'm surprised that there wasn't some way to build what they needed using off-the-shelf components and lower their risk."[9]



**Scope verification** involves formal acceptance of the completed project scope by the stakeholders. This acceptance is often achieved by a customer inspection and then sign-off on key deliverables. To receive formal acceptance of the project scope, the project team must develop clear documentation of the project's products and procedures to evaluate if they were completed correctly and satisfactorily. Recall from Chapter 4 that configuration management specialists identify and document the functional and physical characteristics of the project's products, record and report the changes, and audit the products to verify conformance to requirements. To minimize scope changes, it is crucial to do a good job of configuration management and verifying project scope.

The project management plan, requirements documentation, the requirements traceability matrix, and validated deliverables are the main inputs for scope verification. The main tool for performing scope verification is inspection. The customer, sponsor, or user inspects the work after it is delivered. The main outputs of scope verification are accepted deliverables, change requests, and project document updates. For example, suppose Kim's team members deliver upgraded computers to users as part of the IT Upgrade Project. Several users might complain because the computers did not include special keyboards they need for medical reasons. Appropriate people would review this change request and take appropriate corrective action, such as getting sponsor approval for purchasing the special keyboards.

## CONTROLLING SCOPE

As discussed in the section of Chapter 4 on integrated change control, change is inevitable on projects, especially changes to the scope of information technology projects. Scope control involves controlling changes to the project scope. Users often are not exactly sure how they want screens to look or what functionality they will really need to improve business performance. Developers are not exactly sure how to interpret user requirements, and they also have to deal with constantly changing technologies.

The goal of scope control is to influence the factors that cause scope changes, assure changes are processed according to procedures developed as part of integrated change control, and manage changes when they occur. You cannot do a good job of controlling scope if you do not first do a good job of collecting requirements, defining scope, and verifying scope. How can you prevent scope creep when you have not agreed on the work to be performed and your sponsor hasn't verified that the proposed work was acceptable? You also need to develop a process for soliciting and monitoring changes to project scope.

Stakeholders should be encouraged to suggest changes that will benefit the overall project and discouraged from suggesting unnecessary changes.

The project management plan, work performance data, requirements documentation, requirements traceability matrix, and organizational process assets are the main inputs to scope control. An important tool for performing scope control is variance analysis. **Variance** is the difference between planned and actual performance. For example, if a supplier was supposed to deliver five special keyboards and you received only four, the variance would be one keyboard. The outputs of scope control include work performance measurements, organizational process assets updates, change requests, project management plan updates, and project document updates.



Table 1-2 in Chapter 1 lists the top ten factors that help information technology projects succeed. Four of these ten factors are related to scope verification and control: user involvement, clear business objectives, minimized or clearly defined scope, and firm basic requirements. To avoid project failures, therefore, it is crucial for information technology project managers and their teams to work on improving user input and reducing incomplete and changing requirements and specifications.

## BEST PRACTICE

As seen from the examples of What Went Wrong?, companies should follow these best practices to avoid major scope problems:

1. Keep the scope realistic. Don't make projects so large that they can't be completed. Break large projects down into a series of smaller ones.
2. Involve users in project scope management. Assign key users to the project team and give them ownership of requirements definition and scope verification.
3. Use off-the-shelf hardware and software whenever possible. Many IT people enjoy using the latest and greatest technology, but business needs, not technology trends, must take priority.
4. Follow good project management processes. As described in this chapter and others, there are well-defined processes for managing project scope and others aspects of projects.

Managing the scope of software development is often very difficult, but frameworks such as IBM's Rational Unified Process® (RUP) can help. For example, RUP describes a set of principles that characterize best practices in the creation, deployment, and evolution of software-intensive systems as follows:

- Adapt the process
- Balance competing stakeholder priorities
- Collaborate across teams
- Demonstrate value iteratively
- Elevate the level of abstraction
- Focus continuously on quality[10]

The following sections provide more suggestions for improving scope management on information technology projects.

## Suggestions for Improving User Input



Lack of user input leads to problems with managing scope creep and controlling change. How can you manage this important issue? Following are suggestions for improving user input:

- Develop a good project selection process for information technology projects. Insist that all projects have a sponsor from the user organization. The sponsor should not be someone in the information technology department, nor should the sponsor be the project manager. Make project information, including the project charter, project management plan, project scope statement, WBS, and WBS dictionary, easily available in the organization. Making basic project information available will help avoid duplication of effort and ensure that the most important projects are the ones on which people are working.
- Have users on the project team. Some organizations require project managers to come from the business area of the project instead of the information technology group. Some organizations assign co-project managers to information technology projects, one from information technology and one from the main business group. Users should be assigned full-time to large information technology projects and part-time to smaller projects. A key success factor in Northwest Airline's ResNet project (see the companion Web site to read the entire case study for this project) was training reservation agents—the users—in how to write programming code for their new reservation system. Because the sales agents had intimate knowledge of the business, they provided excellent input and actually created most of the software.
- Have regular meetings with defined agendas. Meeting regularly sounds obvious, but many information technology projects fail because the project team members do not have regular interaction with users. They assume they understand what users need without getting direct feedback. To encourage this interaction, users should sign off on key deliverables presented at meetings.
- Deliver something to project users and sponsors on a regular basis. If it is some sort of hardware or software, make sure it works first.
- Do not promise to deliver what cannot be delivered in a particular time frame. Make sure the project schedule allows enough time to produce the deliverables.
- Co-locate users with the developers. People often get to know each other better by being in close proximity. If the users cannot be physically moved to be near developers during the entire project, they could set aside certain days for co-location.

## Suggestions for Reducing Incomplete and Changing Requirements

Some requirement changes are expected on information technology projects, but many projects have too many changes to their requirements, especially during later stages of the project life cycle when it is more difficult to implement them. The following are suggestions for improving the requirements process:

- Develop and follow a requirements management process that includes procedures for initial requirements determination. (See the Suggested Readings on the companion Web site by Wiegers and Robertson for detailed information on managing requirements.)
- Employ techniques such as prototyping, use case modeling, and Joint Application Design to understand user requirements thoroughly. **Prototyping** involves developing a working replica of the system or some aspect of the system. These working replicas may be throwaways or an incremental component of the deliverable system. Prototyping is an effective tool for gaining an understanding of requirements, determining the feasibility of requirements, and resolving user interface uncertainties. **Use case modeling** is a process for identifying and modeling business events, who initiated them, and how the system should respond to them. It is an effective tool for understanding requirements for information systems. **Joint Application Design (JAD)** uses highly organized and intensive workshops to bring together project stakeholders—the sponsor, users, business analysts, programmers, and so on—to jointly define and design information systems. These techniques also help users become more active in defining system requirements.
- Put all requirements in writing and keep them current and readily available. Several tools are available to automate this function. For example, a type of software called a requirements management tool aids in capturing and maintaining requirements information, provides immediate access to the information, and assists in establishing necessary relationships between requirements and information created by other tools.
- Create a requirements management database for documenting and controlling requirements. Computer Aided Software Engineering (CASE) tools or other technologies can assist in maintaining a repository for project data. A CASE tool's database can also be used to document and control requirements.
- Provide adequate testing to verify that the project's products perform as expected. Conduct testing throughout the project life cycle. Chapter 8, Project Quality Management, includes more information on testing.
- Use a process for reviewing requested requirements changes from a systems perspective. For example, ensure that project scope changes include associated cost and schedule changes. Require approval by appropriate stakeholders. It is crucial for the project manager to lead the team in their focus on achieving approved scope goals and not get side-tracked into doing additional work. For example, in his book, *Alpha Project Managers*, Andy Crowe tried to uncover what the "best" or "alpha" project managers do differently from other project managers. One of these alpha project managers explained how he learned an important lesson about scope control:

  > "Toward the end of some projects I've worked on, the managers made their teams work these really long hours. After the second or third time this happened, I just assumed that this was the way things worked. Then I got to work with a manager who planned

everything out really well and ran the team at a good pace the whole time, and we kept on schedule. When the customer found out that things were on schedule, he kept trying to increase the scope, but we had a good manager this time, and she wouldn't let him do it without adjusting the baselines. That was the first time I was on a project that finished everything on time and on budget, and I was amazed at how easy she made it look."[11]



- Emphasize completion dates. For example, a project manager at Farmland Industries, Inc. in Kansas City, Missouri, kept her 15-month, $7 million integrated supply-chain project on track by setting the project deadline. She says, "May 1 was the drop-dead date, and everything else was backed into it. Users would come to us and say they wanted something, and we'd ask them what they wanted to give up to get it. Sticking to the date is how we managed scope creep."[12]
- Allocate resources specifically for handling change requests. For example, Peeter Kivestu and his ResNet team at Northwest Airlines knew that users would request enhancements to the reservations system they were developing. They provided a special function key on the ResNet screen for users to submit their requests, and the project included three full-time programmers to handle these requests. Users made over 11,000 enhancement requests. The managers who sponsored the four main software applications had to prioritize the software enhancement requests and decide as a group what changes to approve. The three programmers then implemented as many items as they could, in priority order, given the time they had. Although they only implemented 38 percent of the requested enhancements, they were the most important ones, and the users were very satisfied with the system and process. (You can find a detailed description of the ResNet project on the companion Web site for this text.)

## USING SOFTWARE TO ASSIST IN PROJECT SCOPE MANAGEMENT

Project managers and their teams can use several types of software to assist in project scope management. As shown in several of the figures and tables in this chapter, you can use word processing software to create scope-related documents, and most people use spreadsheet or presentation software to develop various charts, graphs, and matrixes related to scope management. Mind-mapping software can be useful in developing a WBS. Project stakeholders also transmit project scope management information using various types of communication software such as e-mail and assorted Web-based applications.

Project management software helps you develop a WBS, which serves as a basis for creating Gantt charts, assigning resources, allocating costs, and so on. You can also use the templates that come with various project management software products to help you create a WBS for your project. (See the section on project scope management in Appendix A for detailed information on using Project 2007 and the companion Web site for information on templates related to project scope management.)

You can also use many types of specialized software to assist in project scope management. Many information technology projects use special software for requirements management, prototyping, modeling, and other scope-related work. Because scope is such a crucial part of project management, there are many software products available to assist in managing project scope.

Project scope management is very important, especially on information technology projects. After selecting projects, organizations must collect the requirements and define the scope of the work, break down the work into manageable pieces, verify the scope with project stakeholders, and manage changes to project scope. Using the basic project management concepts, tools, and techniques discussed in this chapter can help you successfully manage project scope.



## CASE WRAP-UP

Kim Nguyen reviewed guidelines for creating WBSs provided by her company and other sources. She had a meeting with the three team leaders for her project to get their input on how to proceed. They reviewed several sample documents and decided to have major groupings for their project based on updating the inventory database, acquiring the necessary hardware and software, installing the hardware and software, and performing project management. After they decided on a basic approach, Kim led a meeting with the entire project team of 12 people, with some attending virtually. She reviewed the project charter and stakeholder register, described the basic approach they would use to collect requirements and define the project scope, and reviewed sample WBSs. Kim opened the floor for questions, which she answered confidently. She then let each team leader work with his or her people to start writing the detailed scope statement and their sections of the WBS and WBS dictionary. Everyone participated in the meeting, sharing their individual expertise and openly asking questions. Kim could see that the project was off to a good start.

## Chapter Summary

Project scope management includes the processes required to ensure that the project addresses all the work required, and only the work required, to complete the project successfully. The main processes include collecting requirements, defining scope, creating the WBS, verifying scope, and controlling scope.



The first step in project scope management is collecting requirements, a crucial part of many IT projects. It is important to review the project charter and meet with key stakeholders listed in the stakeholder register when collecting requirements. The main outputs of this process are requirements documentation, a requirements management plan, and a requirements traceability matrix.

A project scope statement is created in the scope definition process. This document often includes a product scope description, product user acceptance criteria, detailed information on all project deliverables, and information on project boundaries, constraints, and assumptions. There are often several versions of the project scope statement to keep scope information detailed and up-to-date.

A work breakdown structure (WBS) is a deliverable-oriented grouping of the work involved in a project that defines the total scope of the project. The WBS forms the basis for planning and managing project schedules, costs, resources, and changes. You cannot use project management software without first creating a good WBS. A WBS dictionary is a document that describes detailed information about each WBS item. A good WBS is often difficult to create because of the complexity of the project. There are several approaches for developing a WBS, including using guidelines, the analogy approach, the top-down approach, the bottom-up approach, and mind mapping.

Verifying scope involves formal acceptance of the project scope by the stakeholders. Controlling scope involves controlling changes to the project scope.

Poor project scope management is one of the key reasons projects fail. For information technology projects, it is important for good project scope management to have strong user involvement, a clear statement of requirements, and a process for managing scope changes.

There are many software products available to assist in project scope management. The WBS is a key concept in properly using project management software since it provides the basis for entering tasks.

## Quick Quiz

1. ________________ refer(s) to all the work involved in creating the products of the project and the processes used to create them.
    a. Deliverables
    b. Milestones
    c. Scope
    d. Product development

2. Which tool or technique for collecting requirements is often the most expensive and time consuming?
   a. interviews
   b. focus groups
   c. surveys
   d. observation
3. A ___________ is a deliverable-oriented grouping of the work involved in a project that defines the total scope of the project.
   a. scope statement
   b. WBS
   c. WBS dictionary
   d. work package
4. What approach to developing a WBS involves writing down or drawing ideas in a nonlinear format?
   a. top-down
   b. bottom-up
   c. analogy
   d. mind mapping
5. Assume you have a project with major categories called planning, analysis, design, and testing. What level of the WBS would these items fall under?
   a. 0
   b. 1
   c. 2
   d. 3
6. Which of the following is not a best practice that can help in avoiding scope problems on information technology projects?
   a. Keep the scope realistic
   b. Use off-the-shelf hardware and software whenever possible
   c. Follow good project management processes
   d. Don't involve too many users in scope management
7. What major restaurant chain terminated a large project after spending $170 million on it, primarily because they realized the project scope was too much to handle?
   a. Burger King
   b. Pizza Hut
   c. McDonald's
   d. Taco Bell

8. Scope ____________ is often achieved by a customer inspection and then sign-off on key deliverables.
    a. verification
    b. validation
    c. completion
    d. close-out



9. Which of the following is not a suggestion for improving user input?
    a. Develop a good project selection process for information technology projects
    b. Have users on the project team
    c. Co-locate users with developers
    d. Only have meetings as needed, not on a regular basis
10. Project management software helps you develop a ____________, which serves as a basis for creating Gantt charts, assigning resources, and allocating costs.
    a. project plan
    b. schedule
    c. WBS
    d. deliverable

### Quick Quiz Answers

1. c; 2. a; 3. b; 4. d; 5. c; 6. d; 7. c; 8. a; 9. d; 10. c

## Discussion Questions

1. What is involved in project scope management, and why is good project scope management so important on information technology projects?
2. What is involved in collecting requirements for a project? Why is it often such a difficult thing to do?
3. Discuss the process of defining project scope in more detail as a project progresses, going from information in a project charter to a project scope statement, WBS, and WBS dictionary.
4. Describe different ways to develop a WBS and explain why it is often so difficult to do.
5. What is the main technique used for verifying scope? Give an example of scope verification on a project.
6. Using examples in this book or online, describe a project that suffered from scope creep. Could it have been avoided? How? Can scope creep be a good thing? When?
7. Why do you need a good WBS to use project management software? What other types of software can you use to assist in project scope management?

## Exercises

1. You are working on a project to develop a new or enhanced system to help people at your college, university, or organization to find jobs. The system must be tailored to your student

or work population and be very easy to use. Write a two-page paper describing how you would collect requirements for this system and include at least five requirements in a requirements traceability matrix.

2. Use PowerPoint, Visio, or similar software to create a WBS in chart form (similar to an organizational chart—see the sample in the top of Figure 5-4). Assume the Level 2 categories are initiating, planning, executing, monitoring and controlling, and closing. Under the executing section, include Level 3 categories of analysis, design, prototyping, testing, implementation, and support. Assume the support category includes Level 4 items called training, documentation, user support, and enhancements.
3. Create the same WBS described in Exercise 2 using Project 2007, indenting categories appropriately. Use the outline numbering feature to display the outline numbers (click Tools on the menu bar, click Options, and then click Show outline number). For example, your WBS should start with 1.0 Initiating. Do not enter any durations or dependencies. See Appendix A or Project 2007's Help for instructions on creating a WBS. Print the resulting Gantt chart on one page, being sure to display the entire Task Name column.
4. Create a WBS for one of the following projects:
   - Introducing self-checkout registers at your school's bookstore
   - Updating 50 laptops from Project 2003 to Project 2007
   - Providing a new Internet Cafe onsite at your organization

   Decide on all of the Level 2 categories for the WBS. Draw a mind map (or use mind-mapping software, if desired) and break down the work to at least the fourth level for one of the WBS items. Enter the WBS into Project 2007 and print out the Gantt chart. Do not enter any durations or dependencies. Make notes of questions you had while completing this exercise.
5. Review a template file in the Microsoft Project 2007 templates folder, from Microsoft's "Project templates" link at *http://office.microsoft.com/project*, or from another source. What do you think about the WBS? Write a two-page paper summarizing your analysis, providing at least three suggestions for improving the WBS.
6. Read one of the suggested readings on the companion Web site or find an article related to project scope management. Write a two-page summary of the article, its key conclusions, and your opinion.



## Running Case

Managers at Manage Your Health, Inc. (MYH) selected Tony Prince as the project manager for the Recreation and Wellness Intranet Project. The schedule goal is six months, and the budget is $200,000. Tony had previous project management and systems analysis experience within the company, and he was an avid sports enthusiast. Tony was starting to put the project team together. Tony knew he would have to develop a survey to solicit input from all employees about this new system and make sure it was very user-friendly.

Recall from Chapter 4 that this application would include the following capabilities:

- Allow employees to register for company-sponsored recreational programs, such as soccer, softball, bowling, jogging, walking, and other sports.

- Allow employees to register for company-sponsored classes and programs to help them manage their weight, reduce stress, stop smoking, and manage other health-related issues.
- Track data on employee involvement in these recreational and health-management programs.
- Offer incentives for people to join the programs and do well in them (e.g., incentives for achieving weight goals, winning sports team competitions, etc.).



Assume that MYH would not need to purchase any additional hardware or software for the project.

## Tasks

1. Document your approach for collecting requirements for this project in a two-page paper, and include at least five requirements in a requirements traceability matrix.
2. Develop a first version of a project scope statement for the project. Use the template provided on the companion Web site for this text and the example in Chapter 3 as guides. Be as specific as possible in describing product characteristics and requirements, as well as all of the project's deliverables. Be sure to include testing and training as part of the project scope.
3. Develop a work breakdown structure for the project. Break down the work to Level 3 or Level 4, as appropriate. Use the template on the companion Web site and samples in this text as guides. Print the WBS in list form as a Word file. Be sure the WBS is based on the project charter (created in the Chapter 4 Running Case), the project scope statement created in Task 2 above, and other relevant information.
4. Use the WBS you developed in Task 3 above to begin creating a Gantt chart in Project 2007 for the project. Use the outline numbering feature to display the outline numbers (click Tools on the menu bar, click Options, and then click Show outline number). Do not enter any durations or dependencies. Print the resulting Gantt chart on one page, being sure to display the entire Task Name column.
5. Develop a strategy for scope verification and change control for this project. Write a two-page paper summarizing key points of the strategy.

## Companion Web Site

Visit the companion Web site for this text (*www.cengage.com/mis/schwalbe*) to access:

- References cited in the text and additional suggested readings for each chapter
- Template files
- Lecture notes
- Interactive quizzes
- Podcasts
- Links to general project management Web sites
- And more

See the Preface of this text for additional information on accessing the companion Web site.

## Key Terms

**analogy approach** — creating a WBS by using a similar project's WBS as a starting point

**bottom-up approach** — creating a WBS by having team members identify as many specific tasks related to the project as possible and then grouping them into higher level categories



**decomposition** — subdividing project deliverables into smaller pieces

**deliverable** — a product, such as a report or segment of software code, produced as part of a project

**Joint Application Design (JAD)** — using highly organized and intensive workshops to bring together project stakeholders—the sponsor, users, business analysts, programmers, and so on—to jointly define and design information systems

**project scope management** — the processes involved in defining and controlling what work is or is not included in a project

**project scope statement** — a document that includes, at a minimum, a description of the project, including its overall objectives and justification, detailed descriptions of all project deliverables, and the characteristics and requirements of products and services produced as part of the project

**prototyping** — developing a working replica of the system or some aspect of the system to help define user requirements

**requirement** — a condition or capability that must be met or possessed by a system, product, service, result, or component to satisfy a contract, standard, specification, or other formal document

**requirements management plan** — a plan that describes how project requirements will be analyzed, documented, and managed

**requirements traceability matrix (RTM)** — a table that lists requirements, various attributes of each requirement, and the status of the requirements to ensure that all requirements are addressed

**scope** — all the work involved in creating the products of the project and the processes used to create them

**scope baseline** — the approved project scope statement and its associated WBS and WBS dictionary

**scope creep** — the tendency for project scope to keep getting bigger

**top-down approach** — creating a WBS by starting with the largest items of the project and breaking them into their subordinate items

**use case modeling** — a process for identifying and modeling business events, who initiated them, and how the system should respond to them

**variance** — the difference between planned and actual performance

**WBS dictionary** — a document that describes detailed information about each WBS item

**work breakdown structure (WBS)** — a deliverable-oriented grouping of the work involved in a project that defines the total scope of the project

**work package** — a task at the lowest level of the WBS

## End Notes



[1] Karl Weigers, *Software Requirements, Second Edition*, (Microsoft Press: Microsoft, 2003), p. 7.

[2] Accept Software, "Success Story: Genesys Connects with Accept to Maintain Competitive Edge, *AcceptSoftware.com* (accessed August 7, 2008).

[3] This particular structure is based on a sample Project 98 file. See *www.microsoft.com* for additional template files.

[4] Mindjet Visual Thinking, "About Mind Maps," *Mindjet.com* (2002).

[5] David I. Cleland, *Project Management: Strategic Design and Implementation*, 2nd ed. (New York: McGraw-Hill, 1994).

[6] Geoffrey James, "Information Technology fiascoes . . . and how to avoid them," *Datamation* (November 1997).

[7] Paul McDougall, "8 Expensive IT Blunders," *InformationWeek* (Oct. 16, 2006).

[8] Geoffrey James, "Information Technology fiascoes . . . and how to avoid them," *Datamation* (November 1997).

[9] Marc L. Songini, "21st Century Insurance apps in limbo despite $100M investment," *ComputerWorld* (December 6, 2002).

[10] Per Kroll and Walker Royce, "Key principles of business-driven development," IBM's DeveloperWorks Rational Library, (October 15, 2005).

[11] Andy Crowe, *Alpha Project Managers: What the Top 2% Know That Everyone Else Does Not*, (Kennesaw, GA: Velociteach Press), 2006, p. 46–47.

[12] Julia King, "IS reins in runaway projects," *ComputerWorld* (September 24, 1997).

CHAPTER 6

# PROJECT TIME MANAGEMENT

## LEARNING OBJECTIVES

**After reading this chapter, you will be able to:**

- Understand the importance of project schedules and good project time management
- Define activities as the basis for developing project schedules
- Describe how project managers use network diagrams and dependencies to assist in activity sequencing
- Understand the relationship between estimating resources and project schedules
- Explain how various tools and techniques help project managers perform activity duration estimating
- Use a Gantt chart for planning and tracking schedule information, find the critical path for a project, and describe how critical chain scheduling and the Program Evaluation and Review Technique (PERT) affect schedule development
- Discuss how reality checks and discipline are involved in controlling and managing changes to the project schedule
- Describe how project management software can assist in project time management and review words of caution before using this software

## OPENING CASE

Sue Johnson was the project manager for a consulting company contracted to provide a new online registration system at a local college. This system absolutely had to be operational by May 1, so students could use it to register for the fall semester. Her company's contract had a stiff penalty clause if the system was not ready by then, and Sue and her team would get nice bonuses for doing a good job on this project and meeting the schedule. Sue knew that it was her responsibility to meet the schedule and manage scope, cost, and quality expectations. She and her team developed a detailed schedule and network diagram to help organize the project.

Developing the schedule turned out to be the easy part; keeping the project on track was more difficult. Managing people issues and resolving schedule conflicts were two of the bigger challenges. Many of the customers' employees took unplanned vacations and missed or rescheduled project review meetings. These changes made it difficult for the project team to follow their planned schedule for the system because they had to have customer sign-off at various stages of the systems development life cycle. One senior programmer on Sue's project team quit, and she knew it would take extra time for a new person to get up to speed. It was still early in the project, but Sue knew they were falling behind. What could she do to meet the operational date of May 1?

## THE IMPORTANCE OF PROJECT SCHEDULES

Many information technology projects are failures in terms of meeting scope, time, and cost projections. Managers often cite delivering projects on time as one of their biggest challenges and the main cause of conflict.

Perhaps part of the reason schedule problems are so common is that time is easily and simply measured. You can debate scope and cost overruns and make actual numbers appear closer to estimates, but once a project schedule is set, anyone can quickly estimate schedule performance by subtracting the original time estimate from how long it really took to complete the project. People often compare planned and actual project completion times without taking into account approved changes in the project. Time is also the one variable that has the least amount of flexibility. Time passes no matter what happens on a project.

Individual work styles and cultural differences may also cause schedule conflicts. You will learn in Chapter 9, Project Human Resource Management, about the Myers Briggs Type Indicator. One dimension of this team-building tool deals with peoples' attitudes toward structure and deadlines. Some people prefer detailed schedules and emphasize task completion. Others prefer to keep things open and flexible. Different cultures and even entire countries have different attitudes about schedules. For example, some countries close businesses for several hours every afternoon to have siestas. Others countries may have different religious or secular holidays at certain times of the year when not much work will be done. Cultures may also have different perceptions of work ethic—some may value hard work and strict schedules while others may value the ability to remain relaxed and flexible.

In contrast to the 2002 Salt Lake City Winter Olympic Games (see Chapter 4's Media Snapshot) or the 2008 Beijing Summer Olympic Games, planning and scheduling was not well implemented for the 2004 Summer Olympic Games held in Athens, Greece. Many articles were written before the opening ceremonies predicting that the facilities would not be ready in time. "With just 162 days to go to the opening of the Athens Olympics, the Greek capital is still not ready for the expected onslaught.... By now 22 of the 30 Olympic projects were supposed to be finished. This week the Athens Olympic Committee proudly announced 19 venues would be finished by the end of next month. That's a long way off target."[1]

However, many people were pleasantly surprised by the amazing opening ceremonies, beautiful new buildings, and state-of-the-art security and transportation systems in Athens. For example, traffic flow, considered a major pre-Games hurdle, was superb. One spectator at the games commented on the prediction that the facilities would not be ready in time, "Athens proved them all wrong.... It has never looked better."[2] The Greeks even made fun of critics by having construction workers pretend to still be working as the ceremonies began. Athens Olympics chief Gianna Angelopoulos-Daskalaki deserves a gold medal for her performance in leading the many people involved in organizing the 2004 Olympic Games. Unfortunately, the Greek government suffered a huge financial deficit since the games cost more than twice the planned budget.



With all these possibilities for schedule conflicts, it's important to use good project time management so that project managers can help improve performance in this area. **Project time management**, simply defined, involves the processes required to ensure timely completion of a project. Achieving timely completion of a project, however, is by no means simple. There are six main processes involved in project time management:

1. *Defining activities* involves identifying the specific activities that the project team members and stakeholders must perform to produce the project deliverables. An **activity** or **task** is an element of work normally found on the work breakdown structure (WBS) that has an expected duration, a cost, and resource requirements. The main outputs of this process are an activity list, activity attributes, and milestone list.
2. *Sequencing activities* involves identifying and documenting the relationships between project activities. The main outputs of this process include project schedule network diagrams and project document updates.
3. *Estimating activity resources* involves estimating how many **resources**—people, equipment, and materials—a project team should use to perform project activities. The main outputs of this process are activity resource requirements, a resource breakdown structure, and project document updates.
4. *Estimating activity durations* involves estimating the number of work periods that are needed to complete individual activities. Outputs include activity duration estimates and project document updates.

5. *Developing the schedule* involves analyzing activity sequences, activity resource estimates, and activity duration estimates to create the project schedule. Outputs include a project schedule, a schedule baseline, schedule data, and project document updates.
6. *Controlling the schedule* involves controlling and managing changes to the project schedule. Outputs include work performance measurements, organizational process assets updates, change requests, project management plan updates, and project document updates.



Figure 6-1 summarizes these processes and outputs, showing when they occur in a typical project.

**Planning**
Process: **Define activities**
Outputs: Activity list, activity attributes, milestone list
Process: **Sequence activities**
Outputs: Project schedule network diagrams, project document updates
Process: **Estimate activity resources**
Outputs: Activity resource requirements, resource breakdown structure, project document updates
Process: **Estimate activity durations**
Outputs: Activity duration estimates, project document updates
Process: **Develop schedule**
Outputs: Project schedule, schedule baseline, schedule data, project document updates

**Monitoring and Controlling**
Process: **Control schedule**
Outputs: Work performance measurements, organizational process assets updates, change requests, project management plan updates, project document updates

**Project Start** → **Project Finish**

**FIGURE 6-1** Project time management summary

You can improve project time management by performing these processes and by using some basic project management tools and techniques. Every manager is familiar with some form of scheduling, but most managers have not used several of the tools and techniques unique to project time management, such as Gantt charts, network diagrams, and critical path analysis.

## DEFINING ACTIVITIES

Project schedules grow out of the basic documents that initiate a project. The project charter often mentions planned project start and end dates, which serve as the starting points for a more detailed schedule. The project manager starts with the project charter and develops a project scope statement and WBS, as discussed in Chapter 5, Project Scope Management. The project charter should also include some estimate of how much money will be allocated to the project. Using this information with the main inputs for defining activities—the scope baseline (the scope statement, WBS, and WBS dictionary), enterprise environmental factors, and organizational process assets—the project manager and project team begin developing a detailed list of activities, their attributes, and a milestone list.

The **activity list** is a tabulation of activities to be included on a project schedule. The list should include the activity name, an activity identifier or number, and a brief description of the activity. The **activity attributes** provide more schedule-related information about each activity, such as predecessors, successors, logical relationships, leads and lags, resource requirements, constraints, imposed dates, and assumptions related to the activity. The activity list and activity attributes should be in agreement with the WBS and WBS dictionary. Information is added to the activity attributes as it becomes available, such as logical relationships and resource requirements that are determined in later processes. Many project teams use an automated system to keep track of all of this activity-related information.



A **milestone** on a project is a significant event that normally has no duration. It often takes several activities and a lot of work to complete a milestone, but the milestone itself is like a marker to help in identifying necessary activities. Milestones are also useful tools for setting schedule goals and monitoring progress. For example, milestones on a project like the one in the opening case might include completion and customer sign-off of documents, such as design documents and test plans; completion of specific products, such as software modules or installation of new hardware; and completion of important process-related work, such as project review meetings, tests, and so on. Not every deliverable or output created for a project is really a milestone. Milestones are the most important and visible ones. For example, the term milestone is used in several contexts, such as in child development. Parents and doctors check for milestones, such as a child first rolling over, sitting, crawling, walking, talking, and so on. You will learn more about milestones later in this chapter.

Activity information is a required input to the other time management processes. You cannot determine activity sequencing, resources, or durations, develop the schedule, or control the schedule until you have a good understanding of project activities.

Recall the triple constraint of project management—balancing scope, time, and cost goals—and note the order of these items. Ideally, the project team and key stakeholders first define the project scope, then the time or schedule for the project, and then the project's cost. The order of these three items reflects the basic order of the first four processes in project time management: defining activities (further defining the scope), sequencing activities (further defining the time), and estimating activity resources and activity durations (further defining the time and cost). These four project time management processes are the basis for creating a project schedule.

The goal of the defining activities is to ensure that the project team has complete understanding of all the work they must do as part of the project scope so they can start scheduling the work. For example, a WBS item might be "Produce study report." The project team would have to understand what that means before they can make schedule-related decisions. How long should the report be? Does it require a survey or extensive research to produce it? What skill level does the report writer need to have? Further defining that task will help the project team determine how long it will take to do and who should do it.

The WBS is often dissected further as the project team members further define the activities required for performing the work. For example, the task "Produce study report" might be broken down into several subtasks describing the steps involved in producing the report, such as developing a survey, administering the survey, analyzing the survey results, performing research, writing a draft report, editing the report, and finally producing the report. This process of progressive elaboration, as mentioned as an attribute of a project in Chapter 1, is sometimes called "rolling wave planning."



As stated earlier, activities or tasks are elements of work performed during the course of a project; they have expected durations, costs, and resource requirements. Defining activities also results in supporting detail to document important product information as well as assumptions and constraints related to specific activities. The project team should review the activity list and activity attributes with project stakeholders before moving on to the next step in project time management. If they do not review these items, they could produce an unrealistic schedule and deliver unacceptable results. For example, if a project manager simply estimated that it would take one day for the "Produce study report" task and had an intern or trainee write a 10-page report to complete that task, the result could be a furious customer who expected extensive research, surveys, and a 100-page report. Clearly defining the work is crucial to all projects. If there are misunderstandings about activities, requested changes may be required.

In the opening case, Sue Johnson and her project team had a contract and detailed specifications for the college's new online registration system. They also had to focus on meeting the May 1 date for an operational system so the college could start using the new system for the new semester's registration. To develop a project schedule, Sue and her team had to review the contract, detailed specifications, and desired operational date, create an activity list, activity attributes, and milestone list. After developing more detailed definitions of project activities, Sue and her team would review them with their customers to ensure that they were on the right track.

## WHAT WENT WRONG?

At the U.S. Federal Bureau of Investigation (FBI), poor time management was one of the reasons behind the failure of Trilogy, a "disastrous, unbelievably expensive piece of vaporware, which was more than four years in the (un)making. The system was supposed to enable FBI agents to integrate intelligence from isolated information silos within the Bureau."[3] In May 2006, the Government Accounting Agency said that the Trilogy project failed at its core mission of improving the FBI's investigative abilities and was plagued with missed milestones and escalating costs.

The FBI rushed to develop the new system, beginning in 2001, in response to the attacks on September 11. The need for new software was obvious to former FBI agent, David J. Williams, who recalls joining a roomful of agents shortly after 9/11 to help with intelligence. Agents were wearing out the casters on their chairs by sliding back and forth between 20 old computer terminals, many of which were attached to various databases containing different information.

*continued*

Congressional hearings revealed numerous problems with the project. The requirements for the project were very loosely defined, there were many leadership changes throughout the project, and several contracts had no scheduled milestones for completion or penalties if work was late. The new system was finally completed in 2006 and cost more than $537 million—more than a year late and $200 million over budget. The FBI replaced Trilogy with a new system called Sentinel and began training employees to use it in May 2007. Although that was about a month behind schedule, director Robert Mueller told a Senate panel that he was optimistic about the $425 million project's progress.[4]



## SEQUENCING ACTIVITIES

After defining project activities, the next step in project time management is sequencing them or determining their dependencies. Inputs to the activity sequencing process include the activity list and attributes, project scope statement, milestone list, and organizational process assets. It involves evaluating the reasons for dependencies and the different types of dependencies.

### Dependencies

A **dependency** or **relationship** relates to the sequencing of project activities or tasks. For example, does a certain activity have to be finished before another one can start? Can the project team do several activities in parallel? Can some overlap? Determining these relationships or dependencies between activities has a significant impact on developing and managing a project schedule.

There are three basic reasons for creating dependencies among project activities:

- **Mandatory dependencies** are inherent in the nature of the work being performed on a project. They are sometimes referred to as hard logic. For example, you cannot test code until after the code is written.
- **Discretionary dependencies** are defined by the project team. For example, a project team might follow good practice and not start the detailed design of a new information system until the users sign off on all of the analysis work. Discretionary dependencies are sometimes referred to as soft logic and should be used with care since they may limit later scheduling options.
- **External dependencies** involve relationships between project and non-project activities. The installation of a new operating system and other software may depend on delivery of new hardware from an external supplier. Even though the delivery of the new hardware may not be in the scope of the project, you should add an external dependency to it because late delivery will affect the project schedule.

As with activity definition, it is important that project stakeholders work together to define the activity dependencies that exist on their project. If you do not define the sequence of activities, you cannot use some of the most powerful schedule tools available to project managers: network diagrams and critical path analysis.

## Network Diagrams

Network diagrams are the preferred technique for showing activity sequencing. A **network diagram** is a schematic display of the logical relationships among, or sequencing of, project activities. Some people refer to network diagrams as project schedule network diagrams or PERT charts. PERT is described later in this chapter. Figure 6-2 shows a sample network diagram for Project X, which uses the arrow diagramming method (ADM) or activity-on-arrow (AOA) approach.



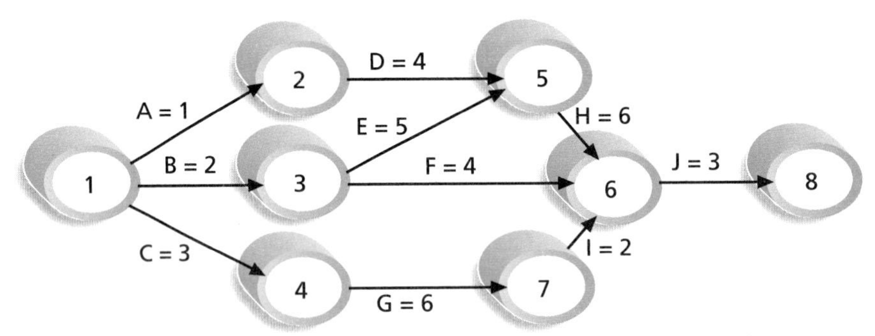


*Note*: Assume all durations are in days; A=1 means Activity A has a duration of 1 day.

**FIGURE 6-2** Activity-on-arrow (AOA) network diagram for Project X

Note the main elements on this network diagram. The letters A through J represent activities with dependencies that are required to complete the project. These activities come from the WBS and activity definition process described earlier. The arrows represent the activity sequencing or relationships between tasks. For example, Activity A must be done before Activity D; Activity D must be done before Activity H, and so on.

The format of this network diagram uses the **activity-on-arrow (AOA)** approach or the **arrow diagramming method (ADM)**—a network diagramming technique in which activities are represented by arrows and connected at points called nodes to illustrate the sequence of activities. A **node** is simply the starting and ending point of an activity. The first node signifies the start of a project, and the last node represents the end of a project.

Keep in mind that the network diagram represents activities that must be done to complete the project. It is not a race to get from the first node to the last node. *Every* activity on the network diagram must be completed in order for the project to finish. It is also important to note that not every single item on the WBS needs to be on the network diagram; only activities with dependencies need to be shown on the network diagram. However, some people like to have start and end milestones and to list every activity. It is a matter of preference. For projects with hundreds of activities, it might be simpler to include only activities with dependencies on a network diagram, especially on large projects. Sometimes it is enough to put summary tasks on a network diagram or to break down the project into several smaller network diagrams.

Assuming you have a list of the project activities and their start and finish nodes, follow these steps to create an AOA network diagram:

1. Find all of the activities that start at Node 1. Draw their finish nodes, and draw arrows between Node 1 and each of those finish nodes. Put the activity letter or name on the associated arrow. If you have a duration estimate, write that next to the activity letter or name, as shown in Figure 6-2. For example, A = 1 means that the duration of Activity A is one day, week, or other standard unit of time. Also be sure to put arrowheads on all arrows to signify the direction of the relationships.
2. Continue drawing the network diagram, working from left to right. Look for bursts and merges. **Bursts** occur when two or more activities follow a single node. A **merge** occurs when two or more nodes precede a single node. For example, in Figure 6-2, Node 1 is a burst since it goes into Nodes 2, 3, and 4. Node 5 is a merge preceded by Nodes 2 and 3.
3. Continue drawing the AOA network diagram until all activities are included on the diagram.
4. As a rule of thumb, all arrowheads should face toward the right, and no arrows should cross on an AOA network diagram. You may need to redraw the diagram to make it look presentable.

Even though AOA or ADM network diagrams are generally easy to understand and create, a different method is more commonly used: the precedence diagramming method. The **precedence diagramming method (PDM)** is a network diagramming technique in which boxes represent activities. It is particularly useful for visualizing certain types of time relationships.

Figure 6-3 illustrates the types of dependencies that can occur among project activities. After you determine the reason for a dependency between activities (mandatory,

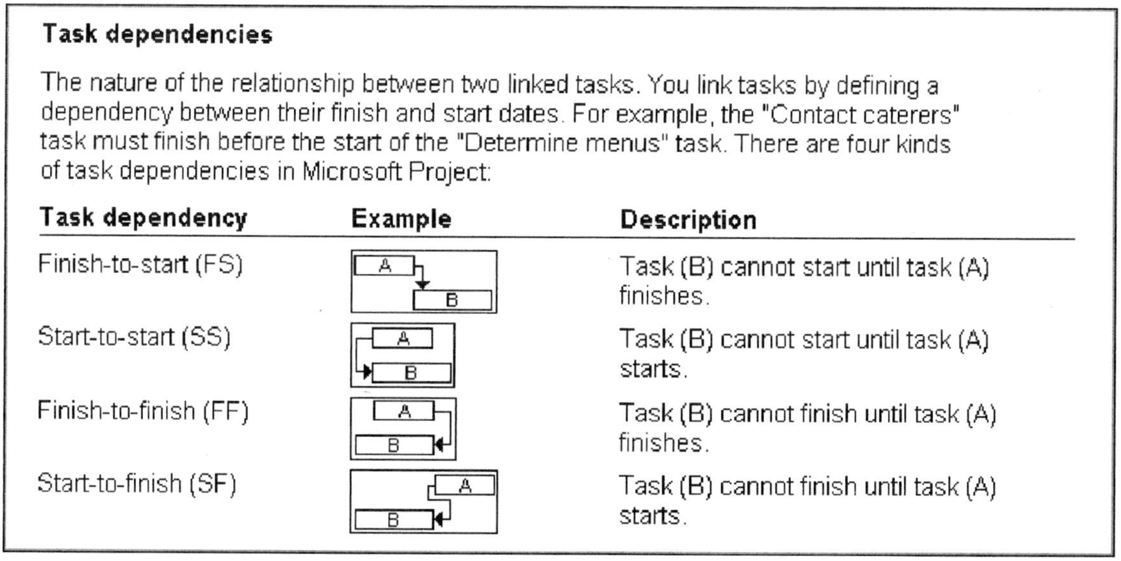

**Task dependencies**

The nature of the relationship between two linked tasks. You link tasks by defining a dependency between their finish and start dates. For example, the "Contact caterers" task must finish before the start of the "Determine menus" task. There are four kinds of task dependencies in Microsoft Project:

| Task dependency | Example | Description |
| --- | --- | --- |
| Finish-to-start (FS) | A B | Task (B) cannot start until task (A) finishes. |
| Start-to-start (SS) | A B | Task (B) cannot start until task (A) starts. |
| Finish-to-finish (FF) | A B | Task (B) cannot finish until task (A) finishes. |
| Start-to-finish (SF) | A B | Task (B) cannot finish until task (A) starts. |

**FIGURE 6-3** Task dependency types

discretionary, or external), you must determine the type of dependency. Note that the terms activity and task are used interchangeably, as are relationship and dependency. See Appendix A to learn how to create dependencies in Microsoft Project 2007. The four types of dependencies or relationships between activities include:



- **Finish-to-start dependency:** a relationship where the "from" activity or predecessor must finish before the "to" activity or successor can start. For example, you cannot provide user training until after software, or a new system, has been installed. Finish-to-start is the most common type of relationship, or dependency, and AOA network diagrams use only finish-to-start dependencies.
- **Start-to-start dependency:** a relationship in which the "from" activity cannot start until the "to" activity or successor is started. For example, on several information technology projects, a group of activities all start simultaneously, such as the many tasks that occur when a new system goes live.
- **Finish-to-finish dependency:** a relationship where the "from" activity must be finished before the "to" activity can be finished. One task cannot finish before another finishes. For example, quality control efforts cannot finish before production finishes, although the two activities can be performed at the same time.
- **Start-to-finish dependency:** a relationship where the "from" activity must start before the "to" activity can be finished. This type of relationship is rarely used, but it is appropriate in some cases. For example, an organization might strive to stock raw materials just in time for the manufacturing process to begin. A delay in the manufacturing process starting should delay completion of stocking the raw materials. Another example would be a babysitter who wants to finish watching a young child but is dependent on the parent arriving. The parent must show up or "start" before the babysitter can finish his or her oversight.

Figure 6-4 illustrates Project X using the precedence diagramming method. Notice that the activities are placed inside boxes, which represent the nodes on this diagram. Arrows show the relationships between activities. This figure was created using Microsoft Project, which automatically places additional information inside each node. Each task box includes the start and finish date, labeled Start and Finish, the task ID number, labeled ID, the task's duration, labeled Dur, and the names of resources, if any, assigned to the task, labeled Res. The border of the boxes for tasks on the critical path appears automatically in red in the Microsoft Project network diagram view. In Figure 6-4, the boxes for critical tasks have a thicker border.

The precedence diagramming method is used more often than AOA network diagrams and offers a number of advantages over the AOA technique. First, most project management software uses the precedence diagramming method. Second, the precedence diagramming method avoids the need to use dummy activities. **Dummy activities** have no duration and no resources but are occasionally needed on AOA network diagrams to show logical relationships between activities. They are represented with dashed arrow lines, and have zero for the duration estimate. Third, the precedence diagramming method shows different dependencies among tasks, whereas AOA network diagrams use only finish-to-start dependencies. You will learn more about activity sequencing using Project 2007 in Appendix A.

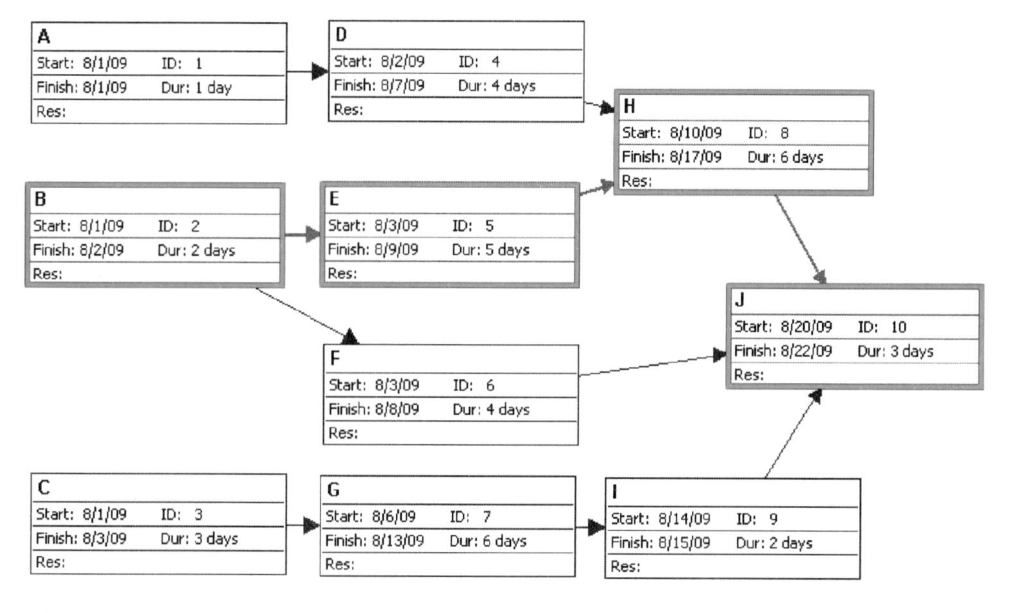


**FIGURE 6-4** Precedence diagramming method (PDM) network diagram for Project X



## ESTIMATING ACTIVITY RESOURCES

Before you can estimate the duration for each activity, you must have a good idea of the quantity and type of resources (people, equipment, and materials) that will be assigned to each activity. The nature of the project and the organization will affect resource estimating. Expert judgment, an analysis of alternatives, estimating data, and project management software are tools available to assist in resource estimating. It is important that the people who help determine what resources are necessary include people who have experience and expertise in similar projects and with the organization performing the project.

Important questions to answer in activity resource estimating include:

- How difficult will it be to do specific activities on this project?
- Is there anything unique in the project's scope statement that will affect resources?
- What is the organization's history in doing similar activities? Has the organization done similar tasks before? What level of personnel did the work?
- Does the organization have people, equipment, and materials that are capable and available for performing the work? Are there any organizational policies that might affect the availability of resources?
- Does the organization need to acquire more resources to accomplish the work? Would it make sense to outsource some of the work? Will outsourcing increase or decrease the amount of resources needed and when they'll be available?

A project's activity list, activity attributes, resource calendars or availability, enterprise environmental factors, and organizational process assets (such as policies regarding staffing and outsourcing) are all important inputs to answering these questions. During the early phases of a project, the project team may not know which specific people, equipment, and materials will be available. For example, they might know from past projects that there will be a mix of experienced and inexperienced programmers working on a project. They might also have information available that approximates the number of people or hours it normally takes to perform specific activities.



It is important to thoroughly brainstorm and evaluate alternatives related to resources, especially on projects that involve people from multiple disciplines and companies. Since most projects involve many human resources and the majority of costs are for salaries and benefits, it is often effective to solicit ideas from different people to help develop alternatives and address resource-related issues early in a project. The resource estimates should also be updated as more detailed information becomes available.

The main outputs of the resource estimating process include a list of activity resource requirements, a resource breakdown structure, and project document updates. For example, if junior employees will be assigned to many activities, the project manager might request that additional activities, time, and resources be approved to help train and mentor those employees. In addition to providing the basis for estimating activity durations, estimating activity resources provides vital information for project cost estimating (Chapter 7), project human resource management (Chapter 9), project communications management (Chapter 10), project risk management (Chapter 11), and project procurement management (Chapter 12). For example, a **resource breakdown structure** is a hierarchical structure that identifies the project's resources by category and type. Resource categories might include analysts, programmers, and testers. Under programmers, there might be types of programmers, such as Java programmers or COBOL programmers. This information would be helpful in determining resource costs, acquiring resources, and so on.

## ESTIMATING ACTIVITY DURATIONS

After working with key stakeholders to define activities, determine their dependencies, and estimate their resources, the next process in project time management is to estimate the duration of activities. It is important to note that **duration** includes the actual amount of time worked on an activity *plus* elapsed time. For example, even though it might take one workweek or five workdays to do the actual work, the duration estimate might be two weeks to allow extra time needed to obtain outside information. The resources assigned to a task will also affect the task duration estimate.

Do not confuse duration with **effort**, which is the number of workdays or work hours required to complete a task. A duration estimate of one day could be based on eight hours of work or eighty hours of work. Duration relates to the time estimate, not the effort estimate. Of course, the two are related, so project team members must document their assumptions when creating duration estimates and update the estimates as the project progresses. The people who will actually do the work, in particular, should have a lot of say in these duration estimates, since they are the ones whose performance will be evaluated on meeting them. If scope changes occur on the project, the duration estimates should be updated to reflect those changes. It is also helpful to review similar projects and seek the advice of experts in estimating activity durations.

There are several inputs to activity duration estimating. The activity list, activity attributes, activity resource requirements, resource calendars, project scope statement, enterprise environmental factors, and organizational process assets all include information that affect duration estimates. In addition to reviewing past project information, the team should also review the accuracy of the duration estimates thus far on the project. For example, if they find that all of their estimates have been much too long or short, the team should update the estimates to reflect what they have learned. One of the most important considerations in making activity duration estimates is the availability of resources, especially human resources. What specific skills do people need to do the work? What are the skill levels of the people assigned to the project? How many people are expected to be available to work on the project at any one time?



The outputs of activity duration estimating include activity duration estimates and project document updates. Duration estimates are often provided as a discrete number, such as four weeks, or as a range, such as three to five weeks, or as a three-point estimate. A **three-point estimate** is an estimate that includes an optimistic, most likely, and pessimistic estimate, such as three weeks for the optimistic, four weeks for the most likely, and five weeks for the pessimistic estimate. The optimistic estimate is based on a best-case scenario, while the pessimistic estimate is based on a worst-case scenario. The most likely estimate, as it sounds, is an estimate based on a most likely or expected scenario. A three-point estimate is required for performing PERT estimates, as described later in this chapter, and for performing Monte Carlo simulations, described in Chapter 11, Project Risk Management. Other duration estimating techniques include analogous and parametric estimating and reserve analysis, as described in Chapter 7, Project Cost Management. Expert judgment is also an important tool for developing good activity duration estimates.

## DEVELOPING THE SCHEDULE

Schedule development uses the results of all the preceding project time management processes to determine the start and end dates of the project. There are often several iterations of all the project time management processes before a project schedule is finalized. The ultimate goal of developing the schedule is to create a realistic project schedule that provides a basis for monitoring project progress for the time dimension of the project. The main outputs of this process are the project schedule, a schedule baseline, project document updates, and schedule data. Some project teams create a computerized model to create a network diagram, enter resource requirements and availability by time period, and adjust other information to quickly generate alternative schedules. See Appendix A for information on using Project 2007 to assist in schedule development.

Several tools and techniques assist in schedule development:

- A Gantt chart is a common tool for displaying project schedule information.
- Critical path analysis is a very important tool for developing and controlling project schedules.
- Critical chain scheduling is a technique that focuses on limited resources when creating a project schedule.
- PERT analysis is a means for considering schedule risk on projects.

The following sections provide samples of each of these tools and techniques and discuss their advantages and disadvantages. (See the *PMBOK® Guide, Fourth Edition* to learn how these main techniques and others are broken into additional categories.)

## Gantt Charts

**Gantt charts** provide a standard format for displaying project schedule information by listing project activities and their corresponding start and finish dates in a calendar format. Gantt charts are sometimes referred to as bar charts since the activities' start and end dates are shown as horizontal bars. Figure 6-5 shows a simple Gantt chart for Project X created with Microsoft Project. Figure 6-6 shows a Gantt chart that is more sophisticated based on a software launch project from a template provided by Microsoft. Recall that the activities on the Gantt chart should coincide with the activities on the WBS, which should coincide with the activity list and milestone list. Notice that the software launch project's Gantt chart contains milestones, summary tasks, individual task durations, and arrows showing task dependencies.



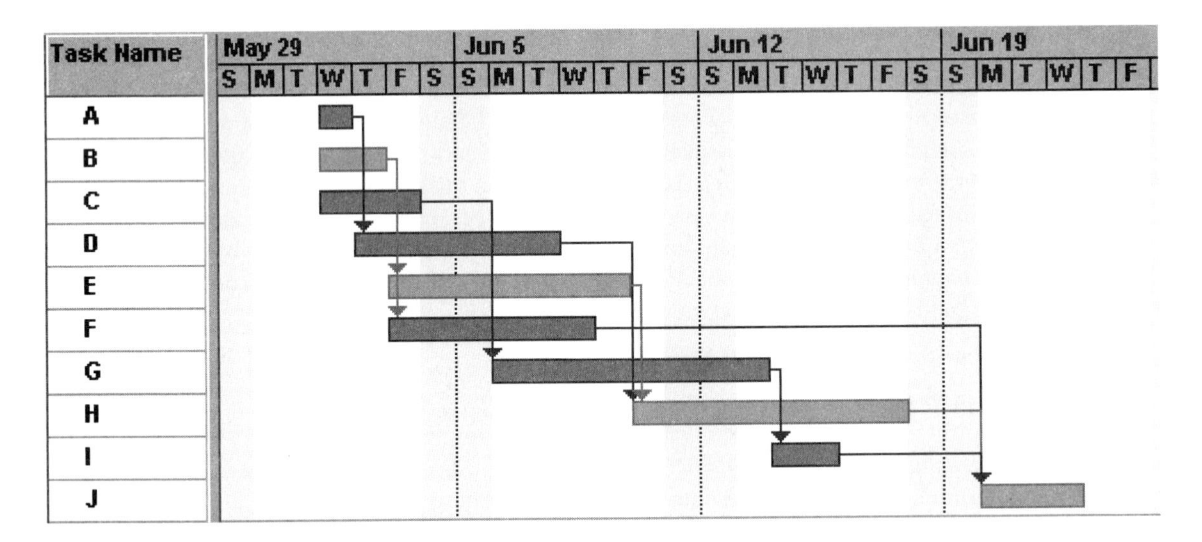


**FIGURE 6-5** Gantt chart for Project X

Notice the different symbols on the software launch project's Gantt chart (Figure 6-6):

- The black diamond symbol represents a milestone. In Figure 6-6, Task 1, "Marketing Plan distributed," is a milestone that occurs on March 17. Tasks 3, 4, 8, 9, 14, 25, 27, 43, and 45 are also milestones. For very large projects, top managers might want to see only milestones on a Gantt chart. Microsoft Project allows you to filter information displayed on a Gantt chart so you can easily show specific tasks, such as milestones.
- The thick black bars with arrows at the beginning and end represent summary tasks. For example, Activities 12 through 15—"Develop creative briefs," "Develop concepts," "Creative concepts," and "Ad development"—are all subtasks of the summary task called "Advertising," Task 11. WBS activities are referred to as tasks and subtasks in most project management software.

- The light gray horizontal bars such as those found in Figure 6-6 for Tasks 5, 6, 7, 10, 12, 13, 15, 22, 24, 26, and 44, represent the duration of each individual task. For example, the light gray bar for Subtask 5, "Packaging," starts in mid-February and extends until early May.
- Arrows connecting these symbols show relationships or dependencies between tasks. Gantt charts often do not show dependencies, which is their major disadvantage. If dependencies have been established in Microsoft Project, they are automatically displayed on the Gantt chart.



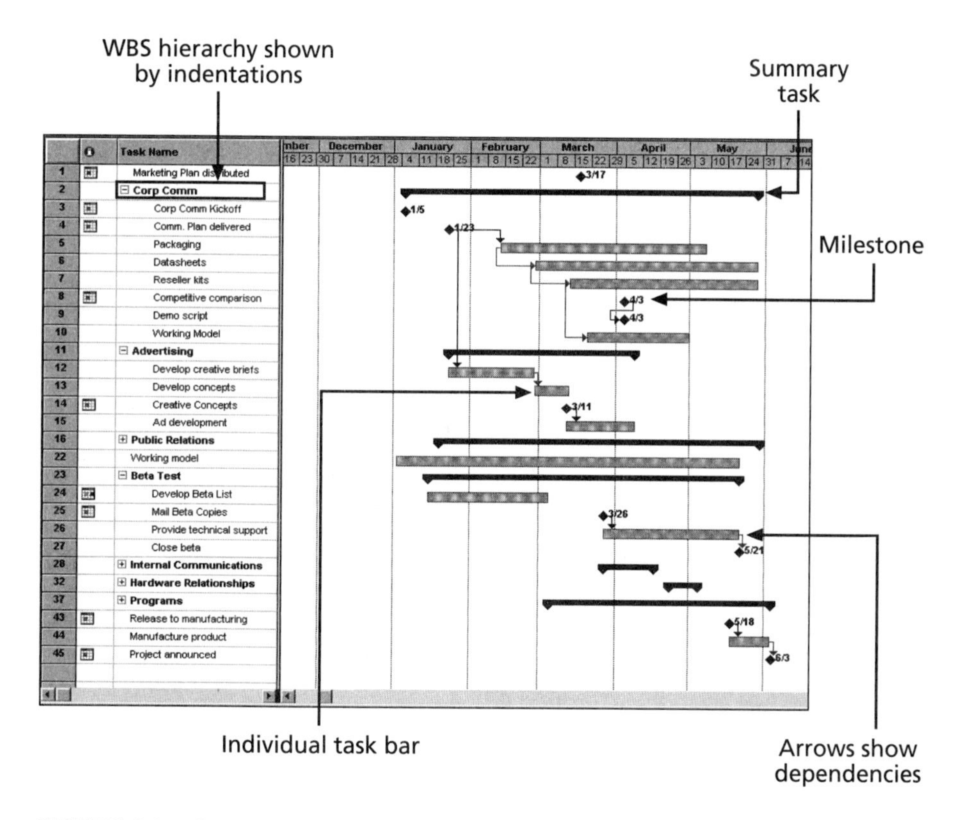


**FIGURE 6-6** Gantt chart for software launch project

### Adding Milestones to Gantt Charts

Milestones can be a particularly important part of schedules, especially for large projects. Many people like to focus on meeting milestones, so you can create milestones to emphasize important events or accomplishments on projects. Normally, you create milestones by entering tasks with zero duration. In Microsoft Project, you can also mark any task as a milestone by checking the appropriate box in the Advanced tab of the Task Information dialog box. The duration of the task will not change to zero, but the Gantt chart will show the milestone symbol to represent that task based on its start date. See Appendix A for more information.

To make milestones meaningful, some people use the SMART criteria to help define them. The **SMART criteria** are guidelines suggesting that milestones should be:

- Specific
- Measurable
- Assignable
- Realistic
- Time-framed



For example, distributing a marketing plan is specific, measurable, and assignable if everyone knows what should be in the marketing plan, how it should be distributed, how many copies should be distributed and to whom, and who is responsible for the actual delivery. Distributing the marketing plan is realistic and able to be time-framed if it is an achievable event and scheduled at an appropriate time.

### BEST PRACTICE

Schedule risk is inherent in the development of complex systems. Luc Richard, the founder of *www.projectmangler.com*, suggests that project managers can reduce schedule risk through project milestones, a best practice that involves identifying and tracking significant points or achievements in the project. The five key points of using project milestones include the following:

1. Define milestones early in the project and include them in the Gantt chart to provide a visual guide.
2. Keep milestones small and frequent.
3. The set of milestones must be all-encompassing.
4. Each milestone must be binary, meaning it is either complete or incomplete.
5. Carefully monitor the critical path.

Additional best practices Richard recommends for software development projects include the following:

- Monitor the project's progress and revise the plan.
- Build on a solid base; that means developing a system with less than a .1 percent defect rate.
- Assign the right people to the right tasks. Put the best developers on the critical tasks.
- Start with high-risk tasks.
- "Don't boil the ocean." In other words, if the entire project consists of high-risk tasks, then the project itself is high-risk and bound for failure.
- Integrate early and often, and follow practices like the daily build process.[5] (A daily build involves compiling the latest version of a software program each day to ensure that all required dependencies are present and that the program is tested to avoid introducing new bugs.)

### Using Tracking Gantt Charts to Compare Planned and Actual Dates

You can use a special form of a Gantt chart to evaluate progress on a project by showing actual schedule information. Figure 6-7 shows a **Tracking Gantt chart**—a Gantt chart that compares planned and actual project schedule information. The planned schedule dates for activities are called the **baseline dates**, and the entire approved planned schedule is called the **schedule baseline**. The Tracking Gantt chart includes columns (hidden in Figure 6-7) labeled "Start" and "Finish" to represent actual start and finish dates for each task, as well as columns labeled "Baseline Start" and "Baseline Finish" to represent planned start and finish dates for each task. In this example, the project is completed, but several tasks missed their planned start and finish dates.



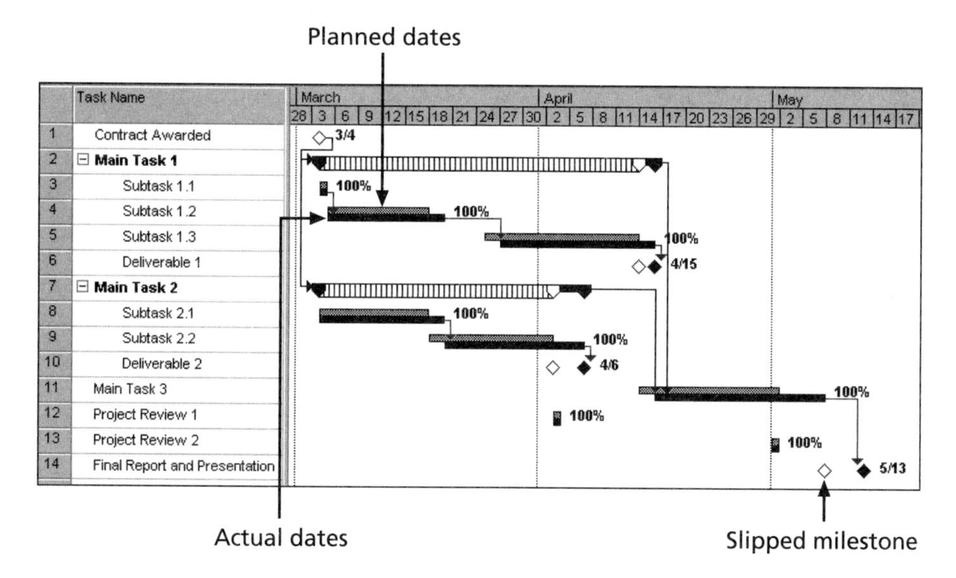


**FIGURE 6-7** Sample Tracking Gantt chart

To serve as a progress evaluation tool, a Tracking Gantt chart uses a few additional symbols:

- Notice that the Gantt chart in Figure 6-7 often shows two horizontal bars for tasks. The top horizontal bar represents the planned or baseline duration for each task. The bar below it represents the actual duration. Subtasks 1.2 and 1.3 illustrate this type of display. If these two bars are the same length, and start and end on the same dates, then the actual schedule was the same as the planned schedule for that task. This scheduling occurred for Subtask 1.1, where the task started and ended as planned on March 4. If the bars do not start and end on the same dates, then the actual schedule differed from the planned or baseline schedule. If the top horizontal bar is shorter than the bottom one, the task took longer than planned, as you can see for Subtask 1.2. If the top horizontal bar is longer than the bottom one, the task took less time than planned. A striped horizontal bar, illustrated by Main Tasks 1 and 2, represents the planned duration for summary tasks. The black bar adjoining it shows progress for summary tasks. For example, Main Task 2 clearly shows the actual duration took longer than what was planned.
- A white diamond on the Tracking Gantt chart represents a **slipped milestone**. A slipped milestone means the milestone activity was actually completed later than originally planned. For example, the last task provides an example of a

slipped milestone since the final report and presentation were completed later than planned.

- Percentages to the right of the horizontal bars display the percentage of work completed for each task. For example, 100 percent means the task is finished, 50 percent means the task is still in progress and is 50 percent completed.

A Tracking Gantt chart is based on the percentage of work completed for project tasks or the actual start and finish dates. It allows the project manager to monitor schedule progress on individual tasks and the whole project. For example, Figure 6-7 shows that this project is completed. It started on time, but it finished a little late, on May 13 (5/13) versus May 8.



The main advantage of using Gantt charts is that they provide a standard format for displaying planned and actual project schedule information. In addition, they are easy to create and understand. The main disadvantage of Gantt charts is that they do not *usually* show relationships or dependencies between tasks. If Gantt charts are created using project management software and tasks are linked, then the dependencies *will* be displayed, but differently than they would be displayed on a network diagram. However, whether you prefer viewing dependencies on a Gantt chart or network diagram is a matter of personal preference.

## Critical Path Method

Many projects fail to meet schedule expectations. **Critical path method (CPM)**—also called **critical path analysis**—is a network diagramming technique used to predict total project duration. This important tool will help you combat project schedule overruns. A **critical path** for a project is the series of activities that determine the *earliest* time by which the project can be completed. It is the *longest* path through the network diagram and has the least amount of slack or float. **Slack** or **float** is the amount of time an activity may be delayed without delaying a succeeding activity or the project finish date. There are normally several tasks done in parallel on projects, and most projects have multiple paths through a network diagram. The longest path or path containing the critical tasks is what is driving the completion date for the project. You are not finished with the project until you have finished *all* the tasks.

### Calculating the Critical Path

To find the critical path for a project, you must first develop a good network diagram, which, in turn, requires a good activity list based on the WBS. Once you create a network diagram, you must also estimate the duration of each activity to determine the critical path. Calculating the critical path involves adding the durations for all activities on each path through the network diagram. The longest path is the critical path.

Figure 6-8 shows the AOA network diagram for Project X again. Note that you can use either the AOA or precedence diagramming method to determine the critical path on projects. Figure 6-8 shows all of the paths—a total of four—through the network diagram. Note that each path starts at the first node (1) and ends at the last node (8) on the AOA network diagram. This figure also shows the length or total duration of each path through the network diagram. These lengths are computed by adding the durations of each activity on the path. Since path B-E-H-J at 16 days has the longest duration, it is the critical path for the project.

What does the critical path really mean? The critical path shows the shortest time in which a project can be completed. Even though the critical path is the *longest* path, it

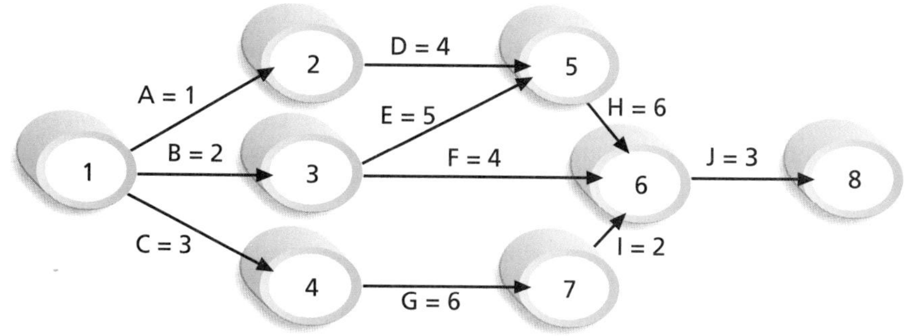


*Note*: Assume all durations are in days.

| | | |
|---|---|---|
| Path 1: | A-D-H-J | Length = 1+4+6+3 = 14 days |
| Path 2: | B-E-H-J | Length = 2+5+6+3 = 16 days |
| Path 3: | B-F-J | Length = 2+4+3 = 9 days |
| Path 4: | C-G-I-J | Length = 3+6+2+3 = 14 days |

Since the critical path is the longest path through the network diagram, Path 2, B-E-H-J, is the critical path for Project X.

**FIGURE 6-8** Determining the critical path for Project X

represents the *shortest* time it takes to complete a project. If one or more of the activities on the critical path takes longer than planned, the whole project schedule will slip *unless* the project manager takes corrective action.

Project teams can be creative in managing the critical path. For example, Joan Knutson, a well-known author and speaker in the project management field, often describes how a gorilla helped Apple computer complete a project on time. Team members worked in an area with cubicles, and whoever was in charge of a task currently on the critical path had a big, stuffed gorilla on top of his or her cubicle. Everyone knew that person was under the most time pressure, so they tried not to distract him or her. When a critical task was completed, the person in charge of the next critical task received the gorilla.

### Growing Grass Can Be on the Critical Path

People are often confused about what the critical path is for a project or what it really means. Some people think the critical path includes the most critical activities. However, the critical path is concerned only with the time dimension of a project. The fact that its name includes the word *critical* does not mean that it includes all critical activities. For example, Frank Addeman, Executive Project Director at Walt Disney Imagineering, explained in a keynote address at a PMI-ISSIG Professional Development Seminar that growing grass was on the critical path for building Disney's Animal Kingdom theme park! This 500-acre park required special grass for its animal inhabitants, and some of the grass took years to grow. Another misconception is that the critical path is the shortest path through the network diagram. In some areas, such as transportation modeling, similar network diagrams are drawn in which identifying the shortest path is the goal. For a project, however, each task or activity must be done in order to complete the project. It is not a matter of choosing the shortest path.

Other aspects of critical path analysis may cause confusion. Can there be more than one critical path on a project? Does the critical path ever change? In the Project X example, suppose that Activity A has a duration estimate of three days instead of one day. This new duration estimate would make the length of Path 1 equal to 16 days. Now the project has two longest paths of equal duration, so there are two critical paths. Therefore, there *can* be more than one critical path on a project. Project managers should closely monitor performance of activities on the critical path to avoid late project completion. If there is more than one critical path, project managers must keep their eyes on all of them.



The critical path on a project can change as the project progresses. For example, suppose everything is going as planned at the beginning of the project. In this example, suppose Activities A, B, C, D, E, F, and G all start and finish as planned. Then suppose Activity I runs into problems. If Activity I takes more than four days, it will cause path C-G-I-J to be longer than the other paths, assuming they progress as planned. This change would cause path C-G-I-J to become the new critical path. Therefore, the critical path can change on a project.

### Using Critical Path Analysis to Make Schedule Trade-Offs

It is important to know what the critical path is throughout the life of a project so the product manager *can* make trade-offs. If the project manager knows that one of the tasks on the critical path is behind schedule, he or she needs to decide what to do about it. Should the schedule be renegotiated with stakeholders? Should more resources be allocated to other items on the critical path to make up for that time? Is it okay if the project finishes behind schedule? By keeping track of the critical path, the project manager and his or her team take a proactive role in managing the project schedule.

A technique that can help project managers make schedule trade-offs is determining the free slack and total slack for each project activity. **Free slack** or **free float** is the amount of time an activity can be delayed without delaying the early start date of any immediately following activities. The **early start date** for an activity is the earliest possible time an activity can start based on the project network logic. **Total slack** or **total float** is the amount of time an activity can be delayed from its early start without delaying the planned project finish date.

Project managers calculate free slack and total slack by doing a forward and backward pass through a network diagram. A **forward pass** determines the early start and early finish dates for each activity. The **early finish date** for an activity is the earliest possible time an activity can finish based on the project network logic. The project start date is equal to the early start date for the first network diagram activity. Early start plus the duration of the first activity is equal to the early finish date of the first activity. It is also equal to the early start date of each subsequent activity unless an activity has multiple predecessors. When an activity has multiple predecessors, its early start date is the latest of the early finish datesof those predecessors. For example, Tasks D and E immediately precede Task H in Figure 6-8. The early start date for Task H, therefore, is the early finish date of Task E, since it occurs later than the early finish date of Task D. A **backward pass** through the network diagram determines the late start and late finish dates for each activity in a similar fashion. The **late start date** for an activity is the latest possible time an activity might begin without delaying the project finish date. The **late finish date** for an activity is the latest possible time an activity can be completed without delaying the project finish date.

Project managers can determine the early and late start and finish dates of each activity by hand. For example, Figure 6-9 shows a simple network diagram with three tasks, A, B, and C. Tasks A and B both precede Task C. Assume all duration estimates are in days. Task A has an estimated duration of 5 days, Task B has an estimated duration of 10 days, and Task C has an estimated duration of 7 days. There are only two paths through this small network diagram: path A-C has a duration of 12 days (5+7), and path B-C has a duration of 17 days (10+7). Since path B-C is longest, it is the critical path. There is no float or slack on this path, so the early and late start and finish dates are the same. However, Task A has 5 days of float or slack. Its early start date is day 0, and its late start date is day 5. Its early finish date is day 5, and its late finish date is day 10. Both the free and total float amounts for Task A are the same at 5 days.



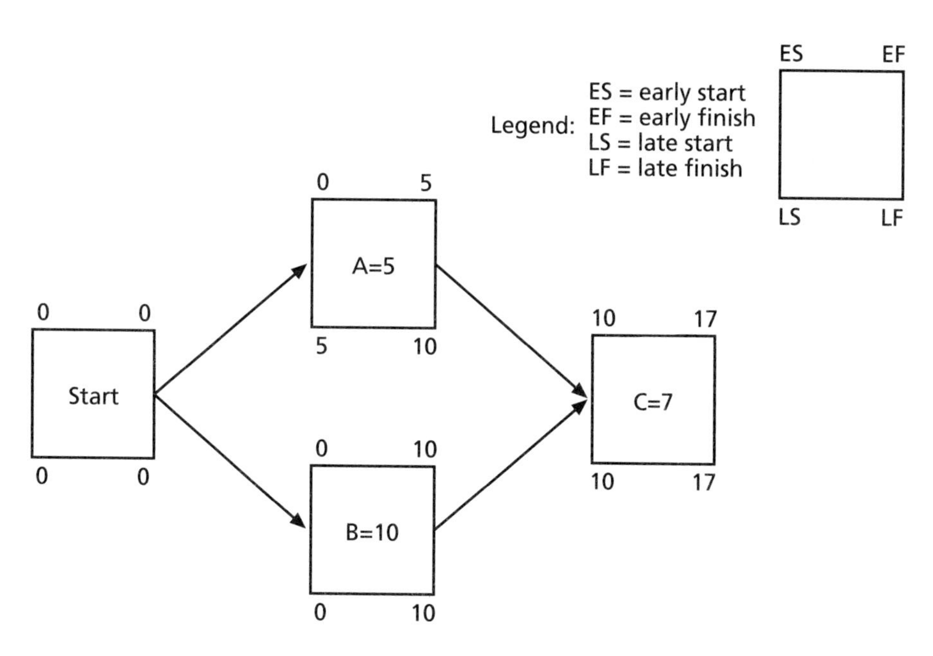


**FIGURE 6-9** Calculating early and late start and finish dates

A much faster and easier way to determine early and late start and finish dates and free and total slack amounts for activities is by using project management software. Table 6-1 shows the free and total slack for all activities on the network diagram for Project X after entering the data from Figure 6-8 and assuming Tasks A, B, and C started on August 3, 2009. (The network diagram is shown in Figure 6-4.) The data in this table was created by selecting the Schedule Table view in Microsoft Project. Knowing the amount of float or slack allows project managers to know whether the schedule is flexible and how flexible it might be. For example, at 7 days (7d), Task F has the most free and total slack. The most slack on any other activity is only 2 days (2d). Understanding how to create and use slack information provides a basis for negotiating (or not negotiating) project schedules. See the Help information in Microsoft Project or research other resources for more detailed information on calculating slack.

**TABLE 6-1** Free and total float or slack for Project X

| Task Name | Start | Finish | Late Start | Late Finish | Free Slack | Total Slack |
|---|---|---|---|---|---|---|
| A | 8/3/09 | 8/3/09 | 8/5/09 | 8/5/09 | 0d | 2d |
| B | 8/3/09 | 8/4/09 | 8/3/09 | 8/4/09 | 0d | 0d |
| C | 8/3/09 | 8/5/09 | 8/5/09 | 8/7/09 | 0d | 2d |
| D | 8/4/09 | 8/7/09 | 8/6/09 | 8/11/09 | 2d | 2d |
| E | 8/5/09 | 8/11/09 | 8/5/09 | 8/11/09 | 0d | 0d |
| F | 8/5/09 | 8/10/09 | 8/14/09 | 8/17/09 | 7d | 7d |
| G | 8/6/09 | 8/13/09 | 8/10/09 | 8/17/09 | 0d | 2d |
| H | 8/12/09 | 8/19/09 | 8/12/09 | 8/19/09 | 0d | 0d |
| I | 8/14/09 | 8/17/09 | 8/18/09 | 8/19/09 | 2d | 2d |
| J | 8/20/09 | 8/24/09 | 8/20/09 | 8/24/09 | 0d | 0d |



### Using the Critical Path to Shorten a Project Schedule

It is common for stakeholders to want to shorten a project schedule estimate. A project team may have done their best to develop a project schedule by defining activities, determining sequencing, and estimating resources and durations for each activity. The results of this work may have shown that the project team needs ten months to complete the project. The sponsor might ask if the project can be done in eight or nine months. Rarely do people ask the project team to take longer than they suggested. By knowing the critical path, the project manager and his or her team can use several duration compression techniques to shorten the project schedule. One technique is to reduce the duration of activities on the critical path. The project manager can shorten the duration of critical path activities by allocating more resources to those activities or by changing their scope.

Recall that Sue Johnson in the opening case was having schedule problems with the online registration project because several users missed important project review meetings and one of the senior programmers quit. If Sue and her team created a realistic project schedule, produced accurate duration estimates, and established dependencies between tasks, they could analyze their status in terms of meeting the May 1 deadline. If some activities on the critical path had already slipped and they did not build in extra time at the end of the project, then they would have to take corrective actions to finish the project on time. Sue could request that her company or the college provide more people to work on the project in an effort to make up time. She could also request that the scope of activities be reduced to complete the project on time. Sue could also use project time management techniques, such as crashing or fast tracking, to shorten the project schedule.

**Crashing** is a technique for making cost and schedule trade-offs to obtain the greatest amount of schedule compression for the least incremental cost. For example, suppose one of the items on the critical path for the online registration project was entering course data for the new semester into the new system. If this task is yet to be done and was originally estimated to take two weeks based on the college providing one part-time data entry clerk, Sue could suggest that the college have the clerk work full time to finish the task in one week instead of two. This change would not cost Sue's company more money, and it could shorten the project end date by one week. If the college could not meet this request, Sue could consider hiring a temporary data entry person for one week to help get the task done faster. By focusing on tasks on the critical path that could be done more quickly for no extra cost or a small cost, the project schedule can be shortened.



The main advantage of crashing is shortening the time it takes to finish a project. The main disadvantage of crashing is that it often increases total project costs. You will learn more about project costs in Chapter 7, Project Cost Management.

Another technique for shortening a project schedule is fast tracking. **Fast tracking** involves doing activities in parallel that you would normally do in sequence. For example, Sue Johnson's project team may have planned not to start any of the coding for the online registration system until *all* of the analysis was done. Instead, they could consider starting some coding activity before the analysis is completed. The main advantage of fast tracking, like crashing, is that it can shorten the time it takes to finish a project. The main disadvantage of fast tracking is that it can end up lengthening the project schedule since starting some tasks too soon often increases project risk and results in rework.

### Importance of Updating Critical Path Data

In addition to finding the critical path at the beginning of a project, it is important to update the schedule with actual data. After the project team completes activities, the project manager should document the actual durations of those activities. He or she should also document revised estimates for activities in progress or yet to be started. These revisions often cause a project's critical path to change, resulting in a new estimated completion date for the project. Again, proactive project managers and their teams stay on top of changes so they can make informed decisions and keep stakeholders informed of, and involved in, major project decisions.

## Critical Chain Scheduling

Another technique that addresses the challenge of meeting or beating project finish dates is an application of the Theory of Constraints called critical chain scheduling. The **Theory of Constraints (TOC)** is a management philosophy developed by Eliyahu M. Goldratt and discussed in his books *The Goal* and *Critical Chain*. The Theory of Constraints is based on the fact that, like a chain with its weakest link, any complex system at any point in time often has only one aspect or constraint that limits its ability to achieve more of its goal. For the system to attain any significant improvements, that constraint must be identified, and the whole system must be managed with it in mind. **Critical chain scheduling** is a method of scheduling that considers limited resources when creating a project schedule and includes buffers to protect the project completion date.

An important concept in critical chain scheduling is the availability of scarce resources. Some projects cannot be done unless a particular resource is available to work on one or several tasks. For example, if a television station wants to produce a show centered around a particular celebrity, it must first check the availability of that celebrity. As another example, if a particular piece of equipment is needed full time to complete each of two tasks that were originally planned to occur simultaneously, critical chain scheduling acknowledges that you must either delay one of those tasks until the equipment is available or find another piece of equipment in order to meet the schedule. Other important concepts related to critical chain scheduling include multitasking and time buffers.



**Multitasking** occurs when a resource works on more than one task at a time. This situation occurs frequently on projects. People are assigned to multiple tasks within the same project or different tasks on multiple projects. For example, suppose someone is working on three different tasks, Task 1, Task 2, and Task 3, for three different projects, and each task takes 10 days to complete. If the person did not multitask, and instead completed each task sequentially, starting with Task 1, then Task 1 would be completed after day 10, Task 2 would be completed after day 20, and Task 3 would be completed after day 30, as shown in Figure 6-10a. However, because many people in this situation try to please all three people who need their tasks completed, they often work on the first task for some time, then the second, then the third, then go back to the first task, and so on, as shown in Figure 6-10b. In this example, the tasks were all half-done one at a time, then completed one at a time. Task 1 is now completed at the end of day 20 instead of day 10, Task 2 is completed at the end of day 25 instead of day 20, and Task 3 is still completed on day 30. This example illustrates how multitasking can delay task completions. Multitasking also often involves wasted setup time, which increases total duration.

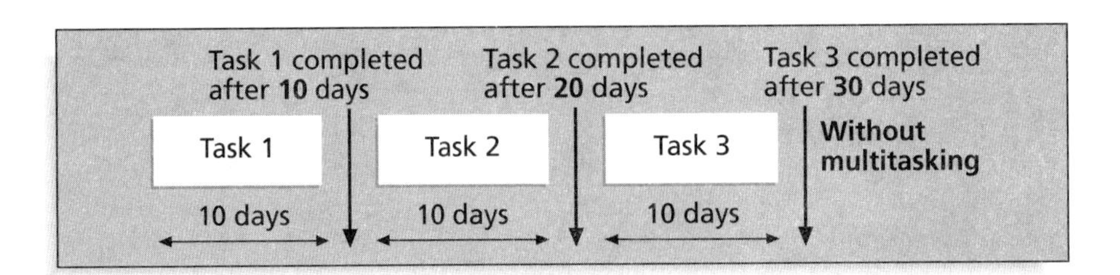


**FIGURE 6-10a** Three tasks without multitasking

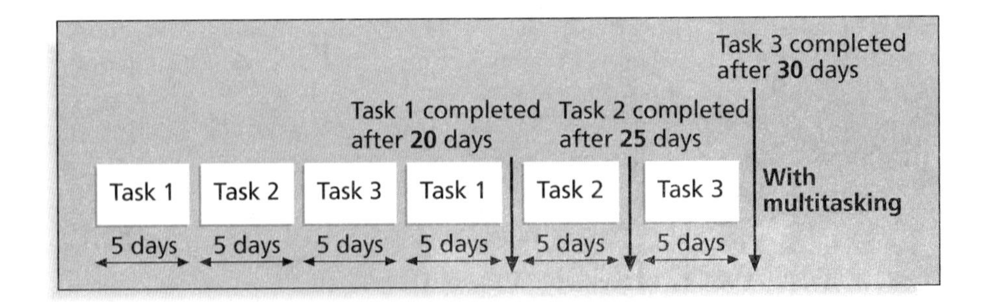


**FIGURE 6-10b** Three tasks with multitasking

Critical chain scheduling assumes that resources do not multitask or at least minimize multitasking. Someone should not be assigned to two tasks simultaneously on the same project when critical chain scheduling is in effect. Likewise, critical chain theory suggests that projects be prioritized so people working on more than one project at a time know which tasks take priority. Preventing multitasking avoids resource conflicts and wasted setup time caused by shifting between multiple tasks over time.



An essential concept to improving project finish dates with critical chain scheduling is to change the way people make task estimates. Many people add a safety or **buffer**, which is additional time to complete a task, to an estimate to account for various factors. These factors include the negative effects of multitasking, distractions and interruptions, fear that estimates will be reduced, Murphy's Law, and so on. **Murphy's Law** states that if something can go wrong, it will. Critical chain scheduling removes buffers from individual tasks and instead creates a **project buffer**, which is additional time added before the project's due date. Critical chain scheduling also protects tasks on the critical chain from being delayed by using **feeding buffers**, which consist of additional time added before tasks on the critical chain that are preceded by non-critical-path tasks.

Figure 6-11 provides an example of a network diagram constructed using critical chain scheduling. Note that the critical chain accounts for a limited resource, X, and the schedule includes use of feeding buffers and a project buffer in the network diagram. The tasks marked with an X are part of the critical chain, which can be interpreted as being the critical path using this technique. The task estimates in critical chain scheduling should be shorter than traditional estimates because they do not include their own buffers. Not having task buffers should mean less occurrence of **Parkinson's Law**, which states that work expands to fill the time allowed. The feeding and project buffers protect the date that really needs to be met—the project completion date.

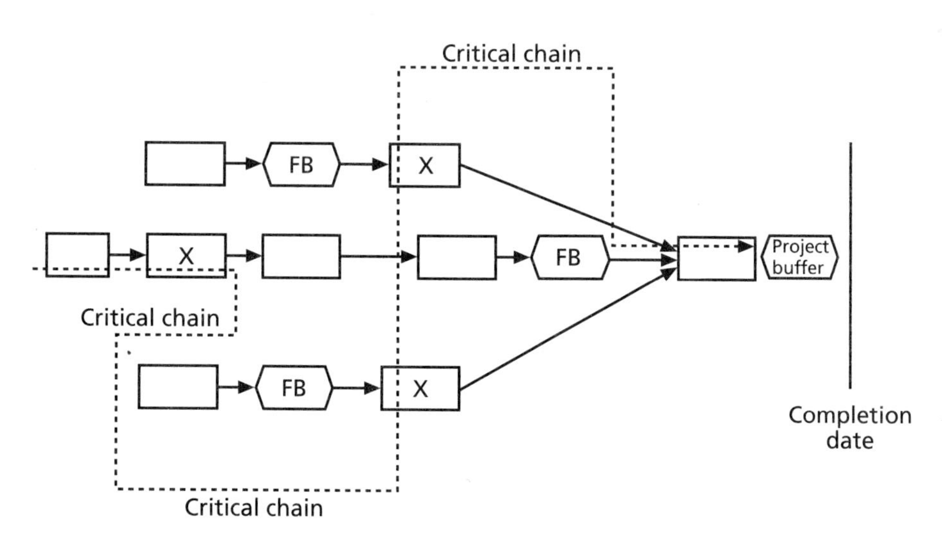


**FIGURE 6-11** Example of critical chain scheduling[6]

Several organizations have reported successes with critical chain scheduling. For example, the U.S. health care industry is learning to think differently by using the Theory of Constraints.

> "Consider a relatively simple system of a physician's office or clinic. The steps in the process could be patients checking in, filling out forms, having vital signs taken by a nurse, seeing the physician, seeing the nurse for a prescribed procedure such as vaccination, and so forth. These steps could take place in a simple linear sequence or chain ... This process or chain only can produce an average of eight per hour. The chain is only as strong as its weakest link and the rate of the slowest resource in this example, the weakest link, is eight. This is true regardless of how fast each of the other resources can process individually, how much work is stuffed into the pipeline, or how complex the process or set of interconnected processes is to complete. Moreover, improving the performance of any link besides the constraint does nothing to improve the system as a whole."[7]



As you can see, critical chain scheduling is a fairly complicated yet powerful tool that involves critical path analysis, resource constraints, and changes in how task estimates are made in terms of buffers. Some consider critical chain scheduling one of the most important new concepts in the field of project management.

## Program Evaluation and Review Technique (PERT)

Another project time management technique is the **Program Evaluation and Review Technique (PERT)**—a network analysis technique used to estimate project duration when there is a high degree of uncertainty about the individual activity duration estimates. PERT applies the critical path method (CPM) to a weighted average duration estimate. This approach was developed about the same time as CPM, in the late 1950s, and also uses network diagrams, which are still sometimes referred to as PERT charts.

PERT uses **probabilistic time estimates**—duration estimates based on using optimistic, most likely, and pessimistic estimates of activity durations—instead of one specific or discrete duration estimate, as CPM does. In other words, PERT uses a three-point estimate, as described earlier. To use PERT, you calculate a weighted average for the duration estimate of each project activity using the following formula:

$$\text{PERT weighted average} = \frac{\text{optimistic time} + 4 * \text{most likely time} + \text{pessimistic time}}{6}$$

By using the **PERT weighted average** for each activity duration estimate, the total project duration estimate takes into account the risk or uncertainty in the individual activity estimates.

Suppose Sue Johnson's project team in the opening case used PERT to determine the schedule for the online registration system project. They would have to collect numbers for the optimistic, most likely, and pessimistic duration estimates for each project activity. Suppose one of the activities was to design an input screen for the system. Someone might estimate that it would take about two weeks or 10 workdays to do this activity. Without using PERT, the duration estimate for that activity would be 10 workdays. Using PERT, the project team would also need to estimate the pessimistic and optimistic times for completing this activity. Suppose an optimistic estimate is that the input screen can be designed in eight workdays, and a pessimistic time estimate is 24 workdays. Applying the PERT formula, you get the following:

$$\text{PERT weighted average} = \frac{8 \text{ workdays } + 4 * 10 \text{ workdays} + 24 \text{ workdays}}{6}$$
$$= 12 \text{ work days}$$

Instead of using the most likely duration estimate of 10 workdays, the project team would use 12 workdays when doing critical path analysis. These additional two days could really help the project team in getting the work completed on time.



The main advantage of PERT is that it attempts to address the risk associated with duration estimates. Since many projects exceed schedule estimates, PERT may help in developing schedules that are more realistic. PERT's main disadvantages are that it involves more work than CPM since it requires several duration estimates, and there are better probabilistic methods for assessing schedule risk (see the information on Monte Carlo simulations in Chapter 11, Project Risk Management).

## CONTROLLING THE SCHEDULE

The final process in project time management is controlling the schedule. Like scope control, schedule control is a portion of the integrated change control process under project integration management. The goal of schedule control is to know the status of the schedule, influence the factors that cause schedule changes, determine that the schedule has changed, and manage changes when they occur.

The main inputs to schedule control are the project management plan, project schedule, work performance data, and organizational process assets. Some of tools and techniques include:

- Progress reports
- A schedule change control system, operated as part of the integrated change control system described in Chapter 4, Project Integration Management
- A scheduling tool and/or project management software, such as Project 2007 or similar software
- Schedule comparison bar charts, such as the Tracking Gantt chart
- Variance analysis, such as analyzing float or slack
- What-if scenario analysis, which can be done manually or with the aid of software
- Adjusting leads and lags
- Schedule compression, such as crashing and fast-tracking described earlier in this chapter
- Performance measurement, such as earned value, described in Chapter 7, Project Cost Management
- Resource leveling, as described in Chapter 9, Project Human Resource Management

The main outputs of schedule control include work performance measurements, organizational process assets updates, such as lessons-learned reports related to schedule control, change requests, project management plan updates, and project document updates.

There are many issues involved in controlling changes to project schedules. It is important first to ensure that the project schedule is realistic. Many projects, especially in information technology, have very unrealistic schedule expectations. It is also important to use discipline and leadership to emphasize the importance of following and meeting project schedules. Although the various tools and techniques assist in developing and managing project schedules, project managers must handle several people-related issues to keep projects on track. "Most projects fail because of people issues, not from failure to draw a good PERT chart."[8] Project managers can perform a number of reality checks that will help them manage changes to project schedules. Several soft skills can help project managers to control schedule changes.



## Reality Checks on Scheduling and the Need for Discipline

It is important for projects to have realistic schedule goals and for project managers to use discipline to help meet those goals. One of the first reality checks a project manager should make is to review the draft schedule usually included in the project charter. Although this draft schedule might include only a project start and end date, the project charter sets some initial schedule expectations for the project. Next, the project manager and his or her team should prepare a more detailed schedule and get stakeholders' approval. To establish the schedule, it is critical to get involvement and commitment from all project team members, top management, the customer, and other key stakeholders.

Several experts have written about the lack of realistic schedule estimates for information technology projects. Ed Yourdon, a well-known and respected software development expert, describes "death march" projects as projects doomed for failure from the start, due to unclear and/or unrealistic expectations, especially for meeting time constraints.[9] For example, in 2008, twenty months after a project was started to overhaul and integrate more than 300 key IT systems for the city of Portland, Oregon, the city changed IT consultants as the cost and schedule for the project continued to increase. Robert Stoll, a Portland attorney representing Ariston Consulting & Technologies, Inc. in the termination negotiations with the city, said Ariston did the best it could do with the information it was given by the city. From Ariston's point of view, he said, city officials who prepared the project requirements and put it out for bid weren't familiar enough with their overall IT systems and needs, making the goal for the work an impossible target to hit—a death march project. "It's sort of garbage in, garbage out, if you know what I mean," Stoll said. "I certainly don't think that Ariston made any mistakes."[10]

Experts in chaos theory suggest that project managers must account not only for project risks based upon their inaccuracies in estimation, but also for the complex behavior of the organization brought on by multiple independent project managers who also must estimate project risks. The interdependencies of such a complex system will typically result in a generally predictable inefficiency. They propose that this inefficiency can be addressed by arranging for additional resources to be committed to the project. Using an analogy related to the behavior of traffic systems, they suggest that project managers schedule resources on the project so that no single resource is utilized more than 75 percent. The actual inefficiency estimate can and should be assessed for each individual organization, but typically will vary between 70 and 80 percent.[11] It is very important, therefore, to set realistic project schedules and allow for contingencies throughout the life of a project.

Another type of reality check comes from progress meetings with stakeholders. The project manager is responsible for keeping the project on track, and key stakeholders like to stay informed, often through high-level periodic reviews. Managers like to see progress made on projects approximately each month. Project managers often illustrate progress with a Tracking Gantt chart showing key deliverables and activities. The project manager needs to understand the schedule, including why activities are or are not on track, and take a proactive approach to meeting stakeholder expectations. It is also important to verify schedule progress, as Sue Johnson from the opening case discovered (see the Case Wrap-Up). Just because a team member says a task was completed on time does not always mean that it was. Project managers must review the actual work and develop a good relationship with team members to ensure work is completed as planned or changes are reported as needed.



Top management hates surprises, so the project manager must be clear and honest in communicating project status. By no means should project managers create the illusion that the project is going fine when, in fact, it is having serious problems. When serious conflicts arise that could affect the project schedule, the project manager must alert top management and work with them to resolve the conflicts.

Project managers must also use discipline to control project schedules. Several information technology project managers have discovered that setting firm dates for key project milestones helps minimize schedule changes. It is very easy for scope creep to raise its ugly head on information technology projects. Insisting that important schedule dates be met and that proper planning and analysis be completed up front helps everyone focus on doing what is most important for the project. This discipline results in meeting project schedules.

## WHAT WENT RIGHT?

A recent example of using discipline and strong user involvement to meet project schedules comes from a winner of Poland's Project Excellence Award in 2007. Mittal Steel Poland is part of ArcelorMittal which is the world's number one steel company, with 320,000 employees in more than 60 countries. The main goal of Mittal Steel Poland's Implementation of SAP System project was to unify IT systems with other Mittal branches in Europe and to improve business and financial processes.[12] Research by AMR found that 46 percent of ERP licenses are unused, leaving many users with unnecessary support and maintenance bills. Derek Prior, research director at AMR Research, identified three things the most successful SAP implementation projects do to deliver business benefits:

- "Form a global competence centre. You need a team, including IT professionals and business analysts, who understand the intricacies of how the software is set up in your company and the business processes it enables.
- Identify super-users for each location. These need to be business managers keen to use the software to its full capacity who also have the respect of their local business peers so they can spread the message on what the software offers.
- You need ongoing involvement of managers in business processes so they feel they own these processes."[13]

# USING SOFTWARE TO ASSIST IN PROJECT TIME MANAGEMENT

Several types of software are available to assist in project time management. Software for facilitating communications helps project managers exchange schedule-related information with project stakeholders. Decision support models can help project managers analyze various trade-offs that can be made related to schedule issues. However, project management software, such as Microsoft Project 2007, was designed specifically for performing project management tasks. You can use project management software to draw network diagrams, determine the critical path for a project, create Gantt charts, and report, view, and filter specific project time management information.



Many projects involve hundreds of tasks with complicated dependencies. After you enter the necessary information, project management software automatically generates a network diagram and calculates the critical path(s) for the project. It also highlights the critical path in red on the network diagram. Project management software also calculates the free and total float or slack for all activities. For example, the data in Table 6-1 was created using Microsoft Project by selecting the schedule table view from the menu bar. Using project management software eliminates the need to perform cumbersome calculations manually and allows for "what if" analysis as activity duration estimates or dependencies change. Recall that knowing which activities have the most slack gives the project manager an opportunity to reallocate resources or make other changes to compress the schedule or help keep it on track.

Project 2007 easily creates Gantt charts and Tracking Gantt charts, which make tracking actual schedule performance versus the planned or baseline schedule much easier. It is important, however, to enter actual schedule information in a timely manner in order to benefit from using the Tracking Gantt chart feature. Some organizations use e-mail or other communications software to send up-to-date task and schedule information to the person responsible for updating the schedule. He or she can then quickly authorize these updates and enter them directly into the project management software. This process provides an accurate and up-to-date project schedule in Gantt chart form.

Project 2007 also includes many built-in reports, views, and filters to assist in project time management. For example, a project manager can quickly run a report to list all tasks that are to start soon. He or she could then send out a reminder to the people responsible for these tasks. If the project manager were presenting project schedule information to top management, he or she could create a Gantt chart showing only summary tasks or milestones. You can also create custom reports, views, tables, and filters. See Appendix A to learn how to use the project time management features of Project 2007.

## Words of Caution on Using Project Management Software

Many people misuse project management software because they do not understand the concepts behind creating a network diagram, determining the critical path, or setting a schedule baseline. They might also rely too heavily on sample files or templates in developing their own project schedules. Understanding the underlying concepts (even being able to work with the tools manually) is critical to successful use of project management software, as is understanding the specific needs of your project.

Many top managers, including software professionals, have made blatant errors using various versions of Microsoft Project and similar tools. For example, one top manager did not know about establishing dependencies among project activities and entered every single start and end date for hundreds of activities. When asked what would happen if the project started a week or two late, she responded that she would have to reenter all of the dates.

This manager did not understand the importance of establishing relationships among the tasks. Establishing relationships among tasks allows the software to update formulas automatically when the inputs change. If the project start date slips by one week, the project management software will update all the other dates automatically, as long as they are not hard-coded into the software. (Hard-coding involves entering all activity dates manually instead of letting the software calculate them based on durations and relationships.) If one activity cannot start before another ends, and the first activity's actual start date is two days late, the start date for the succeeding activity will automatically be moved back two days. To achieve this type of functionality, tasks that have relationships must be linked in project management software.

Another top manager on a large information technology project did not know about setting a baseline in Microsoft Project. He spent almost one day every week copying and pasting information from Microsoft Project into a spreadsheet and using complicated "IF" statements to figure out what activities were behind schedule. He had never received any training on Microsoft Project and did not know about many of its capabilities. To use any software effectively, users must have adequate training in the software and an understanding of its underlying concepts.

Many project management software programs also come with templates or sample files. It is very easy to use these files without considering unique project needs. For example, a project manager for a software development project can use the Microsoft Project Software Development template file, the files containing information from similar projects done in the past, or sample files purchased from other companies. All of these files include suggested tasks, durations, and relationships. There are benefits to using templates or sample files, such as less setup time and a reality check if the project manager has never managed this type of project before. However, there are also drawbacks to this approach. There are many assumptions in these template files that may not apply to the project, such as a design phase that takes three months to complete or the performance of certain types of testing. Project managers and their teams may over-rely on templates or sample files and ignore unique concerns for their particular projects.

## CASE WRAP-UP

It was now March 15, just a month and a half before the new online registration system was supposed to go live. The project was in total chaos. Sue Johnson thought she could handle all of the conflicts that kept appearing on the project, and she was too proud to admit to her top management or the college president that things were not going well. She spent a lot of time preparing a detailed schedule for the project, and she thought she was using their project management software well enough to keep up with project status. However, the five main programmers on the project all figured out a way to generate automatic updates for their tasks every week, saying that everything was completed as planned. They paid very little attention to the actual plan and hated filling out status information. Sue did not verify most of their work to check that it was actually completed. In addition, the head of the Registrar's Office was uninterested in the project and delegated sign-off responsibility to one of his clerks who really did not understand the entire registration process. When Sue and her team started testing the new system, she learned they were using last year's course data. Using last year's course data caused additional problems because the college was moving from quarters to semesters in the new semester. How could they have missed that requirement? Sue hung her head in shame as she walked into a meeting with her manager to ask for help. She learned the hard way how difficult it was to keep a project on track. She wished she had spent more time talking face-to-face with key project stakeholders, especially her programmers and the Registrar's Office representatives, to verify that the project was on schedule and that the schedule was updated accurately.

## Chapter Summary

Project time management is often cited as the main source of conflict on projects. Most information technology projects exceed time estimates. The main processes involved in project time management include defining activities, sequencing activities, estimating activity resources, estimating activity durations, developing the schedule, and controlling the schedule.

Defining activities involves identifying the specific activities that must be done to produce the project deliverables. It usually results in a more detailed WBS.



Sequencing activities determines the relationships or dependencies between activities. Three reasons for creating relationships are that they are mandatory based on the nature of the work, discretionary based on the project team's experience, or external based on non-project activities. Activity sequencing must be done in order to use critical path analysis.

Network diagrams are the preferred technique for showing activity sequencing. The two methods used for creating these diagrams are the arrow diagramming method and the precedence diagramming method. There are four types of relationships between tasks: finish-to-start, finish-to-finish, start-to-start, and start-to-finish.

Estimating activity resources involves determining the quantity and type of resources (people, equipment, and materials) that will be assigned to each activity. The nature of the project and the organization will affect resource estimating.

Estimating activity durations creates estimates for the amount of time it will take to complete each activity. These time estimates include the actual amount of time worked plus elapsed time.

Developing the schedule uses results from all of the other project time management processes to determine the start and end dates for the project. Project managers often use Gantt charts to display the project schedule. Tracking Gantt charts show planned and actual schedule information.

The critical path method predicts total project duration. The critical path for a project is the series of activities that determines the earliest completion date for the project. It is the longest path through a network diagram. If any activity on the critical path slips, the whole project will slip unless the project manager takes corrective action.

Crashing and fast tracking are two techniques for shortening project schedules. Project managers and their team members must be careful about accepting unreasonable schedules, especially for information technology projects.

Critical chain scheduling is an application of the Theory of Constraints (TOC) that uses critical path analysis, resource constraints, and buffers to help meet project completion dates.

The Program Evaluation and Review Technique (PERT) is a network analysis technique used to estimate project duration when there is a high degree of uncertainty about the individual activity duration estimates. It uses optimistic, most likely, and pessimistic estimates of activity durations. PERT is seldom used today.

Controlling the schedule is the final process in project time management. Even though scheduling techniques are very important, most projects fail because of people issues, not from a poor network diagram. Project managers must involve all stakeholders in the schedule development process. It is critical to set realistic project schedules and use discipline to meet schedule goals.

Project management software can assist in project scheduling if used properly. With project management software, you can avoid the need to perform cumbersome calculations manually and perform "what if" analysis as activity duration estimates or dependencies change. Many people misuse project management software because they do not understand the concepts behind creating a network diagram, determining the critical path, or setting a schedule baseline. Project managers must also avoid over-relying on sample files or templates when creating their unique project schedules.



## Quick Quiz

1. What is the first process in planning a project schedule?
   a. defining milestones
   b. defining activities
   c. estimating activity resources
   d. sequencing activity sequencing
2. Predecessors, successors, logical relationships, leads and lags, resource requirements, constraints, imposed dates, and assumptions are all examples of _______ .
   a. items in an activity list
   b. items on a Gantt chart
   c. milestone attributes
   d. activity attributes
3. As the project manager for a software development project, you are helping to develop the project schedule. You decide that writing code for a system should not start until users sign off on the analysis work. What type of dependency is this?
   a. technical
   b. mandatory
   c. discretionary
   d. external
4. You cannot start editing a technical report until someone else completes the first draft. What type of dependency does this represent?
   a. finish-to-start
   b. start-to-start
   c. finish-to-finish
   d. start-to-finish
5. Which of the following statements is false?
   a. A resource breakdown structure is a hierarchical structure that identifies the project's resources by category and type.
   b. Duration and effort are synonymous terms.
   c. A three-point estimate is an estimate that includes an optimistic, most likely, and pessimistic estimate.
   d. A Gantt chart is a common tool for displaying project schedule information.

6. What symbol on a Gantt chart represents a slipped milestone?
   a. a black arrow
   b. a white arrow
   c. a black diamond
   d. a white diamond
7. What type of diagram shows planned and actual project schedule information?
   a. a network diagram
   b. a Gantt chart
   c. a Tracking Gantt chart
   d. a milestone chart
8. ____________ is a network diagramming technique used to predict total project duration.
   a. PERT
   b. A Gantt chart
   c. Critical path method
   d. Crashing
9. Which of the following statements is false?
   a. "Growing grass" was on the critical path for a large theme park project.
   b. The critical path is the series of activities that determine the earliest time by which a project can be completed.
   c. A forward pass through a project network diagram determines the early start and early finish dates for each activity.
   d. Fast tracking is a technique for making cost and schedule trade-offs to obtain the greatest amount of schedule compression for the least incremental cost.
10. ____________ is a method of scheduling that considers limited resources when creating a project schedule and includes buffers to protect the project completion date.
    a. Parkinson's Law
    b. Murphy's Law
    c. Critical path analysis
    d. Critical chain scheduling

### Quick Quiz Answers

1. b; 2. d; 3. c; 4. a; 5. b; 6. d; 7. c; 8. a; 9. d; 10. d

## Discussion Questions

1. Why do you think schedule issues often cause the most conflicts on projects?
2. Why is defining activities the first process involved in project time management?
3. Why is it important to determine activity sequencing on projects? Discuss diagrams you have seen that are similar to network diagrams. Describe their similarities and differences.

4. How does activity resource estimating affect estimating activity durations?
5. Explain the difference between estimating activity durations and estimating the effort required to perform an activity.
6. Explain the following schedule development tools and concepts: Gantt charts, critical path method, PERT, and critical chain scheduling.
7. How can you minimize or control changes to project schedules?



8. List some of the reports you can generate with Project 2007 to assist in project time management.
9. Why is it difficult to use project management software well?

## Exercises

1. Using Figure 6-2, enter the activities, their durations (in days), and their relationships in Project 2007. Use a project start date of August 1, 2009. View the network diagram. Does it look like Figure 6-4? Print the network diagram on one page. Return to the Gantt Chart view. Click View on the menu bar, select Table: Entry, and then click Schedule to re-create Table 6-1. You may need to move the split bar to the right to reveal all of the table columns. (See Appendix  for detailed information on using Project 2007.) Write a few paragraphs explaining what the network diagram and schedule table show concerning Project X's schedule.
2. Consider Table 6-2, Network Diagram Data for a Small Project. All duration estimates or estimated times are in days; and the network proceeds from Node 1 to Node 9.

**TABLE 6-2** Network diagram data for a small project

| Activity | Initial Node | Final Node | Estimated Duration |
|---|---|---|---|
| A | 1 | 2 | 2 |
| B | 2 | 3 | 2 |
| C | 2 | 4 | 3 |
| D | 2 | 5 | 4 |
| E | 3 | 6 | 2 |
| F | 4 | 6 | 3 |
| G | 5 | 7 | 6 |
| H | 6 | 8 | 2 |
| I | 6 | 7 | 5 |
| J | 7 | 8 | 1 |
| K | 8 | 9 | 2 |

a. Draw an AOA network diagram representing the project. Put the node numbers in circles and draw arrows from node to node, labeling each arrow with the activity letter and estimated time.

b. Identify all of the paths on the network diagram and note how long they are, using Figure 6-8 as a guide for how to represent each path.

c. What is the critical path for this project and how long is it?

d. What is the shortest possible time it will take to complete this project?

e. Review the online tutorials for Gantt and PERT charts by Mark Kelly, available on the companion Web site or directly at *www.mckinnonsc.vic.edu.au/vceit/ganttpert*. Write a one-page paper with your answers to the questions in the tutorials. Also include any questions you had in doing the tutorials.

3. Consider Table 6-3, Network Diagram Data for a Large Project. All duration estimates or estimated times are in weeks; and the network proceeds from Node 1 to Node 8.

**TABLE 6-3** Network diagram data for a large project

| Activity | Initial Node | Final Node | Estimated Duration |
|---|---|---|---|
| A | 1 | 2 | 10 |
| B | 1 | 3 | 12 |
| C | 1 | 4 | 8 |
| D | 2 | 3 | 4 |
| E | 2 | 5 | 8 |
| F | 3 | 4 | 6 |
| G | 4 | 5 | 4 |
| H | 4 | 6 | 8 |
| I | 5 | 6 | 6 |
| J | 5 | 8 | 12 |
| K | 6 | 7 | 8 |
| L | 7 | 8 | 10 |

a. Draw an AOA network diagram representing the project. Put the node numbers in circles and draw arrows from node to node, labeling each arrow with the activity letter and estimated time.

b. Identify all of the paths on the network diagram and note how long they are, using Figure 6-8 as a guide for how to represent each path.

c. What is the critical path for this project and how long is it?

d. What is the shortest possible time it will take to complete this project?

4. Enter the information from Exercise 2 into Project 2007. View the network diagram and task schedule table to see the critical path and float or slack for each activity. Print the Gantt chart and network diagram views and the task schedule table. Write a short paper that interprets this information for someone unfamiliar with project time management.



5. Enter the information from Exercise 3 into Project 2007. View the network diagram and task schedule table to see the critical path and float or slack for each activity. Print the Gantt chart and network diagram views and the task schedule table. Write a short paper that interprets this information for someone unfamiliar with project time management.

6. You have been asked to determine a rough schedule for a nine-month Billing System Conversion project, as part of your job as a consultant to a Fortune 500 firm. The firm's old system was written in COBOL on a mainframe computer, and the maintenance costs are prohibitive. The new system will run on an off-the-shelf application. You have identified several high-level activities that must be done in order to initiate, plan, execute, control, and close the project. Table 6-4 shows your analysis of the project's tasks and schedule so far.

**TABLE 6-4** COBOL conversion project schedule

| Tasks | Mar | Apr | May | Jun | Jul | Aug | Sep | Oct | Nov |
|---|---|---|---|---|---|---|---|---|---|
| **Initiating** | | | | | | | | | |
| Develop project charter | | | | | | | | | |
| Meet with stakeholders | | | | | | | | | |
| **Planning** | | | | | | | | | |
| Create detailed WBS and schedule | | | | | | | | | |
| Estimate project costs | | | | | | | | | |
| Create project team | | | | | | | | | |
| Create communication plan | | | | | | | | | |
| Organize a comprehensive project plan | | | | | | | | | |
| **Executing** | | | | | | | | | |
| Award and manage contract for software conversion | | | | | | | | | |
| Install new software on servers | | | | | | | | | |
| Install new hardware and software on clients' machines | | | | | | | | | |
| Test new billing system | | | | | | | | | |
| Train users on new system | | | | | | | | | |
| **Controlling** | | | | | | | | | |
| Closing | | | | | | | | | |

a. Using the information in Table 6-4, draw horizontal bars to illustrate when you think each task would logically start and end. Then use Project 2007 to create a Gantt chart and network diagram based on this information.

b. Identify at least two milestones that could be included under each of the process groups in Table 6-4. Then write a detailed description of each of these milestones that meets the SMART criteria.

7. Interview someone who uses some of the techniques discussed in this chapter. How does he or she feel about network diagrams, critical path analysis, Gantt charts, critical chain scheduling, using project management software, and managing the people issues involved in project time management? Write a two-page paper describing the responses.

8. Review two different articles about critical chain scheduling. Write a two-page paper describing how this technique can help improve project schedule management.



## Running Case

Tony Prince is the project manager for the Recreation and Wellness Intranet Project, and team members include you, a programmer/analyst and aspiring project manager; Patrick, a network specialist; Nancy, a business analyst; and Bonnie, another programmer/analyst. Other people supporting the project from other departments are Yusaff from human resources, and Cassandra from finance. Assume these are the only people who can be assigned and charged to work on project activities. Recall that your schedule and cost goals are to complete the project in six months for under $200,000.

## Tasks

1. Review the WBS and Gantt chart you created for Chapter 5, Task 3. Propose three to five additional activities you think should be added to help you estimate resources and durations. Write a one-page paper describing these new activities.

2. Identify at least eight milestones for this project. Write a one-page paper describing each milestone using the SMART criteria. Discuss how determining these milestones might add additional activities or tasks to the Gantt chart. Remember that milestones normally have no duration, so you must have tasks that will lead to completing the milestone.

3. Using the Gantt chart you created for Chapter 5, Task 3, and the new activities and milestones you proposed in Tasks 1 and 2 above, create a new Gantt chart using Project 2007. Estimate the task durations and enter dependencies, as appropriate. Remember that your schedule goal for the project is six months. Print the Gantt chart and network diagram, each on one page.

4. Write a one-page paper summarizing how you would assign people to each activity. Include a table or matrix listing how many hours each person would work on each task. These resource assignments should make sense given the duration estimates made in Task 3 above. Remember that duration estimates are not the same as effort estimates since they include elapsed time.

## Key Terms



**activity** — an element of work, normally found on the WBS, that has an expected duration and cost, and expected resource requirements; also called a task

**activity attributes** — information about each activity, such as predecessors, successors, logical relationships, leads and lags, resource requirements, constraints, imposed dates, and assumptions related to the activity

**activity list** — a tabulation of activities to be included on a project schedule

**activity-on-arrow (AOA)** — a network diagramming technique in which activities are represented by arrows and connected at points called nodes to illustrate the sequence of activities; also called arrow diagramming method (ADM)

**arrow diagramming method (ADM)** — a network diagramming technique in which activities are represented by arrows and connected at points called nodes to illustrate the sequence of activities; also called activity-on-arrow (AOA)

**backward pass** — a project network diagramming technique that determines the late start and late finish dates for each activity in a similar fashion

**baseline dates** — the planned schedule dates for activities in a Tracking Gantt chart

**buffer** — additional time to complete a task, added to an estimate to account for various factors

**burst** — when a single node is followed by two or more activities on a network diagram

**crashing** — a technique for making cost and schedule trade-offs to obtain the greatest amount of schedule compression for the least incremental cost

**critical chain scheduling** — a method of scheduling that takes limited resources into account when creating a project schedule and includes buffers to protect the project completion date

**critical path** — the series of activities in a network diagram that determines the earliest completion of the project; it is the longest path through the network diagram and has the least amount of slack or float

**critical path method (CPM)** or **critical path analysis** — a project network analysis technique used to predict total project duration

**dependency** — the sequencing of project activities or tasks; also called a relationship

**discretionary dependencies** — sequencing of project activities or tasks defined by the project team and used with care since they may limit later scheduling options

**dummy activities** — activities with no duration and no resources used to show a logical relationship between two activities in the arrow diagramming method of project network diagrams

**duration** — the actual amount of time worked on an activity *plus* elapsed time

**early finish date** — the earliest possible time an activity can finish based on the project network logic

**early start date** — the earliest possible time an activity can start based on the project network logic

**effort** — the number of workdays or work hours required to complete a task

**external dependencies** — sequencing of project activities or tasks that involve relationships between project and non-project activities

**fast tracking** — a schedule compression technique in which you do activities in parallel that you would normally do in sequence

**feeding buffers** — additional time added before tasks on the critical path that are preceded by non-critical-path tasks

**finish-to-finish dependency** — a relationship on a project network diagram where the "from" activity must be finished before the "to" activity can be finished

**finish-to-start dependency** — a relationship on a project network diagram where the "from" activity must be finished before the "to" activity can be started

**float** — the amount of time a project activity may be delayed without delaying a succeeding activity or the project finish date; also called slack

**forward pass** — a network diagramming technique that determines the early start and early finish dates for each activity



**free slack (free float)** — the amount of time an activity can be delayed without delaying the early start of any immediately following activities

**Gantt chart** — a standard format for displaying project schedule information by listing project activities and their corresponding start and finish dates in a calendar format; sometimes referred to as bar charts

**late finish date** — the latest possible time an activity can be completed without delaying the project finish date

**late start date** — the latest possible time an activity may begin without delaying the project finish date

**mandatory dependencies** — sequencing of project activities or tasks that are inherent in the nature of the work being done on the project

**merge** — when two or more nodes precede a single node on a network diagram

**milestone** — a significant event that normally has no duration on a project; serves as a marker to help in identifying necessary activities, setting schedule goals, and monitoring progress

**multitasking** — when a resource works on more than one task at a time

**Murphy's Law** — principle that if something can go wrong, it will

**network diagram** — a schematic display of the logical relationships or sequencing of project activities

**node** — the starting and ending point of an activity on an activity-on-arrow diagram

**Parkinson's Law** — principle that work expands to fill the time allowed

**PERT weighted average** — $\frac{\text{optimistic time} + 4 * \text{most likely time} + \text{pessimistic time}}{6}$

**Precedence Diagramming Method (PDM)** — a network diagramming technique in which boxes represent activities

**probabilistic time estimates** — duration estimates based on using optimistic, most likely, and pessimistic estimates of activity durations instead of using one specific or discrete estimate

**Program Evaluation and Review Technique (PERT)** — a project network analysis technique used to estimate project duration when there is a high degree of uncertainty with the individual activity duration estimates

**project buffer** — additional time added before the project's due date

**project time management** — the processes required to ensure timely completion of a project

**relationship** — the sequencing of project activities or tasks; also called a dependency

**resource breakdown structure** — a hierarchical structure that identifies the project's resources by category and type

**resources** — people, equipment, and materials

**schedule baseline** — the approved planned schedule for the project

**slack** — the amount of time a project activity may be delayed without delaying a succeeding activity or the project finish date; also called float

**slipped milestone** — a milestone activity that is completed later than planned

**SMART criteria** — guidelines to help define milestones that are **s**pecific, **m**easurable, **a**ssignable, **r**ealistic, and **t**ime-framed

**start-to-finish dependency** — a relationship on a project network diagram where the "from" activity cannot start before the "to" activity is finished

**start-to-start dependency** — a relationship on a project network diagram in which the "from" activity cannot start until the "to" activity starts



**task** — an element of work, normally found on the WBS, that has an expected duration and cost, and expected resource requirements; also called an activity

**Theory of Constraints (TOC)** — a management philosophy that states that any complex system at any point in time often has only one aspect or constraint that is limiting its ability to achieve more of its goal

**three-point estimate** — an estimate that includes an optimistic, most likely, and pessimistic estimate

**total slack (total float)** — the amount of time an activity may be delayed from its early start without delaying the planned project finish date

**Tracking Gantt chart** — a Gantt chart that compares planned and actual project schedule information

## End Notes


[1] Kelly, Fran, "The World Today – Olympic planning schedule behind time," *ABC Online* (March 4, 2004).

[2] Weiner, Jay and Rachel Blount, "Olympics are safe but crowds are sparse," *Minneapolis Star Tribune* (August 22, 2004), A9.

[3] Roberts, Paul, "Frustrated contractor sentenced for hacking FBI to speed deployment," *InfoWorld Tech Watch*, (July 6, 2006).

[4] Losey, Stephen, "FBI to begin training employees on Sentinel case management system," FederalTimes.com (April 26, 2007).

[5] Richard, Luc K. "Reducing Schedule Risk, Parts 1 and 2," Gantthead.com (November 10, 2003 and January 31, 2005).

[6] Goldratt, Eliyahu, *Critical Chain* (Great Barrington, MA: The North River Press), 1997, p. 218.

[7] Breen, Anne M., Tracey Burton-Houle, and David C. Aron, "Applying the Theory of Constraints in Health Care: Part 1 - The Philosophy," *Quality Management in Health Care*, Volume 10, Number 3 (Spring 2002) (*www.goldratt.com/for-cause/applyingtocinhcpt1fco.htm*).

[8] Bolton, Bart, "IS Leadership," *ComputerWorld* (May 19, 1997).

[9] Yourdon, Ed, *Death March, Second Edition*, Upper Saddle River, NJ: Prentice Hall (2003).

[10] Weiss, Todd, "Portland IT overhaul runs into delays, excessive costs," ComputerWorld (May 15, 2008).

[11] Monson, Robert, *Conquering Chaos*, Unpublished Manuscript, Saint Paul, Minnesota (2008).

[12] International Project Management Association, "The best managed projects in the world 2007 awarded," www.ipma.ch (2008).

[13] ComputerWeekly.com, " IT rises to the challenge of standardisation of ERP as Mittal goes on a buying spree," (November 29, 2005).

CHAPTER 7

# PROJECT COST MANAGEMENT

## LEARNING OBJECTIVES

**After reading this chapter, you will be able to:**

- Understand the importance of project cost management
- Explain basic project cost management principles, concepts, and terms
- Discuss different types of cost estimates and methods for preparing them
- Understand the processes involved in cost budgeting and preparing a cost estimate and budget for an information technology project
- Understand the benefits of earned value management and project portfolio management to assist in cost control
- Describe how project management software can assist in project cost management

## OPENING CASE

Juan Gonzales was a systems analyst and network specialist for a major Mexican city's waterworks department. He enjoyed helping the city develop its infrastructure. His next career objective was to become a project manager so he could have even more influence. One of his colleagues invited him to attend an important project review meeting for large government projects, including the Surveyor Pro project, in which Juan was most interested. The Surveyor Pro project was a concept for developing a sophisticated information system that included expert systems, object-oriented databases, and wireless communications. The system would provide instant, graphical information to government surveyors to help them do their jobs. For example, after a surveyor touched a map on the screen of a handheld device, the system would prompt him or her for the type of information needed for that area. This system would help in planning and implementing many projects, from laying fiber-optic cable to installing water lines.

Juan was very surprised, however, when the majority of the meeting was spent discussing cost-related issues. The government officials were reviewing many existing projects to evaluate their performance to date and the potential impact on their budgets before discussing funding for any new projects. Juan did not understand many of the terms and charts the presenters were showing. What was this "earned value" term they kept referring to? How were they estimating what it would cost to complete projects or how long it would take? Juan thought he would learn more about the new technologies the Surveyor Pro project would use, but he discovered that the cost estimate and projected benefits were of most interest to the government officials at the meeting. It also seemed as if a lot of effort would go toward detailed financial studies before any technical work could even start. Juan wished he had taken some accounting and finance courses so he could understand the acronyms and concepts people were discussing. Although Juan had a degree in electrical engineering, he had no formal education and little experience in finance. If Juan could understand information systems and networks, he was confident that he could understand financial issues on projects, too. He jotted down questions to discuss with his colleagues after the meeting.



## THE IMPORTANCE OF PROJECT COST MANAGEMENT

Information technology projects have a poor track record in meeting budget goals. The Standish Group's CHAOS studies reported an average cost **overrun**—the additional percentage or dollar amount by which actual costs exceed estimates—for unsuccessful IT projects ranged from 180 percent in 1994 to 56 percent in 2004. Although academic researchers question the validity of these numbers, more rigorous, scientifically reviewed studies acknowledge the problem of cost overruns for IT projects. For example, three separate surveys of software project cost overruns done by Jenkins, Phan, and Bergeron in 1984, 1988, and 1992, respectively, found that the average cost overrun for all of the projects in their survey samples (not just unsuccessful projects) was 33–34 percent.[1] Obviously, there is room for improvement in meeting cost goals for IT projects. This chapter describes important concepts in project cost management, particularly, creating good estimates and using earned value management (EVM) to assist in cost control.

There is no shortage of examples of information technology projects that suffered from poor cost management. The U.S. Internal Revenue Service (IRS) and other government agencies continue to provide examples of how not to manage costs.

- The IRS managed a series of project failures in the 1990s that cost taxpayers more than $50 billion a year—roughly as much money as the annual net profit of the entire computer industry in those years.[2]
- In 2006, the IRS was again in the news for a botched upgrade to its fraud-detection software. The IRS planned to launch the system in January, in time for the 2006 tax season and one year *after* the original implementation date, but that did not happen. The U.S. government estimated that the lack of a functioning anti-fraud system cost $318 million in fraudulent refunds that didn't get caught.[3]
- A 2008 Government Accountability Office (GAO) report stated that more than 400 U.S. government agency IT projects, worth an estimated $25 billion, suffer from poor planning and underperformance. U.S. Senator Tom Carper of Delaware said the IT projects are redundant, lack clear goals, and are managed by unqualified individuals.[4]

Another example that shows the challenges of managing project costs was the United Kingdom's National Health Service (NHS) IT modernization program. Called "the greatest IT disaster in history" by one London columnist, this ten-year program, which started in 2002, was created to provide an electronic patient records system, appointment booking, and a prescription drug system in England and Wales. Britain's Labor government estimates that the program will eventually cost more than $55 billion, *a $26 billion overrun*. The program has been plagued by technical problems due to incompatible systems, resistance from physicians who say they were not adequately consulted about system features, and arguments among contractors about who's responsible for what.[5] A government audit in June 2006 found the program, one of the largest civilian IT projects undertaken worldwide, was progressing despite high-profile problems. In an effort to reduce cost overruns, the NHS program will no longer pay for products until delivery, shifting some financial responsibility to prime contractors, including BT Group, Accenture, and Fujitsu Services.[6]



## What Is Cost?

A popular cost accounting textbook states, "Accountants usually define cost as a resource sacrificed or foregone to achieve a specific objective."[7] Webster's dictionary defines cost as "something given up in exchange." Costs are often measured in monetary amounts, such as dollars, that must be paid to acquire goods and services. (For convenience, the examples in this chapter use dollars for monetary amounts, but monetary amounts could be in any currency.) Because projects cost money and consume resources that could be used elsewhere, it is very important for project managers to understand project cost management.

Many information technology professionals, however, often react to cost overrun information with a smirk. They know that many of the original cost estimates for information technology projects are low to begin with or based on very unclear project requirements, so naturally there will be cost overruns. Not emphasizing the importance of realistic project cost estimates from the outset is only one part of the problem. In addition, many information technology professionals think preparing cost estimates is a job for accountants. On the contrary, preparing good cost estimates is a very demanding, important skill that many professionals need to acquire.

Another perceived reason for cost overruns is that many information technology projects involve new technology or business processes. Any new technology or business process is untested and has inherent risks. Thus, costs grow and failures are to be expected, right? Wrong. Using good project cost management can change this false perception.



## What Is Project Cost Management?

Recall from Chapter 1 that the triple constraint of project management involves balancing scope, time, and cost goals. Chapters 5 and 6 discuss project scope and time management, and this chapter describes project cost management. **Project cost management** includes the processes required to ensure that a project team completes a project within an approved budget. Notice two crucial phrases in this definition: "a project" and "approved budget." Project managers must make sure *their* projects are well defined, have accurate time and cost estimates, and have a realistic budget that *they* were involved in approving. It is the project manager's job to satisfy project stakeholders while continuously striving to reduce and control costs. There are three project cost management processes:

1. *Estimating costs* involves developing an approximation or estimate of the costs of the resources needed to complete a project. The main outputs of the cost estimating process are activity cost estimates, basis of estimates, and project document updates. A cost management plan is created as part of integration management when creating the project management plan. It should include information related to the level of accuracy for estimates, variance thresholds for monitoring cost performance, reporting formats, and other related information.
2. *Determining the budget* involves allocating the overall cost estimate to individual work items to establish a baseline for measuring performance. The main outputs of the cost budgeting process are a cost performance baseline, project funding requirements, and project document updates.
3. *Controlling costs* involves controlling changes to the project budget. The main outputs of the cost control process are work performance measurements, budget forecasts, organizational process asset updates, change requests, project management plan updates, and project document updates.

Figure 7-1 summarizes these processes and outputs, showing when they occur in a typical project.

To understand each of the project cost management processes, you must first understand the basic principles of cost management. Many of these principles are not unique to project management; however, project managers need to understand how these principles relate to their specific projects.

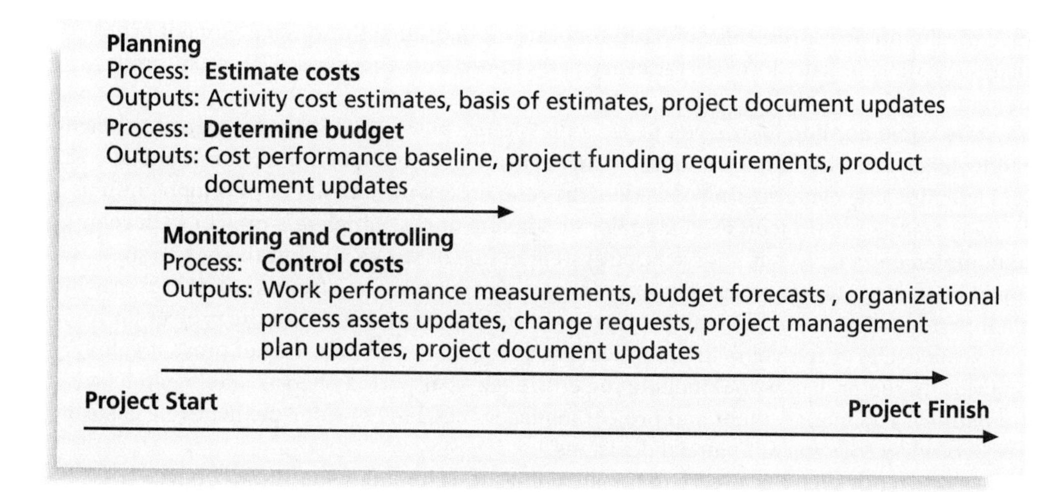




**FIGURE 7-1** Project cost management summary

## BASIC PRINCIPLES OF COST MANAGEMENT

Many information technology projects are never initiated because information technology professionals do not understand the importance of basic accounting and finance principles. Important concepts such as net present value analysis, return on investment, and payback analysis were discussed in Chapter 4, Project Integration Management. Likewise, many projects that are started never finish because of cost management problems. Most members of an executive board have a better understanding of and are more interested in financial terms than information technology terms. Therefore, information technology project managers need to be able to present and discuss project information in financial terms as well as in technical terms. In addition to net present value analysis, return on investment, and payback analysis, project managers must understand several other cost management principles, concepts, and terms. This section describes general topics such as profits, life cycle costing, cash flow analysis, tangible and intangible costs and benefits, direct costs, sunk costs, learning curve theory, and reserves. Another important topic and one of the key tools and techniques for controlling project costs—earned value management—is described in detail in the section on cost control.

**Profits** are revenues minus expenditures. To increase profits, a company can increase revenues, decrease expenses, or try to do both. Most executives are more concerned with profits than with other issues. When justifying investments in new information systems and technology, it is important to focus on the impact on profits, not just revenues or expenses. Consider an e-commerce application that you estimate will increase revenues for a $100 million company by 10 percent. You cannot measure the potential benefits of the e-commerce application without knowing the profit margin. **Profit margin** is the ratio of revenues to profits. If revenues of $100 generate $2 in profits, there is a 2 percent profit margin. If the company loses $2 for every $100 in revenue, there is a –2 percent profit margin.

**Life cycle costing** allows you to see a big-picture view of the cost of a project throughout its life cycle. This helps you develop an accurate projection of a project's financial costs and benefits. Life cycle costing considers the total cost of ownership, or development plus support costs, for a project. For example, a company might complete a project to develop and implement a new customer service system in one or two years, but the new system could be in place for ten years. Project managers, with assistance from financial experts in their organizations, should create estimates of the costs and benefits of the project for its entire life cycle, or ten years in the preceding example. Recall that the net present value analysis for the project would include the entire ten-year period of costs and benefits (see Chapter 4). Top management and project managers need to consider the life cycle costs of projects when they make financial decisions.



Organizations have a history of not spending enough money in the early phases of information technology projects, which impacts total cost of ownership. For example, it is much more cost-effective to spend money on defining user requirements and doing early testing on information technology projects than to wait for problems to appear after implementation. Recall from Chapter 5 that it can cost over 100 times more to correct a defect on a software project that is found late in the project than to fix it early.

Since organizations depend on reliable information technology, there are also huge costs associated with downtime. For example, Table 7-1 summarizes the average cost of a minute of downtime for different IT applications. Costs include the cost to bring the system back up, staff cost to make up for the lost work in production during the system downtime, and direct and indirect lost revenue.

**TABLE 7-1** Costs of downtime for IT applications[8]

| Type of IT Application | Cost/Minute |
|---|---|
| Securities trading | $73,000 |
| Enterprise Requirements Planning (ERP) | $14,800 |
| Order processing | $13,300 |
| Electronic commerce | $12,600 |
| Supply chain | $11,500 |
| Point of sale (POS) | $ 4,700 |
| Automatic teller machine (ATM) | $ 3,600 |
| E-mail | $ 1,900 |

The Standish Group International, "Trends in IT Value," (www.standishgroup.com) (2008).

## WHAT WENT RIGHT?

Many organizations use IT to reduce operational costs. For example, a recent study shows that technology has decreased the costs (adjusted for inflation) associated with processing an automated teller machine (ATM) transaction from a financial institution:

- In 1968, the average cost was $5.
- In 1978, the cost went down to $1.50
- In 1988, the cost was just a nickel.
- In 1998, it only cost a penny.
- In 2008, the cost was just half a penny![9]

Another cost-cutting strategy has been inspired by the global emphasis on improving the environment. Investing in green IT and other initiatives has helped both the environment and companies' bottom lines. Michael Dell, CEO of Dell, said he aimed to make his company "carbon neutral" in 2008. "The computer giant is looking to zero-out its carbon emissions through a number of initiatives, such as offering small businesses and consumers curbside recycling of their old computers, stuffing small recycling bags with free postage into new printer-ink cartridge boxes, and operating a 'Plant a Tree for Me' program."[10] (Dell did reach his goal. Visit *www.dell.com/earth* for more information.)



**Cash flow analysis** is a method for determining the estimated annual costs and benefits for a project and the resulting annual cash flow. Project managers must conduct cash flow analysis to determine net present value. Most consumers understand the basic concept of cash flow. If they do not have enough money in their wallets or checking/credit accounts, they cannot purchase something. Top management must consider cash flow concerns when selecting projects in which to invest. If top management selects too many projects that have high cash flow needs in the same year, the company will not be able to support all of its projects and maintain its profitability. It is also important to define clearly the year on which the company bases the dollar amounts. For example, if a company bases all costs on 2008 estimates, it would need to account for inflation and other factors when projecting costs and benefits in future-year dollars.

Tangible and intangible costs and benefits are categories for determining how definable the estimated costs and benefits are for a project. **Tangible costs or benefits** are those costs or benefits that an organization can easily measure in dollars. For example, suppose the Surveyor Pro project described in the opening case included a preliminary feasibility study. If a company completed this study for $100,000, the tangible cost of the study is $100,000. If Juan's government estimated that it would have cost $150,000 to do the study itself, the tangible benefits of the study would be $50,000 if it assigned the people who would have done the study to other projects. Conversely, **intangible costs or benefits** are costs or benefits that are difficult to measure in monetary terms. Suppose Juan and a few other people, out of personal interest, spent some time using government-owned computers, books, and other resources to research areas related to the study. Although their hours and the government-owned materials were not billed to the project, they could be considered intangible costs. Intangible benefits for projects often include items like goodwill, prestige, and general statements of improved productivity that an organization cannot easily translate into dollar amounts. Because intangible costs and benefits are difficult to quantify, they are often harder to justify.

**Direct costs** are costs that can be directly related to producing the products and services of the project. You can attribute direct costs directly to a certain project. For example, the salaries of people working full time on the project and the cost of hardware and software purchased specifically for the project are direct costs. Project managers should focus on direct costs, since they can control them.



**Indirect costs** are costs that are not directly related to the products or services of the project, but are indirectly related to performing the project. For example, the cost of electricity, paper towels, and so on in a large building housing a thousand employees who work on many projects would be indirect costs. Indirect costs are allocated to projects, and project managers have very little control over them.

**Sunk cost** is money that has been spent in the past. Consider it gone, like a sunken ship that can never be returned. When deciding what projects to invest in or continue, you should not include sunk costs. For example, in the opening case, suppose Juan's office had spent $1 million on a project over the past three years to create a geographic information system, but they never produced anything valuable. If his government were evaluating what projects to fund next year and someone suggested that they keep funding the geographic information system project because they had already spent $1 million on it, he or she would be incorrectly making sunk cost a key factor in the project selection decision. Many people fall into the trap of considering how much money has been spent on a failing project and, therefore, hate to stop spending money on it. This trap is similar to gamblers not wanting to stop gambling because they have already lost money. Sunk costs should be forgotten.

**Learning curve theory** states that when many items are produced repetitively, the unit cost of those items decreases in a regular pattern as more units are produced. For example, suppose the Surveyor Pro project would potentially produce 1,000 handheld devices that could run the new software and access information via satellite. The cost of the first handheld device or unit would be much higher than the cost of the thousandth unit. Learning curve theory should help estimate costs on projects involving the production of large quantities of items. Learning curve theory also applies to the amount of time it takes to complete some tasks. For example, the first time a new employee performs a specific task, it will probably take longer than the tenth time that employee performs a very similar task.

**Reserves** are dollars included in a cost estimate to mitigate cost risk by allowing for future situations that are difficult to predict. **Contingency reserves** allow for future situations that may be partially planned for (sometimes called **known unknowns**) and are included in the project cost baseline. For example, if an organization knows it has a 20 percent rate of turnover for information technology personnel, it should include contingency reserves to pay for recruiting and training costs for information technology personnel. **Management reserves** allow for future situations that are unpredictable (sometimes called **unknown unknowns**). For example, if a project manager gets sick for two weeks or an important supplier goes out of business, management reserve could be set aside to cover the resulting costs.

# ESTIMATING COSTS

Project managers must take cost estimates seriously if they want to complete projects within budget constraints. After developing a good resource requirements list, project managers and their project teams must develop several estimates of the costs for these resources. Recall from Chapter 6 that an important process in project time management is estimating activity resources, which provides a list of activity resource requirements. For example, if an activity on a project is to perform a particular type of test, the list of activity resource requirements would describe the skill level of the people needed to perform the test, the number of people and hours suggested to perform the test, the need for special software or equipment, and so on. All of this information is required to develop a good cost estimate. This section describes various types of cost estimates, tools and techniques for estimating costs, typical problems associated with information technology cost estimates, and a detailed example of a cost estimate for an information technology project.



## Types of Cost Estimates

One of the main outputs of project cost management is a cost estimate. Project managers normally prepare several types of cost estimates for most projects. Three basic types of estimates include the following:

- A **rough order of magnitude (ROM) estimate** provides an estimate of what a project will cost. ROM estimates can also be referred to as a ballpark estimate, a guesstimate, a swag, or a broad gauge. This type of estimate is done very early in a project or even before a project is officially started. Project managers and top management use this estimate to help make project selection decisions. The timeframe for this type of estimate is often three or more years prior to project completion. A ROM estimate's accuracy is typically −50 percent to +100 percent, meaning the project's actual costs could be 50 percent below the ROM estimate or 100 percent above. For example, the actual cost for a project with a ROM estimate of $100,000 could range between $50,000 to $200,000. For information technology project estimates, this accuracy range is often much wider. Many information technology professionals automatically double estimates for software development because of the history of cost overruns on information technology projects.
- A **budgetary estimate** is used to allocate money into an organization's budget. Many organizations develop budgets at least two years into the future. Budgetary estimates are made one to two years prior to project completion. The accuracy of budgetary estimates is typically −10 percent to +25 percent, meaning the actual costs could be 10 percent less or 25 percent more than the budgetary estimate. For example, the actual cost for a project with a budgetary estimate of $100,000 could range between $90,000 to $125,000.
- A **definitive estimate** provides an accurate estimate of project costs. Definitive estimates are used for making many purchasing decisions for which accurate estimates are required and for estimating final project costs. For example, if a project involves purchasing 1,000 personal computers from an outside supplier in the next three months, a definitive estimate would be required to aid in

evaluating supplier proposals and allocating the funds to pay the chosen supplier. Definitive estimates are made one year or less prior to project completion. A definitive estimate should be the most accurate of the three types of estimates. The accuracy of this type of estimate is normally −5 percent to +10 percent, meaning the actual costs could be 5 percent less or 10 percent more than the definitive estimate. For example, the actual cost for a project with a definitive estimate of $100,000 could range between $95,000 to $110,000. Table 7-2 summarizes the three basic types of cost estimates.



**TABLE 7-2** Types of cost estimates

| Type of Estimate | When Done | Why Done | How Accurate |
|---|---|---|---|
| **Rough Order of Magnitude (ROM)** | Very early in the project life cycle, often 3–5 years before project completion | Provides estimate of cost for selection decisions | −50% to +100% |
| **Budgetary** | Early, 1–2 years out | Puts dollars in the budget plans | −10% to +25% |
| **Definitive** | Later in the project, less than 1 year out | Provides details for purchases, estimates actual costs | −5% to +10% |

The number and type of cost estimates vary by application area. For example, the Association for the Advancement of Cost Engineering (AACE) International identifies five types of cost estimates for construction projects: order of magnitude, conceptual, preliminary, definitive, and control. The main point is that estimates are usually done at various stages of a project and should become more accurate as time progresses.

In addition to creating cost estimates, it is also important to provide supporting details for the estimates. The supporting details include the ground rules and assumptions used in creating the estimate, a description of the project (scope statement, WBS, and so on) used as a basis for the estimate, and details on the cost estimation tools and techniques used to create the estimate. These supporting details should make it easier to prepare an updated estimate or similar estimate as needed.

A **cost management plan** is a document that describes how the organization will manage cost variances on the project. For example, if a definitive cost estimate provides the basis for evaluating supplier cost proposals for all or part of a project, the cost management plan describes how to respond to proposals that are higher or lower than the estimates. Some organizations assume that a cost proposal within 10 percent of the estimate is acceptable and only negotiate items that are more than 10 percent higher or 20 percent lower than the estimated costs. The cost management plan is part of the overall project management plan described in Chapter 4, Project Integration Management.

Another important consideration in preparing cost estimates is labor costs, because a large percentage of total project costs are often labor costs. Many organizations estimate the number of people or hours they need by department or skill over the life cycle of a project. For example, when Northwest Airlines developed initial cost estimates for its reservation system project, ResNet, it determined the maximum number of people it could assign to the project each year by department. Table 7-3 shows this information. (Figure 9-7 in Chapter 9, Project Human Resource Management, provides similar resource information in graphical form, where the number of resources are provided by job category, such as business analyst, programmer, and so on.) Note the small number of contractors Northwest Airlines planned to use. Labor costs are often much higher for contractors, so it is important to distinguish between internal and external resources. (See the companion Web site for this text to read the detailed case study on ResNet, including cost estimates.)



**TABLE 7-3** Maximum departmental headcounts by year

| Department | Year 1 | Year 2 | Year 3 | Year 4 | Year 5 | Totals |
|---|---|---|---|---|---|---|
| Information systems | 24 | 31 | 35 | 13 | 13 | 116 |
| Marketing systems | 3 | 3 | 3 | 3 | 3 | 15 |
| Reservations | 12 | 29 | 33 | 9 | 7 | 90 |
| Contractors | 2 | 3 | 1 | 0 | 0 | 6 |
| Totals | 41 | 66 | 72 | 25 | 23 | 227 |

## Cost Estimation Tools and Techniques

As you can imagine, developing a good cost estimate is difficult. Fortunately, there are several tools and techniques available to assist in creating one. Commonly used tools and techniques include analogous cost estimating, bottom-up estimating, parametric modeling, the cost of quality, project management estimating software, vendor bid analysis, and reserve analysis.

**Analogous estimates**, also called **top-down estimates**, use the actual cost of a previous, similar project as the basis for estimating the cost of the current project. This technique requires a good deal of expert judgment and is generally less costly than others are, but it is also less accurate. Analogous estimates are most reliable when the previous projects are similar in fact, not just in appearance. In addition, the groups preparing cost estimates must have the needed expertise to determine whether certain parts of the project will be more or less expensive than analogous projects. For example, estimators often try to find a similar project and then customize/modify it for known differences. However, if the project to be estimated involves a new programming language or working with a new type of hardware or network, the analogous estimate technique could easily result in too low an estimate.

**Bottom-up estimates** involve estimating individual work items or activities and summing them to get a project total. It is sometimes referred to as activity-based costing. The size of the individual work items and the experience of the estimators drive the accuracy of the estimates. If a detailed WBS is available for a project, the project manager could have each person responsible for a work package develop his or her own cost estimate for that work package, or at least an estimate of the amount of resources required. Someone in the financial area of an organization often provides resource cost rates, such as labor rates or costs per pound of materials, which can be entered into project management software to calculate costs. The software automatically calculates information to create cost estimates for each level of the WBS and finally for the entire project. See Appendix A's section on project cost management for detailed information on entering resource costs and assigning resources to tasks to create a bottom-up estimate using Project 2007. Using smaller work items increases the accuracy of the cost estimate because the people assigned to do the work develop the cost estimate instead of someone unfamiliar with the work. The drawback with bottom-up estimates is that they are usually time-intensive and therefore expensive to develop.



**Parametric modeling** uses project characteristics (parameters) in a mathematical model to estimate project costs. A parametric model might provide an estimate of $50 per line of code for a software development project based on the programming language the project is using, the level of expertise of the programmers, the size and complexity of the data involved, and so on. Parametric models are most reliable when the historical information that was used to create the model is accurate, the parameters are readily quantifiable, and the model is flexible in terms of the size of the project. For example, in the 1980s, engineers at McDonnell Douglas Corporation (now part of Boeing) developed a parametric model for estimating aircraft costs based on a large historical database. The model included the following parameters: the type of aircraft (fighter aircraft, cargo aircraft, or passenger aircraft), how fast the plane would fly, the thrust-to-weight ratio of the engine, the estimated weights of various parts of the aircraft, the number of aircraft produced, the amount of time available to produce them, and so on. In contrast to this sophisticated model, some parametric models involve very simple heuristics or rules of thumb. For example, a large office automation project might use a ballpark figure of $10,000 per workstation based on a history of similar office automation projects developed during the same time period. Parametric models that are more complicated are usually computerized. See the Suggested Readings on the companion Web site for examples of parametric models, such as the COCOMO II model. In practice, many people find that using a combination or hybrid approach involving analogous, bottom up, and/or parametric modeling provides the best cost estimates.

Other considerations to make when preparing cost estimates are how much to include in reserves, as described earlier, the cost of quality, described in Chapter 8, Project Quality Management, and other cost estimating methods such as vendor bid analysis, as described in Chapter 12, Project Procurement Management. Using software to assist in cost estimating is described later in this chapter.

## Typical Problems with Information Technology Cost Estimates

Although there are many tools and techniques to assist in creating project cost estimates, many information technology project cost estimates are still very inaccurate, especially those involving new technologies or software development. Tom DeMarco, a well-known author on software development, suggests four reasons for these inaccuracies and some ways to overcome them.[11]

- *Estimates are done too quickly*. Developing an estimate for a large software project is a complex task requiring a significant amount of effort. Many estimates must be done quickly and before clear system requirements have been produced. For example, the Surveyor Pro project described in the opening case involves a lot of complex software development. Before fully understanding what information surveyors really need in the system, someone would have to create a ROM estimate and budgetary estimates for this project. Rarely are the more precise, later estimates less than the earlier estimates for information technology projects. It is important to remember that estimates are done at various stages of the project, and project managers need to explain the rationale for each estimate.
- *Lack of estimating experience*. The people who develop software cost estimates often do not have much experience with cost estimation, especially for large projects. There is also not enough accurate, reliable project data available on which to base estimates. If an organization uses good project management techniques and develops a history of keeping reliable project information, including estimates, it should help improve the organization's estimates. Enabling information technology people to receive training and mentoring on cost estimating will also improve cost estimates.
- *Human beings are biased toward underestimation*. For example, senior information technology professionals or project managers might make estimates based on their own abilities and forget that many junior people will be working on a project. Estimators might also forget to allow for extra costs needed for integration and testing on large information technology projects. It is important for project managers and top management to review estimates and ask important questions to make sure the estimates are not biased.
- *Management desires accuracy*. Management might ask for an estimate, but really wants a more accurate number to help them create a bid to win a major contract or get internal funding. This problem is similar to the situation discussed in Chapter 6, Project Time Management, in which top managers or other stakeholders want project schedules to be shorter than the estimates. It is important for project managers to help develop good cost and schedule estimates and to use their leadership and negotiation skills to stand by those estimates.



It is also important to be cautious with initial estimates. Top management never forgets the first estimate and rarely, if ever, remembers how approved changes affect the estimate. It is a never-ending and crucial process to keep top management informed about revised cost estimates. It should be a formal process, albeit a possibly painful one.

## Sample Cost Estimate

One of the best ways to learn how the cost estimating process works is by studying sample cost estimates. Every cost estimate is unique, just as every project is unique. You can see a short sample cost estimate in Chapter 3 for JWD Consulting's Project Management

Intranet Site project. You can also view the ResNet cost estimate on the companion Web site for this text.

This section includes a step-by-step approach for developing a cost estimate for the Surveyor Pro project described in the opening case. Of course, it is much shorter and simpler than a real cost estimate would be, but it illustrates a process to follow and uses several of the tools and techniques described earlier. For more detailed information on creating a cost estimate, see the NASA Cost Estimating Handbook and other references provided in the Suggested Readings on the companion Web site.



Before beginning any cost estimate, you must first gather as much information as possible about the project and ask how the organization plans to use the cost estimate. If the cost estimate will be the basis for contract awards and performance reporting, it should be a definitive estimate and as accurate as possible, as described earlier.

It is also important to clarify the ground rules and assumptions for the estimate. For the Surveyor Pro project cost estimate, these include the following:

- This project was preceded by a detailed study and proof of concept to show that it was possible to develop the hardware and software needed by surveyors and link the new devices to existing information systems. The proof of concept project produced a prototype handheld device and much of the software to provide basic functionality and link to the Global Positioning Systems (GPS) and other government databases used by surveyors. There is some data available to help estimate future labor costs, especially for the software development. There is also some data to help estimate the cost of the handheld devices.
- The main goal of this project is to produce 100 handheld devices, continue developing the software (especially the user interface), test the new system in the field, and train 100 surveyors in selected cities on how to use the new system. A follow-up contract is expected for a much larger number of devices based on the success of this project.
- There is a WBS for the project, as shown below:
  1. Project management
  2. Hardware
     - 2.1 Handheld devices
     - 2.2 Servers
  3. Software
     - 3.1 Licensed software
     - 3.2 Software development
  4. Testing
  5. Training and support
  6. Reserves
- Costs must be estimated by WBS and by month. The project manager will report progress on the project using earned value analysis, which requires this type of estimate.
- Costs will be provided in U.S. dollars. Since the project length is one year, inflation will not be included.
- The project will be managed by the government's project office. There will be a part-time project manager and four team members assigned to the project. The team members will help manage various parts of the project and provide

their expertise in the areas of software development, training, and support. Their total hours will be allocated as follows: 25 percent to project management, 25 percent to software development, 25 percent to training and support, and 25 percent to non-project work.

- The project involves purchasing the handheld devices from the same company that developed the prototype device. Based on producing 100 devices, the cost rate is estimated to be $600 per unit. The project will require four additional servers to run the software required for the devices and for managing the project.
- The project requires purchased software licenses for accessing the GPS and three other external systems. Software development includes developing a graphical user interface for the devices, an online help system, and a new module for tracking surveyor performance using the device.
- Testing costs should be low due to the success of the prototype project. An estimate based on 10 percent multiplied by the total hardware and software estimates should be sufficient.
- Training will include instructor-led classes in five different locations. The project team believes it will be best to outsource most of the training, including developing course materials, holding the sessions, and providing help desk support for three months as the surveyors start using their devices in the field.
- Because there are several risks related to this project, include 20 percent of the total estimate as reserves.
- You must develop a computer model for the estimate, making it easy to change several inputs, such as the number of labor hours for various activities or labor rates.



Fortunately, the project team can easily access cost estimates and actual information from similar projects. There is a great deal of information available from the proof of concept project, and the team can also talk to contractor personnel from the past project to help them develop the estimate. There are also some computer models available, such as a software-estimating tool based on function points.

Since the estimate must be provided by WBS and by month, the team first reviews a draft of the project schedule and makes further assumptions, as needed. They decide first to estimate the cost of each WBS item and then determine when the work will be performed, even though costs may be incurred at different times than the work is performed. Their budget expert has approved this approach for the estimate. Below are further assumptions and information for estimating the costs for each WBS category:

1. *Project management*: Estimate based on compensation for the part-time project manager and 25 percent of team members' time. The budget expert for this project suggested using a labor rate of $100/hour for the project manager and $75/hour for each team member, based on working an average of 160 hours per month, full time. Therefore, the total hours for the project manager under this category are 960 ($160/2 * 12 = 960$). Costs are also included for the four project team members working 25 percent of their time each, or, a total of 160 hours per month for all project personnel ($160 * 12 = 1920$). An additional amount will

be added for all contracted labor, estimated by multiplying 10 percent of their total estimates for software development and testing costs (10% * ($594,000 + $69,000)).

2. *Hardware*
   2.1 *Handheld devices*: 100 devices estimated by contractor at $600 per unit.
   2.2 *Servers*: Four servers estimated at $4,000 each, based on recent server purchases.



3. *Software*
   3.1 *Licensed software:* License costs will be negotiated with each supplier. Since there is a strong probability of large future contracts and great publicity if the system works well, costs are expected to be lower than usual. A cost of $200/handheld device will be used.
   3.2 *Software development:* This estimate will use two approaches: a labor estimate and a function point estimate. The higher estimate will be used. If the estimates are more than 20 percent apart, a third approach to providing the estimate will be required. The supplier who did the development on the proof of concept project will provide the labor estimate input, and local technical experts will make the function point estimates.
4. *Testing*: Based on similar projects, testing will be estimated as 10 percent of the total hardware and software cost.
5. *Training and support*: Based on similar projects, training will be estimated on a per-trainee basis, plus travel costs. The cost per trainee (100 total) will be $500, and travel will be $700/day/person for the instructors and project team members. It is estimated that there will be a total of 12 travel days. Labor costs for the project team members will be added to this estimate to assist in training and providing support after the training. The labor hours estimate for team members is 1,920 hours total.
6. *Reserves*: As directed, reserves will be estimated at 20 percent of the total estimate.

The project team then develops a cost model using the above information. Figure 7-2 shows a spreadsheet that summarizes the costs by WBS item based on the above information. Notice that the WBS items are listed in the first column, and sometimes the items are broken down into more detail based on how the costs are estimated. For example, the project management category includes three items to calculate costs for the project manager, the project team members, and the contractors, since all of these people will perform some project management activities. Also notice that there are columns for entering the number of units or hours and the cost per unit or hour. Several items are estimated using this approach. There are also some short comments within the estimate, such as reserves being 20 percent of the total estimate. Also notice that you can easily change several input variables, such as number of hours or cost per hour, to quickly change the estimate.

There is also an asterisk by the software development item in Figure 7-2, referring to another reference for detailed information on how this more complicated estimate was made. Recall that one of the assumptions was that software development must be estimated using two approaches, and that the higher estimate be used as long as both estimates were no more than 20 percent apart. The labor estimate in this case was slightly higher than the function

**Surveyor Pro Project Cost Estimate Created October 5**

| | # Units/Hrs. | Cost/Unit/Hr. | Subtotals | WBS Level 2 Totals | % of Total |
|---|---|---|---|---|---|
| WBS Items | | | | | |
| **1. Project Management** | | | | **$306,300** | **20%** |
| Project manager | 960 | $100 | $96,000 | | |
| Project team members | 1920 | $75 | $144,000 | | |
| Contractors (10% of software development and testing) | | | $66,300 | | |
| **2. Hardware** | | | | **$76,000** | **5%** |
| 2.1 Handheld devices | 100 | $600 | $60,000 | | |
| 2.2 Servers | 4 | $4,000 | $16,000 | | |
| **3. Software** | | | | **$614,000** | **40%** |
| 3.1 Licensed software | 100 | $200 | $20,000 | | |
| 3.2 Software development* | | | $594,000 | | |
| **4. Testing (10% of total hardware and software costs)** | | | $69,000 | **$69,000** | **5%** |
| **5. Training and Support** | | | | **$202,400** | **13%** |
| Trainee cost | 100 | $500 | $50,000 | | |
| Travel cost | 12 | $700 | $8,400 | | |
| Project team members | 1920 | $75 | $144,000 | | |
| **6. Reserves (20% of total estimate)** | | | $253,540 | **$253,540** | **17%** |
| **Total project cost estimate** | | | | **$1,521,240** | |

* See software development estimate

**FIGURE 7-2** Surveyor Pro project cost estimate

point estimate, so that number was used ($594,000 versus $562,158). Details are provided in Figure 7-3 on how the function point estimate was made. As you can see, there are many assumptions made in producing the function point estimate. Again, by putting the information into a cost model, you can easily change several inputs to adjust the estimate. See the referenced article and other sources for more information on function point estimates.

It is very important to have several people review the project cost estimate. It is also helpful to analyze the total dollar value as well as the percentage of the total amount for each major WBS category. For example, a senior executive could quickly look at the Surveyor Pro project cost estimate and decide if the numbers are reasonable and the assumptions are well documented. In this case, the government had budgeted $1.5 million for the project, so the estimate was right in line with that amount. The WBS Level 2 items (project management, hardware, software, testing, etc.) also seemed to be at appropriate percentages of the total cost based on similar past projects. In some cases, a project team might also be asked to provide a range estimate for each item instead of one discrete amount. For example, they might estimate that the testing costs will be between $60,000 and $80,000 and document their assumptions in determining those values. It is also import to update cost estimates, especially if any major changes occur on a project.

After the total cost estimate is approved, the team can then allocate costs for each month based on the project schedule and when costs will be incurred. Many organizations also require that the estimated costs be allocated into certain budget categories, as described in the next section.

Surveyor Pro Software Development Estimate Created October 5

| 1. Labor Estimate | # Units/Hrs. | Cost/Unit/Hr. | Subtotals | Calculations |
|---|---|---|---|---|
| Contractor labor estimate | 3000 | $150 | $450,000 | 3000 * 150 |
| Project team member estimate | 1920 | $75 | $144,000 | 1920 * 75 |
| **Total labor estimate** | | | **$594,000** | Sum above two values |
| | | | | |
| **2. Function point estimate**** | Quantity | Conversion Factor | Function Points | **Calculations** |
| External inputs | 10 | 4 | 40 | 10 * 4 |
| External interface files | 3 | 7 | 21 | 3 * 7 |
| External outputs | 4 | 5 | 20 | 4 * 5 |
| External queries | 6 | 4 | 24 | 6 * 4 |
| Logical internal tables | 7 | 10 | 70 | 7 * 10 |
| **Total function points** | | | **175** | Sum above function point values |
| Java 2 languange equivalency value | | | 46 | Assumed value from reference |
| Source lines of code (SLOC) estimate | | | 8,050 | 175 * 46 |
| Productivity×KSLOC^Penalty (in months) | | | 29.28 | 3.13 * 8.05^1.072 (see reference) |
| Total labor hours (160 hours/month) | | | 4,684.65 | 29.28 * 160 |
| Cost/labor hour ($120/hour) | | | $120 | Assumed value from budget expert |
| **Total function point estimate** | | | **$562,158** | 4684.65 * 120 |

**Approach based on paper by William Roetzheim, "Estimating Software Costs," Cost Xpert Group, Inc. (2003) using the COCOMO II default linear productivity factor (3.13) and penalty factor (1.072).

**FIGURE 7-3** Surveyor Pro software development estimate



## DETERMINING THE BUDGET

Determining the project budget involves allocating the project cost estimate to individual work items over time. These work items are based on the activities in the work breakdown structure for the project. The activity cost estimates, basis of estimates, scope baseline, project schedule, resource calendars, contracts, and organizational process assets are all inputs for determining the budget. The main goal of the cost budgeting process is to produce a cost baseline for measuring project performance and project funding requirements. It may also result in project document updates, such as items being added, removed, or modified to the scope statement or project schedule.

For example, the Surveyor Pro project team could use the cost estimate from Figure 7-2 along with the project schedule and other information to allocate costs for each month. Figure 7-4 provides an example of a cost baseline for this project. Again, it's important for the team to document assumptions they made when developing the cost baseline and have several experts review it.

Most organizations have a well-established process for preparing budgets. For example, many organizations require budget estimates to include the number of full-time equivalent (FTE) staff, often referred to as headcount, for each month of the project. This number provides the basis for estimating total compensation costs each year. Many organizations also want to know the amount of money projected to be paid to suppliers for their labor costs or other purchased goods and services. Other common budget categories include travel, depreciation, rents/leases, and other supplies and expenses. It is important to understand these budget categories before developing an estimate to make sure data is collected accordingly. Organizations use this information to track costs across projects and

**Surveyor Pro Project Cost Baseline Created October 10***

| WBS Items | 1 | 2 | 3 | 4 | 5 | 6 | 7 | 8 | 9 | 10 | 11 | 12 | Totals |
|---|---|---|---|---|---|---|---|---|---|---|---|---|---|
| 1. Project Management | | | | | | | | | | | | | |
| 1.1 Project manager | 8,000 | 8,000 | 8,000 | 8,000 | 8,000 | 8,000 | 8,000 | 8,000 | 8,000 | 8,000 | 8,000 | 8,000 | 96,000 |
| 1.2 Project team members | 12,000 | 12,000 | 12,000 | 12,000 | 12,000 | 12,000 | 12,000 | 12,000 | 12,000 | 12,000 | 12,000 | 12,000 | 144,000 |
| 1.3 Contractors | | 6,027 | 6,027 | 6,027 | 6,027 | 6,027 | 6,027 | 6,027 | 6,027 | 6,027 | 6,027 | 6,027 | 66,300 |
| 2. Hardware | | | | | | | | | | | | | |
| 2.1 Handheld devices | | | | 30,000 | 30,000 | | | | | | | | 60,000 |
| 2.2 Servers | | | | 8,000 | 8,000 | | | | | | | | 16,000 |
| 3. Software | | | | | | | | | | | | | |
| 3.1 Licensed software | | | | 10,000 | 10,000 | | | | | | | | 20,000 |
| 3.2 Software development | | 60,000 | 60,000 | 80,000 | 127,000 | 127,000 | 90,000 | 50,000 | | | | | 594,000 |
| 4. Testing | | | 6,000 | 8,000 | 12,000 | 15,000 | 15,000 | 13,000 | | | | | 69,000 |
| 5. Training and Support | | | | | | | | | | | | | |
| 5.1 Trainee cost | | | | | | | | | 50,000 | | | | 50,000 |
| 5.2 Travel cost | | | | | | | | | 8,400 | | | | 8,400 |
| 5.3 Project team members | | | | | | | 24,000 | 24,000 | 24,000 | 24,000 | 24,000 | 24,000 | 144,000 |
| 6. Reserves | | | | 10,000 | 10,000 | 30,000 | 30,000 | 60,000 | 40,000 | 40,000 | 30,000 | 3,540 | 253,540 |
| Totals | 20,000 | 86,027 | 92,027 | 172,027 | 223,027 | 198,027 | 185,027 | 173,027 | 148,427 | 90,027 | 80,027 | 53,567 | 1,521,240 |

*See the lecture slides for this chapter on the companion Web site for a larger view of this and other figures in this chapter. Numbers are rounded, so some totals appear to be off.

**FIGURE 7-4** Surveyor Pro project cost baseline



non-project work and look for ways to reduce costs. They also use the information for legal and tax purposes.

In addition to providing input for budgetary estimates, cost budgeting provides a cost baseline. A **cost baseline** is a time-phased budget that project managers use to measure and monitor cost performance. Estimating costs for each major project activity over time provides project managers and top management with a foundation for project cost control, as described in the next section. See Appendix A for information on using Project 2007 for cost control.

Cost budgeting, as well as requested changes or clarifications, may result in updates to the cost management plan, a subsidiary part of the project management plan. Cost budgeting also provides information for project funding requirements. For example, some projects have all funds available when the project begins, but others must rely on periodic funding to avoid cash flow problems. If the cost baseline shows that more funds are required in certain months than are expected to be available, the organization must make adjustments to avoid financial problems.

## MEDIA SNAPSHOT

Anyone who has run for a public office or worked on a political campaign knows how expensive they can be. Barack Obama and his campaign leaders understood the potential of today's news media and technology, especially for raising money and finding volunteers for his highly successful campaign to become the 44th President of the U.S.

- The Obama campaign used 16 different online social platforms, including Facebook, LinkedIn (business networkers), MySpace (youth), YouTube (video), Fickr (images), Digg (social bookmarking), Twitter (mobile), BlackPlanet (African-

*continued*

Americans), Eons (baby boomers), GLEE (gay and lesbians), and MiGente (Latinos) to interact with people from various backgrounds. Sources say 80 percent of all contributions originated from these social networks, and some say 90 percent of all contributions (which totaled over $600 million) were less than $100.[12]

- In a *60 Minutes* episode shortly after the election, campaign leaders discussed some of the details of the campaign. David Axelrod, Obama's chief strategist, recalled, "When we started the campaign, we met around a table like this. And there was just a handful of us. You know, we started with nothing. And Barack said to us, 'I want this to be a grassroots campaign. I wanna reinvigorate our democracy. First of all I think that's the only way we can win and secondly I want to rekindle some idealism that together we can get things done in this country.'"[13]
- The Web site My.BarackObama was created to develop an online community with over a million members. Users could get access to the tools they needed to effectively organize on a local level to help elect Obama. For example, the site helped users find local events and groups, contact undecided voters in their areas, and share their stories on blogs. Obama spread his message of change and invigorated millions of people to support him.



## CONTROLLING COSTS

Controlling project costs includes monitoring cost performance, ensuring that only appropriate project changes are included in a revised cost baseline, and informing project stakeholders of authorized changes to the project that will affect costs. The project management plan, project funding requirements, work performance data, and organizational process assets are inputs for controlling costs. Outputs of this process are work performance measurements, budget forecasts, organizational process asset updates, change requests, project management plan updates, and product document updates.

Several tools and techniques assist in project cost control. As shown in Appendix A, Project 2007 has many cost management features to help you enter budgeted costs, set a baseline, enter actuals, calculate variances, and run various cost reports. In addition to using software, however, there must be some change control system to define procedures for changing the cost baseline. This cost control change system is part of the integrated change control system described in Chapter 4, Project Integration Management. Since many projects do not progress exactly as planned, new or revised cost estimates are often required, as are estimates to evaluate alternate courses of action. Performance review meetings can be a powerful tool for helping to control project costs. People often perform better when they know they must report on their progress. Another very important tool for cost control is performance measurement. Although many general accounting approaches are available for measuring cost performance, earned value management (EVM) is a very powerful cost control technique that is unique to the field of project management.

## Earned Value Management

**Earned value management (EVM)** is a project performance measurement technique that integrates scope, time, and cost data. Given a cost performance baseline, project managers and their teams can determine how well the project is meeting scope, time, and cost goals by entering actual information and then comparing it to the baseline. A **baseline** is the original project plan plus approved changes. Actual information includes whether or not a WBS item was completed or approximately how much of the work was completed, when the work actually started and ended, and how much it actually cost to do the completed work.



In the past, earned value management was used primarily on large government projects. Today, however, more and more companies are realizing the value of using this tool to help control costs. In fact, a discussion by several academic experts in earned value management and a real practitioner revealed the need to clarify how to actually calculate earned value. Brenda Taylor, a senior project manager for P2 Project Management Solutions in Johannesburg, South Africa, questioned calculating earned value by simply multiplying the planned value to date by a percentage complete value. She suggested using the rate of performance instead, as described below.

Earned value management involves calculating three values for each activity or summary activity from a project's WBS.

1. The **planned value (PV)**, also called the budget, is that portion of the approved total cost estimate planned to be spent on an activity during a given period. Table 7-3 shows an example of earned value calculations. Suppose a project included a summary activity of purchasing and installing a new Web server. Suppose further that, according to the plan, it would take one week and cost a total of $10,000 for the labor hours, hardware, and software involved. The planned value (PV) for that activity that week is, therefore, $10,000.
2. The **actual cost (AC)** is the total direct and indirect costs incurred in accomplishing work on an activity during a given period. For example, suppose it actually took two weeks and cost $20,000 to purchase and install the new Web server. Assume that $15,000 of these actual costs were incurred during Week 1 and $5,000 was incurred during Week 2. These amounts are the actual cost (AC) for the activity each week.
3. The **earned value (EV)** is an estimate of the value of the physical work actually completed. It is based on the original planned costs for the project or activity and the rate at which the team is completing work on the project or activity to date. The **rate of performance (RP)** is the ratio of actual work completed to the percentage of work planned to have been completed at any given time during the life of the project or activity. For example, suppose the server installation was halfway completed by the end of Week 1. The rate of performance would be 50 percent (50/100) because by the end of Week 1, the planned schedule reflects that the task should be 100 percent complete and only 50 percent of that work has been completed. In Table 7-4, the earned value estimate after one week is therefore $5,000.[14]

**TABLE 7-4** Earned value calculations for one activity after Week 1

| Activity | Week 1 |
|---|---|
| Earned Value (EV) | 5,000 |
| Planned Value (PV) | 10,000 |
| Actual cost (AC) | 15,000 |
| Cost variance (CV) | −10,000 |
| Schedule variance (SV) | −5,000 |
| Cost performance index (CPI) | 33% |
| Schedule performance index (SPI) | 50% |



The earned value calculations in Table 7-4 are carried out as follows:

$$EV = 10,000 * 50\% = 5,000$$

$$CV = 5,000 - 15,000 = -10,000$$

$$SV = 5,000 - 10,000 = -5,000$$

$$CPI = 5,000/15,000 = 33\%$$

$$SPI = 5,000/10,000 = 50\%$$

Table 7-5 summarizes the formulas used in earned value management. Note that the formulas for variances and indexes start with EV, the earned value. Variances are calculated by subtracting the actual cost or planned value from EV, and indexes are calculated by

**TABLE 7-5** Earned value formulas

| Term | Formula |
|---|---|
| Earned value (EV) | EV = PV to date * RP |
| Cost variance (CV) | CV = EV – AC |
| Schedule variance (SV) | SV = EV – PV |
| Cost performance index (CPI) | CPI = EV/AC |
| Schedule performance index (SPI) | SPI = EV/PV |
| Estimate at completion (EAC) | EAC = BAC/CPI |
| Estimated time to complete | Original time estimate/SPI |

dividing EV by the actual cost or planned value. After you total the EV, AC, and PV data for all activities on a project, you can use the CPI and SPI to project how much it will cost and how long it will take to finish the project based on performance to date. Given the budget at completion and original time estimate, you can divide by the appropriate index to calculate the estimate at completion (EAC) and estimated time to complete, assuming performance remains the same. There are no standard acronyms for the term estimated time to complete or original time estimate.

**Cost variance (CV)** is the earned value minus the actual cost. If cost variance is a negative number, it means that performing the work cost more than planned. If cost variance is a positive number, it means that performing the work cost less than planned.



**Schedule variance (SV)** is the earned value minus the planned value. A negative schedule variance means that it took longer than planned to perform the work, and a positive schedule variance means that it took less time than planned to perform the work.

The **cost performance index (CPI)** is the ratio of earned value to actual cost and can be used to estimate the projected cost of completing the project. If the cost performance index is equal to one, or 100 percent, then the planned and actual costs are equal—the costs are exactly as budgeted. If the cost performance index is less than one or less than 100 percent, the project is over budget. If the cost performance index is greater than one or more than 100 percent, the project is under budget.

The **schedule performance index (SPI)** is the ratio of earned value to planned value and can be used to estimate the projected time to complete the project. Similar to the cost performance index, a schedule performance index of one, or 100 percent, means the project is on schedule. If the schedule performance index is greater than one or 100 percent, then the project is ahead of schedule. If the schedule performance index is less than one or 100 percent, the project is behind schedule.

Note that in general, *negative numbers for cost and schedule variance indicate problems in those areas*. Negative numbers mean the project is costing more than planned or taking longer than planned. Likewise, *CPI and SPI less than one or less than 100 percent also indicate problems*.

The cost performance index can be used to calculate the **estimate at completion (EAC)**—an estimate of what it will cost to complete the project based on performance to date. Similarly, the schedule performance index can be used to calculate an estimated time to complete the project.

You can graph earned value information to track project performance. Figure 7-5 shows an earned value chart for a one-year project after five months. Note that the actual cost and earned value lines end at five months, since that is the point in time where the data is collected or estimated. The chart includes three lines and two points, as follows:

- Planned value (PV), the cumulative planned amounts for all activities by month. Note that the planned value line extends for the estimated length of the project and ends at the BAC point.
- Actual cost (AC), the cumulative actual amounts for all activities by month.
- Earned value (EV), the cumulative earned value amounts for all activities by month.
- **Budget at completion (BAC)**, the original total budget for the project, or $100,000 in this example. The BAC point is plotted on the chart at the original time estimate of 12 months.

- Estimate at completion (EAC), estimated to be $122,308, in this example. This number is calculated by taking the BAC, or $100,000 in this case, and dividing by the CPI, which, in this example, was 81.761 percent. This EAC point is plotted on the chart at the estimated time to complete of 12.74 months. This number is calculated by taking the original time estimate, or 12 months in this case, and dividing by the SPI, which in this example was 94.203 percent.



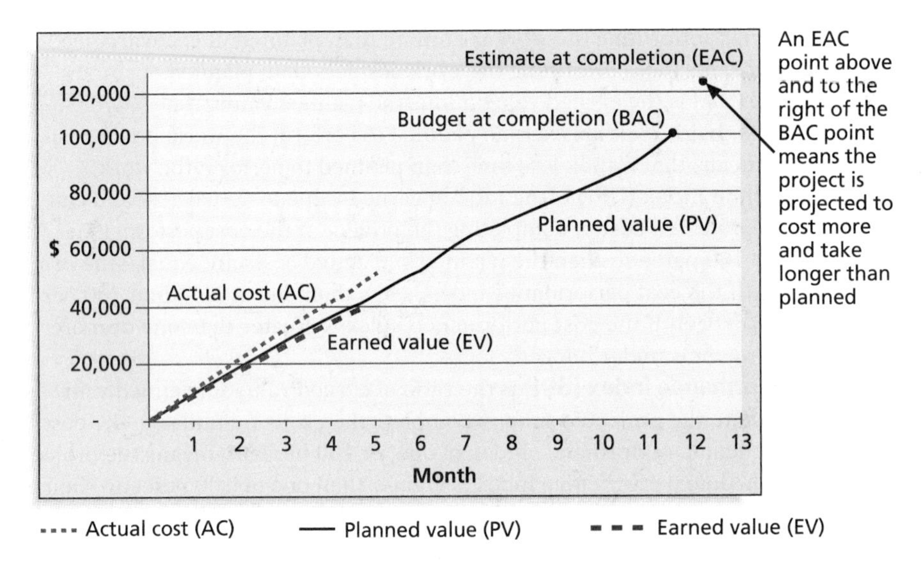


**FIGURE 7-5** Earned value chart for project after five months

Viewing earned value information in chart form helps you visualize how the project is performing. For example, you can see the planned performance by looking at the planned value line. If the project goes as planned, it will finish in 12 months and cost $100,000. Notice in the example in Figure 7-5 that the actual cost line is always right on or above the earned value line. When the actual cost line is right on or above the earned value line, costs are equal to or more than planned. The planned value line is pretty close to the earned value line, just slightly higher in the last month. This relationship means that the project has been right on schedule until the last month, when the project fell behind schedule.

Top managers overseeing multiple projects often like to see performance information in a graphical form such as the earned value chart in Figure 7-5. For example, in the opening case, the government officials were reviewing earned value charts and estimates at completion for several different projects. Earned value charts allow you to see how projects are performing quickly. If there are serious cost and schedule performance problems, top management may decide to terminate projects or take other corrective action. The estimates at completion (EAC) are important inputs to budget decisions, especially if total funds are limited. Earned value management is an important technique because, when used effectively, it helps top management and project managers evaluate progress and make sound management decisions.

If earned value management is such a powerful cost control tool, then why doesn't every organization use it? Why do many government projects require it, but many commercial projects don't? Two reasons why organizations do not widely use earned value management are its focus on tracking actual performance versus planned performance and the importance of percentage completion data in making calculations. Many projects, particularly information technology projects, do not have good planning information, so tracking performance against a plan might produce misleading information. Several cost estimates are usually made on information technology projects, and keeping track of the most recent cost estimate and the actual costs associated with it could be cumbersome. In addition, estimating percentage completion of tasks might produce misleading information. What does it really mean to say that a task is actually 75 percent complete after three months? Such a statement is often not synonymous with saying the task will be finished in one more month or after spending an additional 25 percent of the planned budget.



To make earned value management simpler to use, organizations can modify the level of detail and still reap the benefits of the technique. For example, you can use percentage completion data such as 0 percent for items not yet started, 50 percent for items in progress, and 100 percent for completed tasks. As long as the project is defined in enough detail, this simplified percentage completion data should provide enough summary information to allow managers to see how well a project is doing overall. You can get very accurate total project performance information using these simple percentage complete amounts. For example, using simplified percentage complete amounts for a one-year project with weekly reporting and an average task size of one week, you can expect about a 1 percent error rate.[15]

You can also only enter and collect earned value data at summary levels of the WBS. Quentin Fleming, author of the book *Earned Value Project Management*,[16] often gives presentations about earned value management. Many people express their frustration in trying to collect such detailed information. Quentin explains that you do not have to collect information at the work package level to use earned value management. It is most important to have a deliverable-oriented WBS, and many WBS items can summarize several subdeliverables. For example, you might have a WBS for a house that includes items for each room in the house. Just collecting earned value data for each room would provide meaningful information instead of trying to collect the detailed information on each component of the room, such as flooring, furniture, lighting, and so on.

It is important to remember that the heart and soul of EVM are estimates. The entire EVM process begins with an estimate and when the estimate is off, all the calculations will be off. Before an organization attempts to use EVM, it must learn to develop good estimates. Earned value management is the primary method available for integrating performance, cost, and schedule data. It can be a powerful tool for project managers and top management to use in evaluating project performance. Project management software, such as Project 2007, includes tables for collecting earned value data and reports that calculate variance information. Project 2007 also allows you to easily produce the earned value chart, similar to the one in Figure 7-5, without importing the data into Microsoft Excel. See the project cost management section of Appendix A for an example of using earned value management and the Suggested Readings on the companion Web site for more information. Another approach to evaluating the performance of multiple projects is project portfolio management.

## Project Portfolio Management



As mentioned in Chapter 1, many organizations now collect and control an entire suite of projects or investments as one set of interrelated activities in one place—a portfolio. Project managers need to understand how their projects fit into the bigger picture, and they need to help their organizations make wise investment decisions. Many project managers also want to move on to manage larger projects, become program managers, then vice presidents, and eventually CEOs. Understanding project portfolio management, therefore, is important for project and organizational success.

There can be a portfolio for information technology projects, for example, and portfolios for other types of projects. An organization can view project portfolio management as having five levels, from simplest to most complex, as follows:

1. Put all your projects in one database.
2. Prioritize the projects in your database.
3. Divide your projects into two or three budgets based on type of investment, such as utilities or required systems to keep things running, incremental upgrades, and strategic investments.
4. Automate the repository.
5. Apply modern portfolio theory, including risk-return tools that map project risk on a curve.

For example, Jane Walton, the project portfolio manager for information technology projects at Schlumberger, saved the company $3 million in one year by organizing the organization's 120 information technology projects into a portfolio. Manufacturing companies used project portfolio management in the 1960s, and Walton anticipated the need to justify investments in information technology projects just as managers have to justify capital investment projects. She found that 80 percent of the organization's projects overlapped, and 14 separate projects were trying to accomplish the same thing. Other managers, such as Douglas Hubbard, president of a consulting firm, see the need to use project portfolio management, especially for information technology projects. Hubbard suggests, "IT investments are huge, risky investments. It's time we do this."[17]

Project portfolio managers can start by using spreadsheet software to develop and manage project portfolios, or they can use sophisticated software designed to help manage project portfolios. Several software tools available today help project portfolio managers summarize earned value and project portfolio information, as described in the following section. Although many organizations have adopted project portfolio management tools and techniques (including portfolio management software) for information technology projects, they are not following best practices to truly reap the benefits, as described in the Best Practice feature below.

## BEST PRACTICE

A global survey released by Borland Software in 2006 suggests that many organizations are still at a low-level of maturity in terms of how they define project goals, allocate resources, and measure overall success of their information technology portfolios. Approximately 54 percent of survey respondents were from the Americas, 32 percent were from Asia Pacific, and 14 percent were from Europe, the Middle East and Africa. Some of the findings include the following:



- Only 22 percent of survey respondents reported that their organization either effectively or very effectively uses a project plan for managing projects.
- Only 17 percent have either rigorous or very rigorous processes for project plans, which include developing a baseline and estimating schedule, cost, and business impact of projects.
- Only 20 percent agreed their organizations monitor portfolio progress and coordinate across inter-dependent projects.
- The majority of respondents agreed their organization has no business impact assessment for completed projects, and success is measured only at the project level, based on performance against schedule and budget.
- Just two percent of survey respondents felt their organization was very effective at measuring performance of the overall portfolio.

According to Branndon Stewart, director of Information Technology Management and Governance products at Borland, "The most successful organizations are taking a holistic view of focusing, managing, and measuring their IT efforts with an integrated combination of best practice processes, training and technology. Unfortunately, most organizations today still aren't taking this approach.... IT leaders understand the value of a balanced portfolio aligned with business objectives, but most lack a well-defined and consistent process for managing the origination, evaluation, and execution of IT investments." Stewart continued, "Portfolio management enables IT to make fact-based investment decisions in unison with business stakeholders, thus ensuring alignment, improving visibility, and shifting the burden of investment decisions from the CIO to all stakeholders."[18]

## USING PROJECT MANAGEMENT SOFTWARE TO ASSIST IN PROJECT COST MANAGEMENT

Most organizations use software to assist in various activities related to project cost management. Spreadsheets are a common tool for cost estimating, cost budgeting, and cost control. Many companies also use more sophisticated and centralized financial applications software to provide important cost-related information to accounting and finance personnel. This section focuses specifically on how you can use project management software in cost management. Appendix A includes a section on using the cost management features in Project 2007.

Project management software can be a very helpful tool during each project cost management process. It can help you study overall project information or focus on tasks that are over a specified cost limit. You can use the software to assign costs to resources and tasks, prepare cost estimates, develop cost budgets, and monitor cost performance. Project 2007 has several standard cost reports: cash flow, budget, over budget tasks, over budget resources, and earned value reports. For several of these reports, you must enter percentage completion information and actual costs, just as you need this information when manually calculating earned value or other analyses.



Although Microsoft Project 2007 has a fair amount of cost management features, many information technology project managers use other tools to manage cost information; they do not know that they can use Project 2007 for cost management or they simply do not track costs based on a WBS, as most project management software does. As with most software packages, users need training to use the software effectively and understand the features available. Instead of using dedicated project management software for cost management, some information technology project managers use company accounting systems; others use spreadsheet software to achieve more flexibility. Project managers who use other software often do so because these other systems are more generally accepted in their organizations and more people know how to use them. In order to improve project cost management, several companies have developed methods to link data between their project management software and their main accounting software systems.

Many organizations are starting to use software to organize and analyze all types of project data into project portfolios and across the entire enterprise. Enterprise or project portfolio management tools integrate information from multiple projects to show the projects' status, or health. See Figure 1-5 in Chapter 1 for a sample. A 2008 study by the Gantry Group measured the return on investment of implementing portfolio management software by IT departments. They estimated savings of 6.5 percent of the average annual IT budget by the end of year one. They also found that using project portfolio management software had the following benefits:

- Improved the annual average project timeliness by 45.2 percent
- Reduced IT management time spent on project status reporting by 43 percent, reclaiming 3.8 hours of each manager's time per week
- Reduced IT management time spent on IT labor capitalization reporting by 55 percent, recouping 3.6 hours per report
- Decreased the time to achieve financial sign-off for new IT projects by 20.4 percent, or 8.4 days[19]

As with using any software, however, managers must make sure that the data is accurate and up-to-date and ask pertinent questions before making any major decisions.

## CASE WRAP-UP

After talking to his colleagues about the meeting, Juan had a better idea of the importance of project cost management. He understood the value of doing detailed studies before making major expenditures on new projects, especially after learning about the high cost of correcting defects late in a project. He also learned how important it is to develop good cost estimates and keep costs on track. He really enjoyed seeing how the cost estimate for the Surveyor Pro project was developed and was eager to learn more about various estimating tools and techniques.

At the meeting, government officials cancelled several projects when the project managers showed how poorly the projects were performing and admitted that they did not do much planning and analysis early in the projects. Juan knew that he could not focus on just the technical aspects of projects if he wanted to move ahead in his career. He began to wonder whether several projects the city was considering were really worth the taxpayers' money. Issues of cost management added a new dimension to Juan's job.

## Chapter Summary

Project cost management is a traditionally weak area of information technology projects. Information technology project managers must acknowledge the importance of cost management and take responsibility for understanding basic cost concepts, cost estimating, budgeting, and cost control.

Project managers must understand several basic principles of cost management in order to be effective in managing project costs. Important concepts include profits and profit margins, life cycle costing, cash flow analysis, sunk costs, and learning curve theory.



Estimating costs is a very important part of project cost management. There are several types of cost estimates, including rough order of magnitude (ROM), budgetary, and definitive. Each type of estimate is done during different stages of the project life cycle, and each has a different level of accuracy. There are several tools and techniques for developing cost estimates, including analogous estimating, bottom-up estimating, parametric modeling, and computerized tools. A detailed example of a project cost estimate is provided, illustrating how to apply several of these concepts.

Determining the budget involves allocating costs to individual work items over time. It is important to understand how particular organizations prepare budgets so estimates are made accordingly.

Controlling costs includes monitoring cost performance, reviewing changes, and notifying project stakeholders of changes related to costs. Many basic accounting and finance principles relate to project cost management. Earned value management is an important method used for measuring project performance. Earned value management integrates scope, cost, and schedule information. Project portfolio management allows organizations to collect and control an entire suite of projects or investments as one set of interrelated activities.

Several software products assist with project cost management. Project 2007 has many cost management features, including earned value management. Enterprise project management software and portfolio management software can help managers evaluate data on multiple projects.

## Quick Quiz

1. ____________________ is a resource sacrificed or foregone to achieve a specific objective or something given up in exchange.
   a. Money
   b. Liability
   c. Trade
   d. Cost
2. What is the main goal of project cost management?
   a. to complete a project for as little cost as possible
   b. to complete a project within an approved budget
   c. to provide truthful and accurate cost information on projects
   d. to ensure that an organization's money is used wisely

3. Which of the following is not a key output of project cost management?
    a. activity cost estimates
    b. a cost management plan
    c. updates to the project management plan
    d. a cost performance baseline
4. If a company loses $5 for every $100 in revenue for a certain product, what is the profit margin for that product?
    a. –5 percent
    b. 5 percent
    c. –$5
    d. $5
5. ________________ reserves allow for future situations that are unpredictable.
    a. Contingency
    b. Financial
    c. Management
    d. Baseline
6. You are preparing a cost estimate for a building based on its location, purpose, number of square feet, and other characteristics. What cost estimating technique are you using?
    a. parametric
    b. analogous
    c. bottom-up
    d. top-down
7. ________________ involves allocating the project cost estimate to individual work items over time.
    a. Reserve analysis
    b. Life cycle costing
    c. Project cost budgeting
    d. Earned value analysis
8. ________________ is a project performance measurement technique that integrates scope, time, and cost data.
    a. Reserve analysis
    b. Life cycle costing
    c. Project cost budgeting
    d. Earned value analysis

9. If the actual cost for a WBS item is $1500 and its earned value was $2000, what is its cost variance, and is it under or over budget?
   a. the cost variance is –$500, which is over budget
   b. the cost variance is –$500, which is under budget
   c. the cost variance is $500, which is over budget
   d. the cost variance is $500, which is under budget



10. If a project is halfway completed and its schedule performance index is 110 percent and its cost performance index is 95 percent, how is it progressing?
    a. it is ahead of schedule and under budget
    b. it is ahead of schedule and over budget
    c. it is behind schedule and under budget
    d. it is behind schedule and over budget

### Quick Quiz Answers

1. d; 2. b; 3. b; 4. a; 5. c; 6. a; 7. c; 8. d; 9. d; 10. b

## Discussion Questions

1. Discuss why many information technology professionals may overlook project cost management and how this might affect completing projects within budget.
2. Explain some of the basic principles of cost management, such as profits, life cycle costs, tangible and intangible costs and benefits, direct and indirect costs, reserves, and so on.
3. Give examples of when you would prepare rough order of magnitude (ROM), budgetary, and definitive cost estimates for an information technology project. Give an example of how you would use each of the following techniques for creating a cost estimate: analogous, parametric, and bottom-up.
4. Explain what happens during the process to determine the project budget.
5. Explain how earned value management (EVM) can be used to control costs and measure project performance and speculate as to why it is not used more often. What are some general rules of thumb for deciding if cost variance, schedule variance, cost performance index, and schedule performance index numbers are good or bad?
6. What is project portfolio management? Can project managers use it with earned value management?
7. Describe several types of software that project managers can use to support project cost management.

## Exercises

1. Given the following information for a one-year project, answer the following questions. Recall that PV is the planned value, EV is the earned value, AC is the actual cost, and BAC is the budget at completion.

   PV = $ 23,000
   EV = $ 20,000
   AC = $ 25,000
   BAC = $ 120,000

   a. What is the cost variance, schedule variance, cost performance index (CPI), and schedule performance index (SPI) for the project?
   b. How is the project doing? Is it ahead of schedule or behind schedule? Is it under budget or over budget?
   c. Use the CPI to calculate the estimate at completion (EAC) for this project. Is the project performing better or worse than planned?
   d. Use the schedule performance index (SPI) to estimate how long it will take to finish this project.
   e. Sketch the earned value chart based for this project, using Figure 7-5 as a guide.
2. Create a cost estimate/model for building a new, state-of-the-art multimedia classroom for your organization within the next six months. The classroom should include 20 high-end personal computers with appropriate software for your organization, a network server, Internet access for all machines, an instructor station, and a projection system. Be sure to include personnel costs associated with the project management for this project. Document the assumptions you made in preparing the estimate and provide explanations for key numbers.
3. Research online information, textbooks, and local classes for using the cost management features of Project 2007 or other software tools, including project portfolio management software (e.g., CA Clarity PPM, Primavera, Daptiv PPM, etc.). Also review the project cost management section of Appendix A. Ask three people in different information technology organizations that use project management software if they use the cost management features of the software and in what ways they use them. Write a brief report of what you learned from your research.
4. Read the article by William Roetzheim on the companion Web site on estimating software costs. Find one or two other references on using SLOC and function points. In your own words, explain how to estimate software development costs using both of these approaches in a two-page paper.
5. Create a spreadsheet to calculate your projected total costs, total revenues, and total profits for giving a seminar on cost estimating. Below are some of your assumptions:
   - You will charge $600 per person for a two-day class.
   - You estimate that 30 people will register for and attend the class, but you want to change this input.
   - Your fixed costs include $500 total to rent a room for both days, setup fees of $400 for registration, and $300 for designing a postcard for advertising.

- You will not include any of your labor costs for this estimate, but you estimate that you will spend at least 150 hours developing materials, managing the project, and giving the actual class. You would like to know what your time is worth given different scenarios.
- You will order 5,000 postcards, mail 4,000, and distribute the rest to friends and colleagues.
- Your variable costs include the following:
  a. –$5 per person for registration plus four percent of the class fee per person to handle credit card processing; assume everyone pays by credit card.
  b. $.40 per postcard for printing if you order 5,000 or more.
  c. $.25 per postcard for mailing and postage.
  d. $25 per person for beverages and lunch.
  d. $30 per person for class handouts.



Be sure to have input cells for any variables that might change, such as the cost of postage, handouts, and so on. Calculate your profits based on the following number of people who attend: 10, 20, 30, 40, 50, and 60. In addition, calculate what your time would be worth per hour based on the number of students. Try to use the Excel data table feature showing the profits based on the number of students. If you are unfamiliar with data tables, just repeat the calculations for each possibility of 10, 20, 30, 40, 50, and 60 students. Print your results on one page, highlighting the profits for each scenario and what your time is worth.

## Running Case

Tony Prince and his team working on the Recreation and Wellness Intranet Project have been asked to refine the existing cost estimate for the project so they can evaluate supplier bids and have a solid cost baseline for evaluating project performance. Recall that your schedule and cost goals are to complete the project in six months for under $200,000.

## Tasks

1. Prepare and print a one-page cost model for the project, similar to the one provided in Figure 7-2. Use the following WBS, and be sure to document assumptions you make in preparing the cost model. Assume a labor rate of $100/hour for the project manager and $60/hour for other project team members. Assume that none of the work is outsourced, labor costs for users are not included, and there are no additional hardware costs. The total estimate should be $200,000.

   1. Project management
   2. Requirements definition
   3. Web site design

      3.1 Registration for recreational programs

      3.2 Registration for classes and programs

      3.3 Tracking system

      3.4 Incentive system

4. Web site development
    4.1 Registration for recreational programs
    4.2 Registration for classes and programs
    4.3 Tracking system
    4.4 Incentive system
5. Testing
6. Training, roll out, and support



2. Using the cost model you created above, prepare a cost baseline by allocating the costs by WBS for each month of the project.
3. Assume you have completed three months of the project. The BAC was $200,000 for this six-month project. Also assume the following:

    PV = $ 120,000

    EV = $ 100,000

    AC = $ 90,000

    a. What is the cost variance, schedule variance, cost performance index (CPI), and schedule performance index (SPI) for the project?
    b. How is the project doing? Is it ahead of schedule or behind schedule? Is it under budget or over budget?
    c. Use the CPI to calculate the estimate at completion (EAC) for this project. Is the project performing better or worse than planned?
    d. Use the schedule performance index (SPI) to estimate how long it will take to finish this project.
    e. Sketch an earned value chart using the above information. See Figure 7-5 as a guide.

## Companion Web Site

Visit the companion Web site for this text at *www.cengage.com/mis/schwalbe* to access:

- References cited in the text and additional suggested readings for each chapter
- Template files
- Lecture notes
- Interactive quizzes
- Podcasts
- Links to general project management Web sites
- And more

See the Preface of this text for additional information on accessing the companion Web site.

## Key Terms

**actual cost (AC)** — the total of direct and indirect costs incurred in accomplishing work on an activity during a given period.

**analogous estimates** — a cost estimating technique that uses the actual cost of a previous, similar project as the basis for estimating the cost of the current project, also called top-down estimates.

**baseline** — the original project plan plus approved changes.



**bottom-up estimates** — a cost estimating technique based on estimating individual work items and summing them to get a project total.

**budget at completion (BAC)** — the original total budget for a project.

**budgetary estimate** — a cost estimate used to allocate money into an organization's budget.

**cash flow analysis** — a method for determining the estimated annual costs and benefits for a project.

**contingency reserves** — dollars included in a cost estimate to allow for future situations that may be partially planned for (sometimes called **known unknowns**) and are included in the project cost baseline.

**controlling costs** — controlling changes to the project budget.

**cost baseline** — a time-phased budget that project managers use to measure and monitor cost performance.

**cost management plan** — a document that describes how cost variances will be managed on the project.

**cost performance index (CPI)** — the ratio of earned value to actual cost; can be used to estimate the projected cost to complete the project.

**cost variance (CV)** — the earned value minus the actual cost.

**definitive estimate** — a cost estimate that provides an accurate estimate of project costs.

**determining the budget** — allocating the overall cost estimate to individual work items to establish a baseline for measuring performance.

**direct costs** — costs that can be directly related to producing the products and services of the project.

**earned value (EV)** — an estimate of the value of the physical work actually completed.

**earned value management (EVM)** — a project performance measurement technique that integrates scope, time, and cost data.

**estimate at completion (EAC)** — an estimate of what it will cost to complete the project based on performance to date.

**estimating costs** — developing an approximation or estimate of the costs of the resources needed to complete the project.

**indirect costs** — costs that are not directly related to the products or services of the project, but are indirectly related to performing the project.

**intangible costs or benefits** — costs or benefits that are difficult to measure in monetary terms.

**known unknowns** — dollars included in a cost estimate to allow for future situations that may be partially planned for (sometimes called **contingency reserves**) and are included in the project cost baseline.

**learning curve theory** — a theory that states that when many items are produced repetitively, the unit cost of those items normally decreases in a regular pattern as more units are produced.

**life cycle costing** — considers the total cost of ownership, or development plus support costs, for a project.

**management reserves** — dollars included in a cost estimate to allow for future situations that are unpredictable (sometimes called unknown unknowns).

**overrun** — the additional percentage or dollar amount by which actual costs exceed estimates.

**parametric modeling** — a cost-estimating technique that uses project characteristics (parameters) in a mathematical model to estimate project costs.

**planned value (PV)** — that portion of the approved total cost estimate planned to be spent on an activity during a given period.



**profit margin** — the ratio between revenues and profits.

**profits** — revenues minus expenses.

**project cost management** — the processes required to ensure that the project is completed within the approved budget.

**rate of performance (RP)** — the ratio of actual work completed to the percentage of work planned to have been completed at any given time during the life of the project or activity.

**reserves** — dollars included in a cost estimate to mitigate cost risk by allowing for future situations that are difficult to predict.

**rough order of magnitude (ROM) estimate** — a cost estimate prepared very early in the life of a project to provide a rough idea of what a project will cost.

**schedule performance index (SPI)** — the ratio of earned value to planned value; can be used to estimate the projected time to complete a project.

**schedule variance (SV)** — the earned value minus the planned value.

**sunk cost** — money that has been spent in the past.

**tangible costs or benefits** — costs or benefits that can be easily measured in dollars.

**top-down estimates** — a cost estimating technique that uses the actual cost of a previous, similar project as the basis for estimating the cost of the current project, also called analogous estimates.

**unknown unknowns** — dollars included in a cost estimate to allow for future situations that are unpredictable (sometimes called management reserves).

## End Notes


[1] Magne Jørgensen and Kjetil Moløkken, "How Large Are Software Cost Overruns? A Review of the 1994 CHAOS Report," Simula Research Laboratory (2006).

[2] Geoffrey James, "IT Fiascoes ... and How to Avoid Them," *Datamation* (November 1997).

[3] Paul McDougall, "8 Expensive IT Blunders," *InformationWeek* (October 16, 2006).

[4] Roy Mark,"GAO: Billions Wasted on Federal IT Projects," eWeek.com (July 31, 2008).

[5] Paul McDougall, "U.K. Health System IT Upgrade Called A 'Disaster'" *InformationWeek* (June 5, 2006).

[6] Jeremy Kirk, "Datacenter failure pinches U.K. health service," IDG News Service (August 1, 2006).

[7] Charles T. Horngren, George Foster, and Srikanti M. Datar, *Cost Accounting*, 8th ed. Englewood Cliffs, NJ: Prentice-Hall, 1994.

[8] The Standish Group International, "Trends in IT Value," (www.standishgroup.com) (2008).

[9] Ibid.

[10] Dawn Kawamoto, "Dell's green goal for 2008," CNET (September 27, 2007).

[11] Tom DeMarco, *Controlling Software Projects*. New York: Yourdon Press, 1982.



[12] David Krejci, "Message received," StarTribune, Minneapolis, MN (November 5, 2008).

[13] Steve Kroft, "Obama's Inner Circle Shares Inside Story," CBS News (November 9, 2008).

[14] Brenda Taylor, P2 Senior Project Manager, P2 Project Management Solutions, Johannesburg, South Africa, E-mail, 2004.

[15] Daniel M. Brandon, Jr., "Implementing Earned Value Easily and Effectively," *Project Management Journal* (June 1998) 29 (2), p. 11–18.

[16] Quentin W. Fleming and Joel M. Koppelman, *Earned Value Project Management, Third Edition* (Project Management Institute, 2006).

[17] Scott Berinato, "Do the Math," *CIO Magazine* (October 1, 2001).

[18] Borland Software, "Organizations Making Progress with IT Management and Governance but Still Face Significant Challenges according to Borland Survey," Borland Press Release (August 28, 2006).

[19] Michael Kringsman, "ROI Study: Product Portfolio Management Yields Results," *ZDNet* (March 7, 2008).

CHAPTER 8

# PROJECT QUALITY MANAGEMENT

## LEARNING OBJECTIVES

**After reading this chapter, you will be able to:**

- Understand the importance of project quality management for information technology products and services
- Define project quality management and understand how quality relates to various aspects of information technology projects
- Describe quality planning and its relationship to project scope management
- Discuss the importance of quality assurance
- Explain the main outputs of the quality control process
- Understand the tools and techniques for quality control, such as the Seven Basic Tools of Quality, statistical sampling, Six Sigma, and testing
- Summarize the contributions of noteworthy quality experts to modern quality management
- Describe how leadership, the cost of quality, organizational influences, expectations, cultural differences, and maturity models relate to improving quality in information technology projects
- Discuss how software can assist in project quality management

## OPENING CASE

A large medical instruments company just hired Scott Daniels, a senior consultant from a large consulting firm, to lead a project to resolve the quality problems with the company's new Executive Information System (EIS). A team of internal programmers and analysts worked with several company executives to develop this new system. Many executives were hooked on the new, user-friendly EIS. They loved the way the system allowed them to track sales of various medical instruments quickly and easily by product, country, hospital, and sales representative. After successfully testing the new EIS with several executives, the company decided to make the system available to all levels of management.



Unfortunately, several quality problems developed with the new EIS after a few months of operation. People were complaining that they could not get into the Web-based system. The system started going down a couple of times a month, and the response time was reportedly getting slower. Users complained when they could not access information within a few seconds. Several people kept forgetting how to log in to the system, thus increasing the number of calls to the company's help desk. There were complaints that some of the reports in the system gave inconsistent information. How could a summary report show totals that were not consistent with a detailed report on the same information? The executive sponsor of the EIS wanted the problems fixed quickly and accurately, so he decided to hire an expert in quality from outside the company whom he knew from past projects. Scott Daniels' job was to lead a team of people from both the medical instruments company and his own firm to identify and resolve quality-related issues with the EIS and to develop a plan to help prevent quality problems from happening on future projects.

# THE IMPORTANCE OF PROJECT QUALITY MANAGEMENT

Most people have heard jokes about what cars would be like if they followed a development history similar to that of computers. A well-known joke that has been traveling around the Internet goes as follows:

> At a recent computer exposition (COMDEX), Bill Gates, the founder and CEO of Microsoft Corporation, stated: "If General Motors had kept up with technology like the computer industry has, we would all be driving $25 cars that got 1,000 miles to the gallon." In response to Gates' comments, General Motors issued a press release stating: "If GM had developed technology like Microsoft, we would all be driving cars with the following characteristics:
>
> 1. For no reason whatsoever your car would crash twice a day.
> 2. Every time they repainted the lines on the road, you would have to buy a new car.
> 3. Occasionally, your car would die on the freeway for no reason, and you would just accept this, restart, and drive on.
> 4. Occasionally, executing a maneuver such as a left turn would cause your car to shut down and refuse to restart, in which case you would have to reinstall the engine.

5. Only one person at a time could use the car, unless you bought "Car95" or "CarNT." But then you would have to buy more seats.
6. Macintosh would make a car that was powered by the sun, reliable, five times as fast, and twice as easy to drive, but would run on only five percent of the roads.
7. The oil, water temperature, and alternator warning lights would be replaced by a single "general car default" warning light.
8. New seats would force everyone to have the same size hips.
9. The airbag system would say "Are you sure?" before going off.
10. Occasionally, for no reason whatsoever, your car would lock you out and refuse to let you in until you simultaneously lifted the door handle, turned the key, and grabbed hold of the radio antenna.
11. GM would require all car buyers to also purchase a deluxe set of Rand McNally road maps (now a GM subsidiary), even though they neither need them nor want them. Attempting to delete this option would immediately cause the car's performance to diminish by 50 percent or more. Moreover, GM would become a target for investigation by the Justice Department.
12. Every time GM introduced a new model car, buyers would have to learn how to drive all over again because none of the controls would operate in the same manner as the old car.
13. You would press the Start button to shut off the engine.[1]



Most people simply accept poor quality from many information technology products. So what if your computer crashes a couple of times a month? Just make sure you back up your data. So what if you cannot log in to the corporate intranet or the Internet right now? Just try a little later when it is less busy. So what if the latest version of your word-processing software was shipped with several known bugs? You like the software's new features, and all new software has bugs. Is quality a real problem with information technology projects?

Yes, it is! Information technology is not just a luxury available in some homes, schools, or offices. Companies throughout the world provide employees with access to computers. The majority of people in the United States use the Internet, and usage in other countries continues to grow rapidly. It took only five years for 50 million people to use the Internet compared to 25 years for 50 million people to use telephones. Many aspects of our daily lives depend on high-quality information technology products. Food is produced and distributed with the aid of computers; cars have computer chips to track performance; children use computers to help them learn in school; corporations depend on technology for many business functions; and millions of people rely on technology for entertainment and personal communications. Many information technology projects develop mission-critical systems that are used in life-and-death situations, such as navigation systems on aircraft and computer components built into medical equipment. Financial institutions and their customers also rely on high-quality information systems. Customers get very upset when systems provide inaccurate financial data or reveal information to unauthorized users that could lead to identity theft. When one of these systems does not function correctly, it is much more than a slight inconvenience, as described in the following What Went Wrong? examples.

- In 1981, a small timing difference caused by a computer program change created a 1 in 67 chance that the space shuttle's five on-board computers would not synchronize. The error caused a launch abort.[2]
- In 1986, two hospital patients died after receiving fatal doses of radiation from a Therac 25 machine. A software problem caused the machine to ignore calibration data.[3]
- In one of the biggest software errors in banking history, Chemical Bank mistakenly deducted about $15 million from more than 100,000 customer accounts. The problem resulted from a single line of code in an updated computer program that caused the bank to process every withdrawal and transfer at its automated teller machines (ATMs) twice. For example, a person who withdrew $100 from an ATM had $200 deducted from his or her account, though the receipt indicated only a withdrawal of $100. The mistake affected 150,000 transactions.[4]
- In August 2008, the Privacy Rights Clearinghouse, a nonprofit consumer advocacy group in San Diego, stated on their Web site that more than 236 million data records of U.S. residents have been exposed due to security breaches since January 2005. For example, Bank of America Corp. lost computer data tapes containing personal information of 1.2 million federal employees, including some U.S. senators; Ameritrade lost a back-up computer tape containing the personal information of 200,000 online trading customers; and the U.S. Air Force confirmed that personal data of 33,000 officers and enlisted personnel were hacked from an online system.[5]



Before you can improve the quality of information technology projects, it is important to understand the basic concepts of project quality management.

## WHAT IS PROJECT QUALITY MANAGEMENT?

Project quality management is a difficult knowledge area to define. The International Organization for Standardization (ISO) defines **quality** as "the totality of characteristics of an entity that bear on its ability to satisfy stated or implied needs" (ISO8042:1994) or "the degree to which a set of inherent characteristics fulfils requirements." (ISO9000:2000). Many people spent many hours coming up with these definitions, yet they are still very vague. Other experts define quality based on conformance to requirements and fitness for use. **Conformance to requirements** means the project's processes and products meet written specifications. For example, if the project scope statement requires delivery of 100 computers with specific processors, memory, and so on, you could easily check whether suitable computers had been delivered. **Fitness for use** means a product can be used as it was intended. If these computers were delivered without monitors or keyboards and were left in boxes on the customer's shipping dock, the customer might not be satisfied because the computers would not be fit for use. The customer may have assumed that the delivery included monitors and keyboards, unpacking the computers, and installation so they would be ready to use.

The purpose of **project quality management** is to ensure that the project will satisfy the needs for which it was undertaken. Recall that project management involves meeting or exceeding stakeholder needs and expectations. The project team must develop good relationships with key stakeholders, especially the main customer for the project, to understand what quality means to them. *After all, the customer ultimately decides if quality is acceptable*. Many technical projects fail because the project team focuses only on meeting the written requirements for the main products being produced and ignores other stakeholder needs and expectations for the project. For example, the project team should know what successfully delivering 100 computers means to the customer.



Quality, therefore, must be on an equal level with project scope, time, and cost. If a project's stakeholders are not satisfied with the quality of the project management or the resulting products of the project, the project team will need to adjust scope, time, and cost to satisfy the stakeholder. Meeting only written requirements for scope, time, and cost is not sufficient. To achieve stakeholder satisfaction, the project team must develop a good working relationship with all stakeholders and understand their stated or implied needs.

Project quality management involves three main processes:

1. *Planning quality* includes identifying which quality standards are relevant to the project and how to satisfy those standards. Incorporating quality standards into project design is a key part of quality planning. For an information technology project, quality standards might include allowing for system growth, planning a reasonable response time for a system, or ensuring that the system produces consistent and accurate information. Quality standards can also apply to information technology services. For example, you can set standards for how long it should take to get a reply from a help desk or how long it should take to ship a replacement part for a hardware item under warranty. The main outputs of quality planning are a quality management plan, quality metrics, quality checklists, a process improvement plan, and project document updates. A **metric** is a standard of measurement. Examples of common metrics include failure rates of products produced, availability of goods and services, and customer satisfaction ratings.
2. *Performing quality assurance* involves periodically evaluating overall project performance to ensure that the project will satisfy the relevant quality standards. The quality assurance process involves taking responsibility for quality throughout the project's life cycle. Top management must take the lead in emphasizing the roles all employees play in quality assurance, especially senior managers' roles. The main outputs of this process are organizational process asset updates, change requests, project management plan updates, and project document updates.
3. *Performing quality control* involves monitoring specific project results to ensure that they comply with the relevant quality standards while identifying ways to improve overall quality. This process is often associated with the technical tools and techniques of quality management, such as Pareto charts, quality control charts, and statistical sampling. You will learn more about these tools and

techniques later in this chapter. The main outputs of quality control include quality control measurements, validated changes, validated deliverables, organizational process asset updates, change requests, project management plan updates, and project document updates.

Figure 8-1 summarizes these processes and outputs, showing when they occur in a typical project.



**Planning**
Process: **Plan quality**
Outputs: Quality management plan, quality metrics, quality checklists, process improvement plan, and project document updates

**Executing**
Process: **Perform quality assurance**
Outputs: Organizational process asset updates, change requests, project management plan updates, and project document updates

**Monitoring and Controlling**
Process: **Perform quality control**
Outputs: Quality control measurements, validated changes, validated deliverables, organizational process asset updates, change requests, project management plan updates, and project document updates

**Project Start** **Project Finish**

**FIGURE 8-1** Project quality management summary

## PLANNING QUALITY

Project managers today have a vast knowledge base of information related to quality, and the first step to ensuring project quality management is planning. **Quality planning** implies the ability to anticipate situations and prepare actions that bring about the desired outcome. The current thrust in modern quality management is the prevention of defects through a program of selecting the proper materials, training and indoctrinating people in quality, and planning a process that ensures the appropriate outcome. In project quality planning, it is important to identify relevant quality standards, such as ISO standards described later in this chapter, for each unique project and to design quality into the products of the project and the processes involved in managing the project.

**Design of experiments** is a quality planning technique that helps identify which variables have the most influence on the overall outcome of a process. Understanding which variables affect outcome is a very important part of quality planning. For example, computer chip designers might want to determine which combination of materials and equipment will produce the most reliable chips at a reasonable cost. You can also apply design of experiments to project management issues such as cost and schedule trade-offs. For example, junior programmers or consultants cost less than senior programmers or consultants, but you cannot expect them to complete the same level of work in the same amount of time. An appropriately designed experiment to compute project costs and durations for various combinations of junior and senior programmers or consultants can allow you to determine an optimal mix of personnel, given limited resources. Refer to the section on the Taguchi method later in this chapter for more information.

Quality planning also involves communicating the correct actions for ensuring quality in a format that is understandable and complete. In quality planning for projects, it is important to describe important factors that directly contribute to meeting the customer's requirements. Organizational policies related to quality, the particular project's scope statement and product descriptions, and related standards and regulations are all important input to the quality planning process.



As mentioned in the discussion of project scope management (see Chapter 5), it is often difficult to completely understand the performance dimension of information technology projects. Even if the development of hardware, software, and networking technology would stand still for a while, it is often difficult for customers to explain exactly what they want in an information technology project. Important scope aspects of information technology projects that affect quality include functionality and features, system outputs, performance, and reliability and maintainability.

- **Functionality** is the degree to which a system performs its intended function. **Features** are the system's special characteristics that appeal to users. It is important to clarify what functions and features the system *must* perform, and what functions and features are *optional*. In the EIS example in the opening case, the mandatory functionality of the system might be that it allows users to track sales of specific medical instruments by predetermined categories such as the product group, country, hospital, and sales representative. Mandatory features might be a graphical user interface with icons, menus, online help, and so on.
- **System outputs** are the screens and reports the system generates. It is important to define clearly what the screens and reports look like for a system. Can the users easily interpret these outputs? Can users get all of the reports they need in a suitable format?
- **Performance** addresses how well a product or service performs the customer's intended use. To design a system with high quality performance, project stakeholders must address many issues. What volumes of data and transactions should the system be capable of handling? How many simultaneous users should the system be designed to handle? What is the projected growth rate in the number of users? What type of equipment must the system run on? How fast must the response time be for different aspects of the system under different circumstances? For the EIS in the opening case, several of the quality problems appear to relate to performance issues. The system is failing a couple of times a month, and users are unsatisfied with the response time. The project team may not have had specific performance requirements or tested the system under the right conditions to deliver the expected performance. Buying faster hardware might address these performance issues. Another performance

problem that might be more difficult to fix is the fact that some of the reports are generating inconsistent results. This could be a software quality problem that may be difficult and costly to correct since the system is already in operation.

- **Reliability** is the ability of a product or service to perform as expected under normal conditions. In discussing reliability for information technology projects, many people use the term IT service management. See the Suggested Readings on the companion Web site, such as ISO/IEC 20000, which is based on the Information Technology Infrastructure Library (ITIL).
- **Maintainability** addresses the ease of performing maintenance on a product. Most information technology products cannot reach 100 percent reliability, but stakeholders must define what their expectations are. For the EIS, what are the normal conditions for operating the system? Should reliability tests be based on 100 people accessing the system at once and running simple queries? Maintenance for the EIS might include uploading new data into the system or performing maintenance procedures on the system hardware and software. Are the users willing to have the system be unavailable several hours a week for system maintenance? Providing help desk support could also be a maintenance function. How fast a response do users expect for help desk support? How often can users tolerate system failure? Are the stakeholders willing to pay more for higher reliability and fewer failures?



These aspects of project scope are just a few of the requirement issues related to quality planning. Project managers and their teams need to consider all of these project scope issues in determining quality goals for the project. The main customers for the project must also realize their role in defining the most critical quality needs for the project and constantly communicate these needs and expectations to the project team. Since most information technology projects involve requirements that are not set in stone, it is important for all project stakeholders to work together to balance the quality, scope, time, and cost dimensions of the project. *Project managers, however, are ultimately responsible for quality management on their projects.*

Project managers should be familiar with basic quality terms, standards, and resources. For example, the International Organization for Standardization (ISO) provides information based on inputs from 157 different countries. They have an extensive Web site (*www.iso.org*), which is the source of ISO 9000 and more than 17,000 international standards for business, government and society. If you're curious where the acronym came from, the word "iso" comes from the Greek, meaning "equal." IEEE also provides many standards related to quality and has detailed information on their Web site (*www.ieee.org*).

## PERFORMING QUALITY ASSURANCE

It is one thing to develop a plan for ensuring quality on a project; it is another to ensure delivery of quality products and services. **Quality assurance** includes all of the activities related to satisfying the relevant quality standards for a project. Another goal of quality assurance is continuous quality improvement.

Many companies understand the importance of quality assurance and have entire departments dedicated to this area. They have detailed processes in place to make sure their products and services conform to various quality requirements. They also know they must produce those products and services at competitive prices. To be successful in today's competitive business environment, successful companies develop their own best practices and evaluate other organizations' best practices to continuously improve the way they do business.

Top management and project managers can have the greatest impact on the quality of projects by doing a good job of quality assurance. The importance of leadership in improving information technology project quality is discussed in more detail later in this chapter.



Several tools used in quality planning can also be used in quality assurance. Design of experiments, as described under quality planning, can also help ensure and improve product quality. **Benchmarking** generates ideas for quality improvements by comparing specific project practices or product characteristics to those of other projects or products within or outside the performing organization. For example, if a competitor has an EIS with an average down time of only one hour a week, that might be a benchmark for which to strive. Fishbone or Ishikawa diagrams, as described later in this chapter, can assist in ensuring and improving quality by finding the root causes of quality problems.

An important tool for quality assurance is a quality audit. A **quality audit** is a structured review of specific quality management activities that help identify lessons learned that could improve performance on current or future projects. In-house auditors or third parties with expertise in specific areas can perform quality audits, and quality audits can be scheduled or random. Industrial engineers often perform quality audits by helping to design specific quality metrics for a project and then applying and analyzing the metrics throughout the project. For example, the Northwest Airlines ResNet project (available on the companion Web site for this text) provides an excellent example of using quality audits to emphasize the main goals of a project and then track progress in reaching those goals. The main objective of the ResNet project was to develop a new reservation system to increase direct airline ticket sales and reduce the time it took for sales agents to handle customer calls. The measurement techniques for monitoring these goals helped ResNet's project manager and project team supervise various aspects of the project by focusing on meeting those goals. Measuring progress toward increasing direct sales and reducing call times also helped the project manager justify continued investments in ResNet.

## PERFORMING QUALITY CONTROL

Many people only think of quality control when they think of quality management. Perhaps it is because there are many popular tools and techniques in this area. Before describing some of these tools and techniques, it is important to distinguish quality control from quality planning and quality assurance.

Although one of the main goals of **quality control** is to improve quality, the main outcomes of this process are acceptance decisions, rework, and process adjustments.

- **Acceptance decisions** determine if the products or services produced as part of the project will be accepted or rejected. If they are accepted, they are considered to be validated deliverables. If project stakeholders reject some of the products or services produced as part of the project, there must be rework. For example, the executive who sponsored development of the EIS in the opening case was obviously not satisfied with the system and hired an outside consultant, Scott Daniels, to lead a team to address and correct the quality problems.

- **Rework** is action taken to bring rejected items into compliance with product requirements or specifications or other stakeholder expectations. Rework often results in requested changes and validated defect repair, resulting from recommended defect repair or corrective or preventive actions. Rework can be very expensive, so the project manager must strive to do a good job of quality planning and quality assurance to avoid this need. Since the EIS did not meet all of the stakeholders' expectations for quality, the medical instruments company was spending additional money for rework.
- **Process adjustments** correct or prevent further quality problems based on quality control measurements. Process adjustments are often found by using quality control measurements, and they often result in updates to the quality baseline, organization process assets, and the project management plan. For example, Scott Daniels, the consultant in the opening case, might recommend that the medical instruments company purchase a faster server for the EIS to correct the response-time problems. This change would require changes to the project management plan since it would require more work to be done related to the project. The company also hired Scott to develop a plan to help prevent future information technology project quality problems.



## TOOLS AND TECHNIQUES FOR QUALITY CONTROL

Quality control includes many general tools and techniques. This section describes the Seven Basic Tools of Quality, statistical sampling, and Six Sigma—and discusses how they can be applied to information technology projects. The section concludes with a discussion on testing, since information technology projects use testing extensively to ensure quality.

The following seven tools are known as the Seven Basic Tools of Quality:

1. *Cause-and-effect diagrams*: **Cause-and-effect diagrams** trace complaints about quality problems back to the responsible production operations. In other words, they help you find the root cause of a problem. They are also known as **fishbone** or **Ishikawa diagrams**, named after their creator, Kaoru Ishikawa. You can also use the technique known as the **5 whys**, where you repeatedly ask the question "Why?" (five is a good rule of thumb) to help peel away the layers of symptoms that can lead to the root cause of a problem. These symptoms can be branches on the cause-and-effect diagram.

   Figure 8-2 provides an example of a cause-and-effect diagram that Scott Daniels, the consultant in the opening case, might create to uncover the root cause of the problem of users not being able to log in to the EIS. Notice that it resembles the skeleton of a fish, hence the name fishbone diagram.
   This fishbone diagram lists the main areas that could be the cause of the problem: the EIS system's hardware, the user's hardware or software, or the user's

training. This figure describes two of these areas, the individual user's hardware and training, in more detail. For example, using the 5 whys, you could first ask why the users cannot get into the system, then why they keep forgetting their passwords, why they did not reset their passwords, why they did not check a box to save a password, and so on. The root cause of the problem would have a significant impact on the actions taken to solve the problem. If many users could not get into the system because their computers did not have enough memory, the solution might be to upgrade memory for those computers. If many users could not get into the system because they forgot their passwords, there might be a much quicker, less expensive solution.



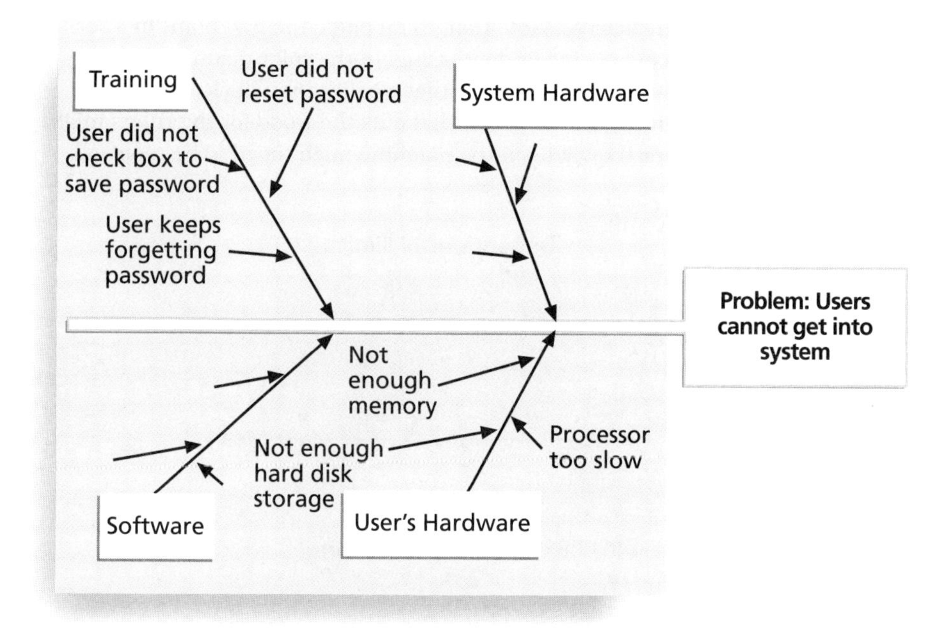


**FIGURE 8-2** Sample cause-and-effect diagram

2. *Control charts*: A **control chart** is a graphic display of data that illustrates the results of a process over time. Control charts allow you to determine whether a process is in control or out of control. When a process is in control, any variations in the results of the process are created by random events. Processes that are in control do not need to be adjusted. When a process is out of control, variations in the results of the process are caused by nonrandom events. When a process is out of control, you need to identify the causes of those nonrandom events and adjust the process to correct or eliminate them. For example, Figure 8-3 provides an example of a control chart for a process that manufactures 12-inch rulers. Assume that these are wooden rulers created by machines on an assembly line. Each point on the chart represents a length measurement for a ruler that comes off the assembly line. The scale on the vertical axis goes from 11.90 to 12.10. These numbers represent the lower and upper

specification limits for the ruler. In this case, this would mean that the customer for the rulers has specified that all rulers purchased must be between 11.90 and 12.10 inches long, or 12 inches plus or minus 0.10 inches. The lower and upper control limits on the quality control chart are 11.91 and 12.09 inches, respectively. This means the manufacturing process is designed to produce rulers between 11.91 and 12.09 inches long. Looking for and analyzing patterns in process data is an important part of quality control. You can use quality control charts and the seven run rule to look for patterns in data. The **seven run rule** states that if seven data points in a row are all below the mean, above the mean, or are all increasing or decreasing, then the process needs to be examined for non-random problems. In Figure 8-3, data points that violate the seven run rule are starred. Note that you include the first point in a series of points that are all increasing or decreasing. In the ruler manufacturing process, these data points may indicate that a calibration device may need adjustment. For example, the machine that cuts the wood for the rulers might need to be adjusted or the blade on the machine might need to be replaced.



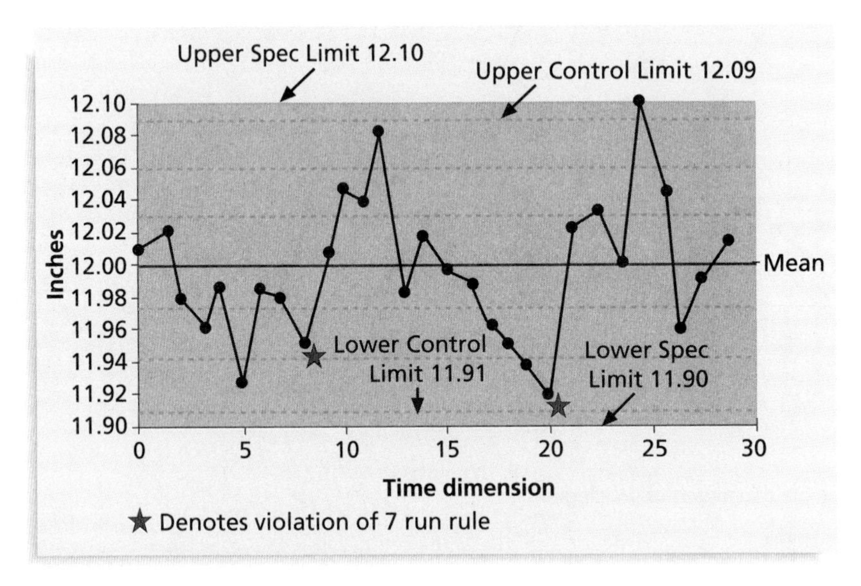


**FIGURE 8-3** Sample control chart

3. *Run chart*: A **run chart** displays the history and pattern of variation of a process over time. It is a line chart that shows data points plotted in the order in which they occur. You can use run charts to perform trend analysis to forecast future outcomes based on historical results. For example, trend analysis can help you analyze how many defects have been identified over time and see if there are trends. Figure 8-4 shows a sample run chart, charting the number of defects each month for three different types of defects. Notice that you can easily see the patterns of Defect 1 continuing to increase over time, Defect 2 decreasing the first several months and then holding steady, and Defect 3 fluctuating each month.

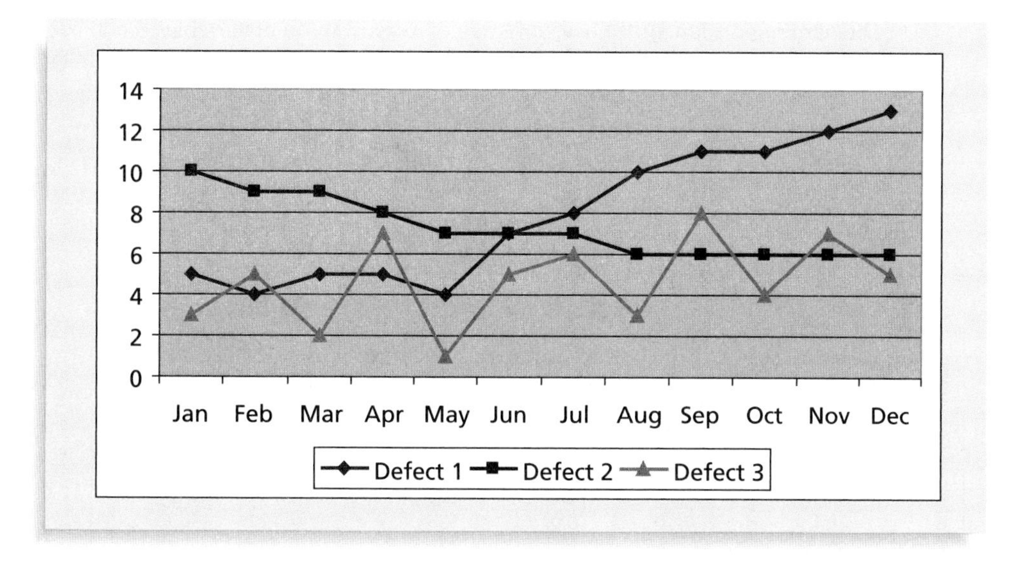


**FIGURE 8-4** Sample run chart



4. *Scatter diagram*: A **scatter diagram** helps to show if there is a relationship between two variables. The closer data points are to a diagonal line, the more closely the two variables are related. For example, Figure 8-5 provides a sample scatter diagram that Scott Daniels might create to compare user satisfaction ratings of the EIS system to the age of respondents to see if there is a relationship. They might find that younger users are less satisfied with the system, for example, and make decisions based on that finding.

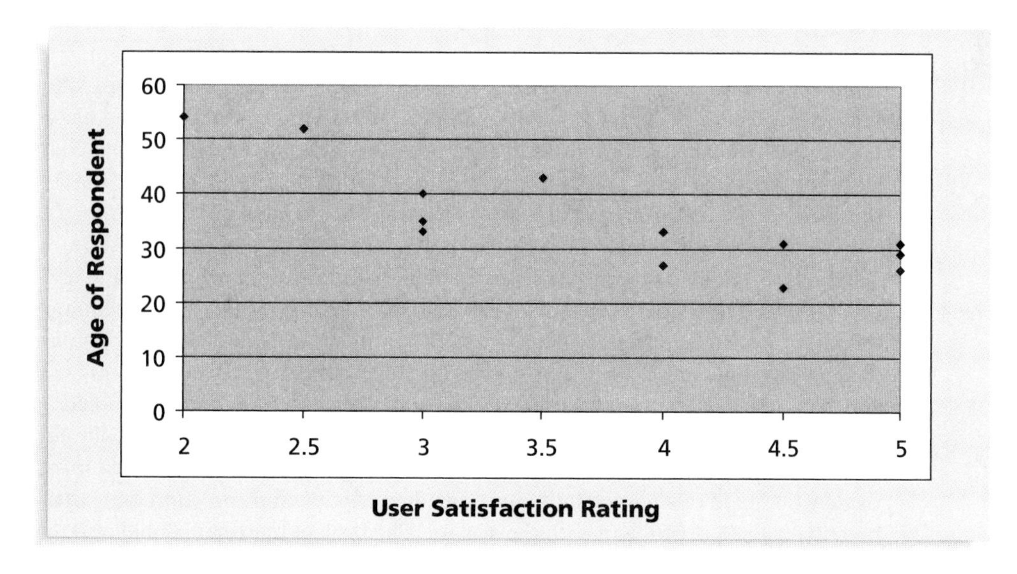


**FIGURE 8-5** Sample scatter diagram

5. *Histograms*: A **histogram** is a bar graph of a distribution of variables. Each bar represents an attribute or characteristic of a problem or situation, and the height of the bar represents its frequency. For example, Scott Daniels might ask the Help Desk to create a histogram to show how many total complaints they received each week related to the EIS system. Figure 8-6 shows a sample histogram.



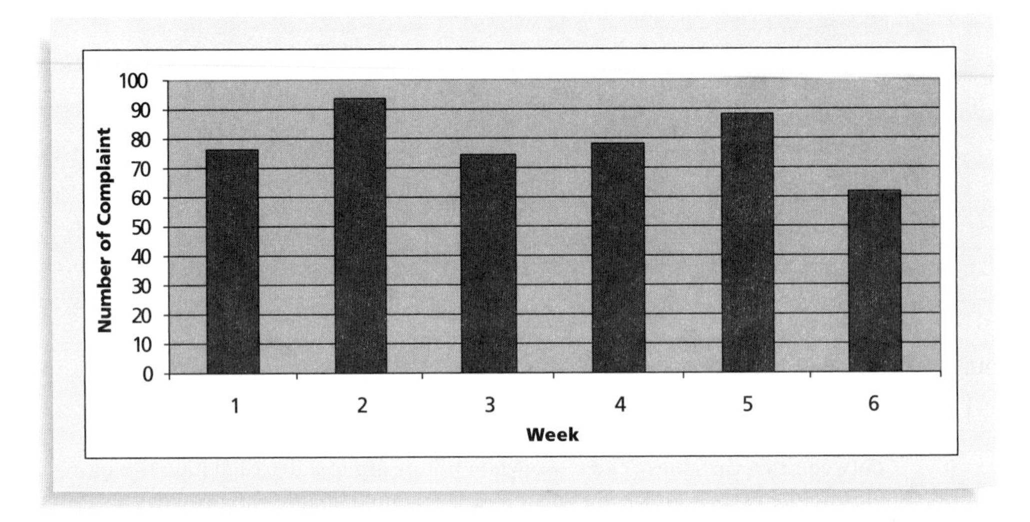


**FIGURE 8-6** Sample histogram

6. *Pareto charts*: A **Pareto chart** is a histogram that can help you identify and prioritize problem areas. The variables described by the histogram are ordered by frequency of occurrence. Pareto charts help you identify the vital few contributors that account for most quality problems in a system. **Pareto analysis** is sometimes referred to as the 80-20 rule, meaning that 80 percent of problems are often due to 20 percent of the causes. For example, suppose there was a detailed history of user complaints about the EIS. The project team could create a Pareto chart based on that data, as shown in Figure 8-7. Notice that login problems are the most frequent user complaint, followed by the system locking up, the system being too slow, the system being hard to use, and the reports being inaccurate. The first complaint accounts for 55 percent of the total complaints. The first and second complaints have a cumulative percentage of almost 80 percent, meaning these two areas account for 80 percent of the complaints. Therefore, the company should focus on making it easier to log in to the system to improve quality, since the majority of complaints fall under that category. The company should also address why the system locks up. Because Figure 8-7 shows that inaccurate reports are a problem that is rarely mentioned, the project manager should investigate who made this complaint before spending a lot of effort on addressing that potentially critical problem with the system. The project manager should also find out if complaints about the system being too slow were actually due to the user not being able to log in or the system locking up.

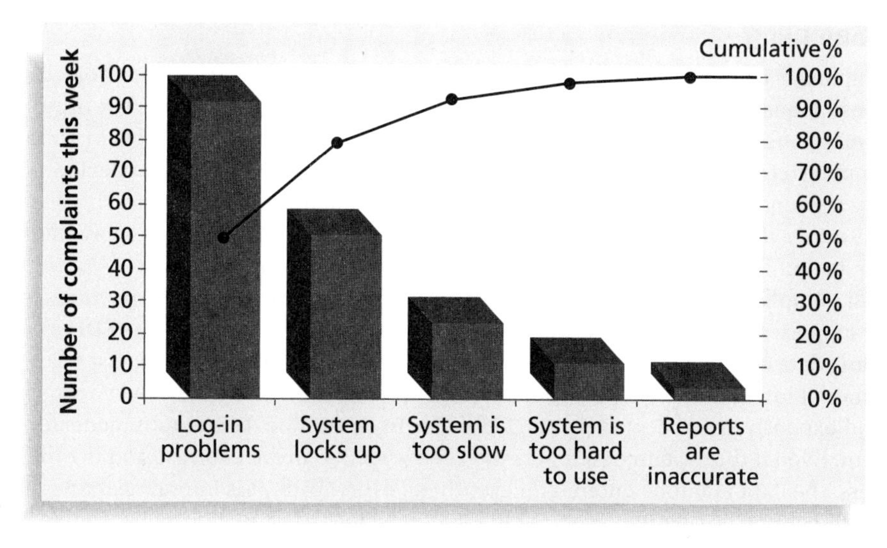


**FIGURE 8-7** Sample Pareto chart



7. *Flowcharts*: **Flowcharts** are graphic displays of the logic and flow of processes that help you analyze how problems occur and how processes can be improved. They show activities, decision points, and the order of how information is processed. Figure 8-8 provides a simple example of a flowchart that shows the process a project team might use for accepting or rejecting deliverables.

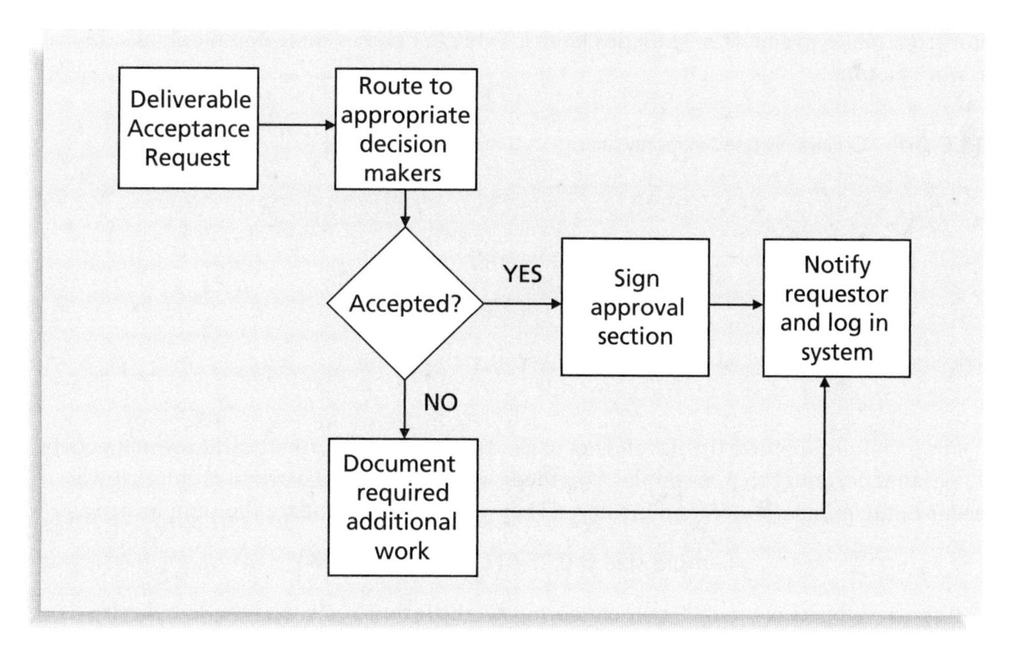


**FIGURE 8-8** Sample flowchart

## Statistical Sampling

Statistical sampling is a key concept in project quality management. Members of a project team who focus on quality control must have a strong understanding of statistics, but other project team members need to understand only the basic concepts. These concepts include statistical sampling, certainty factor, standard deviation, and variability. Standard deviation and variability are fundamental concepts for understanding quality control charts. This section briefly describes these concepts and describes how a project manager might apply them to information technology projects. Refer to statistics texts for additional details.



**Statistical sampling** involves choosing part of a population of interest for inspection. For example, suppose a company wants to develop an electronic data interchange (EDI) system for handling data on invoices from all of its suppliers. Assume also that in the past year, the total number of invoices was 50,000 from 200 different suppliers. It would be very time consuming and expensive to review every single invoice to determine data requirements for the new system. Even if the system developers did review all 200 invoice forms from the different suppliers, the data might be entered differently on every form. It is impractical to study every member of a population, such as all 50,000 invoices, so statisticians have developed techniques to help determine an appropriate sample size. If the system developers used statistical techniques, they might find that by studying only 100 invoices, they would have a good sample of the type of data they would need in designing the system.

The size of the sample depends on how representative you want the sample to be. A simple formula for determining sample size is:

$$\text{Sample size} = .25 * (\text{certainty factor} / \text{acceptable error})^2$$

The certainty factor denotes how certain you want to be that the data sampled will not include variations that do not naturally exist in the population. You calculate the certainty factor from tables available in statistics books. Table 8-1 shows some commonly used certainty factors.

**TABLE 8-1** Commonly used certainty factors

| Desired Certainty | Certainty Factor |
| --- | --- |
| 95% | 1.960 |
| 90% | 1.645 |
| 80% | 1.281 |

For example, suppose the developers of the EDI system described earlier would accept a 95 percent certainty that a sample of invoices would contain no variation unless it was present in the population of total invoices. They would then calculate the sample size as:

$$\text{Sample size} = 0.25 * (1.960 / .05)^2 = 384$$

If the developers wanted 90 percent certainty, they would calculate the sample size as:

$$\text{Sample size} = 0.25 * (1.645 / .10)^2 = 68$$

If the developers wanted 80 percent certainty, they would calculate the sample size as:

$$\text{Sample size} = 0.25 * (1.281 / .20)^2 = 10$$

Assume the developers decide on 90 percent for the certainty factor. Then they would need to examine 68 invoices to determine the type of data the EDI system would need to capture.

## Six Sigma



The work of many project quality experts contributed to the development of today's Six Sigma principles. There has been some confusion in the past few years about the term Six Sigma. This section summarizes recent information about this important concept and explains how organizations worldwide use Six Sigma principles to improve quality, decrease costs, and better meet customer needs.

In their book, *The Six Sigma Way*, authors Peter Pande, Robert Neuman, and Roland Cavanagh define **Six Sigma** as "a comprehensive and flexible *system* for achieving, sustaining and maximizing business success. Six Sigma is uniquely driven by close understanding of customer needs, disciplined use of facts, data, and statistical analysis, and diligent attention to managing, improving, and reinventing business processes."[6]

Six Sigma's target for perfection is the achievement of no more than 3.4 defects, errors, or mistakes per million opportunities. This target number is explained in more detail later in this section. An organization can apply the Six Sigma principles to the design and production of a product, a help desk, or other customer-service process.

Projects that use Six Sigma principles for quality control normally follow a five-phase improvement process called **DMAIC** (pronounced de-MAY-ick), which stands for Define, Measure, Analyze, Improve, and Control. DMAIC is a systematic, closed-loop process for continued improvement that is scientific and fact based. The following are brief descriptions of each phase of the DMAIC improvement process:

1. *Define*: Define the problem/opportunity, process, and customer requirements. Important tools used in this phase include a project charter, a description of customer requirements, process maps, and Voice of the Customer (VOC) data. Examples of VOC data include complaints, surveys, comments, and market research that represent the views and needs of the organization's customers.
2. *Measure*: Define measures, then collect, compile, and display data. Measures are defined in terms of defects per opportunity, as defined later in this section.
3. *Analyze*: Scrutinize process details to find improvement opportunities. A project team working on a Six Sigma project, normally referred to as a Six Sigma team, investigates and verifies data to prove the suspected root causes of quality problems and substantiates the problem statement. An important tool in this phase is the fishbone or Ishikawa diagram, as described earlier in this chapter.
4. *Improve*: Generate solutions and ideas for improving the problem. A final solution is verified with the project sponsor, and the Six Sigma team develops a plan to pilot test the solution. The Six Sigma team reviews the results of the

pilot test to refine the solution, if needed, and then implements the solution where appropriate.

5. *Control*: Track and verify the stability of the improvements and the predictability of the solution. Control charts are one tool used in the control phase, as described later in this chapter.

### How Is Six Sigma Quality Control Unique?



How does using Six Sigma principles differ from using previous quality control initiatives? Many people remember other quality initiatives from the past few decades such as Total Quality Management (TQM) and Business Process Reengineering (BPR), to name a few. The origins of many Six Sigma principles and tools are found in these previous initiatives; however, there are several new ideas included in Six Sigma principles that help organizations improve their competitiveness and bottom-line results. Below are a few of these principles:

- Using Six Sigma principles is an organization-wide commitment. CEOs, top managers, and all levels of employees in an organization that embraces Six Sigma principles (often referred to as a *Six Sigma organization*) have seen remarkable improvements due to its use. There are often huge training investments, but these investments pay off as employees practice Six Sigma principles and produce higher quality goods and services at lower costs.
- Six Sigma training normally follows the "belt" system, similar to a karate class in which students receive different color belts for each training level. In Six Sigma training, those in the Yellow Belt category receive the minimum level of training, which is normally two to three full days of Six Sigma training for project team members who work on Six Sigma projects on a part-time basis. Those in the Green Belt category usually participate in two to three full weeks of training. Those in the Black Belt category normally work on Six Sigma projects full-time and attend four to five full weeks of training. Project managers are often Black Belts. The Master Black Belt category describes experienced Black Belts who act as technical resources and mentors to people with lower-level belts.
- Organizations that successfully implement Six Sigma principles have the ability and willingness to adopt two seemingly contrary objectives at the same time. Authors James Collins and Jerry Porras describe this as the "We can do it all" or "Genius of the And" approach in their book, *Built to Last*.[7] For example, Six Sigma organizations believe that they can be creative *and* rational, focus on the big picture *and* minute details, reduce errors *and* get things done faster, and make customers happy *and* make a lot of money.
- Six Sigma is not just a program or a discipline to organizations that have benefited from it. Six Sigma is an operating philosophy that is customer-focused and strives to drive out waste, raise levels of quality, and improve financial performance at *breakthrough* levels. A Six Sigma organization sets high goals and uses the DMAIC improvement process to achieve extraordinary quality improvements.

Many organizations do some of what now fits under the definition of Six Sigma, and many Six Sigma principles are not brand-new. What *is* new is its ability to bring together many different themes, concepts, and tools into a coherent management process that can be used on an organization-wide basis.

## WHAT WENT RIGHT?

Many organizations have reported dramatic improvements as a result of implementing Six Sigma principles.

- Motorola, Inc. pioneered the adoption of Six Sigma in the 1980s. Its reason for developing and implementing Six Sigma was clear: to stay in business. Japanese competitors were putting several U.S. and European companies out of business. Using Six Sigma provided Motorola with a simple, consistent way to track and compare performance to customer requirements and meet ambitious quality goals in terms of defect reduction. Chairman Bob Galvin set a goal of ten times improvement in defect reduction every two years, or 100 times improvement in four years. Motorola did stay in business, achieving excellent growth and profitability in the 1980s and 1990s. Motorola estimates the cumulative savings based on Six Sigma efforts to be about $14 billion.[8]
- Allied Signal/Honeywell, a technology and manufacturing company, began several quality improvement activities in the early 1990s. By 1999, the company reported saving more than $600 million a year, thanks to training in and application of Six Sigma principles. Six Sigma teams reduced the costs of reworking defects and applied the same principles to designing new products, like aircraft engines. They reduced the time it took from design to certification of engines from 42 to 33 months. Allied Signal/Honeywell credits Six Sigma with a 6 percent productivity increase in 1998 and record profit margins of 13 percent. As one of the company's directors said, "[Six Sigma] changed the way we think and the way we communicate. We never used to talk about the process or the customer; now they're part of our everyday conversation."[9]
- In an effort to improve the efficiency of the Specials Radiology Department, Baptist St. Anthony's Hospital in Amarillo, Texas initiated an improvement project in May 2006. Project goals included improving start times for cases, increasing throughput, implementing control mechanisms for continuous improvement, and providing feedback to other departments concerning performance. After implementing the solutions recommended by the project team and measuring results, the percent of delayed cases dropped from 79 percent to 33 percent, delays within the department decreased by 22 percent, and the number of orders missing or needing clarification dropped to zero from 11 percent. The hospital receives orders from more than 500 physicians at an average rate of 250 orders per day, so this one aspect of the project affects more than 90,000 orders per year.[10]



### Six Sigma and Project Selection and Management

Joseph M. Juran stated, "All improvement takes place project by project, and in no other way."[11] Organizations implement Six Sigma by selecting and managing projects. An important part of project management is good project selection.

This statement is especially true for Six Sigma projects. Pande, Neuman, and Cavanagh conducted an informal poll to find out what the most critical and most commonly mishandled activity was in launching Six Sigma, and the unanimous answer was project selection. "It's a pretty simple equation, really: Well-selected and -defined improvement projects equal better, faster results. The converse equation is also simple: Poorly selected and defined projects equal delayed results and frustration."[12]

Organizations must also be careful to apply higher quality where it makes sense. An article in *Fortune* stated that companies that have implemented Six Sigma have not necessarily boosted their stock values. Although GE boasted savings of more than $2 billion in 1999 due to its use of Six Sigma, other companies, such as Whirlpool, could not clearly demonstrate the value of their investments. Why can't all companies benefit from Six Sigma? Because minimizing defects does not matter if an organization is making a product that no one wants to buy. As one of Six Sigma's biggest supporters, Mikel Harry, puts it, "I could genetically engineer a Six Sigma goat, but if a rodeo is the marketplace, people are still going to buy a Four Sigma horse."[13]



As described in Chapter 4, Project Integration Management, there are several methods for selecting projects. However, what makes a project a potential Six Sigma project? First, there must be a quality problem or gap between the current and desired performance. Many projects do not meet this first criterion, such as building a house, merging two corporations, or providing an information technology infrastructure for a new organization. Second, the project should not have a clearly understood problem. Third, the solution should not be predetermined, and an optimal solution should not be apparent.

Once a project is selected as a good candidate for Six Sigma, many project management concepts, tools, and techniques described in this text come into play. For example, Six Sigma projects usually have a business case, a project charter, requirements documents, a schedule, a budget, and so on. Six Sigma projects are done in teams and have sponsors called champions. There are also, of course, project managers, often called *team leaders* in Six Sigma organizations. In other words, Six Sigma projects are simply types of projects that focus on supporting the Six Sigma philosophy by being customer-focused and striving to drive out waste, raise levels of quality, and improve financial performance at breakthrough levels.

### Six Sigma and Statistics

An important concept in Six Sigma is improving quality by reducing variation. The term *sigma* means standard deviation. **Standard deviation** measures how much variation exists in a distribution of data. A small standard deviation means that data clusters closely around the middle of a distribution and there is little variability among the data. A large standard deviation means that data is spread out around the middle of the distribution and there is relatively greater variability. Statisticians use the Greek symbol $\sigma$ (sigma) to represent the standard deviation.

Figure 8-9 provides an example of a **normal distribution**—a bell-shaped curve that is symmetrical regarding the **mean** or average value of the population (the data being analyzed). In any normal distribution, 68.3 percent of the population is within one standard deviation ($1\sigma$) of the mean, 95.5 percent of the population is within two standard deviations ($2\sigma$), and 99.7 percent of the population is within three standard deviations ($3\sigma$) of the mean.

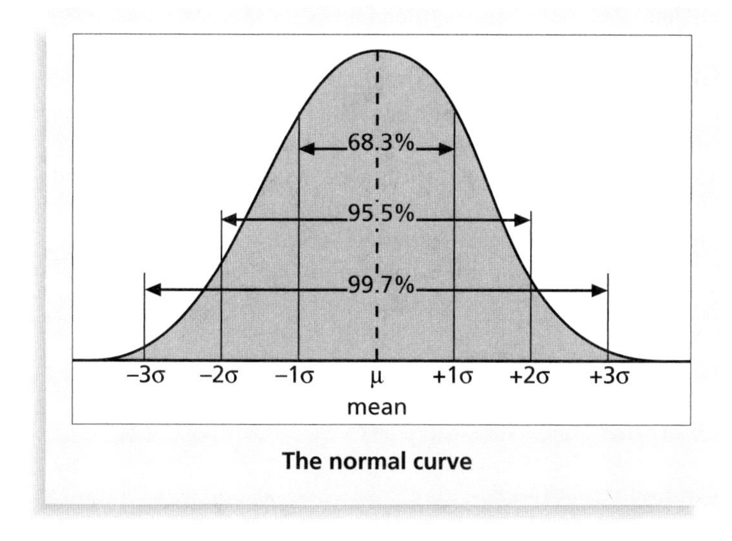


**FIGURE 8-9** Normal distribution and standard deviation



Standard deviation is a key factor in determining the acceptable number of defective units found in a population. Table 8-2 illustrates the relationship between sigma, the percentage of the population within that sigma range, and the number of defective units per billion. Note that this table shows that being plus or minus six sigma in pure statistical terms means only *two* defective units per *billion*. Why, then, is the target for Six Sigma programs 3.4 defects per million opportunities, as stated earlier in this chapter?

**TABLE 8-2** Sigma and defective units

| Specification Range (in +/– Sigmas) | Percent of Population within Range | Defective Units per Billion |
|---|---|---|
| 1 | 68.27 | 317,300,000 |
| 2 | 95.45 | 45,400,000 |
| 3 | 99.73 | 2,700,000 |
| 4 | 99.9937 | 63,000 |
| 5 | 99.999943 | 57 |
| 6 | 99.9999998 | 2 |

Based on Motorola's original work on Six Sigma in the 1980s, the convention used for Six Sigma is a scoring system that accounts for more variation in a process than you would typically find in a few weeks or months of data gathering. In other words, time is an important factor in determining process variations. Table 8-3 shows the Six Sigma conversion table applied to Six Sigma projects. (See the Suggested Reading by Pande, Neuman, and Cavanagh for a detailed description of Six Sigma scoring.) The **yield** represents the number of units handled correctly through the process steps. A **defect** is any instance where the product or service fails to meet customer requirements. Because most products or services have multiple customer requirements, there can be several opportunities to have a defect. For example, suppose a company is trying to reduce the number of errors on customer billing statements. There could be several errors on a billing statement due to a misspelled name, incorrect address, wrong date of service, calculation error, and so on. There might be 100 opportunities for a defect to occur on one billing statement. Instead of measuring the number of defects per unit or billing statement, Six Sigma measures the number of defects based on the number of opportunities.



**TABLE 8-3** Sigma conversion table

| Sigma | Yield | Defects per Million Opportunities (DPMO) |
|---|---|---|
| 1 | 31.0% | 690,000 |
| 2 | 69.2% | 308,000 |
| 3 | 93.3% | 66,800 |
| 4 | 99.4% | 6,210 |
| 5 | 99.97% | 230 |
| 6 | 99.99966% | 3.4 |

As you can see, the Six Sigma conversion table shows that a process operating at six sigma means there are no more than 3.4 defects per million opportunities. However, most organizations today use the term *six sigma project* in a broad sense to describe projects that will help them in achieving, sustaining, and maximizing business success through better business processes.

Another term you may hear being used in the telecommunications industry is **six 9s of quality**. Six 9s of quality is a measure of quality control equal to 1 fault in 1 million opportunities. In the telecommunications industry, it means 99.9999 percent service availability or *30 seconds of down time a year*. This level of quality has also been stated as the target goal for the number of errors in a communications circuit, system failures, or errors in lines of code. To achieve six 9s of quality requires continual testing to find and eliminate errors or enough redundancy and back-up equipment in systems to reduce the overall system failure rate to that low a level.

## Testing

Many information technology professionals think of testing as a stage that comes near the end of information technology product development. Instead of putting serious effort into proper planning, analysis, and design of information technology projects, some organizations rely on testing just before a product ships to ensure some degree of quality. In fact, testing needs to be done during almost every phase of the systems development life cycle, not just before the organization ships or hands over a product to the customer.

Figure 8-10 shows one way of portraying the systems development life cycle. This example includes 17 main tasks involved in a software development project and shows their relationship to each other. For example, every project should start by initiating the project, conducting a feasibility study, and then performing project planning. The figure then shows that the work involved in preparing detailed requirements and the detailed architecture for the system can be performed simultaneously. The oval-shaped phases represent actual tests or tasks, which will include test plans to help ensure quality on software development projects.[14]



Several of the phases in Figure 8-10 include specific work related to testing.

- A **unit test** is done to test each individual component (often a program) to ensure that it is as defect-free as possible. Unit tests are performed before moving onto the integration test.
- **Integration testing** occurs between unit and system testing to test functionally grouped components. It ensures a subset(s) of the entire system works together.
- **System testing** tests the entire system as one entity. It focuses on the big picture to ensure that the entire system is working properly.
- **User acceptance testing** is an independent test performed by end users prior to accepting the delivered system. It focuses on the business fit of the system to the organization, rather than technical issues.

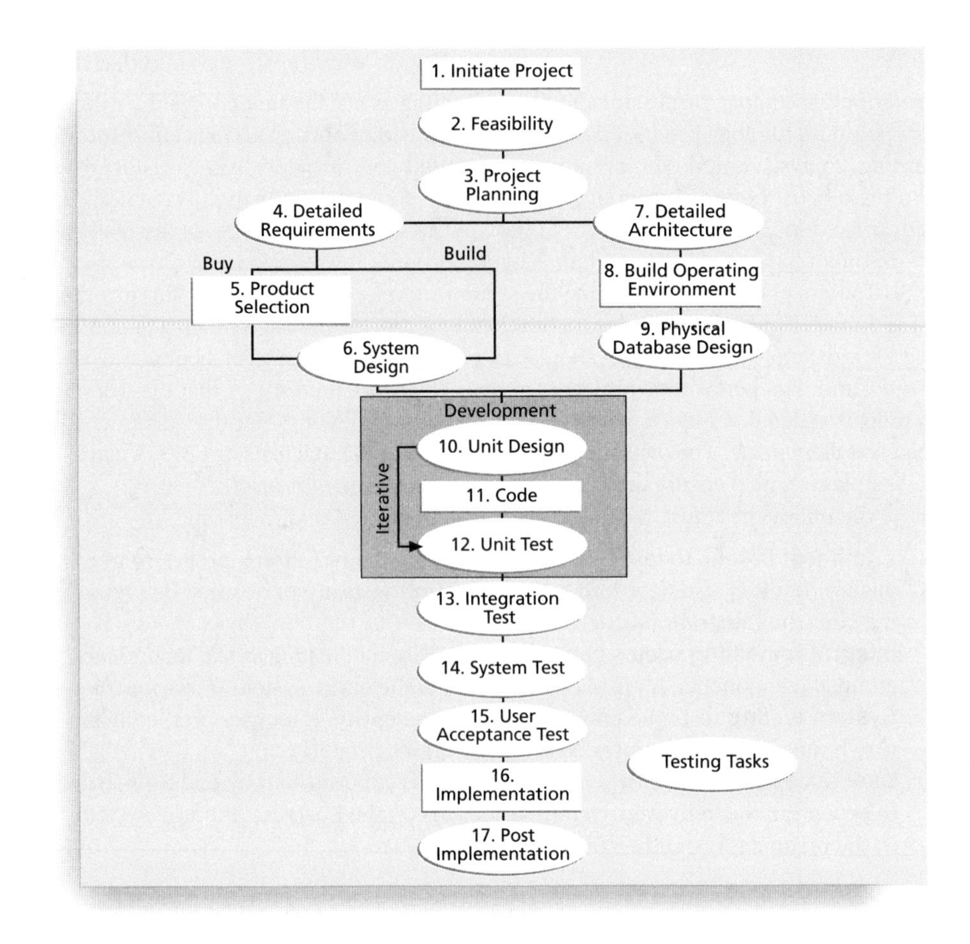


**FIGURE 8-10** Testing tasks in the software development life cycle

To help improve the quality of software development projects, it is important for organizations to follow a thorough and disciplined testing methodology. System developers and testers must also establish a partnership with all project stakeholders to make sure the system meets their needs and expectations and the tests are done properly. As described in the next section, there are tremendous costs involved in failure to perform proper testing.

Testing alone, however, cannot always solve software defect problems, according to Watts S. Humphrey, a renowned expert on software quality and Fellow at Carnegie Mellon's Software Engineering Institute. He believes that the traditional code/test/fix cycle for software development is not enough. As code gets more complex, the number of defects missed by testing increases and becomes the problem not just of testers, but also of paying customers. Humphrey says that, on average, programmers introduce a defect for every nine or ten lines of code, and the finished software, after all testing, contains about five to six defects per thousand lines of code. Although there are many different definitions, Humphrey defines a **software defect** as anything that must be changed before delivery of the program. Testing does not sufficiently prevent software defects because the number of ways to test a complex system is huge. In addition, users will continue to invent new ways to use a system that its developers never considered, so certain functionalities may never have been tested or even included in the system requirements. Humphrey suggests that people rethink the software development process to provide *no* potential defects when you enter system testing. This means that developers must be responsible for providing error-free code at each stage of testing. Humphrey teaches a development process where programmers measure and track the kinds of errors they commit so they can use the data to improve their performance. He also acknowledges that top management must support developers by letting them self-direct their work. Programmers need to be motivated and excited to do high-quality work and have some control over how they do it.[15]



## MODERN QUALITY MANAGEMENT

Modern quality management requires customer satisfaction, prefers prevention to inspection, and recognizes management responsibility for quality. Several noteworthy people helped develop the following theories, tools, and techniques that define modern quality management.[16] The suggestions from these quality experts led to many projects to improve quality and provided the foundation for today's Six Sigma projects. This section summarizes major contributions made by Deming, Juran, Crosby, Ishikawa, Taguchi, and Feigenbaum.

### Deming and His 14 Points for Management

Dr. W. Edwards Deming is known primarily for his work on quality control in Japan. Dr. Deming went to Japan after World War II, at the request of the Japanese government, to assist them in improving productivity and quality. Deming, a statistician and former professor at New York University, taught Japanese manufacturers that higher quality meant greater productivity and lower cost. American industry did not recognize Deming's theories until Japanese manufacturers started producing products that seriously challenged American products, particularly in the auto industry. Ford Motor Company then adopted Deming's quality methods and experienced dramatic improvement in quality and sales thereafter. By the 1980s, after seeing the excellent work coming out of Japan, several U.S. corporations vied for Deming's expertise to help them establish quality improvement programs in their own factories. Many people are familiar with the Deming Prize, an award given to recognize high-quality organizations, and Deming's Cycle for Improvement: plan, do, check, and act. Most Six Sigma principles described earlier are based on the plan-do-check-act model created by Deming. Many are also familiar with Deming's 14 Points for Management, summarized below from Deming's text *Out of the Crisis*[17]:

1. Create constancy of purpose for improvement of product and service.
2. Adopt the new philosophy.
3. Cease dependence on inspection to achieve quality.
4. End the practice of awarding business based on price tag alone. Instead, minimize total cost by working with a single supplier.
5. Improve constantly and forever every process for planning, production, and service.
6. Institute training on the job.

7. Adopt and institute leadership.
8. Drive out fear.
9. Break down barriers between staff areas.
10. Eliminate slogans, exhortations, and targets for the workforce.
11. Eliminate numerical quotas for the workforce and numerical goals for management.
12. Remove barriers that rob people of workmanship. Eliminate the annual rating or merit system.
13. Institute a vigorous program of education and self-improvement for everyone.
14. Put everyone in the company to work to accomplish the transformation.



## Juran and the Importance of Top Management Commitment to Quality

Joseph M. Juran, like Deming, taught Japanese manufacturers how to improve their productivity. U.S. companies later discovered him as well. He wrote the first edition of the *Quality Control Handbook* in 1974, stressing the importance of top management commitment to continuous product quality improvement. In 2000, at the age of 94, Juran published the fifth edition of this famous handbook.[18] He also developed the Juran Trilogy: quality improvement, quality planning, and quality control. Juran stressed the difference between the manufacturer's view of quality and the customer's view. Manufacturers often focus on conformance to requirements, but customers focus on fitness for use. Most definitions of quality now use fitness for use to stress the importance of satisfying stated or implied needs and not just meeting stated requirements or specifications. Juran developed ten steps to quality improvement:

1. Build awareness of the need and opportunity for improvement.
2. Set goals for improvement.
3. Organize to reach the goals (establish a quality council, identify problems, select projects, appoint teams, designate facilitators).
4. Provide training.
5. Carry out projects to solve problems.
6. Report progress.
7. Give recognition.
8. Communicate results.
9. Keep score.
10. Maintain momentum by making annual improvement part of the regular systems and processes of the company.

## Crosby and Striving for Zero Defects

Philip B. Crosby wrote *Quality Is Free* in 1979 and is best known for suggesting that organizations strive for zero defects.[19] He stressed that the costs of poor quality should include all the costs of not doing the job right the first time, such as scrap, rework, lost labor hours and machine hours, customer ill will and lost sales, and warranty costs. Crosby suggested that the cost of poor quality is so understated that companies can profitably spend unlimited amounts of money on improving quality. Crosby developed the following 14 steps for quality improvement:

1. Make it clear that management is committed to quality.
2. Form quality improvement teams with representatives from each department.
3. Determine where current and potential quality problems lie.
4. Evaluate the cost of quality and explain its use as a management tool.
5. Raise the quality awareness and personal concern of all employees.
6. Take actions to correct problems identified through previous steps.
7. Establish a committee for the zero-defects program.
8. Train supervisors to actively carry out their part of the quality improvement program.
9. Hold a "zero-defects day" to let all employees realize that there has been a change.
10. Encourage individuals to establish improvement goals for themselves and their groups.
11. Encourage employees to communicate to management the obstacles they face in attaining their improvement goals.
12. Recognize and appreciate those who participate.
13. Establish quality councils to communicate on a regular basis.
14. Do it all over again to emphasize that the quality improvement program never ends.



Crosby also developed the Quality Management Process Maturity Grid in 1978. This grid can be applied to an organization's attitude toward product usability. For example, the first stage in the grid is ignorance, where people might think they don't have any problems with usability. The final stage is wisdom, where people have changed their attitude so that usability defect prevention is a routine part of their operation.

## Ishikawa's Guide to Quality Control

Kaoru Ishikawa is best known for his 1972 book *Guide to Quality Control*.[20] He developed the concept of quality circles and pioneered the use of cause-and-effect diagrams, as described earlier in this chapter. **Quality circles** are groups of non-supervisors and work leaders in a single company department who volunteer to conduct group studies on how to improve the effectiveness of work in their department. Ishikawa suggested that Japanese managers and workers were totally committed to quality, but that most U.S. companies delegated the responsibility for quality to a few staff members.

## Taguchi and Robust Design Methods

Genichi Taguchi is best known for developing the Taguchi methods for optimizing the process of engineering experimentation. Key concepts in the Taguchi methods are that quality should be designed into the product and not inspected into it and that quality is best achieved by minimizing deviation from the target value. For example, if the target response time for accessing the EIS described in the opening case is half a second, there should be little deviation from this time. By the late 1990s, Taguchi had become, in the words of *Fortune* magazine, "America's new quality hero."[21] Many companies, including Xerox, Ford, Hewlett-Packard, and Goodyear, have recently used Taguchi's Robust Design methods to design high-quality products. **Robust Design methods** focus on eliminating defects by substituting scientific inquiry for trial-and-error methods.

## Feigenbaum and Workers' Responsibility for Quality

Armand V. Feigenbaum developed the concept of total quality control (TQC) in his 1983 book *Total Quality Control: Engineering and Management*.[22] He proposed that the responsibility for quality should rest with the people who do the work. In TQC, product quality is more important than production rates, and workers are allowed to stop production whenever a quality problem occurs.



## Malcolm Baldrige National Quality Award

The **Malcolm Baldrige National Quality Award** originated in 1987 in the U.S. to recognize companies that have achieved a level of world-class competition through quality management. The award was started in honor of Malcolm Baldrige, who was the U.S. secretary of commerce from 1981 until his death in a rodeo accident in July 1987. Baldrige was a proponent of quality management as a key element in improving the prosperity and long-term strength of U.S. organizations. The Malcolm Baldrige National Quality Award is given by the president of the United States to U.S. businesses—manufacturing and service, small and large—and to education and health care organizations. Organizations must apply for the award, and they must be judged outstanding in seven areas: leadership, strategic planning, customer and market focus, information and analysis, human resource focus, process management, and business results. Three awards may be given annually in each of these categories: manufacturing, service, small business and, from 1999, education and health care. The awards recognize achievements in quality and performance and raise awareness about the importance of quality as a competitive edge; the award is not given for specific products or services.

## ISO Standards

The International Organization for Standardization (ISO) is a network of national standards institutes that work in partnership with international organizations, governments, industries, businesses, and consumer representatives. **ISO 9000**, a quality system standard developed by the ISO, is a three-part, continuous cycle of planning, controlling, and documenting quality in an organization. ISO 9000 provides minimum requirements needed for an organization to meet its quality certification standards. The ISO 9000 family of international quality management standards and guidelines has earned a global reputation as the basis for establishing quality management systems. The ISO's Web site (*www.iso.org*) includes testimonials from many business leaders throughout the world explaining the benefits of following these standards.

- A government official in Malaysia believes that the universally accepted ISO 9000 standard can contribute significantly to improving quality and enhancing development of an excellent work culture. ISO 9000 leads to a more systematic management of quality and provides a means of consolidating quality management systems in public service.
- Managers in the United Kingdom say that ISO 9000 has become a firmly established cornerstone of their organizations' drive toward quality improvement. It leads to substantial cost savings and contributes strongly to customer satisfaction.

- A Brazilian newspaper reported that following ISO 9000 helped its home delivery service improve so much that the complaints dropped to a mere 0.06 percent of the total copies distributed.[23]

ISO continues to offer standards to provide a framework for the assessment of software processes. The overall goals of a standard are to encourage organizations interested in improving quality of software products to employ proven, consistent, and reliable methods for assessing the state of their software development processes. They can also use their assessment results as part of coherent improvement programs. One of the outcomes of assessment and consequent improvement programs is reliable, predictable, and continuously improving software processes.



The contributions of several quality experts, quality awards, and quality standards are important parts of project quality management. The Project Management Institute was proud to announce in 1999 that their certification department had become the first certification department in the world to earn ISO 9000 certification, and that the *PMBOK® Guide* had been recognized as an international standard. Emphasizing quality in project management helps ensure that projects produce products or services that meet customer needs and expectations.

## IMPROVING INFORMATION TECHNOLOGY PROJECT QUALITY

In addition to some of the suggestions provided for using good quality planning, quality assurance, and quality control, there are several other important issues involved in improving the quality of information technology projects. Strong leadership, understanding the cost of quality, providing a good workplace to enhance quality, and working toward improving the organization's overall maturity level in software development and project management can all assist in improving quality.

### Leadership

As Joseph M. Juran said in 1945, "It is most important that top management be quality-minded. In the absence of sincere manifestation of interest at the top, little will happen below."[24] Juran and many other quality experts argue that the main cause of quality problems is a lack of leadership.

As globalization continues to increase and customers become more and more demanding, creating quality products quickly at a reasonable price is essential for staying in business. Having good quality programs in place helps organizations remain competitive. To establish and implement effective quality programs, top management must lead the way. A large percentage of quality problems are associated with management, not technical issues. Therefore, top management must take responsibility for creating, supporting, and promoting quality programs.

Motorola provides an excellent example of a high-technology company that truly emphasizes quality. Leadership is one of the factors that helped Motorola achieve its great success in quality management and Six Sigma. Top management emphasized the need to improve quality and helped all employees take responsibility for customer satisfaction. Strategic objectives in Motorola's long-range plans included managing quality improvement in the same way that new products or technologies were managed. Top management stressed the need to develop and use quality standards and provided resources such as staff, training, and customer inputs to help improve quality.

Leadership provides an environment conducive to producing quality. Management must publicly declare the company's philosophy and commitment to quality, implement company-wide training programs in quality concepts and principles, implement measurement programs to establish and track quality levels, and actively demonstrate the importance of quality. When every employee understands and insists on producing high-quality products, then top management has done a good job of promoting the importance of quality.



## The Cost of Quality

The **cost of quality** is the cost of conformance plus the cost of nonconformance. **Conformance** means delivering products that meet requirements and fitness for use. Examples of these costs include the costs associated with developing a quality plan, costs for analyzing and managing product requirements, and costs for testing. The **cost of nonconformance** means taking responsibility for failures or not meeting quality expectations.

A 2002 study by RTI International reported that software bugs cost the U.S. economy $59.6 billion each year, about 0.6 percent of gross domestic product. Most of the costs are borne by software users, and the rest by developers and vendors. RTI International also suggested that more than one third of these costs could be eliminated by an improved testing infrastructure to enable earlier and more effective identification and removal of software defects.[25]

Other studies estimate the costs per hour of down time for systems. For example, Gartner estimated that the hourly cost of downtime for computer networks is about $42,000. Therefore, a company that suffers from a worse-than-average downtime of 175 hours a year can lose more than $7 million per year.[26] The five major cost categories related to quality include:

1. **Prevention cost**: The cost of planning and executing a project so that it is error-free or within an acceptable error range. Preventive actions such as training, detailed studies related to quality, and quality surveys of suppliers and subcontractors fall under this category. Recall from the discussion of cost management (see Chapter 7) that detecting defects in information systems during the early phases of the systems development life cycle is much less expensive than during the later phases. One hundred dollars spent refining user requirements could save millions by finding a defect before implementing a large system. The Year 2000 (Y2K) issue provided a good example of these costs. If organizations had decided during the 1960s, 1970s, and 1980s that all dates would need four characters to represent the year instead of two characters, they would have saved billions of dollars.
2. **Appraisal cost**: The cost of evaluating processes and their outputs to ensure that a project is error-free or within an acceptable error range. Activities such as inspection and testing of products, maintenance of inspection and test equipment, and processing and reporting inspection data all contribute to appraisal costs of quality.

3. **Internal failure cost**: A cost incurred to correct an identified defect before the customer receives the product. Items such as scrap and rework, charges related to late payment of bills, inventory costs that are a direct result of defects, costs of engineering changes related to correcting a design error, premature failure of products, and correcting documentation all contribute to internal failure cost.
4. **External failure cost**: A cost that relates to all errors not detected and not corrected before delivery to the customer. Items such as warranty cost, field service personnel training cost, product liability suits, complaint handling, and future business losses are examples of external failure costs.
5. **Measurement and test equipment costs**: The capital cost of equipment used to perform prevention and appraisal activities.



Many industries tolerate a very low cost of nonconformance, but not the information technology industry. Tom DeMarco is famous for several studies he conducted on the cost of nonconformance in the information technology industry. In the early 1980s, DeMarco found that the average large company devoted more than 60 percent of its software development efforts to maintenance. Around 50 percent of development costs were typically spent on testing and debugging software.[27] Although these percentages may have improved some, they are still very high.

Top management is primarily responsible for the high cost of nonconformance in information technology. Top managers often rush their organizations to develop new systems and do not give project teams enough time or resources to do a project right the first time. To correct these quality problems, top management must create a culture that embraces quality.

**MEDIA SNAPSHOT**

What do Melissa, Anna Kournikova, Code Red, and Sobig have to do with quality and information technology? They are all names of computer viruses that have cost companies millions of dollars. A quality issue faced by computer users around the world is lost productivity due to computer viruses and spam—unsolicited e-mail sent to multiple mailing lists, individuals, or newsgroups. A 2007 study by Nucleus Research Inc. estimated that spam management costs U.S. businesses more than $71 billion annually in lost productivity or $712 per employee. One e-mail security firm estimated that spam accounts for *95 percent* of total e-mail volume worldwide.[28]

Computer viruses are also getting more technically advanced and infectious as they flood systems across the globe. In 2008, Reuters reported that spyware (software installed on a computer without the owner's consent), viruses, and phishing (using e-mail to try to fraudulently get information such as passwords or credit card numbers) cost consumers $7.1 billion in 2007, up from $2 billion the previous year.[29]

## Organizational Influences, Workplace Factors, and Quality

A study done by Tom DeMarco and Timothy Lister produced interesting results related to organizations and relative productivity. Starting in 1984, DeMarco and Lister conducted "Coding War Games" over several years, in which time more than 600 software developers from 92 organizations participated. The games were designed to examine programming quality and productivity over a wide range of organizations, technical environments, and programming languages. The study demonstrated that organizational issues had a much greater influence on productivity than the technical environment or programming languages.

For example, DeMarco and Lister found that productivity varied by a factor of about one to ten across all participants. That is, one team may have finished a coding project in one day while another team took ten days to finish the same project. In contrast, productivity varied by an average of only 21 percent between pairs of software developers from the same organization. If one team from a specific organization finished the coding project in one day, the longest time it took for another team from the same organization to finish the project was 1.21 days.



DeMarco and Lister also found *no correlation* between productivity and programming language, years of experience, or salary. Furthermore, the study showed that providing a dedicated workspace and a quiet work environment were key factors in improving productivity. The results of the study suggest that top managers must focus on workplace factors to improve productivity and quality.[30]

DeMarco and Lister wrote a book titled *Peopleware* in 1987, and the second edition was published in 1999.[31] Their underlying thesis is that major problems with work performance and project failures are not technological but sociological in nature. They suggest minimizing office politics and giving smart people physical space, intellectual responsibility, and strategic direction—and then just *letting* them work. The manager's function is not to *make* people work but to make it possible for people to work by removing political roadblocks.

## Expectations and Cultural Differences in Quality

Many experienced project managers know that a crucial aspect of project quality management is managing expectations. Although many aspects of quality can be clearly defined and measured, many cannot. Different project sponsors, customers, users, and other stakeholders have different expectations about various aspects of projects. It's very important to understand these expectations and manage any conflicts that might occur due to differences in expectations. For example, in the opening case, several users were upset when they could not access information within a few seconds. In the past, it may have been acceptable to have to wait two or three seconds for a system to load, but many of today's computer users expect systems to run much faster. Project managers and their teams must consider quality-related expectations as they define the project scope.

Expectations can also vary based on an organization's culture or geographic region. Anyone who has traveled to different parts of an organization, a country, or the world understands that expectations are not the same everywhere. For example, one department in a company might expect workers to be in their work areas most of the work day and to dress a certain way. Another department in the same company might focus on whether or not workers produce expected results, no matter where they work or how they dress. People working in smaller towns expect little traffic driving to work, while people working in large cities expect traffic to be a problem or rely on mass transit systems.

People working in other countries for the first time are often amazed at different quality expectations. Visitors to other countries may complain about things they take for granted, such as easily making cell phone calls, using a train or subway instead of relying on a car for transportation, or getting up-to-date maps. It's important to realize that different countries are at different stages of development in terms of quality.

## Maturity Models

Another approach to improving quality in software development projects and project management in general is the use of **maturity models**, which are frameworks for helping organizations improve their processes and systems. Maturity models describe an evolutionary path of increasingly organized and systematically more mature processes. Many maturity models have five levels, with the first level describing characteristics of the least organized or mature organizations, and level five describing the characteristics of the most organized and mature organizations. Three popular maturity models include the Software Quality Function Deployment (SQFD) model, the Capability Maturity Model Integration (CMMI), and the project management maturity model.



### Software Quality Function Deployment Model

The **Software Quality Function Deployment (SQFD) model** is an adaptation of the quality function deployment model suggested in 1986 as an implementation vehicle for Total Quality Management (TQM). SQFD focuses on defining user requirements and planning software projects. The result of SQFD is a set of measurable technical product specifications and their priorities. Having clearer requirements can lead to fewer design changes, increased productivity, and, ultimately, software products that are more likely to satisfy stakeholder requirements. The idea of introducing quality early in the design stage was based on Taguchi's emphasis on robust design methods.[32]

### Capability Maturity Model Integration

Another popular maturity model is in continuous development at the Software Engineering Institute at Carnegie Mellon University. The Software Engineering Institute (SEI) is a federally funded research and development center established in 1984 by the U.S. Department of Defense with a broad mandate to address the transition of software engineering technology. The **Capability Maturity Model Integration (CMMI)** is "a process improvement approach that provides organizations with the essential elements of effective processes. It can be used to guide process improvement across a project, a division, or an entire organization. CMMI helps integrate traditionally separate organizational functions, set process improvement goals and priorities, provide guidance for quality processes, and provide a point of reference for appraising current processes."

The capability levels of the CMMI are:

0. *Incomplete*: At this level, a process is either not performed or partially performed. No generic goals exist for this level, and one or more of the specific goals of the process area are not satisfied.
1. *Performed*: A performed process satisfies the specific goals of the process area and supports and enables the work needed to produce work products. Although this capability level can result in improvements, those improvements can be lost over time if they are not institutionalized.

2. *Managed*: At this level, a process has the basic infrastructure in place to support it. The process is planned and executed based on policies and employs skilled people who have adequate resources to produce controlled outputs. The process discipline reflected by this level ensures that existing practices are retained during times of stress.
3. *Defined*: At this maturity level, a process is rigorously defined and the standards, process descriptions, and procedures for a project are tailored from the organization's set of standard processes to suit that particular project.
4. *Quantitatively managed*: At this level, a process is controlled using statistical and other quantitative techniques. The organization establishes quantitative objectives for quality and process performance that are used as criteria in managing the process.[33]
5. *Optimizing*: An optimizing process is improved based on an understanding of the common causes of variation inherent in the process. The focus is on continually improving the range of process performance through incremental and innovative improvements.[34]



Many companies that want to work in the government market have realized that they will not get many opportunities even to bid on projects unless they have a CMMI Level 3. According to one manager, "CMMI is really the future. People who aren't on the bandwagon now are going to find themselves falling behind."[35]

### Project Management Maturity Models

In the late 1990s, several organizations began developing project management maturity models based on the Capability Maturity Model. Just as organizations realized the need to improve their software development processes and systems, they also realized the need to enhance their project management processes and systems for all types of projects.

The PMI Standards Development Program published the Organizational Project Management Maturity Model (OPM3) in December 2003, and the second edition was released in late 2008. More than 200 volunteers from around the world were part of the initial OPM3 team. The model is based on market research surveys that had been sent to more than 30,000 project management professionals and incorporates 180 best practices and more than 2,400 capabilities, outcomes, and key performance indicators.[36] According to John Schlichter, the OPM3 program director, "The standard would help organizations to assess and improve their project management capabilities as well as the capabilities necessary to achieve organizational strategies through projects. The standard would be a project management maturity model, setting the standard for excellence in project, program, and portfolio management best practices, and explaining the capabilities necessary to achieve those best practices."[37]

## BEST PRACTICE

OPM3 provides the following example to illustrate a best practice, capability, outcome, and key performance indicator:

- *Best practice*: Establish internal project management communities
- *Capability*: Facilitate project management activities
- *Outcome*: Local initiatives, meaning the organization develops pockets of consensus around areas of special interest
- *Key performance indicator*: Community addresses local issues

Best practices are organized into three levels: project, program, and portfolio. Within each of those categories, best practices are categorized by four stages of process improvement: standardize, measure, control, and improve. For example, the list that follows contains several best practices listed in OPM3:

- Project best practices:
  - Project Initiation Process Standardization
  - Project Plan Development Process Measurement
  - Project Scope Planning Process Control
  - Project Scope Definition Process Improvement
- Program best practices:
  - Program Activity Definition Process Standardization
  - Program Activity Sequencing Process Measurement
  - Program Activity Duration Estimating Process Control
  - Program Schedule Development Process Improvement
- Portfolio best practices:
  - Portfolio Resource Planning Process Standardization
  - Portfolio Cost Estimating Process Measurement
  - Portfolio Cost Budgeting Process Control
  - Portfolio Risk Management Planning Process Improvement



Several other companies provide similar project management maturity models. The International Institute for Learning, Inc. calls the five levels in its model common language, common processes, singular methodology, benchmarking, and continuous improvement. ESI International Inc.'s model's five levels are called ad hoc, consistent, integrated, comprehensive, and optimizing. Regardless of the names of each level, the goal is clear: organizations want to improve their ability to manage projects. Many organizations are assessing where they stand in terms of project management maturity, just as they did for software development maturity with the SQFD and CMMI maturity models. Organizations are recognizing that they must make a commitment to the discipline of project management to improve project quality.

## USING SOFTWARE TO ASSIST IN PROJECT QUALITY MANAGEMENT

This chapter provides examples of several tools and techniques used in project quality management. Software can be used to assist with several of these tools and techniques. For example, you can create charts and diagrams from many of the Seven Basic Tools of Quality using spreadsheet and charting software. You can use statistical software packages to help you determine standard deviations and perform many types of statistical analyses. You can create Gantt charts using project management software to help you plan and track work related to project quality management. There are also several specialized software products to assist people with managing Six Sigma projects, creating quality control charts, and assessing maturity levels. Project teams need to decide what types of software will help them manage their particular projects.

As you can see, quality itself is a very broad topic, and it is only one of the nine project management knowledge areas. Project managers must focus on defining how quality relates to their specific projects and ensure that those projects will satisfy the needs for which they were undertaken.



## CASE WRAP-UP

Scott Daniels assembled a team to identify and resolve quality-related issues with the EIS and to develop a plan to help the medical instruments company prevent future quality problems. The first thing Scott's team did was to research the problems with the EIS. They created a cause-and-effect diagram similar to the one in Figure 8-2. They also created a Pareto chart (Figure 8-7) to help analyze the many complaints the Help Desk received and documented about the EIS. After further investigation, Scott and his team found out that many of the managers using the system were very inexperienced in using computer systems beyond basic office automation systems. They also found out that most users received no training at all on how to properly access or use the new EIS. There did not appear to be any major problems with the hardware for the EIS or the users' individual computers. The complaints about reports not giving consistent information all came from one manager, who had actually misread the reports, so there were no problems with the way the software was designed. Scott was very impressed with the quality of the entire project, except for the training. Scott reported his team's findings to the project sponsor of the EIS, who was relieved to find out that the quality problems were not as serious as many people feared.

## Chapter Summary

Many news headlines regarding the poor quality of information technology projects demonstrate that quality is a serious issue. Several mission-critical information technology systems have caused deaths, and quality problems in many business systems have resulted in major financial losses.

Customers are ultimately responsible for defining quality. Important quality concepts include satisfying stated or implied stakeholder needs, conforming to requirements, and delivering items that are fit for use.



Project quality management includes planning quality, performing quality assurance, and performing quality control. Quality planning identifies which quality standards are relevant to the project and how to satisfy them. Quality assurance involves evaluating overall project performance to ensure that the project will satisfy the relevant quality standards. Quality control includes monitoring specific project results to ensure that they comply with quality standards and identifying ways to improve overall quality.

There are many tools and techniques related to project quality management. The Seven Basic Tools of Quality include cause-and-effect diagrams, control charts, run charts, scatter diagrams, histograms, Pareto charts, and flowcharts. Statistical sampling helps define a realistic number of items to include in analyzing a population. Six Sigma helps companies improve quality by reducing defects. Standard deviation measures the variation in data. Testing is very important in developing and delivering high-quality information technology products.

Many people contributed to the development of modern quality management. Deming, Juran, Crosby, Ishikawa, Taguchi, and Feigenbaum all made significant contributions to the field. Many organizations today use their ideas, which also influenced Six Sigma principles. The Malcolm Baldrige National Quality Award and ISO 9000 have also helped organizations emphasize the importance of improving quality.

There is much room for improvement in information technology project quality. Strong leadership helps emphasize the importance of quality. Understanding the cost of quality provides an incentive for its improvement. Providing a good workplace can improve quality and productivity. Understanding stakeholders' expectations and cultural differences are also related to project quality management. Developing and following maturity models can help organizations systematically improve their project management processes to increase the quality and success rate of projects.

There are several types of software available to assist in project quality management. It is important for project teams to decide which software will be most helpful for their particular projects.

## Quick Quiz

1. ___________ is the degree to which a set of inherent characteristics fulfils requirements.
   a. Quality
   b. Conformance to requirements
   c. Fitness for use
   d. Reliability



2. What is the purpose of project quality management?
   a. to produce the highest quality products and services possible
   b. to ensure that appropriate quality standards are met
   c. to ensure that the project will satisfy the needs for which it was undertaken
   d. All of the above
3. ___________ generates ideas for quality improvements by comparing specific project practices or product characteristics to those of other projects or products within or outside the performing organization.
   a. Quality audits
   b. Design of experiments
   c. Six Sigma
   d. Benchmarking
4. What tool could you use to see if there is a relationship between two variables?
   a. a cause-and-effect diagram
   b. a control chart
   c. a run chart
   d. a scatter diagram
5. What tool can you use to determine whether a process is in control or out of control?
   a. a cause-and-effect diagram
   b. a control chart
   c. a run chart
   d. a scatter diagram
6. Six Sigma's target for perfection is the achievement of no more than ________ defects, errors, or mistakes per million opportunities.
   a. 6
   b. 9
   c. 3.4
   d. 1

7. The seven run rule states that if seven data points in a row on a control chart are all below the mean, above the means, or all increasing or decreasing, then the process needs to be examined for __________ problems.
   a. random
   b. non-random
   c. Six Sigma
   d. quality
8. What is the preferred order for performing testing on information technology projects?
   a. unit testing, integration testing, system testing, user acceptance testing
   b. unit testing, system testing, integration testing, user acceptance testing
   c. unit testing, system testing, user acceptance testing, integration testing
   d. unit testing, integration testing, user acceptance testing, system testing
9. ______________ is known for his work on quality control in Japan and developed the 14 Points for Management in his text *Out of the Crisis*.
   a. Juran
   b. Deming
   c. Crosby
   d. Ishikawa
10. PMI's OPM3 is an example of a model or framework for helping organizations improve their processes and systems.
   a. benchmarking
   b. Six Sigma
   c. maturity
   d. quality



### Quick Quiz Answers

1. a; 2. c; 3. d; 4. d; 5. b; 6. c; 7. b; 8. a; 9. b; 10. c

## Discussion Questions

1. Discuss some of the examples of poor quality in information technology projects presented in the "What Went Wrong?" section. Could most of these problems have been avoided? Why do you think there are so many examples of poor quality in information technology projects?
2. What are the main processes included in project quality management?
3. How do functionality, system outputs, performance, reliability, and maintainability requirements affect quality planning?
4. What are benchmarks, and how can they assist in performing quality assurance? Describe typical benchmarks associated with a college or university.
5. What are the three main categories of outputs for quality control?

6. Provide examples of when you would use the Seven Basic Tools of Quality on an information technology project.
7. Discuss the history of modern quality management. How have experts such as Deming, Juran, Crosby, and Taguchi affected the quality movement and today's use of Six Sigma?
8. Discuss three suggestions for improving information technology project quality that were not made in this chapter.
9. Describe three types of software that can assist in project quality management.



## Exercises

1. Assume your organization wants to hire new instructors for your project management course. Develop a list of quality standards that you could use in making this hiring decision.
2. Create a Pareto chart based on the information in the following table. First, create a spreadsheet in Excel, using the data in the table. List the most frequent customer problems first. Use the Excel template called pareto_chart and check your entries so your resulting chart looks similar to the one in Figure 8-7. See the companion Web site for help in creating Pareto charts.

| Customer Complaints | Frequency/Week |
| --- | --- |
| Customer is on hold too long | 90 |
| Customer gets transferred to wrong area or cut off | 20 |
| Service rep cannot answer customer's questions | 120 |
| Service rep does not follow through as promised | 40 |

3. To illustrate a normal distribution, shake and roll a pair of dice 30 times and graph the results. It is more likely for someone to roll a 6, 7, or 8 than a 2 or 12, so these numbers should come up more often. To create the graph, use graph paper or draw a grid. Label the x-axis with the numbers 2 through 12. Label the y-axis with the numbers 1 through 10. Fill in the appropriate grid for each roll of the dice. Do your results resemble a normal distribution? Why or why not?
4. Research the criteria for the Malcolm Baldrige National Quality Award or a similar quality award provided by another organization. Investigate a company that has received this award. What steps did the company take to earn this quality award? What are the benefits of earning a quality award? Summarize your findings in a two-page paper.
5. Review the information in this chapter about Six Sigma principles and Six Sigma organizations. Brainstorm ideas for a potential Six Sigma project that could improve quality on your campus, at your workplace, or in your community. Write a two-page paper describing one project idea and explain why it would be a Six Sigma project. Review and discuss how you could use the DMAIC process on this project.
6. Review the concepts in this chapter related to improving the quality of software. Write a two-page paper describing how you could apply these concepts to software development projects.

## Running Case

The Recreation and Wellness Intranet Project team is working hard to ensure that the new system they develop meets expectations. The team has a detailed scope statement, but the project manager, Tony Prince, wants to make sure they're not forgetting any requirements that might affect how different people view the quality of the project. He knows that the project's sponsor and other senior managers are most concerned with getting people to use the system, improve their health, and reduce healthcare costs. System users will want the system to be very user-friendly, informative, fun to use, and fast.



## Tasks

1. Develop a list of quality standards or requirements related to meeting the stakeholder expectations described above. Also provide a brief description of each requirement. For example, a requirement might be that 90 percent of employees have logged into the system within two weeks after the system rolls out.
2. Based on the list created for Task 1, determine how you will measure progress on meeting the requirements. For example, you might have employees log into the system as part of the training program and track who attends the training. You could also build a feature into the system to track usage by user name, department, and so on.
3. After analyzing survey information, you decide to create a Pareto chart to easily see which types of recreational programs and company-sponsored classes most people were interested in. First, create a spreadsheet in Excel, using the data in the table below. List the most frequently requested programs or classes first. Use the Excel template called pareto_chart and check your entries so your resulting chart looks similar to the one in Figure 8-7. See the companion Web site for help in creating Pareto charts.

| Requested Programs/Classes | # of Times Requested |
|---|---|
| Walking program | 7,115 |
| Volleyball program | 2,054 |
| Weight reduction class | 8,875 |
| Stop smoking class | 4,889 |
| Stress reduction class | 1,894 |
| Soccer program | 3,297 |
| Table tennis program | 120 |
| Softball program | 976 |

## Companion Web Site

Visit the companion Web site for this text at *www.cengage.com/mis/schwalbe* to access:

- References cited in the text and additional suggested readings for each chapter
- Template files
- Lecture notes
- Interactive quizzes
- Podcasts
- Links to general project management Web sites
- And more



See the Preface of this text for additional information on accessing the companion Web site.

## Key Terms

**5 whys** — a technique where you repeatedly ask the question "Why?" (five is a good rule of thumb) to help peel away the layers of symptoms that can lead to the root cause of a problem

**acceptance decisions** — decisions that determine if the products or services produced as part of the project will be accepted or rejected

**appraisal cost** — the cost of evaluating processes and their outputs to ensure that a project is error-free or within an acceptable error range

**benchmarking** — a technique used to generate ideas for quality improvements by comparing specific project practices or product characteristics to those of other projects or products within or outside the performing organization

**Capability Maturity Model Integration (CMMI)** — a process improvement approach that provides organizations with the essential elements of effective processes

**cause-and-effect diagram** — diagram that traces complaints about quality problems back to the responsible production operations to help find the root cause. Also known as *fishbone diagram* or *Ishikawa diagram*

**conformance** — delivering products that meet requirements and fitness for use

**conformance to requirements** — the project processes and products meet written specifications

**control chart** — a graphic display of data that illustrates the results of a process over time

**cost of nonconformance** — taking responsibility for failures or not meeting quality expectations

**cost of quality** — the cost of conformance plus the cost of nonconformance

**defect** — any instance where the product or service fails to meet customer requirements

**DMAIC (Define, Measure, Analyze, Improve, Control)** — a systematic, closed-loop process for continued improvement that is scientific and fact based

**design of experiments** — a quality technique that helps identify which variables have the most influence on the overall outcome of a process

**external failure cost** — a cost related to all errors not detected and corrected before delivery to the customer

**features** — the special characteristics that appeal to users

**fishbone diagram** — diagram that traces complaints about quality problems back to the responsible production operations to help find the root cause. Also known as *cause-and-effect diagram* or *Ishikawa diagram*

**flowchart** — graphic display of the logic and flow of processes that helps you analyze how problems occur and how processes can be improved

**fitness for use** — a product can be used as it was intended

**functionality** — the degree to which a system performs its intended function

**histogram** — a bar graph of a distribution of variables

**integration testing** — testing that occurs between unit and system testing to test functionally grouped components to ensure a subset(s) of the entire system works together

**internal failure cost** — a cost incurred to correct an identified defect before the customer receives the product



**Ishikawa diagram** — diagram that traces complaints about quality problems back to the responsible production operations to help find the root cause. Also known as *cause-and-effect diagram* or *fishbone diagram*

**ISO 9000** — a quality system standard developed by the International Organization for Standardization (ISO) that includes a three-part, continuous cycle of planning, controlling, and documenting quality in an organization

**maintainability** — the ease of performing maintenance on a product

**Malcolm Baldrige National Quality Award** — an award started in 1987 to recognize companies that have achieved a level of world-class competition through quality management

**maturity model** — a framework for helping organizations improve their processes and systems

**mean** — the average value of a population

**measurement and test equipment costs** — the capital cost of equipment used to perform prevention and appraisal activities

**metric** — a standard of measurement

**normal distribution** — a bell-shaped curve that is symmetrical about the mean of the population

**Pareto analysis** — identifying the vital few contributors that account for most quality problems in a system

**Pareto chart** — histogram that helps identify and prioritize problem areas

**performance** — how well a product or service performs the customer's intended use

**prevention cost** — the cost of planning and executing a project so that it is error-free or within an acceptable error range

**process adjustments** — adjustments made to correct or prevent further quality problems based on quality control measurements

**project quality management** — ensuring that a project will satisfy the needs for which it was undertaken

**quality** — the totality of characteristics of an entity that bear on its ability to satisfy stated or implied needs or the degree to which a set of inherent characteristics fulfill requirements

**quality assurance** — periodically evaluating overall project performance to ensure that the project will satisfy the relevant quality standards

**quality audit** — structured review of specific quality management activities that helps identify lessons learned and can improve performance on current or future projects

**quality circles** — groups of nonsupervisors and work leaders in a single company department who volunteer to conduct group studies on how to improve the effectiveness of work in their department

**quality control** — monitoring specific project results to ensure that they comply with the relevant quality standards and identifying ways to improve overall quality

**quality planning** — identifying which quality standards are relevant to the project and how to satisfy them

**reliability** — the ability of a product or service to perform as expected under normal conditions

**rework** — action taken to bring rejected items into compliance with product requirements or specifications or other stakeholder expectations

**Robust Design methods** — methods that focus on eliminating defects by substituting scientific inquiry for trial-and-error methods

**run chart** — chart that displays the history and pattern of variation of a process over time



**scatter diagram** — diagram that helps to show if there is a relationship between two variables; also called XY charts

**seven run rule** — if seven data points in a row on a quality control chart are all below the mean, above the mean, or are all increasing or decreasing, then the process needs to be examined for nonrandom problems

**six 9s of quality** — a measure of quality control equal to 1 fault in 1 million opportunities

**Six Sigma** — a comprehensive and flexible system for achieving, sustaining, and maximizing business success that is uniquely driven by close understanding of customer needs, disciplined use of facts, data, statistical analysis, and diligent attention to managing, improving, and reinventing business processes

**software defect** — anything that must be changed before delivery of the program

**Software Quality Function Deployment (SQFD) model** — a maturity model that focuses on defining user requirements and planning software projects

**standard deviation** — a measure of how much variation exists in a distribution of data

**statistical sampling** — choosing part of a population of interest for inspection

**system outputs** — the screens and reports the system generates

**system testing** — testing the entire system as one entity to ensure that it is working properly

**unit test** — a test of each individual component (often a program) to ensure that it is as defect-free as possible

**user acceptance testing** — an independent test performed by end users prior to accepting the delivered system

**yield** — the number of units handled correctly through the development process

## End Notes


[1] This joke was found on hundreds of Web sites and printed in the *Consultants in Minnesota Newsletter*, Independent Computer Consultants Association, December 1998.

[2] *Design News* (February 1988).

[3] *Datamation* (May 1987).

[4] *New York Times* (February 18, 1994).

[5] Robert Gavin, "Subscriber credit data distributed by mistake," *Boston Globe* (February 1, 2006).

[6] Peter S. Pande, Robert P. Neuman, and Roland R. Cavanagh, *The Six Sigma Way* (New York: McGraw-Hill, 2000), p. xi.

[7] James C. Collins and Jerry I. Porras, *Built to Last: Successful Habits of Visionary Companies* (New York: HarperBusiness, 1994).

[8] Peter S. Pande, Robert P. Neuman, and Roland R. Cavanagh, *The Six Sigma Way* (New York: McGraw-Hill, 2000), p. 7.

[9] Ibid. p. 9.

[10] Lewis Brown and Denise Taylor, "Reducing Delayed Starts in Special Lab With Six Sigma," *iSixSigma Healthcare* 5, no. 2 (January 17, 2007).

[11] "What You Need to Know About Six Sigma," *Productivity Digest* (December 2001), p. 38.

[12] Peter S. Pande, Robert P. Neuman, and Roland R. Cavanagh, *The Six Sigma Way* (New York: McGraw-Hill, 2000), p. 137.

[13] Lee Clifford, "Why You Can Safely Ignore Six Sigma," *Fortune* (January 22, 2001), p. 140.

[14] Hollstadt & Associates, Inc., *Software Development Project Life Cycle Testing Methodology User's Manual* (Burnsville: MN, August 1998), p. 13.

[15] Bart Eisenberg, "Achieving Zero-Defects Software," *Pacific Connection* (January 2004).

[16] Harold Kerzner, *Project Management*, 6th ed. (New York: Van Nostrand Reinhold, 1998), p. 1048.

[17] W. Edwards Deming, *Out of the Crisis* (Boston: MIT Press, 1986).

[18] Joseph M. Juran and A. Blanton Godfrey, *Juran's Quality Handbook*, 5th ed. (New York: McGraw-Hill Professional, 2000).

[19] Philip B. Crosby, *Quality Is Free: The Art of Making Quality Certain* (New York: McGraw-Hill, 1979).

[20] Kaoru Ishikawa, *Guide to Quality Control*, 2nd ed. (Asian Productivity Organization, 1986).

[21] Bylinsky, Gene, "How to Bring Out Better Products Faster," *Fortune* (November 23, 1998) p. 238[B].

[22] Armand V. Feigenbaum, *Total Quality Control: Engineering and Management*, 3rd rev. ed. (New York: McGraw-Hill, 1991).

[23] International Organization for Standardization, (Market feedback) *ISO.org* (March 2003).

[24] American Society for Quality (ASQ), "About ASQ: Joseph M. Juran" *ASQ.com* (*www.asq.org/about-asq/who-we-are/bio_juran.html*).

[25] RTI International, "Software Bugs Cost U.S. Economy $59.6 Billion Annually, RTI Study Finds" (July 1, 2002).

[26] Tom Pisello and Bill Quirk, "How to Quantify Downtime," *Network World* (January 4, 2004).

[27] Tom DeMarco, *Quality Controlling Software Projects: Management, Measurement and Estimation* (New Jersey: Prentice Hall PTR Facsimile Edition, 1986).

[28] Thomas Claburn, "Spam Turns 30 and Never Looked Healthier," *Information Week* (May 2, 2008).

[29] Reuters, "Internet Fraud Ignored by Authorities, Study Charges," *PCWorld* (August 16, 2008).

[30] Tom DeMarco and Timothy Lister, *Peopleware: Productive Projects and Teams* (New York: Dorset House, 1987).

[31] Tom DeMarco and Timothy Lister, *Peopleware: Productive Projects and Teams*, 2nd ed. (New York: Dorset House, 1999).

[32] R. R. Yilmaz and Sangit Chatterjee, "Deming and the Quality of Software Development," *Business Horizons* [Foundation for the School of Business at Indiana University] 40, no. 6 (November–December 1997), pp. 51–58.

[33] Software Engineering Institute, Carnegie Mellon, "What is CMMI," (January 2007) (*http://www.sei.cmu.edu/cmmi/general/index.html*).

[34] CMMI Product Team, "CMMI® for Development, Version 1.2," CMU/SEI–2006–TR–008 ESC–TR–2006–008 (August 2006).

[35] Michael Hardy, "Strength in numbers: New maturity ratings scheme wins support among systems integrators" *FCW.com* (March 28, 2004).

[36] John Schlichter, "The Project Management Institute's Organizational Project Management Maturity Model: An Update on the PMI's OPM3 Program" *PMForum.com* (September 2002).

[37] John Schlichter, "The History of OPM3," *Project Management World Today* (June 2003).

CHAPTER 9

# PROJECT HUMAN RESOURCE MANAGEMENT

## LEARNING OBJECTIVES

**After reading this chapter, you will be able to:**

- Explain the importance of good human resource management on projects, including the current state and future implications of the global IT workforce
- Define project human resource management and understand its processes
- Summarize key concepts for managing people by understanding the theories of Abraham Maslow, Frederick Herzberg, David McClelland, and Douglas McGregor on motivation, H. J. Thamhain and D. L. Wilemon on influencing workers, and Stephen Covey on how people and teams can become more effective
- Discuss human resource planning and be able to create a human resource plan, project organizational chart, responsibility assignment matrix, and resource histogram
- Understand important issues involved in project staff acquisition and explain the concepts of resource assignments, resource loading, and resource leveling
- Assist in team development with training, team-building activities, and reward systems
- Explain and apply several tools and techniques to help manage a project team and summarize general advice on managing teams
- Describe how project management software can assist in project human resource management

## OPENING CASE

This was the third time someone from the Information Technology department tried to work with Ben, the head of the F-44 aircraft program. Ben, who had been with the company for almost 30 years, was known for being "rough around the edges" and very demanding. The company was losing money on the F-44 upgrade project because the upgrade kits were not being delivered on time. The Canadian government had written severe late-penalty fees into its contract, and other customers were threatening to take their business elsewhere. Ben blamed it all on the Information Technology department for not letting his staff access the F-44 upgrade project's information system directly so they could work with their customers and suppliers more effectively. The information system was based on very old technology that only a couple of people in the company knew how to use. It often took days or even weeks for Ben's group to get the information they needed.

Ed Davidson, a senior programmer, attended a meeting with Sarah Ellis, an internal information technology business consultant. Sarah was in her early thirties and had advanced quickly in her company, primarily due to her keen ability to work well with all types of people. Sarah's job was to uncover the real problems with the F-44 aircraft program's information technology support, and then develop a solution with Ben and his team. If she found it necessary to invest in more information technology hardware, software, or staff, Sarah would write a business case to justify these investments and then work with Ed, Ben, and his group to implement the suggested solution as quickly as possible. Ben and three of his staff entered the conference room. Ben threw his books on the table and started yelling at Ed and Sarah. Ed could not believe his eyes or ears when Sarah stood nose to nose with Ben and started yelling right back at him.



## THE IMPORTANCE OF HUMAN RESOURCE MANAGEMENT

Many corporate executives have said, "People are our most important asset." People determine the success and failure of organizations and projects. Most project managers agree that managing human resources effectively is one of the toughest challenges they face. Project human resource management is a vital component of project management, especially in the information technology field—in which qualified people are often hard to find and keep. It is important to understand global IT workforce issues and their implications for the future.

### The Global IT Workforce

Although there have been ups and downs in the information technology labor market, there will always be a need for people to develop and maintain hardware, software, networks, and applications of information technology. The global job market for information technology workers is expanding, and the demand for project managers continues to increase. Below are statistics from several recent studies.

The Digital Planet 2008 study estimated that the global marketplace for information and communications technology (ICT) would top $3.7 trillion in 2008 and reach almost $4 trillion by 2011. The marketplace posted a 10.8 percent average annual growth rate in 2007 and 2008. Additional findings from this report include the following:

- Communications products and services represented the largest single category of ICT spending (57 percent) in 2007 with $1.9 trillion;
- Consumers spent 29 percent of ICT dollars worldwide, while spending by business and government accounted for 71 percent;
- In spending by country, the top ten ICT spending countries are, in descending order: the United States, Japan, China, Germany, United Kingdom, France, Italy, Brazil, Canada, and Spain. In 2008, China jumped ahead of Germany, the United Kingdom, and France; and
- The Americas growth in ICT spending will be the slowest of the three broad regions charted in Digital Planet, at 4 percent between 2007 and 2011. The Asia-Pacific region and the Europe, Africa, and Middle East regions will grow annually at 10.5 percent and 5 percent, respectively.[1]



In the U.S. the size of the IT workforce topped 4 million workers for the first time in 2008. Unemployment rates in many information technology occupations are among the lowest in the labor force, at only 2.3 percent. Demand for talent is high, and several organizations cannot grow as desired due to difficulties in hiring and recruiting the people they need.[2] According to the Bureau of Labor Statistics, several IS-related occupations will be among the top 30 fastest-growing occupations in the United States between now and 2016, with network systems/data communications analysts and computer software engineers listed as numbers one and four, respectively.[3]

A 2007 Forrester survey found that project managers remain in high demand but short supply, with 26 percent of IT leaders planning to hire project managers and 59 percent planning to train their current staff in project management. Little had changed since their 2002 survey, when 55 percent of IT leaders identified project management as the missing skill set in their organization. "The reason for the continued emphasis on project management skills is because IT's value to business remains contingent on its ability to deliver projects which meet business requirements both on time and on budget. IT staff accustomed to more technical roles struggle to transition to project management, CIOs argue, and complain that educational institutions are not putting adequate focus on these skills through coursework."[4] Employers like to hire people with a solid record of accomplishment. It helps to have previous experience in a related field and a four-year college degree. Hiring managers say interpersonal skills are the most important soft skill for information technology workers.[5]

These studies highlight the continued need for skilled information technology workers and project managers to lead their projects. As the job market changes, however, people should upgrade their skills to remain marketable and flexible. Many highly qualified information technology workers lost their jobs in the past few years. Workers need to develop a support system and financial reserves to make it through difficult economic times. They also need to consider different types of jobs and industries where they can use their skills and earn a living. Negotiation and presentation skills also become crucial in finding and keeping a good job. Workers need to know how they personally contribute to an organization's bottom line.

## Implications for the Future of IT Human Resource Management

It is crucial for organizations to practice what they preach about human resources. If people truly are their greatest asset, organizations must work to fulfill their human resource needs *and* the needs of individual people in their organizations, regardless of the job market. If organizations want to be successful at implementing information technology projects, they need to understand the importance of project human resource management and take actions to make effective use of people.

Proactive organizations are addressing current and future human resource needs by, for example, improving benefits, redefining work hours and incentives, and finding future workers. Many organizations have changed their benefits policies to meet worker needs. Most workers assume that their companies provide some perks, such as casual dress codes, flexible work hours, and tuition assistance. Other companies might provide on-site day care, fitness club discounts, or matching contributions to retirement savings. Google, the winner of Fortune's 100 Best Companies award in 2007 and 2008, provides employees with free gourmet meals and doctors on site, a swimming spa and corporate gym, beach volleyball, Foosball, videogames, pool tables, ping-pong, roller hockey, and weekly Thank Goodness It's Friday (TGIF) parties! Of course, employees are top-notch workers, and Google receives 1,300 resumes a day.



Other implications for the future of human resource management relate to the hours organizations expect many information technology professionals to work and how they reward performance. Today people brag about the fact that they can work *less* than 40 hours a week or work from home several days a week—not that they work a lot of overtime and haven't had a vacation in years. If companies plan their projects well, they can avoid the need for overtime, or they can make it clear that overtime is optional. Many companies also outsource more and more of their project work, as described in Chapter 12, Project Procurement Management, to manage the fluctuating demand for workers.

Companies can also provide incentives that use performance, not hours worked, as the basis of rewards. If performance can be objectively measured, as it can in many aspects of information technology jobs, then it should not matter where employees do their work or how long it takes them to do it. For example, if a technical writer can produce a high-quality publication at home in one week, the company and the writer are better off than if the company insisted that the writer come to the office and take two weeks to produce the publication. Objective measures of work performance and incentives based on meeting those criteria are important considerations.

### MEDIA SNAPSHOT

People often brag about the productivity of American workers. An article in the *Minneapolis Star Tribune* by Joan Williams and Ariane Hegewisch, however, revealed some interesting facts:

> Here's the dirty little secret: U.S. productivity is No. 1 in the world when productivity is measured as gross domestic product per worker, but our lead vanishes when productivity is measured as GDP per hour worked, according to the Organization for Economic Cooperation and Development, whose members are the world's 30 most

*continued*

developed nations. Productivity per hour is higher in France, with the United States at about the same level as other advanced European economies. As it turns out, the U.S. "productivity advantage" is just another way of saying that we work more hours than workers in any other industrialized country except South Korea. Is that something to brag about? Europeans take an average of six to seven weeks of paid annual leave, compared with just 12 days in the United States. Twice as many American as European workers put in more than 48 hours per week. Particularly sobering is the fact that in two out of three American families with small children in which both parents work, the couples work more than 80 total hours per week, also more than double the European rate.[6]



Economists say that Americans prefer the higher income gained from working extra hours, while Europeans prefer more family time and leisure. However, sociologists have shown that many Americans, especially men, would like to have more family or leisure time. Recent surveys show that many Americans are willing to sacrifice up to a quarter of their salaries in return for more time off!

So why don't Americans work less? Some do need the money, but many others cannot find jobs where they can work fewer hours. Even more Americans fear losing their jobs, healthcare benefits, or promotion opportunities if they work fewer hours or even request the opportunity to do so. As a result, many families who would prefer both parents work less than 40 hours a week end up having one parent working 50 or more hours a week while the other parent is unemployed altogether. Williams and Hegewisch describe this situation as "a squandering of human capital that is typically overlooked in discussions of productivity and GDP. A new and more promising way to fuel economic growth would be to offer good jobs with working hours that enable fathers as well as mothers to maintain an active involvement with family life as well as an active career."[7]

The need to develop future talent in information technology also has important implications for everyone. Who will maintain the systems we have today when the last of the "baby boomer" generation retires? Who will continue to develop new products and services using new technologies that have not yet been developed? Some schools require all students to take computer literacy courses, although most teenagers today already know how to use computers, iPods, cell phones, and other technologies. But are today's children learning the skills they'll need to develop new technologies and work on global teams? As the workforce becomes more diverse, will more women and minorities be ready and willing to enter IT fields? Several colleges, government agencies, and private groups have programs to help recruit more women and minorities into technical fields. Some companies provide options for working parents to help them balance work and family. All of these efforts will help to develop the human resources needed for future information technology projects.

## WHAT WENT WRONG?

A 2006 report by The Conference Board, Corporate Voices for Working Families, Partnership for 21st Century Skills, and the Society for Human Resource Management suggests that entry level workers in the U.S. are ill-prepared for the workplace. More than

*continued*

400 employers across the United States participated in the survey, describing what skills they believe are important for new job entrants and how well they think recently hired graduates exhibit these skills. Employers listed "very important" skills at all education levels and then rated entry level workers as being deficient, adequate, or excellent in each skill.

For high school graduates, respondents listed many deficiencies, as follows. The percentage next to each skill indicates the percentage of respondents who rated graduates as deficient.

- Written Communications—80.9%
- Professionalism/Work Ethic—70.3%
- Critical Thinking/Problem Solving—69.6%
- Oral Communications—52.7%
- Ethics/Social Responsibility—44.1%
- Reading Comprehension—38.4%
- Teamwork/Collaboration—34.6%
- Diversity—27.9%
- Information Technology Application—21.5%
- English Language—21.0%

Two-year college/technical school graduates were listed as deficient in the following seven skills:

- Written Communications—47.3%
- Writing in English—46.4%
- Lifelong Learning/Self Direction—27.9%
- Creativity/Innovation—27.6%
- Critical Thinking/Problem Solving—22.8%
- Oral Communications—21.3%
- Ethics/Social Responsibility—21.0%

Four-year college graduates were listed as deficient in the following three skills:

- Written Communications—27.8%
- Writing in English—26.2%
- Leadership—23.8%

An interesting finding was that 46.3 percent of respondents ranked graduates of four-year colleges as excellent in information technology application, but they were more concerned that college grads were deficient in written communications, writing in English, and leadership.[8]



## WHAT IS PROJECT HUMAN RESOURCE MANAGEMENT?

Project human resource management includes the processes required to make the most effective use of the people involved with a project. Human resource management includes all project stakeholders: sponsors, customers, project team members, support staff, suppliers supporting the project, and so on. Human resource management includes the following four processes:

1. *Developing the human resource plan* involves identifying and documenting project roles, responsibilities, and reporting relationships. The main output of this process is a human resource plan.
2. *Acquiring the project team* involves getting the needed personnel assigned to and working on the project. Key outputs of this process are project staff assignments, resource calendars, and project management plan updates.
3. *Developing the project team* involves building individual and group skills to enhance project performance. Team-building skills are often a challenge for many project managers. The main outputs of this process are team performance assessments and enterprise environmental factors updates.
4. *Managing the project team* involves tracking team member performance, motivating team members, providing timely feedback, resolving issues and conflicts, and coordinating changes to help enhance project performance. Outputs of this process include enterprise environmental factors updates, organizational process assets updates, change requests, and project management plan updates.

Figure 9-1 summarizes these processes and outputs, showing when they occur in a typical project.



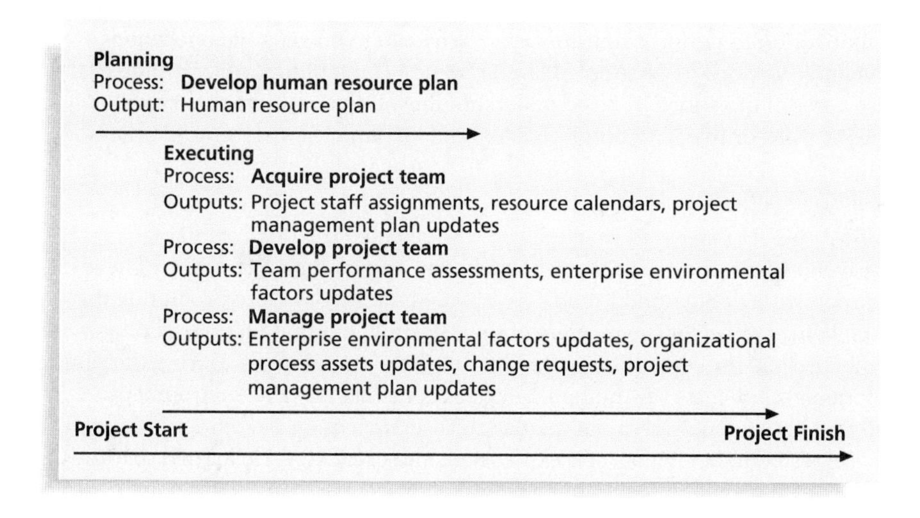


**FIGURE 9-1** Project human resource management summary

Some topics related to human resource management (understanding organizations, stakeholders, and different organizational structures) were introduced in Chapter 2, The Project Management and Information Technology Context. This chapter will expand on some of those topics and introduce other important concepts in project human resource management, including theories about managing people, resource loading, and resource leveling. You will also learn how to use software to assist in project human resource management.

Industrial-organizational psychologists and management theorists have devoted much research and thought to the field of managing people at work. Psychosocial issues that affect how people work and how well they work include motivation, influence and power, and effectiveness. This section will review Abraham Maslow's, Frederick Herzberg's, David McClelland's, and Douglas McGregor's contributions to an understanding of motivation; H. J. Thamhain's and D. L. Wilemon's work on influencing workers and reducing conflict; the effect of power on project teams; and Stephen Covey's work on how people and teams can become more effective. Finally, you will look at some implications and recommendations for project managers.



## Motivation Theories

Psychologists, managers, coworkers, teachers, parents, and most people in general still struggle to understand what motivates people, or, why they do what they do. **Intrinsic motivation** causes people to participate in an activity for their own enjoyment. For example, some people love to read, write, or play an instrument because it makes them feel good. **Extrinsic motivation** causes people to do something for a reward or to avoid a penalty. For example, some young children would prefer *not* to play an instrument, but they do because they receive a reward or avoid a punishment for doing so. Why do some people require no external motivation whatsoever to produce high-quality work while others require significant external motivation to perform routine tasks? Why can't you get someone who is extremely productive at work to do simple tasks at home? Humankind will continue to try to answer these types of questions. A basic understanding of motivational theory will help anyone who has to work or live with other people to understand themselves and others.

### Maslow's Hierarchy of Needs

Abraham Maslow, a highly respected psychologist who rejected the dehumanizing negativism of psychology in the 1950s, is best known for developing a hierarchy of needs. In the 1950s, proponents of Sigmund Freud's psychoanalytic theory were promoting the idea that human beings were not the masters of their destiny and that all their actions were governed by unconscious processes dominated by primitive sexual urges. During the same period, behavioral psychologists saw human beings as controlled by the environment. Maslow argued that both schools of thought failed to recognize unique qualities of human behavior: love, self-esteem, belonging, self-expression, and creativity. He argued that these unique qualities enable people to make independent choices, which gives them full control of their destiny.

Figure 9-2 shows the basic pyramid structure of Maslow's **hierarchy of needs**, which states that people's behaviors are guided or motivated by a sequence of needs. At the bottom of the hierarchy are physiological needs. Once physiological needs are satisfied, safety needs guide behavior. Once safety needs are satisfied, social needs come to the forefront, and so on up the hierarchy. The order of these needs and their relative sizes in the pyramid are significant. Maslow suggests that each level of the hierarchy is a prerequisite for the levels above. For example, it is not possible for a person to consider self-actualization if he or she has not addressed basic needs concerning security and safety. People in an emergency situation, such as a flood or hurricane, are not going to worry about personal growth. Personal survival will be their main motivation. Once a particular need is satisfied, however, it no longer serves as a potent motivator of behavior.

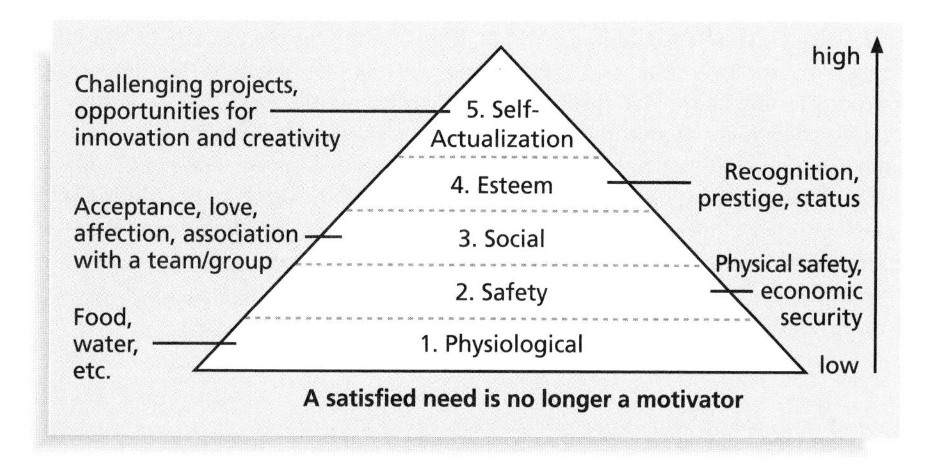

**FIGURE 9-2** Maslow's hierarchy of needs

The bottom four needs in Maslow's hierarchy—physiological, safety, social, and esteem—are referred to as deficiency needs, and the highest level, self-actualization, is considered a growth need. Only after meeting deficiency needs can individuals act upon growth needs. Self-actualized people are problem-focused, have an appreciation for life, are concerned about personal growth, and have the ability to have peak experiences.

Most people working on an information technology project will probably have their basic physiological and safety needs met. If someone has a sudden medical emergency or is laid off from work, however, physiological and safety needs will move to the forefront. To motivate project team members, the project manager needs to understand each person's motivation, especially with regard to social, esteem, and self-actualization or growth needs. Team members new to a company and city might be motivated by social needs. To address social needs, some companies organize gatherings and social events for new workers. Other project members may find these events to be an invasion of personal time they would rather spend with their friends and family or working on an advanced degree.

Maslow's hierarchy conveys a message of hope and growth. People can work to control their own destinies and naturally strive to satisfy higher and higher needs. Successful project managers know they must focus on meeting project goals, but they also know that they must understand team members' personal goals and needs to provide appropriate motivation and maximize team performance.

### Herzberg's Motivation-Hygiene Theory

Frederick Herzberg is best known for distinguishing between motivational factors and hygiene factors when considering motivation in work settings. He called factors that cause job satisfaction motivators and factors that cause dissatisfaction hygiene factors. The term "hygiene" is used in the sense that these factors are considered maintenance factors that are necessary to avoid dissatisfaction but, by themselves, they do not provide satisfaction.

Head of Case Western University's psychology department, Herzberg wrote the book *Work and the Nature of Man* in 1966 and a famous *Harvard Business Review* article, "One More Time: How Do You Motivate Employees?" in 1968.[9] Herzberg analyzed the factors that affected productivity among a sample of 1,685 employees. Popular beliefs at that time were that work output was most improved through larger salaries, more supervision, or a more attractive work environment. According to Herzberg, these hygiene factors would cause dissatisfaction if not present, but would not motivate workers to do more if present. Today, professionals might also expect employers to provide them with health benefits, training, and a computer or other equipment required to perform their jobs. Herzberg found that people were motivated to work mostly by feelings of personal achievement and recognition. Motivators, Herzberg concluded, included achievement, recognition, the work itself, responsibility, advancement, and growth, and are shown in Table 9-1.



**TABLE 9-1** Examples of Herzberg's hygiene factors and motivators

| Hygiene Factors | Motivators |
|---|---|
| Larger salaries | Achievement |
| More supervision | Recognition |
| More attractive work environment | Work itself |
| Computer or other required equipment | Responsibility |
| Health benefits | Advancement |
| Training | Growth |

In his books and articles, Herzberg explained why attempts to use positive factors such as reducing time spent at work, upward spiraling wages, offering fringe benefits, providing human relations and sensitivity training, and so on did not instill motivation. He argued that people want to actualize themselves by being able to use their creativity and work on challenging projects. They need stimuli for their growth and advancement needs, in accordance with Maslow's hierarchy of needs. Factors such as achievement, recognition, responsibility, advancement, and growth produce job satisfaction and are work motivators.

### McClelland's Acquired-Needs Theory

David McClelland proposed that an individual's specific needs are acquired or learned over time and shaped by life experiences. The main categories of acquired needs include achievement, affiliation, and power.[10] Normally one or two of these needs will be dominant in individuals.

- *Achievement*: People with a high need for achievement (nAch) seek to excel and tend to avoid both low-risk and high-risk situations to improve their chances for achieving something worthwhile. Achievers need regular feedback and often prefer to work alone or with other high achievers. Managers should give high achievers challenging projects with achievable goals. Achievers should receive frequent performance feedback, and although money is not an important motivator to them, it is an effective form of feedback.
- *Affiliation*: People with a high need for affiliation (nAff) desire harmonious relationships with other people and need to feel accepted by others. They tend to conform to the norms of their work group and prefer work that involves significant personal interaction. Managers should try to create a cooperative work environment to meet the needs of people with a high need for affiliation.
- *Power*: People with a need for power (nPow) desire either personal power or institutional power. People who need personal power want to direct others and can be seen as bossy. People who need institutional power or social power want to organize others to further the goals of the organization. Management should provide those seeking institutional or social power with the opportunity to manage others, emphasizing the importance of meeting organizational goals.



The Thematic Apperception Test (TAT) is a tool to measure the individual needs of different people using McClelland's categories. The TAT presents subjects with a series of ambiguous pictures and asks them to develop a spontaneous story for each picture, assuming they will project their own needs into the story.

### McGregor's Theory X and Theory Y

Douglas McGregor was one of the great popularizers of a human relations approach to management, and he is best known for developing Theory X and Theory Y. In his research, documented in his 1960 book *The Human Side of Enterprise*, McGregor found that although many managers spouted the right ideas, they actually followed a set of assumptions about worker motivation that he called *Theory X* (sometimes referred to as classical systems theory).[11] People who believe in Theory X assume that workers dislike and avoid work if possible, so managers must use coercion, threats, and various control schemes to get workers to make adequate efforts to meet objectives. They assume that the average worker wants to be directed and prefers to avoid responsibility, has little ambition, and wants security above all else. Research seemed to demonstrate clearly that these assumptions were not valid. McGregor suggested a different series of assumptions about human behavior that he called *Theory Y* (sometimes referred to as human relations theory). Managers who believe in Theory Y assume that individuals do not inherently dislike work, but consider it as natural as play or rest. The most significant rewards are the satisfaction of esteem and self-actualization needs, as described by Maslow. McGregor urged managers to motivate people based on these more valid Theory Y notions.

In 1981, William Ouchi introduced another approach to management in his book *Theory Z: How American Business Can Meet the Japanese Challenge*.[12] Theory Z is based on the Japanese approach to motivating workers, which emphasizes trust, quality, collective decision making, and cultural values. Whereas Theory X and Theory Y emphasize how management views employees, Theory Z also describes how workers perceive management. Theory Z workers, it is assumed, can be trusted to do their jobs to their utmost ability, as long as management can be trusted to support them and look out for their well-being. Theory Z emphasizes things such as job rotation, broadening of skills, generalization versus specialization, and the need for continuous training of workers.

### Thamhain and Wilemon's Influence and Power

Many people working on a project do not report directly to project managers, and project managers often do not have control over project staff who report to them. For example, people are free to change jobs. If they are given work assignments they do not like, many workers will simply quit or transfer to other departments or projects. H. J. Thamhain and D. L. Wilemon investigated the approaches project managers use to deal with workers and how those approaches relate to project success. They identified nine influence bases available to project managers:



1. *Authority*: the legitimate hierarchical right to issue orders
2. *Assignment*: the project manager's perceived ability to influence a worker's later work assignments
3. *Budget*: the project manager's perceived ability to authorize others' use of discretionary funds
4. *Promotion*: the ability to improve a worker's position
5. *Money*: the ability to increase a worker's pay and benefits
6. *Penalty*: the project manager's perceived ability to dispense or cause punishment
7. *Work challenge*: the ability to assign work that capitalizes on a worker's enjoyment of doing a particular task, which taps an intrinsic motivational factor
8. *Expertise*: the project manager's perceived special knowledge that others deem important
9. *Friendship*: the ability to establish friendly personal relationships between the project manager and others[13]

Top management grants authority to the project manager. Assignment, budget, promotion, money, and penalty influence bases may or may not be inherent in a project manager's position. Unlike authority, they are not automatically available to project managers as part of their position. Others' perceptions are important in establishing the usefulness of these influence bases. For example, any manager can influence workers by providing challenging work; the ability to provide challenging work (or take it away) is not a special ability of project managers. In addition, project managers must earn the ability to influence by using expertise and friendship.

Thamhain and Wilemon found that projects were more likely to fail when project managers relied too heavily on using authority, money, or penalty to influence people. *When project managers used work challenge and expertise to influence people, projects were more likely to succeed*. The effectiveness of work challenge in influencing people is consistent with Maslow's and Herzberg's research on motivation. The importance of expertise as a means of influencing people makes sense on projects that involve special knowledge, as in most information technology projects.

Influence is related to the topic of power. **Power** is the potential ability to influence behavior to get people to do things they would not otherwise do. Power has a much stronger connotation than influence, especially since it is often used to force people to change their behavior. There are five main types of power, based on French and Raven's classic study, "The Bases of Social Power."[14]

- **Coercive power** involves using punishment, threats, or other negative approaches to get people to do things they do not want to do. This type of power is similar to Thamhain's and Wilemon's influence category called *penalty*. For example, a project manager can threaten to fire workers or subcontractors to try to get them to change their behavior. If the project manager really has the power to fire people, he or she could follow through on the threat. Recall, however, that influencing using penalties is correlated with unsuccessful projects. Still, coercive power can be very effective in stopping negative behavior. For example, if students tend to hand in assignments late, an instructor can have a policy in his or her syllabus stating late penalties, such as a 20 percent grade reduction for each day an assignment is late.
- **Legitimate power** is getting people to do things based on a position of authority. This type of power is similar to the authority basis of influence. If top management gives project managers organizational authority, project managers can use legitimate power in several situations. They can make key decisions without involving the project team, for example. Overemphasis of legitimate power or authority also correlates with project failure.
- **Expert power** involves using personal knowledge and expertise to get people to change their behavior. If people perceive that project managers are experts in certain situations, they will follow their suggestions. For example, if a project manager has expertise in working with a particular information technology supplier and their products, the project team will be more likely to follow the project manager's suggestions on how to work with that vendor and its products.
- **Reward power** involves using incentives to induce people to do things. Rewards can include money, status, recognition, promotions, special work assignments, or other means of rewarding someone for desired behavior. Many motivation theorists suggest that only certain types of rewards, such as work challenge, achievement, and recognition, truly induce people to change their behavior or work hard.
- **Referent power** is based on an individual's personal charisma. People hold someone with referent power in very high regard and will do what they say based on their regard for the person. People such as Martin Luther King, Jr., John F. Kennedy, and Bill Clinton had referent power. Very few people possess the natural charisma that underlies referent power.



It is important for project managers to understand what types of influence and power they can use in different situations. New project managers often overemphasize their position—their legitimate power or authority influence—especially when dealing with project team members or support staff. They also neglect the importance of reward power or work challenge influence. People often respond much better to a project manager who motivates them with challenging work and provides positive reinforcement for doing a good job. It is important for project managers to understand the basic concepts of influence and power, and to practice using them to their own and their project team's advantage.

## Covey and Improving Effectiveness

Stephen Covey, author of *The 7 Habits of Highly Effective People: Powerful Lessons in Personal Change*,[15] expanded on the work done by Maslow, Herzberg, and others to develop an approach for helping people and teams become more effective. Covey's first three habits of effective people—be proactive, begin with the end in mind, and put first things first—help people achieve a private victory by becoming independent. After achieving independence, people can then strive for interdependence by developing the next three habits—think win/win, seek first to understand then to be understood, and synergize. (**Synergy** is the concept that the whole is equal to more than the sum of its parts.) Finally, everyone can work on Covey's seventh habit—sharpen the saw—to develop and renew their physical, spiritual, mental, and social/emotional selves.



Project managers can apply Covey's seven habits to improve effectiveness on projects, as follows:

1. *Be proactive*. Covey, like Maslow, believes that people have the ability to be proactive and choose their responses to different situations. Project managers must be proactive, anticipate, and plan for problems and inevitable changes on projects. They can also encourage their team members to be proactive in working on their project activities.
2. *Begin with the end in mind*. Covey suggests that people focus on their values, what they really want to accomplish, and how they really want to be remembered in their lives. He suggests writing a mission statement to help achieve this habit. Many organizations and projects have mission statements that help them focus on their main purpose.
3. *Put first things first*. Covey developed a time management system and matrix to help people prioritize their time. He suggests that most people need to spend more time doing things that are important, but not urgent. Important but not urgent activities include planning, reading, and exercising. Project managers need to spend a lot of time working on important and not urgent activities, such as developing various project plans, building relationships with major project stakeholders, and mentoring project team members. They also need to avoid focusing only on important and urgent activities—putting out fires.
4. *Think win/win*. Covey presents several paradigms of interdependence, with think win/win being the best choice in most situations. When you use a win/win paradigm, parties in potential conflict work together to develop new solutions that benefit all parties. Project managers should strive to use a win/win approach in making decisions, but sometimes, especially in competitive situations, they must use a win/lose paradigm.
5. *Seek first to understand, then to be understood*. **Empathic listening** is listening with the intent to understand. It is even more powerful than active listening because you forget your personal interests and focus on truly understanding the other person. To really understand other people, you must learn to focus on others first. When you practice empathic listening, you can begin two-way communication. This habit is critical for project managers so they can really understand their stakeholders' needs and expectations.

6. *Synergize*. In projects, a project team can synergize by creating collaborative products that are much better than a collection of individual efforts. Covey also emphasizes the importance of valuing differences in others to achieve synergy. Synergy is essential to many highly technical projects; in fact, several major breakthroughs in information technology occurred because of synergy. For example, in his Pulitzer Prize–winning book, *The Soul of a New Machine*, Tracy Kidder documented the 1970s synergistic efforts of a team of Data General researchers to create a new 32-bit superminicomputer.[16]
7. *Sharpen the saw*. When you practice sharpening the saw, you take time to renew yourself physically, spiritually, mentally, and socially. The practice of self-renewal helps people avoid burnout. Project managers must make sure that they and their project team have time to retrain, reenergize, and occasionally even relax to avoid burnout.



Douglas Ross, author of *Applying Covey's Seven Habits to a Project Management Career*, related Covey's seven habits to project management. Ross suggests that Habit 5—Seek first to understand, then to be understood—differentiates good project managers from average or poor project managers. People have a tendency to focus on their own agendas instead of first trying to understand other people's points of view. Empathic listening can help project managers and team members find out what motivates different people. Understanding what motivates key stakeholders and customers can mean the difference between project success and project failure. Once project managers and team members begin to practice empathic listening, they can communicate and work together to tackle problems more effectively.[17]

Before you can practice empathic listening, you first have to get people to talk to you. In many cases, you must work on developing a rapport with the other person before he or she will really talk to you. **Rapport** is a relation of harmony, conformity, accord, or affinity. Without rapport, people cannot begin to communicate. For example, in the opening case, Ben was not ready to talk to anyone from the Information Technology department, even if Ed and Sarah were ready to listen. Ben was angry about the lack of support he received from the Information Technology department and bullied anyone who reminded him of that group. Before Sarah could begin to communicate with Ben, she had to establish rapport.

One technique for establishing rapport is mirroring. **Mirroring** is matching certain behaviors of the other person. People tend to like people who are like themselves, and mirroring helps you take on some of the other person's characteristics. It also helps them realize if they are behaving unreasonably, as in this case. You can mirror someone's voice tone and/or tempo, breathing, movements, or body postures. After Ben started yelling at her, Sarah quickly decided to mirror his voice tone, tempo, and body posture. She stood nose to nose with Ben and started yelling right back at him. This action made Ben realize what he was doing, and it also made him notice and respect this colleague from the Information Technology department. Once Ben overcame his anger, he could start communicating his needs. In most cases, such extreme measures are not needed, and mirroring must be used with caution. Done improperly or with the wrong person, it could result in very negative consequences. (This example is based on a true situation where the author was "Sarah.")

Recall the importance of getting users involved in information technology projects. For organizations to be truly effective in information technology project management, they must find ways to help users and developers of information systems work together. It is widely accepted that organizations make better project decisions when business professionals and information technology staff collaborate. It is also widely accepted that this task is easier said than done. Many companies have been very successful in integrating technology and business departments, but many other companies continue to struggle with this issue.

It is important to understand and pay attention to concepts of motivation, influence, power, and improving effectiveness in all project processes. It is likewise important to remember that projects operate within an organizational environment. The challenge comes from applying these theories to the many unique individuals involved in particular projects in particular organizations.



As you can see, there are many important topics related to motivation, influence, power, and effectiveness that are relevant to project management. Projects are done by and for people, so it is important for project managers and team members to understand and practice key concepts related to these topics. Remember that everyone prefers to work with people they like and respect, and it is important to treat others with respect, regardless of their title or position. This is especially true for people in support roles, such as administrative assistants, security guards, or members of the cleaning crew. You never know when you may need their help at some critical point in a project.

## DEVELOPING THE HUMAN RESOURCE PLAN

In order to develop a human resource plan for a project, you must identify and document project roles, responsibilities, skills, and reporting relationships. The human resource plan often includes an organizational chart for the project, detailed information on roles and responsibilities, and a staffing management plan.

Before creating an organizational chart or any part of the human resource plan for a project, top management and the project manager must identify what types of people the project really needs to ensure project success. If the key to success lies in having the best Java programmers you can find, planning should reflect that need. If the real key to success is having a top-notch project manager and team leaders whom people respect in the company, that need should drive human resource planning.

### Project Organizational Charts

Recall from Chapter 2 that the nature of information technology projects often means that project team members come from different backgrounds and possess a wide variety of skill sets. It can be very difficult to manage such a diverse group of people, so it is important to provide a clear organizational structure for a project. After identifying important skills and the types of people needed to staff a project, the project manager should work with top management and project team members to create an organizational chart for the project. Figure 9-3 provides a sample organizational chart for a large information technology project. Note that the project personnel include a deputy project manager, subproject managers, and teams. **Deputy project managers** fill in for project managers in their absence and assist them as needed, which is similar to the role of a vice president. **Subproject managers** are responsible for managing the subprojects into which a large project might be divided. There are often subproject managers that focus on managing various software (S/W) and hardware (H/W) components of large projects. This structure is typical for large projects. With many people working on a project, clearly defining and allocating project work is essential. (Visit the companion Web site for this text to see the project organization chart that Northwest Airlines used for their large ResNet project.) Smaller information technology projects usually do not have deputy project managers or subproject managers. On smaller projects, the project managers might have just team leaders reporting directly to them.



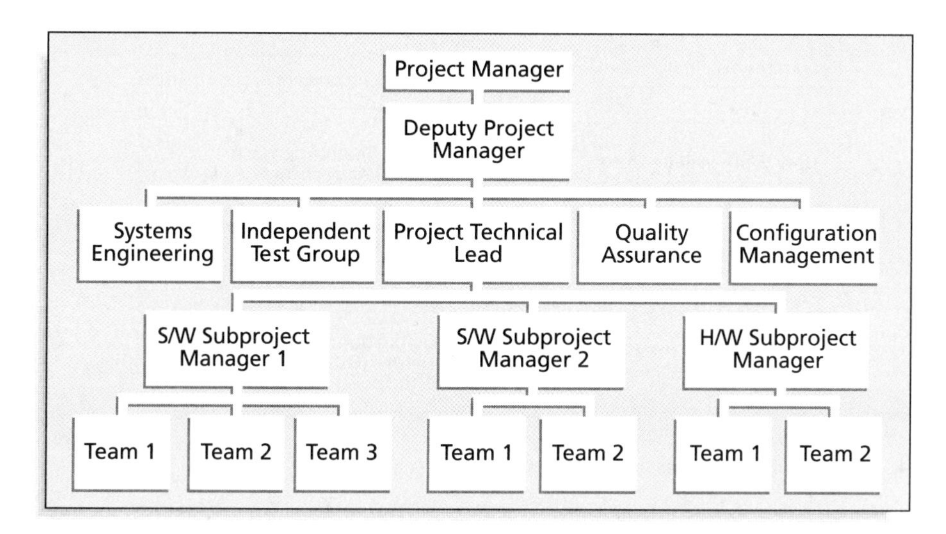


**FIGURE 9-3** Sample organizational chart for a large information technology project

In addition to defining an organizational structure for a project, it is also important to follow a work definition and assignment process. Figure 9-4 provides a framework for defining and assigning work. This process consists of four steps:

1. Finalizing the project requirements
2. Defining how the work will be accomplished
3. Breaking down the work into manageable elements
4. Assigning work responsibilities

The work definition and assignment process is carried out during the proposal and startup phases of a project. Note that the process is iterative, meaning it often takes more than one pass to refine it. A Request for Proposal (RFP) or draft contract often provides the basis for defining and finalizing work requirements, which are then documented in a final contract and technical baseline. If there were not an RFP, then the internal project charter and scope statement would provide the basis for defining and finalizing work requirements, as described in Chapter 5, Project Scope Management. The project team leaders then decide on a technical approach for how to do the work. Should work be broken down using a product-oriented approach or a phased approach? Will the project team outsource some of the work or subcontract to other companies? Once the project team has decided on a technical approach, they develop a work breakdown structure (WBS) to establish manageable elements of work (see Chapter 5, Project Scope Management). They then develop activity definitions to define the work involved in each activity on the WBS further (see Chapter 6, Project Time Management). The last step is assigning the work.



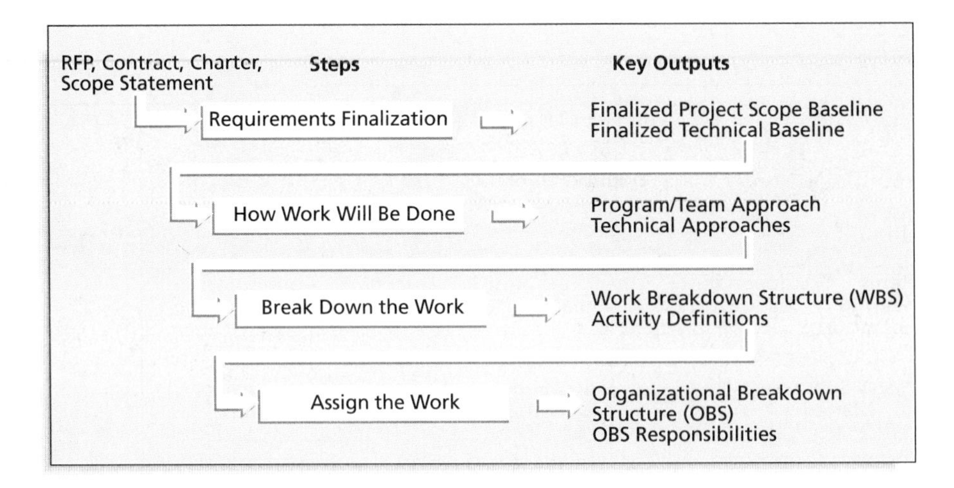


**FIGURE 9-4** Work definition and assignment process

Once the project manager and project team have broken down the work into manageable elements, the project manager assigns work to organizational units. The project manager often bases these work assignments on where the work fits in the organization and uses an organizational breakdown structure to conceptualize the process. An **organizational breakdown structure (OBS)** is a specific type of organizational chart that shows which organizational units are responsible for which work items. The OBS can be based on a general organizational chart and then broken down into more detail, based on specific units within departments in the company or units in any subcontracted companies. For example, OBS categories might include software development, hardware development, training, and so on.

## Responsibility Assignment Matrices

After developing an OBS, the project manager is in a position to develop a responsibility assignment matrix. A **responsibility assignment matrix (RAM)** is a matrix that maps the work of the project as described in the WBS to the people responsible for performing the work as described in the OBS. Figure 9-5 shows an example of a RAM. The RAM allocates work to responsible and performing organizations, teams, or individuals, depending on the desired level of detail. For smaller projects, it would be best to assign individual people to WBS activities. For very large projects, it is more effective to assign the work to organizational units or teams.

WBS activities →

OBS units ↓

| | 1.1.1 | 1.1.2 | 1.1.3 | 1.1.4 | 1.1.5 | 1.1.6 | 1.1.7 | 1.1.8 |
|---|---|---|---|---|---|---|---|---|
| Systems Engineering | R | R P | | | | | R | |
| Software Development | | | R P | | | | | |
| Hardware Development | | | | R P | | | | |
| Test Engineering | P | | | | | | | |
| Quality Assurance | | | | | R P | | | |
| Configuration Management | | | | | | R P | | |
| Integrated Logistics Support | | | | | | | P | |
| Training | | | | | | | | R P |

**R = Responsible organizational unit**
**P = Performing organizational unit**



**FIGURE 9-5** Sample responsibility assignment matrix (RAM)

In addition to using a RAM to assign detailed work activities, you can also use a RAM to define general roles and responsibilities on projects. This type of RAM can include the stakeholders in the project. Figure 9-6 provides a RAM that shows whether stakeholders are accountable or just participants in part of a project, and whether they are required to provide input, review, or sign off on parts of a project. This simple tool can be a very effective way for the project manager to communicate roles and expectations of important stakeholders on projects.

| | Stakeholders | | | | |
|---|---|---|---|---|---|
| **Items** | **A** | **B** | **C** | **D** | **E** |
| Unit Test | S | A | I | I | R |
| Integration Test | S | P | A | I | R |
| System Test | S | P | A | I | R |
| User Acceptance Test | S | P | I | A | R |

A = Accountable
P = Participant
R = Review Required
I = Input Required
S = Sign-off Required

**FIGURE 9-6** RAM showing stakeholder roles

Some organizations use **RACI charts** to show Responsibility (who does the task), Accountability (who signs off on the task or has authority for it), Consultation (who has information necessary to complete the task), and Informed (who needs to be notified of task status/results) roles for project stakeholders. As shown in Table 9-2, a RACI chart lists tasks vertically, individuals or groups horizontally, and each intersecting cell contains an R, A, C, or I. Each task may have multiple R, C, or I entries, but there can be only one A entry to clarify which particular individual or group is accountable for each task. For example, a mechanic is responsible for repairing a car, but the shop owner is accountable for the repairs getting done properly. Note that some people reverse the definitions of responsible and accountable.

**TABLE 9-2** Sample RACI Chart

| | Group A | Group B | Group C | Group D | Group E |
|---|---|---|---|---|---|
| Test Plans | R | A | C | C | I |
| Unit Test | C | I | R | A | I |
| Integration Test | A | R | I | C | C |
| System Test | I | C | A | I | R |
| User Acceptance Test | R | I | C | R | A |



### Staffing Management Plans and Resource Histograms

A **staffing management plan** describes when and how people will be added to and taken off the project team. The level of detail may vary based on the type of project. For example, if an information technology project is projected to need 100 people on average over a year, the staffing management plan would describe the types of people needed to work on the project, such as Java programmers, business analysts, technical writers, and so on, and the number of each type of person needed each month. It would also describe how these resources would be acquired, trained, rewarded, reassigned after the project, and so on. All of these issues are important to meeting the needs of the project, the employees, and the organization.

**WHAT WENT RIGHT?**

Senior managers realize that they must invest in human resources to attract, hire, and retain qualified staff. In addition to providing technical training for IT personnel, several companies have made significant investments in project management training to provide career paths for project managers.

For example, Hewlett-Packard (HP) employed only six registered PMPs in 1997, but by August 2004, it employed more than 1,500 PMPs and was adding 500 more per year.[18] HP even paid for an ad in PMI's Project Network magazine stressing the importance they place on having certified project managers. "At HP, a PMP® [credential] is a stamp of approval. Our major reason for focusing on project management certification is customer-based: We want to make sure we've got the best project managers. Customers across countries and industries ask us, what kind of project managers do you have? What kind of certification do they have? We can tell them that the majority of our project managers are certified. HP values certification. We have four levels of project managers, and the top three require a PMP certification."[19]

IBM Global Business Services organization offers numerous paths to career success, with one dedicated to project management. While most consulting firms offer a single path

*continued*

to a leadership position, IBM has four to allow their people to succeed by focusing on their strengths and interests in one or more disciplines. Their four paths include:

- *Partner*: client management, practice development, engagement sales and delivery
- *Sales Executive*: client relationship and sales
- *Delivery Executive*: project management
- *Distinguished Engineer*: technical capabilities[20]



The staffing management plan often includes a **resource histogram**, which is a column chart that shows the number of resources assigned to a project over time. Figure 9-7 provides an example of a histogram that might be used for a six-month information technology project. Notice that the columns represent the number of people needed in each area—managers, business analysts, programmers, and technical writers. By stacking the columns, you can see the total number of people needed each month. After determining the project staffing needs, the next steps in project human resource management are to acquire the necessary staff and then develop the project team.

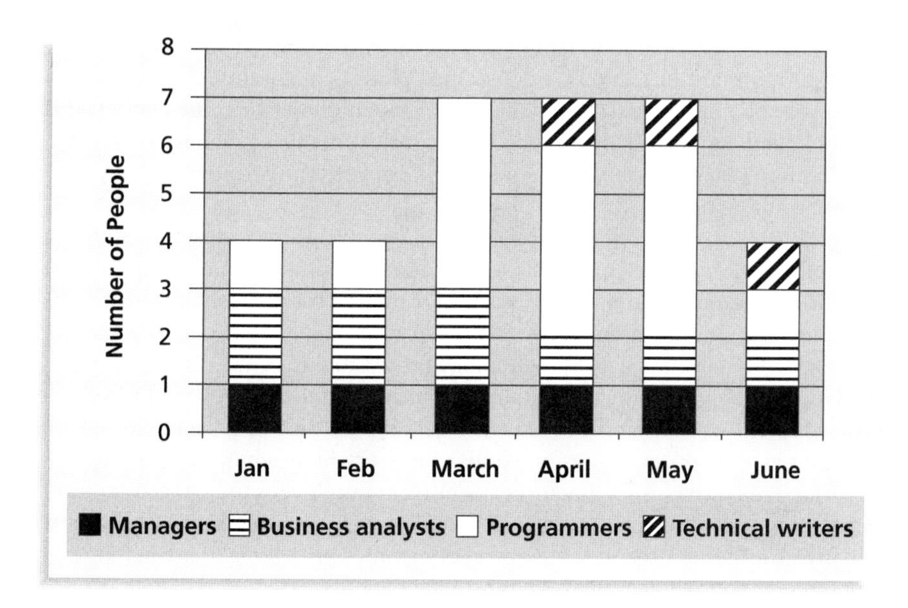


**FIGURE 9-7** Sample resource histogram

## ACQUIRING THE PROJECT TEAM

During the late 1990s, the information technology job market became extremely competitive. It was a seller's market with corporations competing fiercely for a shrinking pool of qualified, experienced information technology professionals. In the early 2000s, the market declined tremendously, so employers could be very selective in recruiting. Today, many organizations again face a shortage of IT staff. Regardless of the current job market, acquiring qualified information technology professionals is critical. There is a saying that the project manager who is the smartest person on the team has done a poor job of recruiting! In addition to recruiting team members, it is also important to assign the appropriate type and number of people to work on projects at the appropriate times. This section addresses important topics related to acquiring the project team: resource assignment, resource loading, and resource leveling.

## Resource Assignment



After developing a staffing management plan, project managers must work with other people in their organizations to assign particular personnel to their projects or to acquire additional human resources needed to staff the project. Project managers with strong influencing and negotiating skills are often good at getting internal people to work on their projects. However, the organization must ensure that people are assigned to the projects that best fit their skills and the needs of the organization. The main outputs of this process are project staff assignments, resource availability information, and updates to the staffing management plan. Many project teams also find it useful to create a project team directory.

Organizations that do a good job of staff acquisition have good staffing plans. These plans describe the number and type of people who are currently in the organization and the number and type of people anticipated to be needed for the project based on current and upcoming activities. An important component of staffing plans is maintaining a complete and accurate inventory of employees' skills. If there is a mismatch between the current mix of people's skills and needs of the organization, it is the project manager's job to work with top management, human resource managers, and other people in the organization to address staffing and training needs.

It is also important to have good procedures in place for hiring subcontractors and recruiting new employees. Since the Human Resource department is normally responsible for hiring people, project managers must work with their human resource managers to address any problems in recruiting appropriate people. It is also a priority to address retention issues, especially for information technology professionals.

One innovative approach to hiring and retaining information technology staff is to offer existing employees incentives for helping recruit and retain personnel. For example, several consulting companies give their employees one dollar for every hour a new person they helped recruit works. This provides an incentive for current employees to help attract new people and to keep both the employees and the people they help to recruit working at their respective company. Another approach that several companies are taking to attract and retain information technology professionals is to provide benefits based on personal need. For example, some people might want to work only four days a week or have the option of working a couple of days a week from home. As it gets more difficult to find good information technology professionals, organizations must become more innovative and proactive in addressing this issue.

Several organizations, publications, and Web sites address the need for good staff acquisition and retention. Enrollment in U.S. computer science and engineering programs has dropped almost in half since 2000, and one-third of U.S. workers will be over the age of 50 by 2010. CIO's researchers suggest that organizations rethink hiring practices and incentives to hire and retain IT talent. For example, if it's important for IT employees to have strong business and communications skills, organizations shouldn't focus primarily on technical skills. A company could require candidates to give a presentation to see how well they communicate and understand the business. Also, organizations should focus on flexibility when negotiating perks. According to the "2006 Compensation and Benefits Report" by Hudson Highland Group, a third of IT workers said they value a flexible work schedule more than other nontraditional benefits. "Employees are more willing to forgo additional cash in order to have a more improved work-life balance," says Peg Buchenroth, Hudson Highland Group's managing director of compensation and benefits.[21]

It is very important to consider the needs of individuals and the organization when making recruiting and retention decisions and to study the best practices of leading companies in these areas. It is also important to address a growing trend in project team members—many of them work in a virtual environment. (See the section on managing the project team for suggestions on working with virtual team members.)

## BEST PRACTICE

Best practices can also be applied to include the best places for people to work. For example, *Fortune* magazine lists the "100 Best Companies to Work For" in the United States every year, with Google taking the honors in 2007 and 2008. *Working Mothers* magazine lists the best companies in the U.S. for women based on benefits for working families. The *Timesonline* (*www.timesonline.co.uk*) provides the London *Sunday Times* list of the "100 Best Companies to Work For," a key benchmark against which U.K. companies can judge their performance as employers. The Great Place to Work Institute, which produces *Fortune* magazine's "100 Best Companies to Work For," uses the same selection methodology for more than 20 international lists, including "Best Companies to Work For" lists in all 15 countries of the European Union, Brazil, Korea, and a number of other countries throughout Latin America and Asia. Companies make these lists based on feedback from their best critics: their own employees. Quotes from employees often show why certain companies made the lists:

- "It is a friendly, courteous, caring hospital. We generally care about our co-workers and our patients. I can always get the help and support that I need to function in this hospital. This goes from the top all the way down to the cleaning people."
- "This is the best place I have ever worked. There's an open door policy. Everyone is allowed to voice their opinion."
- "I get information about everything—profits, losses, problems. Relationships with people are easier here. It's more direct and open."[22]

## Resource Loading

Chapter 6, Project Time Management, described using network diagrams to help manage a project's schedule. One of the problems or dangers inherent in scheduling processes is that they often do not address the issues of resource utilization and availability. (Hence, the development of critical chain scheduling, as described in Chapter 6.) Schedules tend to focus primarily on time rather than on both time and resources, which includes people. An important measure of a project manager's success is how well he or she balances the trade-offs among performance, time, and cost. During a period of crisis, it is occasionally possible to add additional resources—such as additional staff—to a project at little or no cost. Most of the time, however, resolving performance, time, and cost trade-offs entails additional costs to the organization. The project manager's goal must be to achieve project success without increasing the costs or time required to complete the project. The key to accomplishing this goal is effectively managing human resources on the project.

Once people are assigned to projects, there are two techniques available to project managers that help them use project staff most effectively: resource loading and resource leveling. **Resource loading** refers to the amount of individual resources an existing schedule requires during specific time periods. Resource loading helps project managers develop a general understanding of the demands a project will make on the organization's resources, as well as on individual people's schedules. Project managers often use histograms, as described in Figure 9-7, to depict period-by-period variations in resource loading. A histogram can be very helpful in determining staffing needs or in identifying staffing problems.



A resource histogram can also show when work is being overallocated to a certain person or group. **Overallocation** means more resources than are available are assigned to perform work at a given time. For example, Figure 9-8 provides a sample resource histogram created in Microsoft Project. This histogram illustrates how much one individual, Joe Franklin, is assigned to work on the project each week. The percentage numbers on the vertical axis represent the percentage of Joe's available time that is allocated for him to work on the project. The top horizontal axis represents time in weeks. Note that Joe Franklin is overallocated most of the time. For example, for most of March and April and part of May, Joe's work allocation is 300 percent of his available time. If Joe is normally available eight hours per day, this means he would have to work 24 hours a day to meet this staffing projection! Many people don't use the resource assignment features of project management software properly. (See Appendix A for detailed information on using Microsoft Project 2007.)

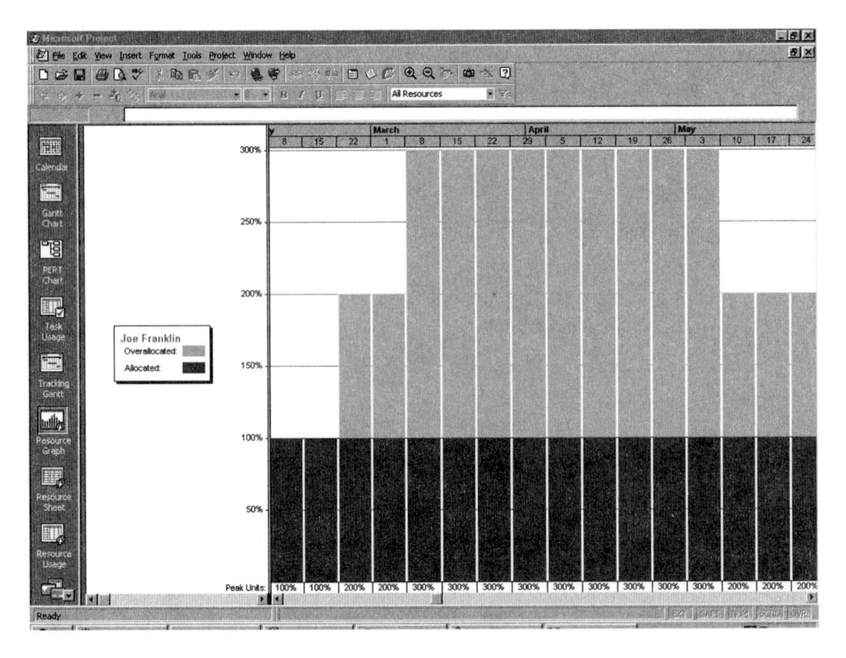


**FIGURE 9-8** Sample histogram showing an overallocated individual

## Resource Leveling

**Resource leveling** is a technique for resolving resource conflicts by delaying tasks. It is a form of network analysis in which resource management concerns drive scheduling decisions (start and finish dates). The main purpose of resource leveling is to create a smoother distribution of resource usage. Project managers examine the network diagram for areas of slack or float, and to identify resource conflicts. For example, you can sometimes remove overallocations by delaying noncritical tasks, which does not result in an overall schedule delay. Other times you will need to delay the project completion date to reduce or remove overallocations. Appendix A has information on using Microsoft Project 2007 to level resources using both of these approaches. You can also view resource leveling as addressing the resource constraints described in critical chain scheduling (see Chapter 6, Project Time Management).



Overallocation is one type of resource conflict. If a certain resource is overallocated, the project manager can change the schedule to remove resource overallocation. If a certain resource is underallocated, the project manager can change the schedule to try to improve the use of the resource. Resource leveling, therefore, aims to minimize period-by-period variations in resource loading by shifting tasks within their slack allowances.

Figure 9-9 illustrates a simple example of resource leveling. The network diagram at the top of this figure shows that Activities A, B, and C can all start at the same time. Activity A has a duration of two days and will take two people to complete; Activity B has a duration of five days and will take four people to complete; and Activity C has a duration of three

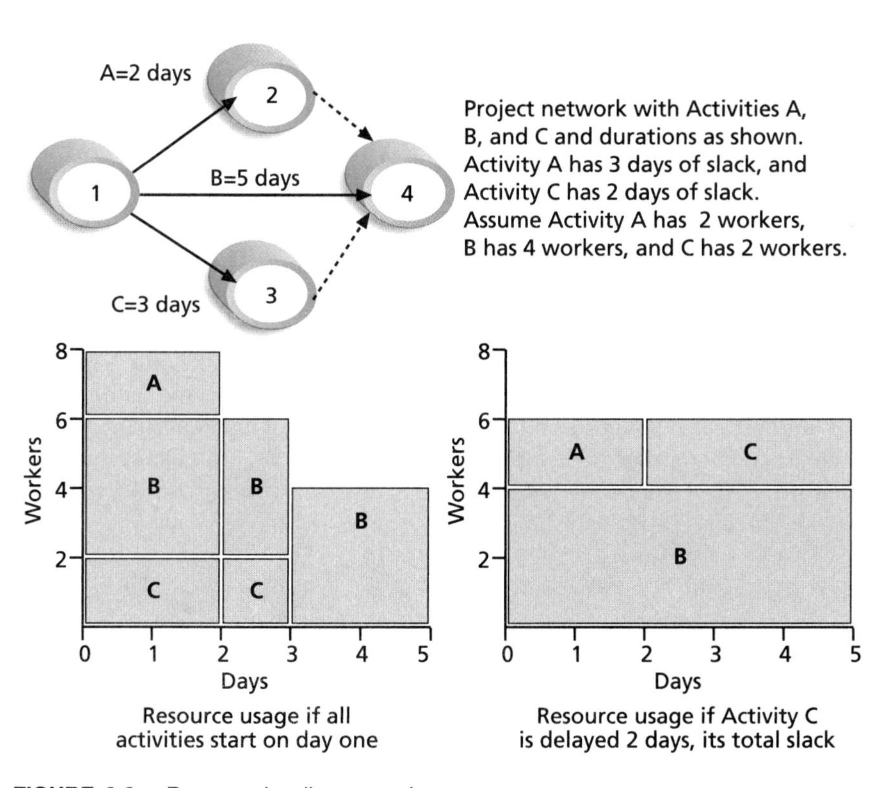


**FIGURE 9-9** Resource leveling example

days and will take two people to complete. The histogram on the lower-left of this figure shows the resource usage if all activities start on day one. The histogram on the lower-right of Figure 9-9 shows the resource usage if Activity C is delayed two days, its total slack allowance. Notice that the lower-right histogram is flat or leveled; that is, its pieces (activities) are arranged to take up the least space (saving days and numbers of workers). You may recognize this strategy from the computer game Tetris, in which you earn points for keeping the falling shapes as level as possible. The player with the most points (most level shape allocation) wins. Resources are also used best when they are leveled.

Resource leveling has several benefits. First, when resources are used on a more constant basis, they require less management. For example, it is much easier to manage a part-time project member who is scheduled to work 20 hours per week on a project for the next three months than it is to manage the same person who is scheduled to work 10 hours one week, 40 the next, 5 the next, and so on.



Second, resource leveling may enable project managers to use a just-in-time inventory type of policy for using subcontractors or other expensive resources. For example, a project manager might want to level resources related to work that must be done by particular subcontractors such as testing consultants. This leveling might allow the project to use four outside consultants full-time to do testing for four months instead of spreading the work out over more time or needing to use more than four people. The latter approach is usually more expensive. Recall from Chapter 6 that crashing and fast-tracking can also be used along with resource concerns to improve a project schedule.

Third, resource leveling results in fewer problems for project personnel and accounting departments. Increasing and decreasing labor levels and particular human resources often produce additional work and confusion. For example, if a person with expertise in a particular area is only assigned to a project two days a week and another person they need to work with is not assigned to the project those same days, they cannot work well together. The Accounting department might complain when subcontractors charge a higher rate for billing less than 20 hours a week on a project. The accountants will remind project managers to strive for getting the lowest rates possible.

Finally, resource leveling often improves morale. People like to have some stability in their jobs. It is very stressful for people not to know from week to week or even day to day what projects they will be working on and with whom they will be working.

Project management software can automatically level resources. However, the project manager must be careful in using the results without making adjustments. Automatic leveling often pushes out the project's completion date. Resources may also be reallocated to work at times that are inappropriate with other constraints. A wise project manager would have one of his or her team members who is proficient in using the project management software ensure that the leveling is done appropriately.

## DEVELOPING THE PROJECT TEAM

Even if a project manager has successfully recruited enough skilled people to work on a project, he or she must ensure that people can work together as a team to achieve project goals. Many information technology projects have had very talented individuals working on them. However, it takes teamwork to complete most projects successfully. The main goal of **team development** is to help people work together more effectively to improve project performance.

Dr. Bruce Tuckman published his four-stage model of team development in 1965 and modified it to include an additional stage in the 1970s. The **Tuckman model** describes five stages of team development:

1. *Forming* involves the introduction of team members, either at the initiation of the team, or as new members are introduced. This stage is necessary, but little work is actually achieved.
2. *Storming* occurs as team members have different opinions as to how the team should operate. People test each other, and there is often conflict within the team.
3. *Norming* is achieved when team members have developed a common working method, and cooperation and collaboration replace the conflict and mistrust of the previous phase.
4. *Performing* occurs when the emphasis is on reaching the team goals, rather than working on team process. Relationships are settled, and team members are likely to build loyalty towards each other. At this stage, the team is able to manage tasks that are more complex and cope with greater change.
5. *Adjourning* involves the break-up of the team after they successfully reach their goals and complete the work.[23]



There is an extensive body of literature on team development. This section will highlight a few important tools and techniques for team development, including training, team-building activities, and reward and recognition systems.

## Training

Project managers often recommend that people take specific training courses to improve individual and team development. For example, Sarah from the opening case had gone through training in emotional intelligence and dealing with difficult people. She was familiar with the mirroring technique and felt comfortable using that approach with Ben. Many other people would not have reacted so quickly and effectively in the same situation. If Ben and Sarah did reach agreement on what actions they could take to resolve the F-44 aircraft program's information technology problems, it might result in a new project to develop and deliver a new system for Ben's group. If Sarah became the project manager for this new project, she would understand the need for special training in interpersonal skills for specific people in her and Ben's departments. Individuals could take special training classes to improve their personal skills. If Sarah thought the whole project team could benefit from taking training together to learn to work as a team, she could arrange for a special team-building session for the entire project team and key stakeholders.

It is very important to provide training in a just-in-time fashion. For example, if Sarah was preparing for a technical assignment where she would need to learn a new programming language, training to deal with difficult people would not help her much. However, the training was very timely for her new consulting position. Many organizations provide e-learning opportunities for their employees so they can learn specific skills at any time and any place. They have also found e-learning to sometimes be more cost-effective than traditional instructor-led training courses. It is important to make sure that the timing and delivery method for the training is appropriate for specific situations and individuals. Organizations have also found that it is often more economical to train current employees in particular areas than it is to hire new people who already possess those skills.

Several organizations that have successfully implemented Six Sigma principles have taken a unique and effective approach to training. They only let high-potential employees attend Six Sigma Black Belt training, which is a substantial investment in terms of time and money. In addition, they do not let employees into a particular Black Belt course until they have had a potential Six Sigma project approved that relates to their current job. Attendees can then apply the new concepts and techniques they learn in the classes to their work settings. High-potential employees feel rewarded by being picked to take this training, and the organization benefits by having these employees implement high-payoff projects because of the training.



## Team-Building Activities

Many organizations provide in-house team-building training activities, and many also use specialized services provided by external companies that specialize in this area. Two common approaches to team-building activities include using physical challenges and psychological preference indicator tools. It is important to understand individual needs, including learning styles, past training, and physical limitations, when determining team-building training options.

Several organizations have teams of people go through certain physically challenging activities to help them develop as a team. Military basic training or boot camps provide one example. Men and women who wish to join the military must first make it through basic training, which often involves several strenuous physical activities such as rappelling off towers, running and marching in full military gear, going through obstacle courses, passing marksmanship training, and mastering survival training. Many organizations use a similar approach by sending teams of people to special locations where they work as a team to navigate white water rapids, climb mountains or rocks, participate in ropes courses, and so on. Research shows that physical challenges often help teams of strangers to work together more effectively, but it can cause already dysfunctional teams to have even more problems.

Even more organizations have teams participate in mental team-building activities in which they learn about themselves, each other, and how to work as a group most effectively. It is important for people to understand and value each other's differences in order to work effectively as a team. Three common exercises used in mental team building include the Myers-Briggs Type Indicator, Wilson Learning Social Styles Profile, and the DISC Profile.

### The Meyers-Briggs Type Indicator

The **Myers-Briggs Type Indicator (MBTI)** is a popular tool for determining personality preferences. During World War II, Isabel B. Myers and Katherine C. Briggs developed the first version of the MBTI based on psychologist Carl Jung's theory of psychological type. The four dimensions of psychological type in the MBTI include:

- *Extrovert/Introvert (E/I)*: This first dimension determines if you are generally extroverted or introverted. The dimension also signifies whether people draw their energy from other people (extroverts) or from inside themselves (introverts).
- *Sensation/Intuition (S/N)*: This second dimension relates to the manner in which you gather information. Sensation (or Sensing) type people take in facts,

details, and reality and describe themselves as practical. Intuitive type people are imaginative, ingenious, and attentive to hunches or intuition. They describe themselves as innovative and conceptual.

- *Thinking/Feeling (T/F)*: This third dimension represents thinking judgment and feeling judgment. Thinking judgment is objective and logical, and feeling judgment is subjective and personal.
- *Judgment/Perception (J/P)*: This fourth dimension concerns people's attitude toward structure. Judgment type people like closure and task completion. They tend to establish deadlines and take them seriously, expecting others to do the same. Perceiving types prefer to keep things open and flexible. They regard deadlines more as a signal to start rather than complete a project and do not feel that work must be done before play or rest begins.[24]



There is much more involved in personality types, and many books are available on this topic. In 1998, David Keirsey published *Please Understand Me II: Temperament Character Intelligence*.[25] This book includes an easy-to-take and interpret test called the *Keirsey Temperament Sorter*, which is a personality type preference test based on the work done by Jung, Myers, and Briggs.

In 1985, an interesting study of the MBTI types of the general population in the United States and information systems (IS) developers revealed some significant contrasts.[26] Both groups of people were most similar in the judgment/perception dimension, with slightly more than half of each group preferring the judgment type ( J). There were significant differences, however, in the other three dimensions. Most people would not be surprised to hear that most information systems developers are introverts. This study found that 75 percent of IS developers were introverts (I), and only 25 percent of the general population were introverts. This personality type difference might help explain some of the problems users have communicating with developers. Another sharp contrast found in the study was that almost 80 percent of IS developers were thinking types (T) compared to 50 percent of the general population. IS developers were also much more likely to be intuitive (N) (about 55 percent) than the general population (about 25 percent). These results fit with Keirsey's classification of NT (Intuitive/Thinking types) people as *rationals*. Educationally, they tend to study the sciences, enjoy technology as a hobby, and pursue systems work. Keirsey also suggests that no more than 7 percent of the general population are NTs. Would you be surprised to know that Bill Gates is classified as a rational?[27]

Project managers can often benefit from knowing their team members' MBTI profiles by adjusting their management styles for each individual. For example, if the project manager is a strong N and one of the team members is a strong S, the project manager should take the time to provide more concrete, detailed explanations when discussing that person's task assignments. Project managers may also want to make sure they have a variety of personality types on their teams. For example, if all team members are strong introverts, it may be difficult for them to work well with users and other important stakeholders who are often extroverts.

The MBTI, like any test, should be used with caution. A 2007 study argues that the lack of progress in improving software development teams is due in part to the inappropriate use of psychological tests and basic misunderstandings of personality theory by those who use them. "Software engineers often complain about those who, in the course of their work, do some programming in support of their professional activities: the claim being that such individuals are not professionals and do not understand the discipline. The same can be said of those who adopt psychological approaches without the relevant qualifications and background."[28]

## The Social Styles Profile

Many organizations also use the Social Styles Profile in team-building activities. Psychologist David Merril, who helped develop the Wilson Learning Social Styles Profile, describes people as falling into four approximate behavioral profiles, or zones. People are perceived as behaving primarily in one of four zones, based on their assertiveness and responsiveness:



- *Drivers* are proactive and task-oriented. They are firmly rooted in the present, and they strive for action. Adjectives to describe drivers include pushy, severe, tough, dominating, harsh, strong-willed, independent, practical, decisive, and efficient.
- *Expressives* are proactive and people-oriented. They are future-oriented and use their intuition to look for fresh perspectives on the world around them. Adjectives to describe expressives include manipulating, excitable, undisciplined, reacting, egotistical, ambitious, stimulating, wacky, enthusiastic, dramatic, and friendly.
- *Analyticals* are reactive and task-oriented. They are past-oriented and strong thinkers. Adjectives to describe analyticals include critical, indecisive, stuffy, picky, moralistic, industrious, persistent, serious, expecting, and orderly.
- *Amiables* are reactive and people-oriented. Their time orientation varies depending on whom they are with at the time, and they strongly value relationships. Adjectives to describe amiables include conforming, unsure, ingratiating, dependent, awkward, supportive, respectful, willing, dependable, and agreeable.[29]

Figure 9-10 shows these four social styles and how they relate to assertiveness and responsiveness. Note that the main determinants of the social style are your levels of assertiveness—if you are more likely to tell people what to do or ask what should be done—and how you respond to tasks—by focusing on the task itself or on the people involved in performing the task.

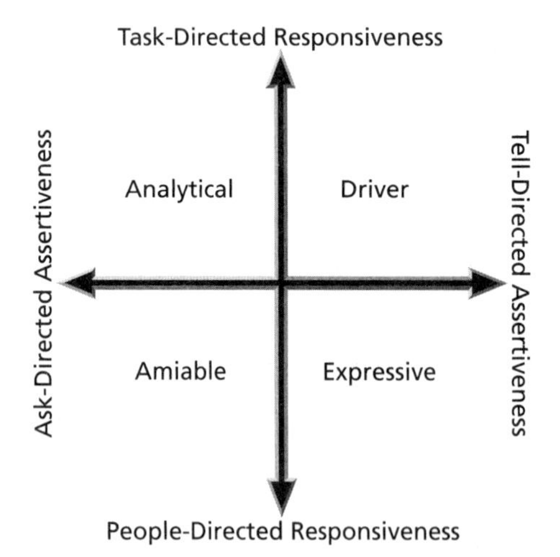


**FIGURE 9-10** Social styles

Knowing the social styles of project stakeholders can help project managers understand why certain people may have problems working together. For example, drivers are often very impatient working with amiables, and analyticals often have difficulties understanding expressives.

### DISC Profiles

Similar to the Social Styles Profile, the DISC Profile uses a four-dimensional model of normal behavior. The four dimensions—Dominance, Influence, Steadiness, and Compliance—provide the basis for the name, DISC. Note that similar terms are sometimes used for some of these letters, such as stability for steadiness or conscientiousness for compliance. The DISC Profile is based on the 1928 work of psychologist William Moulton Marston. The DISC Profile reveals people's behavioral tendencies under certain situations. For example, it reveals how you tend to behave under stress, in conflict, when communicating, when avoiding certain activities, and so on. According to *www.onlinediscprofile.com*, "over 5 million people have taken various forms of the DISC Profile throughout the world. Marston's original work continues to be enhanced by ongoing behavioral research and profiles can be found in more than 50 languages by various publishers of the disc assessment."[30]

Figure 9-11 shows the four dimensions of the DISC Profile model and describes key characteristics of each dimension. Notice that each dimension is also associated with a color and emphasis, such as I, We, You, or It:

- *Dominance*: Represented by red and emphasizes "I," dominance traits include being direct, decisive, assertive, outcome-oriented, competitive, self assured, controlling, and wanting to win

| *It*<br>**Compliance (Blue)**<br>Data driven, risk averse, concerned, works well alone, prefers processes and procedures, not very communicative or social | *I*<br>**Dominance (Red)**<br>Direct, decisive, assertive, outcome oriented, competitive, self assured, takes control, has to win |
|---|---|
| *You*<br>**Steadiness (Green)**<br>Calm, sincere, sympathetic, cooperative, cautious, conflict averse, good listener, wants to maintain stability | *We*<br>**Influence (Yellow)**<br>Persuasive, optimistic , outgoing, verbal, enthusiastic, strives to win others over, leadership through acclimation |

**FIGURE 9-11** The DISC Profile

- *Influence*: Represented by yellow and emphasizes "We," influence traits include being persuasive, optimistic, outgoing, verbal, enthusiastic, striving to win others over, and practicing leadership through acclimation
- *Steadiness*: Represented by green and emphasizes "You," steadiness traits include being calm, sincere, cautious, conflict averse, a good listener, and wanting to maintain stability
- *Compliance*: Represented by blue and emphasizes "It," compliance traits include being data driven, risk averse, concerned, working well alone, preferring processes and procedures, and being not very communicative or social



Like the Social Styles Profile, people in opposite quadrants, such as Dominance and Steadiness or Influence and Compliance, can have problems understanding each other. There are many other team-building activities and tests available. For example, some people like to take Dr. Meredith Belbin's test to help determine which of nine team roles they might prefer. Again, any team-building or personality tool should be used by professionals with caution. In reality, most professionals must be flexible and do whatever work is needed for their teams to succeed. Project managers can use their leadership and coaching skills to help all types of people communicate better with each other and focus on meeting project goals.

### Reward and Recognition Systems

Another important tool for promoting team development is the use of team-based reward and recognition systems. If management rewards teamwork, they will promote or reinforce people to work more effectively in teams. Some organizations offer bonuses, trips, or other rewards to workers that meet or exceed company or project goals. In a project setting, project managers can recognize and reward people who willingly work overtime to meet an aggressive schedule objective or go out of their way to help a teammate. Project managers should not reward people who work overtime just to get extra pay or because of their own poor work or planning.

Project managers must continually assess their team's performance. When they find areas where individuals or the entire team can improve, it's their job to find the best way to develop their people and improve performance.

## MANAGING THE PROJECT TEAM

In addition to developing the project team, the project manager must lead them in performing various project activities. After assessing team performance and related information, the project manager must decide if changes should be requested to the project, or if updates are needed to enterprise environmental factors, organizational process assets, or the project management plan. Project managers must use their soft skills to find the best way to motivate and manage each team member.

### Tools and Techniques for Managing Project Teams

There are several tools and techniques available to assist in managing project teams:

- *Observation and conversation*: It is hard to assess how your team members are performing or how they are feeling about their work if you never see or discuss these issues. Many project managers like to practice "management by walking around" to physically see and hear their team members at work. Informal or formal conversations about how a project is going can provide crucial information. For virtual workers, project managers can still observe and discuss work and personal issues via e-mail, telephone, or other communications media.

- *Project performance appraisals*: Just as general managers provide performance appraisals for their workers, so can project managers. The need for and type of project performance appraisals will vary depending on the length of the project, complexity of the project, organizational policies, contract requirements, and related communications. Even if a project manager does not provide official project performance appraisals for team members, it is still important to provide timely performance feedback. If a team member hands in sloppy or late work, the project manager should determine the reason for this behavior and take appropriate action. Perhaps the team member had a death in the family and could not concentrate. Perhaps the team member was planning to leave the project. The reasons for the behavior would have a strong impact on what action the project manager should take.
- *Conflict management*: Few projects are completed without any conflicts. Some types of conflict are actually desirable on projects, but many are not. As described in Chapter 10, Project Communications Management, there are several ways to handle conflicts. It's important for project managers to understand strategies for handling conflicts and to proactively manage conflict.
- *Issue logs*: Many project managers keep an **issue log** to document, monitor, and track issues that need to be resolved for the project team to work effectively. Issues could include items where people have differences of opinion, situations that need more clarification or investigation, or general concerns that need to be addressed. It is important to acknowledge issues that can hurt team performance and take action to resolve them. The project manager should assign someone to resolve each issue and assign a target date for resolution.
- *Interpersonal skills*: As mentioned in Chapter 1, project managers must possess several interpersonal skills. To effectively manage teams, it is especially important to focus on leadership, influencing, and decision-making skills.



## General Advice on Managing Teams

According to Patrick Lencioni, a well-known author and consultant on teams, "Teamwork remains the one sustainable competitive advantage that has been largely untapped... teamwork is almost always lacking within organizations that fail, and often present within those that succeed."[31] The five dysfunctions of teams are:

1. Absence of trust
2. Fear of conflict
3. Lack of commitment
4. Avoidance of accountability
5. Inattention to results

Lencioni's books provide suggestions for overcoming each of these dysfunctions. For example, he suggests that team members take the Myers-Briggs Type Indicator, as described earlier in this chapter, to help people open up to each other and build trust. To master conflict, he suggests that teams practice having unfiltered, passionate debates about important issues. To achieve commitment, he stresses the importance of getting out all possible ideas, getting people to agree to disagree, but then having them commit to decisions. To embrace accountability, Lencioni stresses the importance of clarifying and focusing on everyone's top priorities. He also suggests that peer pressure and the distaste for letting down a colleague are often better motivators than authoritative intervention. Finally, using some type of scoreboard to focus on team results helps eliminate ambiguity so everyone knows what it means to achieve positive results.

Additional suggestions for ensuring that teams are productive include the following:



- Be patient and kind with your team. Assume the best about people; do not assume that your team members are lazy and careless.
- Fix the problem instead of blaming people. Help people work out problems by focusing on behaviors.
- Establish regular, effective meetings. Focus on meeting project objectives and producing positive results.
- Allow time for teams to go through the basic team-building stages of forming, storming, norming, performing, and adjourning. Don't expect teams to work at the highest performance level right away.
- Limit the size of work teams to three to seven members.
- Plan some social activities to help project team members and other stakeholders get to know each other better. Make the social events fun and not mandatory.
- Stress team identity. Create traditions that team members enjoy.
- Nurture team members and encourage them to help each other. Identify and provide training that will help individuals and the team as a whole become more effective.
- Acknowledge individual and group accomplishments.
- Take additional actions to work with virtual team members. If possible, have a face-to-face or phone meeting at the start of a virtual project or when introducing a virtual team member. Screen people carefully to make sure they can work effectively in a virtual environment. Clarify how virtual team members will communicate.

As you can imagine, team development and management are critical concerns on many information technology projects. Many information technology project managers must break out of their rational/NT preference and focus on empathically listening to other people to address their concerns and create an environment in which individuals and teams can grow and prosper.

## USING SOFTWARE TO ASSIST IN HUMAN RESOURCE MANAGEMENT

Earlier in this chapter, you read that a simple responsibility assignment matrix (Figures 9-5 and 9-6) or histograms (Figures 9-7 and 9-8) are useful tools that can help you effectively manage human resources on projects. You can use several different software packages, including spreadsheets or project management software such as Microsoft Project 2007, to create matrixes and histograms. Many people do not realize that Project 2007 provides a variety of human resource management tools, some of which include assigning and tracking resources, resource leveling, resource usage reports, overallocated resource reports, and to-do lists. You can learn how to use many of these functions and features in Appendix A. (Collaborative software to help people communicate is described in Chapter 10, Project Communications Management.)

You can use Project 2007 to assign resources—including equipment, materials, facilities, or people—to tasks. Project 2007 enables you to allocate individual resources to individual projects or to pool resources and share them across multiple projects. By defining and assigning resources in Project 2007, you can:

- Keep track of the whereabouts of resources through stored information and reports on resource assignments.
- Identify potential resource shortages that could force a project to miss scheduled deadlines and possibly extend the duration of a project.
- Identify underutilized resources and reassign them, which may enable you to shorten a project's schedule and possibly reduce costs.
- Use automated leveling to make level resources easier to manage.



Just as many project management professionals are not aware of the powerful cost-management features of Project 2007, many are also unaware of its powerful human resource management features. The Microsoft Enterprise Project Management Solution provides additional human resource management capabilities. You can also purchase add-in software for Microsoft's products or purchase software from other companies to perform a variety of project human resource management functions. With the aid of this type of software, project managers can have more information available in useful formats to help them decide how to manage human resources most effectively.

Project resource management involves much more than using software to assess and track resource loading, level resources, and so on. People are the most important asset on most projects, and human resources are very different from other resources. You cannot simply replace people the same way that you would replace a piece of equipment. People need far more than a tune-up now and then to keep them performing well. It is essential to treat people with consideration and respect, to understand what motivates them, and to communicate carefully with them. What makes good project managers great is not their use of tools, but rather their ability to enable project team members to deliver the best work they possibly can on a project.

## CASE WRAP-UP

After Sarah yelled right back at Ben, he said, "You're the first person who's had the guts to stand up to me." After that brief introduction, Sarah, Ben, and the other meeting participants had a good discussion about what was really happening on the F-44 upgrade project. Sarah was able to write a justification to get Ben's group special software and support to download key information from the old system so they could manage their project better. When Sarah stood nose to nose with Ben and yelled at him, she used a technique for establishing rapport called mirroring. Although Sarah was by no means a loud and obnoxious person, she saw that Ben was and decided to mirror his behavior and attitude. She put herself in his shoes for a while, and doing so helped break the ice so Sarah, Ben, and the other people at the meeting could really start communicating and working together as a team to solve their problems.

## Chapter Summary

People are the most important assets in organizations and on projects. Therefore, it is essential for project managers to be good human resource managers.

The major processes involved in project human resource management include developing the human resource plan, acquiring the project team, developing the project team, and managing the project team.

Psychosocial issues that affect how people work and how well they work include motivation, influence and power, and effectiveness.

Maslow developed a hierarchy of needs that suggests physiological, safety, social, esteem, and self-actualization needs motivate behavior. Once a need is satisfied, it no longer serves as a motivator.



Herzberg distinguished between motivators and hygiene factors. Hygiene factors such as larger salaries or a more attractive work environment will cause dissatisfaction if not present, but do not motivate workers to do more if present. Achievement, recognition, the work itself, responsibility, and growth are factors that contribute to work satisfaction and motivate workers.

McClelland proposed the acquired-needs theory, suggesting that an individual's needs are acquired or learned over time and shaped by their life experiences. The three types of acquired needs are a need for achievement, a need for affiliation, and a need for power.

McGregor developed Theory X and Theory Y to describe different approaches to managing workers, based on assumptions of worker motivation. Research supports the use of Theory Y, which assumes that people see work as natural and indicates that the most significant rewards are the satisfaction of esteem and self-actualization needs that work can provide. According to Ouchi's Theory Z, workers can be trusted to do their jobs to their utmost ability, as long as management can be trusted to support them and look out for their well-being. Theory Z emphasizes things such as job rotation, broadening of skills, generalization versus specialization, and the need for continuous training of workers.

Thamhain and Wilemon identified nine influence bases available to project managers: authority, assignment, budget, promotion, money, penalty, work challenge, expertise, and friendship. Their research found that project success is associated with project managers who use work challenge and expertise to influence workers. Project failure is associated with using too much influence by authority, money, or penalty.

Power is the potential ability to influence behavior to get people to do things they would not otherwise do. The five main types of power are coercive power, legitimate power, expert power, reward power, and referent power.

Project managers can use Steven Covey's seven habits of highly effective people to help themselves and project teams become more effective. The seven habits include being proactive; beginning with the end in mind; putting first things first; thinking win/win; seeking first to understand, then to be understood; achieving synergy; and sharpening the saw. Using empathic listening is a key skill of good project managers.

Developing the human resource plan involves identifying, assigning, and documenting project roles, responsibilities, and reporting relationships. A responsibility assignment matrix (RAM), staffing management plans, resource histograms, and RACI charts are key tools for defining roles and responsibilities on projects. The main output is a human resource plan.

Acquiring the project team means getting the appropriate staff assigned to and working on the project. This is an important issue in today's competitive environment. Companies must use innovative approaches to find and retain good information technology staff.

Resource loading shows the amount of individual resources an existing schedule requires during specific time frames. Histograms show resource loading and identify overallocation of resources.

Resource leveling is a technique for resolving resource conflicts, such as overallocated resources, by delaying tasks. Leveled resources require less management, lower costs, produce fewer personnel and accounting problems, and often improve morale.

Two crucial skills of a good project manager are team development and team management. Teamwork helps people work more effectively to achieve project goals. Project managers can recommend individual training to improve skills related to teamwork, organize team-building activities for the entire project team and key stakeholders, and provide reward and recognition systems that encourage teamwork. Project managers can use several tools and techniques, including observation and conversation, project performance appraisals, conflict management, issue logs, and interpersonal skills to help them effectively manage their teams.



Spreadsheets and project management software such as Microsoft Project 2007 can help project managers in project human resource management. Software makes it easy to produce responsibility assignment matrixes, create resource histograms, identify overallocated resources, level resources, and provide various views and reports related to project human resource management.

Project human resource management involves much more than using software to facilitate organizational planning and assign resources. What makes good project managers great is their ability to enable project team members to deliver the best work they possibly can on a project.

## Quick Quiz

1. Which of the following is not part of project human resource management?
   a. Resource estimating
   b. Acquiring the project team
   c. Developing the project team
   d. Managing the project team
2. __________ causes people to participate in an activity for their own enjoyment.
   a. Intrinsic motivation
   b. Extrinsic motivation
   c. Self motivation
   d. Social motivation
3. At the bottom of Maslow's pyramid or hierarchy of needs are __________ needs.
   a. self-actualization
   b. esteem
   c. safety
   d. physiological

4. According to McClelland's acquired needs theory, people who desire harmonious relations with other people and need to feel accepted have a high __________ need.
   a. social
   b. achievement
   c. affiliation
   d. extrinsic

5. __________ power is based on a person's individual charisma.
   a. Affiliation
   b. Referent
   c. Personality
   d. Legitimate



6. A __________ maps the work of a project as described in the WBS to the people responsible for performing the work.
   a. project organizational chart
   b. work definition and assignment process
   c. resource histogram
   d. responsibility assignment matrix

7. A staffing management plan often includes a resource __________, which is a column chart that shows the number of resources assigned to the project over time.
   a. chart
   b. graph
   c. histogram
   d. timeline

8. What technique can you use to resolve resource conflicts by delaying tasks?
   a. resource loading
   b. resource leveling
   c. critical path analysis
   d. overallocation

9. What are the five stages in Tuckman's model of team development, in chronological order?
   a. forming, storming, norming, performing, and adjourning
   b. storming, forming, norming, performing, and adjourning
   c. norming, forming, storming, performing, and adjourning
   d. forming, storming, performing, norming, and adjourning

10. Which of the following is not a tool or technique for managing project teams?
    a. observation and conversation
    b. project performance appraisals
    c. issue logs
    d. Social Styles Profile

### Quick Quiz Answers

1. a; 2. a; 3. d; 4. c; 5. b; 6. d; 7. c; 8. b; 9. a; 10. d



## Discussion Questions

1. Discuss the changes in the job market for information technology workers. How does the job market and current state of the economy affect human resource management?
2. Summarize the processes involved in project human resource management.
3. Briefly summarize the works of Maslow, Herzberg, McClelland, McGregor, Ouchi, Thamhain and Wilemon, and Covey. How do their theories relate to project management?
4. Describe situations where it would be appropriate to create a project organizational chart, a responsibility assignment matrix, a RACI chart, and a resource histogram. Describe what these charts or matrices look like.
5. Discuss the difference between resource loading and resource leveling, and provide an example of when you would use each technique.
6. Explain two types of team-building activities described in this chapter.
7. Summarize different tools and techniques project managers can use to help them manage project teams. What can they do to manage virtual team members?
8. How can you use Project 2007 to assist in project human resource management?

## Exercises

1. Your company is planning to launch an important new project starting January 1, which will last one year. You estimate that you will need one full-time project manager, two full-time business analysts for the first six months, two full-time senior programmers for the whole year, four full-time junior programmers for the months of July, August, and September, and one full-time technical writer for the last three months. Use the resource_histogram template file from the companion Web site to create a stacked column chart showing a resource histogram for this project, similar to the one shown in Figure 9-6. Be sure to include a legend to label the types of resources needed. Use appropriate titles and axis labels.
2. Take the MBTI test and research information on this tool. There are several Web sites that have different versions of the test, such as *www.humanmetrics.com*, *www.personalitytype.com*, and *www.keirsey.com*. Write a two-page paper describing your MBTI type and your thoughts on this test as a team-building tool.

3. Summarize three of Covey's habits in your own words and give examples of how these habits would apply to project management. Document your ideas in a two-page paper, and include at least two references.
4. Research recruiting and retention strategies at three different companies. What distinguishes one company from another in this area? Are strategies such as signing bonuses, tuition reimbursement, and business casual dress codes standard for new information technology workers? What strategies appeal most to you? Summarize your ideas in a two-page paper, citing at least three references.
5. Write a two-page paper summarizing the main features of Microsoft Project 2007 that can assist project managers in human resource management. In addition, interview someone who uses Project 2007. Ask him or her if their organization uses any project human resource management features as described in this chapter and Appendix A, and document their reasons for using or not using certain features.
6. Developing good IT project managers is an important issue. Review several of the studies cited in this chapter related to the IT and project management job market and required skills. Also review requirements at your college or university for people entering these fields. Summarize your findings and opinions on this issue in a two- to three-page paper.



## Running Case

Several people working on the Recreation and Wellness Intranet Project are confused about who needs to do what for the testing portion of the project. Recall that the team members include you, a programmer/analyst and aspiring project manager; Patrick, a network specialist; Nancy, a business analyst; and Bonnie, another programmer/analyst. Tony Prince is the project manager, and he has been working closely with managers in other departments to make sure everyone knows what's going on with the project.

1. Prepare a responsibility assignment matrix based on the following information: The main tasks that need to be done for testing include writing a test plan, unit testing, integration testing for each of the main system modules (registration, tracking, and incentives), system testing, and user acceptance testing. In addition to the project team members, there is a team of user representatives available to help with testing, and Tony has also hired an outside consulting firm to help as needed. Prepare a RACI chart to help clarify roles and responsibilities for these testing tasks. Document key assumptions you make in preparing the chart.
2. The people working for the outside consulting firm and user representatives have asked you to create a resource histogram to show how many people you think the project will need for the testing and when. Assume that the consulting firm has junior and senior testers and that the user group has workers and managers. You estimate that you'll need both groups involvement in testing over a period of six weeks. Assume you'll need one senior tester for all six weeks, two junior testers for the last four weeks, two user-group workers for the first week, four user-group workers for the last three weeks, and two user-group managers for the last two weeks. Create a resource histogram, similar to the one in Figure 9-6, based on this information.

3. One of the issues in Tony's issue log is working effectively with the user group during testing. Tony knows that several of his project team members are very introverted and strong thinking types, while several members of the user group are very extroverted and strong feeling types. Write a one-page paper describing options for resolving this issue.

## Companion Web Site

Visit the companion Web site for this text (*www.cengage.com/mis/schwalbe*) to access:

- References cited in the text and additional suggested readings for each chapter
- Template files
- Lecture notes
- Interactive quizzes
- Podcasts
- Links to general project management Web sites
- And more

See the Preface of this text for additional information on accessing the companion Web site.



## Key Terms

**coercive power** — using punishment, threats, or other negative approaches to get people to do things they do not want to do

**deputy project managers** — people who fill in for project managers in their absence and assist them as needed, similar to the role of a vice president

**empathic listening** — listening with the intent to understand

**expert power** — using one's personal knowledge and expertise to get people to change their behavior

**extrinsic motivation** — causes people to do something for a reward or to avoid a penalty

**hierarchy of needs** — a pyramid structure illustrating Maslow's theory that people's behaviors are guided or motivated by a sequence of needs

**intrinsic motivation** — causes people to participate in an activity for their own enjoyment

**issue log** — a tool for managing project teams where the project manager documents, monitors, and tracks issues that need to be resolved in order for the project to run smoothly

**legitimate power** — getting people to do things based on a position of authority

**mirroring** — matching certain behaviors of the other person

**Myers-Briggs Type Indicator (MBTI)** — a popular tool for determining personality preferences

**organizational breakdown structure (OBS)** — a specific type of organizational chart that shows which organizational units are responsible for which work items

**overallocation** — when more resources than are available are assigned to perform work at a given time

**power** — the potential ability to influence behavior to get people to do things they would not otherwise do

**RACI charts** — charts that show Responsibility, Accountability, Consultation, and Informed roles for project stakeholders

**rapport** — a relation of harmony, conformity, accord, or affinity

**referent power** — getting people to do things based on an individual's personal charisma

**resource histogram** — a column chart that shows the number of resources assigned to a project over time

**resource leveling** — a technique for resolving resource conflicts by delaying tasks

**resource loading** — the amount of individual resources an existing schedule requires during specific time periods

**responsibility assignment matrix (RAM)** — a matrix that maps the work of the project as described in the WBS to the people responsible for performing the work as described in the organizational breakdown structure (OBS)



**reward power** — using incentives to induce people to do things

**staffing management plan** — a document that describes when and how people will be added to and taken off a project team

**subproject managers** — people responsible for managing the subprojects that a large project might be broken into

**synergy** — an approach where the whole is greater than the sum of the parts

**team development** — building individual and group skills to enhance project performance

**Tuckman model** — describes five stages of team development: forming, storming, norming, performing, and adjourning

## End Notes

[1] World Information Technology and Services Alliance, "Global ICT Spending Tops $3.5 Trillion," (May 4, 2008).

[2] Eric Chabrow, "Computer Jobs Hit Record High," *CIO Insight* (July 7, 2008).

[3] Bureau of Labor Statistics, "Economic and Employment Projections" (December 4, 2007).

[4] Deb Perelman, "Project Managers in High Demand, Short Supply," *eWeek.com* (March 12, 2007).

[5] Information Technology Association of America (ITAA), "Recovery Slight for IT Job Market in 2004," *ITAA.org* (September 8, 2004).

[6] Joan Williams and Ariane Hegewisch, "Confusing productivity with long workweek," *StarTribune.com* (Minneapolis–St. Paul, Minnesota) (September 6, 2004).

[7] Ibid.

[8] The Conference Board, Corporate Voices for Working Families, Partnership for 21st Century Skills, and Society for Human Resource Management, *"Are They Really Ready to Work? Employers' Perspectives on the Basic Knowledge and Applied Skills of New Entrants to the 21st Century U.S. Workforce"* (2006) (*www.conference-board.org/pdf_free/BED-06-Workforce.pdf*).

[9] Frederick Herzberg, "One More Time: How Do You Motivate Employees?" *Harvard Business Review* (February 1968) pp. 51–62.

[10] David C. McClelland, *The Achieving Society* (New York: Free Press, 1961).

[11] Douglas McGregor, *The Human Side of Enterprise* (New York: McGraw-Hill, 1960).

[12] William Ouchi, *Theory Z: How American Business Can Meet the Japanese Challenge* (New York: Avon Books, 1981).

[13] H. J. Thamhain and D. L. Wilemon, "Building Effective Teams for Complex Project Environments," *Technology Management* 5, no. 2 (May 1999).

[14] John R. French and Bertram H. Raven, "The Bases of Social Power," in D. Cartwright (Ed.), *Studies in Social Power* (Ann Arbor: University of Michigan Press, 1959).

[15] Stephen Covey, *The 7 Habits of Highly Effective People: Powerful Lessons in Personal Change* (New York: Simon & Schuster, 1990).

[16] Kidder, Tracy, *The Soul of a New Machine*. (New York: Modern Library, 1997).



[17] Douglas Ross, "Applying Covey's Seven Habits to a Project Management Career," *PM Network* (April 1996), pp. 26–30.

[18] Rewi, Adrienne, "The Rise of PMP," *PM Network* (October 2004), p. 18.

[19] Ronald L. Kempf, "The Most Universal Three Letters Since URL," *PM Network* (December 2006), p. 59.

[20] IBM Global Business Services, "Career development" (*www-935.ibm.com/services/us/gbs/bus/html/bcs_careers_development.html*).

[21] Stephanie Overby, "How to Hook the Talent You Need," *CIO Magazine* (September 1, 2006) (*www.cio.com/archive/090106/fea_talent.html*).

[22] Great Place to Work Institute, "Best Companies Lists" *GreatPlaceToWork.com*.

[23] Bruce Tuckman and Mary Ann Jensen, "Stages of Small-Group Development Revisited," Group Organization Management.1977; 2: 419–427.

[24] Isabel Myers Briggs, with Peter Myers, *Gifts Differing: Understanding Personality Type* (Palo Alto, CA: Consulting Psychologists Press, 1995).

[25] David Keirsey, *Please Understand Me II: Temperament, Character, Intelligence* (Del Mar, CA: Prometheus Nemesis Book Company, 1998).

[26] Michael L. Lyons, "The DP Psyche," *Datamation* (August 15, 1985).

[27] The information provided here about the Keirsey Temperament Sorter and Keirsey Temperament Theory was compiled from *Keirsey.com*.

[28] Sharon McDonald and Helen M. Edwards, "Who Should Test Whom? Examining the use and abuse of personality tests in software engineering," *Communications of the ACM* 50, no. 1 (January 2007).

[29] Harvey A. Robbins and Michael Finley, *The New Why Teams Don't Work: What Goes Wrong and How to Make It Right* (San Francisco, CA: Berrett-Koehler Publishers, 1999).

[30] John C. Goodman, "DISC, What Is It? Who Created the DISC Model?" *OnlineDiscProfile.com* (2004–2008).

[31] Patrick Lencioni, Overcoming the Five Dysfunctions of a Team: A Field Guide for Leaders, Managers, and Facilitators (San Francisco: Jossey-Bass, 2005), p. 3.

CHAPTER 10

# PROJECT COMMUNICATIONS MANAGEMENT

## LEARNING OBJECTIVES

**After reading this chapter, you will be able to:**

- Understand the importance of good communications on projects
- Discuss the process of identifying stakeholders and how to create a stakeholder register and stakeholder management strategy
- Explain the elements of project communications planning and how to create a communications management plan
- Describe various methods for distributing project information and the advantages and disadvantages of each, discuss the importance of addressing individual communication needs, and calculate the number of communications channels on a project
- Recognize the importance of managing stakeholder expectations
- Understand how reporting performance helps stakeholders stay informed about project progress
- List various methods for improving project communications, such as managing conflicts, running effective meetings, using e-mail and other technologies effectively, and using templates
- Describe how software can enhance project communications management

## OPENING CASE

Peter Gumpert worked his way up the corporate ladder in a large telecommunications company. He was intelligent, competent, and a strong leader, but the new Fiber-optic Undersea Telecommunications program was much larger and more complicated than anything he had previously worked on, let alone managed. This program consisted of several distinct projects, and Peter was in charge of overseeing them all. The changing marketplace for undersea telecommunications systems and the large number of projects involved made communications and flexibility critical concerns for Peter. For missing milestone and completion dates, his company would suffer huge financial penalties ranging from thousands of dollars per day for smaller projects to more than $250,000 per day for larger projects. Many projects depended on the success of other projects, so Peter had to understand and actively manage those critical interfaces.



Peter held several informal and formal discussions with the project managers reporting to him on this program. He worked with them and his project executive assistant, Christine Braun, to develop a communications plan for the program. He was still unsure, however, of the best way to distribute information and manage all of the inevitable changes that would occur. He also wanted to develop consistent ways for all of the project managers to develop their plans and track performance without stifling their creativity and autonomy. Christine suggested that they consider using some new communications technologies to keep important project information up to date and synchronized. Although Peter knew a lot about telecommunications and laying fiber-optic lines, he was not an expert in using information technology to improve the communication process. In fact, that was part of the reason he asked Christine to be his assistant. Could they really develop a process for communicating that would be flexible and easy to use? Time was of the essence as more projects were being added to the Fiber-optics Undersea Telecommunications program every week.

## THE IMPORTANCE OF PROJECT COMMUNICATIONS MANAGEMENT

Many experts agree that the greatest threat to the success of any project, especially information technology projects, is a failure to communicate. Many problems in other knowledge areas, such as an unclear scope or unrealistic schedules, indicate problems with communications. It is crucial for project managers and their teams to make good communications a priority, especially with key stakeholders, like top management.

The information technology field is constantly changing, and these changes bring with them a great deal of technical jargon. When computer professionals have to communicate with non–computer professionals, like most business professionals and senior managers, technical jargon can often complicate matters and confuse those who aren't technically savvy. Even though most people use computers today, the gap between users and developers increases as technology advances. Of course, not every computer professional is a poor communicator, but most people in any field can improve their communication skills.

In addition, most educational systems for information technology graduates promote strong technical skills over strong communication and social skills. Most IT–related degree programs have many technical requirements, but few require courses in communications (speaking, writing, listening), psychology, sociology, and the humanities. People often assume it is easy to pick up these soft skills, but they *are* important skills and, as such, people must learn and develop them.

Many studies have shown that information technology professionals need these soft skills just as much or even more than other skills. You cannot totally separate technical skills and soft skills when working on information technology projects. For projects to succeed, every project team member needs both types of skills and needs to develop them continuously through formal education and on-the-job training.

An article in the *Journal of Information Systems Education* on the importance of communications skills for information technology professionals presented the following conclusion:

> Based on the results of this research we can draw some general conclusions. First, it is evident that IS professionals engage in numerous verbal communication activities that are informal in nature, brief in duration, and with a small number of people at a time. Second, we can infer that most of the communication is indeed verbal in nature but sometimes it is supported by notes or graphs on a board or a handout and also by computer output. Third, it is clear that people expect their peers to listen carefully during a conversation and respond correctly to the issues at hand. Fourth, all IS professionals must be aware of the fact that they will have to engage in some form of informal public speaking. Fifth, it is evident that IS professionals must be able to communicate effectively in order to be successful in their current position, but they must also be able to do so in order to move to higher positions. Since our respondents, on average, seem to have moved throughout their IS career, from lower to higher positions, and they ranked verbal skills more important for their advancement than for their current job, the ability to communicate verbally seems to be a key factor in career advancement.[1]



In a recent study, respondents again stated that non-technical skills are most crucial for IT professionals, even at the entry level:

> The investigation confirms previous findings that non-technical skills are considered most important, especially those pertaining to Personal Attributes and Business Expertise. These soft skills are important as a foundation for all IT positions, and they enhance future learning and productivity as IT professionals advance in their careers. Nevertheless, technical skills are also important, though company dependent to a certain degree (different technical abilities are required for different jobs and organizations). Technical skills reduce the amount of training required of new employees and allow some entry-level personnel to be immediately productive.[2]

This chapter will highlight key aspects of project communications management, provide some suggestions for improving communications, and describe how software can assist in project communications management.

The goal of project communications management is to ensure timely and appropriate generation, collection, dissemination, storage, and disposition of project information. There are five main processes in project communications management:

1. *Identifying stakeholders* involves identifying everyone involved in or affected by the project and determining the best ways to manage relationships with them. The main outputs of this process are a stakeholder register and stakeholder management strategy.

2. *Planning communications* involves determining the information and communications needs of the stakeholders: who needs what information, when will they need it, and how will the information be given to them. The outputs of this process include a communications management plan and project document updates.
3. *Distributing information* involves making needed information available to project stakeholders in a timely manner. The main output of this process is organizational process assets updates. Recall from Chapter 4 that organizational process assets include formal and informal plans, policies, procedures, guidelines, information systems, financial systems, management systems, lessons learned, and historical information that help people understand, follow, and improve business processes in a specific organization.



4. *Managing stakeholder expectations* involves managing communications to satisfy the needs and expectations of project stakeholders and to resolve issues. The outputs of this process are organizational process assets updates, change requests, project management plan updates, and project document updates.
5. *Reporting performance* involves collecting and disseminating performance information, including status reports, progress measurements, and forecasts. The outputs of this process are performance reports, organizational process assets updates, and change requests.

Figure 10-1 summarizes these processes and outputs, showing when they occur in a typical project.

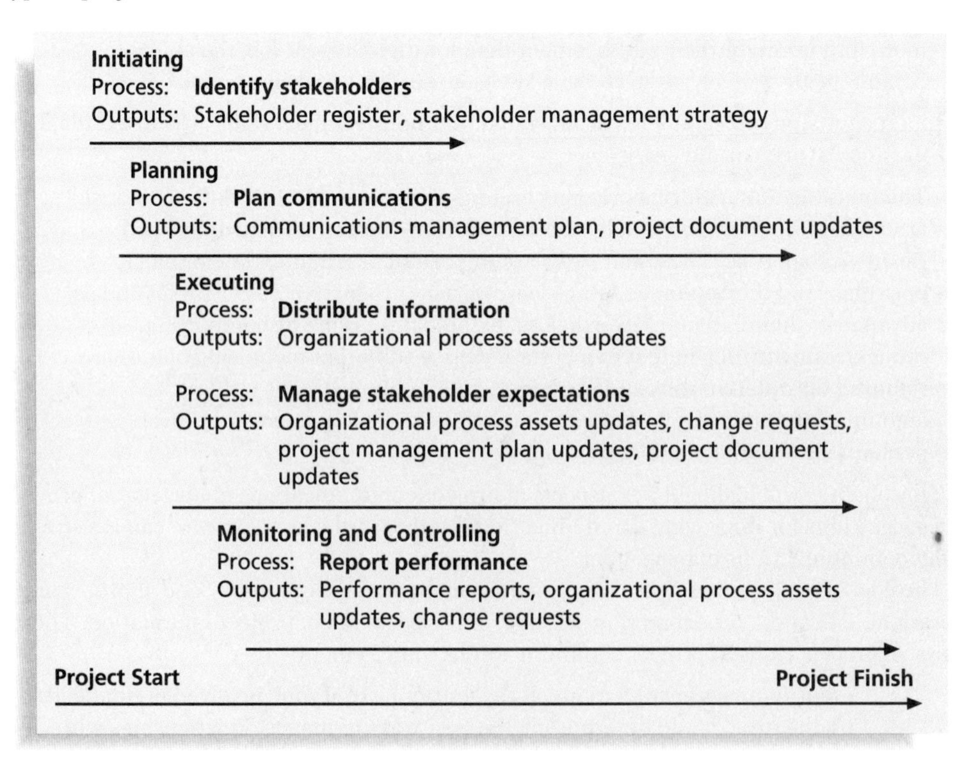


**FIGURE 10-1** Project communications management summary

## IDENTIFYING STAKEHOLDERS

Recall from Chapter 1 that stakeholders are people involved in or affected by project activities and include the project sponsor, project team, support staff, customers, users, suppliers, and even opponents to the project. Also recall that the ultimate goal of project management is to meet or exceed stakeholder needs and expectations from a project. In order to do that, you must first identify who your particular project stakeholders are. Identifying some stakeholders is obvious, but others might be more difficult. For example, there might be competitors outside the organization or even inside the organization who are opposed to the project without the project manager's knowledge. Stakeholders also might change during a project due to employee turnover, partnerships, and so on. It is important to use formal as well as informal communications networks to make sure that all key stakeholders are identified.



A simple way to document basic information on stakeholders is by creating a **stakeholder register**, a public document that includes details related to the identified project stakeholders. These details include the stakeholder's name, position, if he/she is internal or external to the organization, project role, and contact information. It might also include information on stakeholder requirements and expectations. Table 3-4 in Chapter 3 provides an example of a stakeholder register. Since this document is usually available to the public or people outside the project team, project managers should be careful not to include information that might be sensitive, such as how strongly the stakeholder supports the project. Sensitive information is often used in developing the stakeholder management strategy.

A **stakeholder management strategy** is an approach to help increase the support of stakeholders throughout the project. It includes basic information, such as stakeholder names, level of interest in the project, level of influence on the project, and potential management strategies for gaining support or reducing obstacles from that particular stakeholder. Since much of this information can be sensitive, it should be considered confidential. Some project managers do not even write down this information, but they do consider it since stakeholder management is a crucial part of their jobs.

For example, Peter Gumpert, the program manager from the opening case, worked with a few key colleagues to develop a stakeholder management strategy. Since Peter was in charge of several projects with many different stakeholders, it was crucial that he know who his most important stakeholders were and analyze how to work with them. Table 10-1 provides an example of part of Peter's stakeholder management strategy for the Fiber-optic Undersea Telecommunications program. This example also includes items from the stakeholder register, such as the stakeholder position and whether he or she is internal or external to the organization. Notice that the project managers working under Peter are also included in the stakeholder management strategy.

It is important to perform a stakeholder analysis during the initiating process group and to keep the stakeholder register and management strategy updated throughout the life of the project. As stakeholders and other information changes, it is important to update these documents. It is also important to use the stakeholder analysis information when planning communications.

**TABLE 10-1** Sample stakeholder management strategy

| Name | Position | Internal/ External | Level of interest | Level of influence | Potential management strategies |
|---|---|---|---|---|---|
| John Huntz | Project Manager for largest project under Peter | Internal | High | High | John does a great job, but he often upsets other PMs with his harsh approach. Keep him in line and remind him he is part of a bigger team. |
| Carolyn Morris | VP Telecommunications, Peter's boss | Internal | High | High | Carolyn is the first woman VP at our company and still likes to prove herself. Keep her informed of key issues and never surprise her! |
| Subbu Thangi | Dept. Head State of Oregon | External | Low | High | Subbu is in charge of a lot of state issues, like getting permits to install fiber-optic lines. He has a lot on his plate, but he doesn't seem concerned with our projects. Schedule a short, special meeting with him to increase visibility and discuss key issues. |
| Tom Morgan | CEO of major Telecomm. Customer | External | Medium | High | Tom is the sponsor of several of our projects. Give him the status on all of them at once to use his time efficiently. |



## PLANNING COMMUNICATIONS

Because communications is so important on projects, every project should include a **communications management plan**—a document that guides project communications. This plan should be part of the overall project management plan (described in Chapter 4, Project Integration Management). The communications management plan will vary with the needs of the project, but some type of written plan should always be prepared. For example, for small projects, such as the Project Management Intranet Site project described in Chapter 3, the communications management plan can be part of the team contract. For large projects, it should be a separate document. The communications management plan should address the following items:

1. Stakeholder communications requirements
2. Information to be communicated, including format, content, and level of detail
3. Who will receive the information and who will produce it
4. Suggested methods or technologies for conveying the information
5. Frequency of communication
6. Escalation procedures for resolving issues
7. Revision procedures for updating the communications management plan
8. A glossary of common terminology

It is important to know what kinds of information will be distributed to which stakeholders. By analyzing stakeholder communications needs, you can avoid wasting time or money on creating or disseminating unnecessary information.



Table 10-2 provides part of a sample stakeholder communications analysis that shows which stakeholders should get which written communications. Note that the stakeholder communications analysis includes information such as the contact person for the information, when the information is due, and the preferred format for the information. You can create a similar table to show which stakeholders should attend which project meetings. It is always a good idea to include comment sections with these types of tables to record special considerations or details related to each stakeholder, document, meeting, and so on.

**TABLE 10-2** Sample stakeholder communications analysis

| Stakeholders | Document Name | Document Format | Contact Person | Due |
|---|---|---|---|---|
| Customer management | Monthly status report | Hard copy and meeting | Tina Erndt, Tom Silva | First of month |
| Customer business staff | Monthly status report | Hard copy | Julie Grant, Sergey Cristobal | First of month |
| Customer technical staff | Monthly status report | E-mail | Li Chau, Nancy Michaels | First of month |
| Internal management | Monthly status report | Hard copy and meeting | Bob Thomson | First of month |
| Internal business and technical staff | Monthly status report | Intranet | Angie Liu | First of month |
| Training subcontractor | Training plan | Hard copy | Jonathan Kraus | November 1 |
| Software subcontractor | Software implementation plan | E-mail | Najwa Gates | June 1 |

Comments: Put the titles and dates of documents in e-mail headings and have recipients acknowledge receipt.

Having stakeholders review and approve the stakeholder communications analysis will ensure that the information is correct and useful.

Many projects do not include enough initial information on communications. Project managers, top management, and project team members assume using existing communications channels to relay project information is sufficient. The problem with using existing communications channels is that each of these groups (as well as other stakeholders) has different communications needs. Creating some sort of communications management plan and reviewing it with project stakeholders early in a project helps prevent or reduce later communication problems. If organizations work on many projects, developing some consistency in handling project communications helps the organization run smoothly.



Consistent communication helps organizations improve project communications, especially for programs composed of multiple projects. For example, Peter Gumpert, the Fiber-optic Undersea Telecommunications Program manager in the opening case, would benefit greatly from having a communications management plan that all of the project managers who report to him help develop and follow. Since several of the projects have some of the same stakeholders, it is even more important to develop a coordinated communications management plan. For example, if customers receive status reports from Peter's company that have totally different formats and do not coordinate information from related projects within the same company, they will question the ability of Peter's company to manage large programs.

Information regarding the content of essential project communications comes from the work breakdown structure (WBS). In fact, many WBSs include a section for project communications to ensure that reporting key information is a project deliverable. If reporting essential information is an activity defined in the WBS, it becomes even more important to develop a clear understanding of what project information to report, when to report it, how to report it, who is responsible for generating the report, and so on.

## DISTRIBUTING INFORMATION

Getting project information to the right people at the right time and in a useful format is just as important as developing the information in the first place. The stakeholder communications analysis serves as a good starting point for information distribution. Project managers and their teams must decide who receives what information, but they must also decide the best way to distribute the information. Is it sufficient to send written reports for project information? Are meetings alone effective in distributing project information? Are meetings and written communications both required for project information? What is the best way to distribute information to virtual team members?

During project execution, project teams must address important considerations for information distribution, and they often end up updating business processes through improved communications. For example, they might modify policies and procedures, information systems, or incorporate new technologies to improve information distribution. For example, Peter Gumpert, the program manager in the opening case, might decide that providing key people on his projects with handheld wireless devices, such as an iPhone or

BlackBerry, would enhance communications. He would need to request additional funds to provide these devices and training on how to use them.

After answering key questions related to project communications, project managers and their teams must decide the best way to distribute the information. Important considerations for information distribution include the use of technology, formal and informal communications, and the complexity of communications.

## Using Technology to Enhance Information Distribution

Technology can facilitate the process of distributing information, when used properly. Most people and businesses rely on e-mail, instant messaging, Web sites, telephones, cell phones, and other technologies to communicate. Using an internal project management information system, you can organize project documents, meeting minutes, customer requests, and so on, and make them available in an electronic format. You can store this information in local software or make it available on an intranet, an extranet, or the Internet, if the information is not sensitive. Storing templates and samples of project documents electronically can make accessing standard forms easier, thus making the information distribution process easier. It is also important to have backup procedures in place in case something goes wrong with normal communications technologies, as described in the What Went Wrong? You will learn more about using software to assist in project communications management later in this chapter.

### WHAT WENT WRONG?

Telecommunications throughout Asia were severely disrupted on December 26, 2006, after earthquakes off Taiwan damaged undersea cables, slowing Internet services and hindering financial transactions, particularly in the currency market. Property damage was minimal, but six of seven undersea cable systems, which accounted for 90 percent of telecommunications capacity of the region, broke in the quake and its aftershocks. International telephone traffic was restricted from some countries, and Internet access slowed to a crawl. The initial earthquake measured at a magnitude of 7.1 on the Richter scale by the U.S. Geological Survey. Philippine Long Distance Telephone Co. said its Internet service was intermittent, and international phone calls had been affected. Globe Telecom said, "the entire country's telecom services to the United States were disrupted."[3]

PCCW, the largest telephone company in Hong Kong, said on December 27 that it had lost half its Internet capacity because of the broken cables. PCCW reported "normal" voice services on December 29. On December 30, Singapore Telecommunications, the largest Southeast Asian phone operator, said that voice and Internet access was "back to normal." Chunghwa Telecom said that only 1 percent of major domestic customers' dedicated lines were damaged by December 30, compared with about 30 percent immediately after the quake. Operators were rerouting traffic to other lines and satellites. More than 80 percent of calls made through Chunghwa Telecom to the United States, Canada,

*continued*

Japan, Europe, and China could be connected as of December 31, rising from less than 30 percent after the quakes. However, *Bloomberg News* reported on December 31 that most Internet users would continue to experience slow access and suggested that customers minimize nonessential activities that demand large bandwidth over international connections.

"Phone calls and Internet connections are forecast to rise in time for the new year, but we have prepared contingency plans in case anything happens," said Kim Cheol Kee, a spokesman at KT, the biggest South Korean phone and Internet company. The financial impact of the cutoff was limited by the fact that many executives and traders were still away for the holidays. Luckily, markets were quiet and trading was light.[4]



## Formal and Informal Methods for Distributing Information

It is not enough for project team members to submit status reports to their project managers and other stakeholders and assume that everyone who needs to know that information will read the reports. Some technical professionals might assume that submitting the appropriate status reports is sufficient because they are introverts and prefer communicating that way. Occasionally, that approach might work, but many people prefer informal communications. Recall from Chapter 9 that 75 percent of the general population are extroverts, so they enjoy talking to other people. Often, many non-technical professionals—from colleagues to managers—prefer to have a two-way conversation about project information, rather than reading detailed reports, e-mails, or Web pages to try to find pertinent information.

Instead of focusing on getting information by reading technical documents, many colleagues and managers want to know the people working on their projects and develop a trusting relationship with them. They use informal discussions about the project to develop these relationships. Therefore, project managers must be good at nurturing relationships through good communication. Many experts believe that the difference between good project managers and excellent project managers is their ability to nurture relationships and use empathic listening skills, as described in Chapter 9, Project Human Resource Management.

Effective distribution of information depends on project managers and project team members having good communication skills. Communicating includes many different dimensions such as writing, speaking, and listening, and project personnel need to use all of these dimensions in their daily routines. In addition, different people respond positively to different levels or types of communication. For example, a project sponsor may prefer to stay informed through informal discussions held once a week over coffee. The project manager needs to be aware and take advantage of this special communication need. The project sponsor will give better feedback about the project during these informal talks than he or she could give through some other form of communication. Informal conversations allow the project sponsor to exercise his or her role of leadership and provide insights and information that are critical to the success of the project and the organization as a whole. Short face-to-face meetings are often more effective than electronic communications, particularly for sensitive information.

### Distributing Important Information in an Effective and Timely Manner

Many written reports neglect to provide the important information that good managers and technical people have a knack for asking about. For example, it is important to include detailed technical information that will affect critical performance features of products or services the company is producing as part of a project. It is even more important to document any changes in technical specifications that might affect product performance. For example, if the Fiber-Optic Undersea Telecommunications program included a project to purchase and provide special diving gear, and the supplier who provided the oxygen tanks enhanced the tanks so divers could stay under water longer, it would be very important to let other people know about this new capability. The information should not be buried in an attachment with the supplier's new product brochure.



People also have a tendency to not want to report bad news. If the oxygen tank vendor was behind on production, the person in charge of the project to purchase the tanks might wait until the last minute to report this critical information. Oral communication via meetings and informal talks helps bring important information—positive or negative—out into the open.

Oral communication also helps build stronger relationships among project personnel and project stakeholders. People make or break projects, and people like to interact with each other to get a true feeling for how a project is going. Many people cite research that says in a face-to-face interaction, 58 percent of communication is through body language, 35 percent through how the words are said, and a mere 7 percent through the content or words that are spoken. The author of this information (see *Silent Messages* by Albert Mehrabian[5]) was careful to note that these percentages were specific findings for a specific set of variables. Even if the actual percentages are different in verbal project communications today, it is safe to say that it is important to pay attention to more than just the actual words someone is saying. A person's tone of voice and body language say a lot about how they really feel.

Since information technology projects often require a lot of coordination, it is a good idea to have short, frequent meetings. For example, some information technology project managers require all project personnel to attend a "stand-up" meeting every week or even every morning, depending on the project needs. Stand-up meetings have no chairs, and the lack of chairs forces people to focus on what they really need to communicate. If people can't meet face to face, they are often in constant communications via cell phones, e-mail, instant messaging, or other technologies.

To encourage face-to-face, informal communications, some companies have instituted policies that workers cannot use e-mail between certain hours of the business day or even entire days of the week. For example, in the summer of 2004, Jeremy Burton, then vice president of marketing at a large Silicon Valley company, decreed that in his department, Fridays would be e-mail-free. The 240 people in his department had to use the phone or meet face-to-face with people, and violators who did use e-mail were fined.[6] Some companies are also taking advantage of social networking software to increase informal communications. In 2008, Jeremy Burton, now CEO of Serena Software, instituted Facebook Friday. "I told all the employees it's OK on a Friday for everybody to goof off and spend an hour or two on Facebook.... I said 'Go nuts! I dare you to participate, and I bet you'll find out something new about somebody in the company that you never knew before.'"[7] Burton wanted his people to get to know each other better, but he also wanted them to keep up with ever-changing technology so they could continue to develop useful software products. “The subversive message was ‘Guys, the world is a different place and if we’re going to stay relevant we’re going to have to wake up,’” Burton said.[8]

## Selecting the Appropriate Communications Medium

Table 10-3 provides guidelines from Practical Communications, Inc., a communications consulting firm, about how well different types of media, such as hard copy, phone calls,

**TABLE 10-3** Media choice table

| KEY: 1 = EXCELLENT | 2 = ADEQUATE | | 3 = INAPPROPRIATE | | | |
|---|---|---|---|---|---|---|
| How Well Medium Is Suited to: | Hard Copy | Phone Call | Voice Mail | E-mail | Meeting | Web Site |
| Assessing commitment | 3 | 2 | 3 | 3 | 1 | 3 |
| Building consensus | 3 | 2 | 3 | 3 | 1 | 3 |
| Mediating a conflict | 3 | 2 | 3 | 3 | 1 | 3 |
| Resolving a misunderstanding | 3 | 1 | 3 | 3 | 2 | 3 |
| Addressing negative behavior | 3 | 2 | 3 | 2 | 1 | 3 |
| Expressing support/appreciation | 1 | 2 | 2 | 1 | 2 | 3 |
| Encouraging creative thinking | 2 | 3 | 3 | 1 | 3 | 3 |
| Making an ironic statement | 3 | 2 | 2 | 3 | 1 | 3 |
| Conveying a reference document | 1 | 3 | 3 | 3 | 3 | 2 |
| Reinforcing one’s authority | 1 | 2 | 3 | 3 | 1 | 1 |
| Providing a permanent record | 1 | 3 | 3 | 1 | 3 | 3 |
| Maintaining confidentiality | 2 | 1 | 2 | 3 | 1 | 3 |
| Conveying simple information | 3 | 1 | 1 | 1 | 2 | 3 |
| Asking an informational question | 3 | 1 | 1 | 1 | 3 | 3 |
| Making a simple request | 3 | 1 | 1 | 1 | 3 | 3 |
| Giving complex instructions | 3 | 3 | 2 | 2 | 1 | 2 |
| Addressing many people | 2 | 3 or 1* | 2 | 2 | 3 | 1 |

*Depends on system functionality

See Tess Galati, *Email Composition and Communication (EmC2)*. Practical Communications, Inc. (*www.praccom.com*) (2001).

voice mail, e-mail, meetings, and Web sites, are suited to different communication needs. For example, if you were trying to assess commitment of project stakeholders, a meeting would be the most appropriate medium to use. (A face-to-face meeting would be preferable, but a Web conference, where participants can see and hear each other, would also qualify as a meeting.) A phone call would be adequate, but the other media would not be appropriate. Project managers must assess the needs of the organization, the project, and individuals in determining which communication medium to use, and when. They must also be aware of new technologies that can enhance communications and collaboration, as described in the What Went Right?

## WHAT WENT RIGHT?

A 2006 Frost & Sullivan study sponsored by Verizon Business and Microsoft Corp. called "Meetings Around the World: The Impact of Collaboration on Business Performance" found that collaboration is a key driver of overall performance of companies around the world. The impact of collaboration is twice as significant as a company's aggressiveness in pursuing new market opportunities and five times as significant as the external market environment. The study defines collaboration as an interaction between culture and technology such as audio and Web conferencing, e-mail, and instant messaging. The researchers also created a method to specifically measure how collaboration affects business performance.

Of all the collaboration technologies that were studied, three were more commonly present in high-performing companies than in low-performing ones: Web conferencing, audio conferencing, and meeting-scheduler technologies. "This study reveals a powerful new metric business leaders can use to more successfully manage their companies and achieve competitive advantage," said Brian Cotton, a vice president at Frost & Sullivan. "Measuring the quality and capability of collaboration in a given organization presents an opportunity for management to prioritize technology investments, encourage adoption of new tools and open up communications lines for improved collaboration."[9]

The study also showed that there are regional differences in how people in various countries prefer to communicate with one another. These differences highlight an opportunity for greater cultural understanding to improve collaborative efforts around the world. For example,

- American professionals are more likely to enjoy working alone, and they prefer to send e-mail rather than calling a person or leaving a voice mail message. They are also more comfortable with audio, video, and Web conferencing technologies than people of other regions. In addition, they tend to multitask the most when on conference calls.
- Europeans thrive on teamwork more than their counterparts elsewhere and prefer to interact in real time with other people. They are more likely to feel it is irresponsible not to answer the phone, and they want people to call them back rather than leave a voice mail message.
- Professionals in the Asia-Pacific region, more than anywhere else, want to be in touch constantly during the workday. As a result, they find the phone to be an indispensable tool and prefer instant messaging to e-mail.

## Understanding Group and Individual Communication Needs

Many top managers think they can just add more people to a project that is falling behind schedule. Unfortunately, this approach often causes more setbacks because of the increased complexity of communications. In his popular book, *The Mythical Man-Month*, Frederick Brooks illustrates this concept very clearly.[10] People are not interchangeable parts. You cannot assume that a task originally scheduled to take two months of one person's time can be done in one month by two people. A popular analogy is that you cannot take nine women and produce a baby in one month!

In addition to understanding that people are not interchangeable, it is also important to understand individuals' personal preferences for communications. As described in Chapter 9, people have different personality traits, which often affect their communication preferences. For example, if you want to praise a project team member for doing a good job, an introvert would be more comfortable receiving that praise in private while an extrovert would like everyone to hear about his or her good work. An intuitive person would want to understand how something fits into the big picture, while a sensing person would prefer to have more focused, step-by-step details. A strong thinker would want to know the logic behind information, while a feeling person would want to know how the information affects him or her personally as well as other people. Someone who is a judging person would be very driven to meet deadlines with few reminders while a perceiving person would need more assistance in developing and following plans.



Rarely does the receiver interpret a message exactly as the sender intended. Therefore, it is important to provide many methods of communication and an environment that promotes open dialogue. It is important for project managers and their teams to be aware of their own communication styles and preferences and those of other project stakeholders. As described in the previous chapter, many information technology professionals have different personality traits than the general population, such as being more introverted, intuitive, and oriented to thinking (as opposed to feeling). These personality differences can lead to miscommunication with people who are extroverted, sensation-oriented, and feeling types. For example, a user guide written by an information technology professional might not provide the detailed steps most users need. Many users also prefer face-to-face meetings to learn how to use a new system instead of trying to follow a cryptic user guide.

Geographic location and cultural background also affect the complexity of project communications. If project stakeholders are in different countries, it is often difficult or impossible to schedule times for two-way communication during normal working hours. Language barriers can also cause communication problems. The same word may have very different meanings in different languages. Times, dates, and other units of measure are also interpreted differently. People from some cultures also prefer to communicate in ways that may be uncomfortable to others. For example, managers in some countries still do not allow workers of lower ranks or women to give formal presentations. Some cultures also reserve written documents for binding commitments.

## Setting the Stage for Communicating Bad News

It is also important to put information in context, especially if it's bad news. An amusing example of putting bad news in perspective is in the following letter from a college student to her parents. Variations of this letter can be found on numerous Web sites.

*Dear Mom and Dad, or should I say Grandma & Grandpa,*

*Yes, I am pregnant. No, I'm not married yet since Larry, my boyfriend, is out of a job. Larry's employers just don't seem to appreciate the skills he has learned since he quit high school. Larry looks much younger than you, Dad, even though he is three years older. I'm quitting college and getting a job so we can get an apartment before the baby is born. I found a beautiful apartment above a 24-hour auto repair garage with good insulation so the exhaust fumes and noise won't bother us.*

*I'm very happy. I thought you would be too.*

*Love, Ashley*



*P.S. There is no Larry. I'm not pregnant. I'm not getting married. I'm not quitting school, but I am getting a "D" in Chemistry. I just wanted you to have some perspective.*

## Determining the Number of Communications Channels

Another important aspect of information distribution is the number of people involved in a project. As the number of people involved increases, the complexity of communications increases because there are more communications channels or pathways through which people can communicate. There is a simple formula for determining the number of communications channels as the number of people involved in a project increases. You can calculate the number of communications channels as follows:

$$\text{number of communications channels} = \frac{n(n-1)}{2}$$

where $n$ is the number of people involved.

For example, two people have one communications channel: $(2(2 - 1))/2 = 1$. Three people have three channels: $(3(3 - 1))/2 = 3$. Four people have six channels, five people have 10, and so on. Figure 10-2 illustrates this concept. You can see that as the number of people communicating increases above three, the number of communications channels increases rapidly. Project managers should try to limit the size of teams or sub teams to avoid making communications too complex. For example, if three people are working together on one particular project task, they have three communications channels. If you add two more people to their team, you would have ten communications channels, an increase of seven. If you added three more people instead of two, you'd have 12 communication channels. You can see how quickly communications becomes more complex as you increase team size.

Good communicators consider many factors before deciding how to distribute information, including the size of the group, the type of information, and the appropriate communication medium. People tend to overuse e-mail because it is an easy, inexpensive way to send information to a lot of people. When asked why you cannot always send an e-mail to a team of 100 people, just as you would to a team of five, one CIO answered, "As a group increases in size, you have a whole slew of management challenges. Communicating badly exponentially increases the possibility of making fatal mistakes. A large-scale project has a

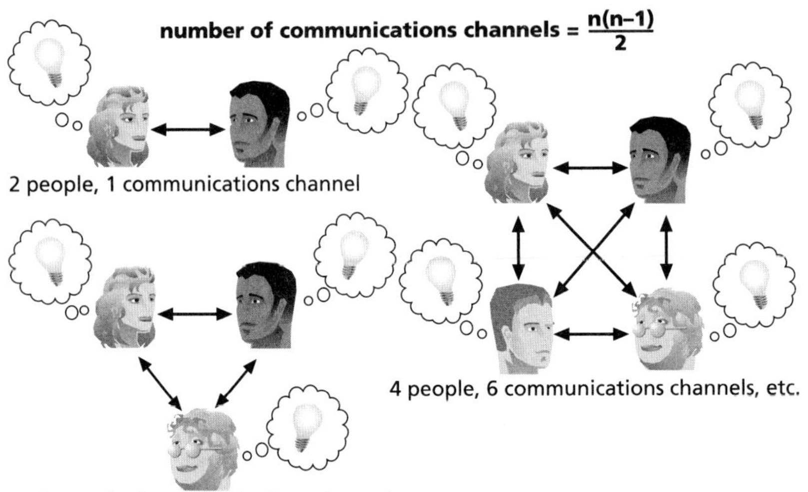


**FIGURE 10-2** The impact of the number of people on communications channels



lot of moving parts, which makes it that much easier to break down. Communication is the oil that keeps everything working properly. It's much easier to address an atmosphere of distrust among a group of five team members than it is with a team of 500 members."[11]

However, there are situations in which you cannot have face-to-face meetings and must e-mail a large group of people. Many information technology professionals work on virtual projects where they never meet their project sponsors, other team members, or other project stakeholders. In a virtual project environment, it is crucial for project managers to develop clear communication procedures. They can and must use e-mail, Web conferencing, instant messaging, discussion threads, project Web sites, and other technologies to communicate most information. They might be able to use phone calls or other media occasionally, but in general, they must rely on good written communications.

As you can see, information distribution involves more than creating and sending status reports or holding periodic meetings. Many good project managers know their personal strengths and weaknesses in this area and surround themselves with people who complement their skills, just as Peter Gumpert in the opening case did in asking Christine to be his assistant. It is good practice to share the responsibility for project communications management with the entire project team.

## MANAGING STAKEHOLDERS

Project managers must understand and work with various stakeholders; therefore, they should specifically address how they can use various communications methods as well as their interpersonal and management skills to satisfy the needs and expectations of project stakeholders. Recall that project success is often measured in different ways. Many studies define project success as meeting project scope, time, *and* cost goals. Many practitioners, however, define project success as satisfying the customer/sponsor, knowing that it's rare to meet scope, time, and cost goals without modifying at least one goal.

Project sponsors can usually rank scope, time, and cost goals in order of importance and provide guidelines on how to balance the triple constraint. This ranking is shown in an **expectations management matrix**, which can help clarify expectations. For example, Table 10-4 shows an expectations management matrix that Peter's project managers from the opening case could use to help manage their key stakeholders. The expectations management matrix includes a list of measures of success as well as priorities, expectations, and guidelines related to each measure. You could add additional measures of success, such as meeting quality expectations, achieving a certain customer satisfaction rating, meeting ROI projections after the project is completed, and so on to the matrix to meet individual project needs.



**TABLE 10-4** Expectations management matrix

| Measure of Success | Priority | Expectations | Guidelines |
|---|---|---|---|
| Scope | 2 | The scope statement clearly defines mandatory requirements and optional requirements. | Focus on meeting mandatory requirements before considering optional ones. |
| Time | 1 | There is no give in the project completion date. Every major deadline must be met, and the schedule is very realistic. | The project sponsor and program manager must be alerted if there are any issues that might affect meeting schedule goals. |
| Cost | 3 | This project is crucial to the organization. If you can clearly justify the need for more funds, they can be made available. | There are strict rules for project expenditures and escalation procedures. Cost is very important, but it takes a back seat to meeting schedule and then scope goals. |
| Quality | 6 | Quality is important, and the expectation is that we follow our well-established processes for testing this system. | All new personnel are required to complete several in-house courses to make sure they understand our quality processes. All corporate quality standards must be followed. |
| Customer Satisfaction | 4 | Our customer expects us to act professionally, answer questions in a timely manner, and work collaboratively with them to get the project done. | All presentations and formal documents provided to the customer must be edited by a tech writer. Everyone should reply to customer requests within 24 hours. |

TABLE 10-4 Expectations management matrix (*continued*)

| Measure of Success | Priority | Expectations | Guidelines |
|---|---|---|---|
| ROI Projections | 5 | The business case for this project projected an ROI of 40% within two years after implementation. | Our finance department will work with the customer to measure the ROI. Meeting/ exceeding this projection will help us bring in future business with this and other customers. |
| Etc. | | | |



Understanding the stakeholders' expectations can help in managing issues. If the project manager knows that cost is not as high a priority as the schedule, he or she will know that it shouldn't be too difficult to ask the project sponsor for needed funds, as long as there is good logic behind the request. Unresolved issues can be a major source of conflict and result in not meeting stakeholder expectations.

## REPORTING PERFORMANCE

Performance reporting keeps stakeholders informed about how resources are being used to achieve project objectives. Work performance information and measurements, forecasted completion dates, quality control measurements, the project management plan, approved change requests, and deliverables are all important inputs to performance reporting. Two key outputs of performance reporting are performance reports and forecasts. Performance reports are normally provided as status reports or progress reports. Many people use the two terms interchangeably, but some people distinguish between them as follows:

- **Status reports** describe where the project stands at a specific point in time. Recall the importance of the triple constraint. Status reports address where the project stands in terms of meeting scope, time, and cost goals. How much money has been spent to date? How long did it take to do certain tasks? Is work being accomplished as planned? Status reports can take various formats depending on the stakeholders' needs.
- **Progress reports** describe what the project team has accomplished during a certain period. Many projects have each team member prepare a monthly or sometimes weekly progress report. Team leaders often create consolidated progress reports based on the information received from team members. A sample template for a monthly progress report is provided later in this chapter.

**Forecasts** predict future project status and progress based on past information and trends. How long will it take to finish the project based on how things are going? How much more money will be needed to complete the project? Project managers can also use earned value management (see Chapter 7, Project Cost Management) to answer these questions by estimating the budget at completion and projected completion date based on how the project is progressing.

Another important technique for performance reporting is the status review meeting. Status review meetings, as described in Chapter 4, Project Integration Management, are a good way to highlight information provided in important project documents, empower people to be accountable for their work, and have face-to-face discussions about important project issues. Many program and project managers hold periodic status review meetings to exchange important project information and motivate people to make progress on their parts of the project. Likewise, many top managers hold monthly or quarterly status review meetings where program and project managers must report overall status information.

Status review meetings sometimes become battlegrounds where conflicts between different parties come to a head. Project managers or higher-level top managers should set ground rules for status review meetings to control the amount of conflict and should work to resolve any potential problems. It is important to remember that project stakeholders should work together to address performance problems.



# SUGGESTIONS FOR IMPROVING PROJECT COMMUNICATIONS

You have seen that good communication is vital to the management and success of information technology projects; you have also learned that project communications management can ensure that essential information reaches the right people at the right time, that feedback and reports are appropriate and useful, and that there is a formalized process of stakeholder management. This section highlights a few areas that all project managers and project team members should consider in their quests to improve project communications. The following text provides guidelines for managing conflict, developing better communication skills, running effective meetings, using e-mail, instant messaging, and collaborative tools effectively, and using templates for project communications.

## Using Communication Skills to Manage Conflict

Most large information technology projects are high-stake endeavors that are highly visible within organizations. They require tremendous effort from team members, are expensive, commandeer significant resources, and can have an extensive impact on the way work is done in an organization. When the stakes are high, conflict is never far away; when the potential for conflict is high, good communication is a necessity.

Chapter 6, Project Time Management, explained that schedule issues cause the most conflicts over the project life cycle and provided suggestions for improving project scheduling. Other conflicts occur over project priorities, staffing, technical issues, administrative procedures, personalities, and cost. It is crucial for project managers to develop and use their human resources and communication skills to help identify and manage conflict on projects. Project managers should lead their teams in developing norms for dealing with various types of conflicts that might arise on their projects. For example, team members should know that disrespectful behavior toward any project stakeholder is inappropriate, and that team members are expected to try to work out small conflicts themselves before elevating them to higher levels. As mentioned earlier in this chapter, escalation procedures should be documented in the communications management plan. Blake and Mouton (1964) delineated five basic modes for handling conflicts: confrontation, compromise, smoothing, forcing, and withdrawal.

1. *Confrontation*. When using the **confrontation mode**, project managers directly face a conflict using a problem-solving approach that allows affected parties to work through their disagreements. This approach is also called the problem-solving mode.
2. *Compromise*. With the **compromise mode**, project managers use a give-and-take approach to resolving conflicts. They bargain and search for solutions that bring some degree of satisfaction to all the parties in a dispute.
3. *Smoothing*. When using the **smoothing mode**, the project manager deemphasizes or avoids areas of differences and emphasizes areas of agreement. This approach is also called accommodating.
4. *Forcing*. The **forcing mode** can be viewed as the win-lose approach to conflict resolution. Project managers exert their viewpoint at the potential expense of another viewpoint. Managers who are very competitive or autocratic in their management style might favor this approach.
5. *Withdrawal*. When using the **withdrawal mode**, project managers retreat or withdraw from an actual or potential disagreement. This approach is also called avoiding and is the least desirable conflict-handling mode.

   More recent studies recognize a sixth conflict-handling mode:

6. *Collaborating*: Using the **collaborating mode**, decision makers incorporate different viewpoints and insights to develop consensus and commitment.

Research indicates that project managers favor using confrontation for conflict resolution over the other modes. The term confrontation may be misleading. This mode really focuses on addressing conflicts using a problem-solving approach. Using Stephen Covey's paradigms of interdependence, this mode focuses on a win-win approach. All parties work together to find the best way to solve the conflict. Other popular approaches to conflict resolution include collaboration and compromise. Successful project managers are less likely to use smoothing, forcing, or withdrawal if they want to make effective decisions.

Project managers must also realize that not all conflict is bad. In fact, conflict can often be good. Conflict often produces important results, such as new ideas, better alternatives, and motivation to work harder and more collaboratively. Project team members may become stagnant or develop **groupthink**—conformance to the values or ethical standards of a group—if there are no conflicting viewpoints on various aspects of a project. Research by Karen Jehn, Professor of Management at Wharton, suggests that task-related conflict, which is derived from differences over team objectives and how to achieve them, often improves team performance. Emotional conflict, however, which stems from personality clashes and misunderstandings, often depresses team performance.[12] Project managers should create an environment that encourages and maintains the positive and productive aspects of conflict.

## Developing Better Communication Skills

Some people seem to be born with great communication skills. Others seem to have a knack for picking up technical skills. It is rare to find someone with a natural ability for both. Both communication and technical skills, however, can be developed. Most information technology professionals enter the field because of their technical skills. Most find, however, that communication skills are the key to advancing in their careers, especially if they want to become good project managers.

Most companies spend a lot of money on technical training for their employees, even when employees might benefit more from communications training. Individual employees are also more likely to enroll voluntarily in classes on the latest technology than those on developing their soft skills. Communication skills training usually includes role-playing activities in which participants learn concepts such as building rapport, as described in Chapter 9, Project Human Resource Management. Training sessions also give participants a chance to develop specific skills in small groups. Training sessions that focus on presentation skills usually use video to record the participants' presentations. Most people are surprised to see some of their mannerisms and enjoy the challenge of improving their skills. A minimal investment in communication and presentation training can have a tremendous payback to individuals, their projects, and their organizations. These skills also have a much longer shelf life than many of the skills learned in technical training courses.



As organizations become more global, they realize that they must also invest in ways to improve communication with people from different countries and cultures. For example, many Americans are raised to speak their minds, while in some other cultures people are offended by outspokenness. Not understanding how to communicate effectively with other cultures and people of diverse backgrounds hurts projects and businesses. Many training courses are available to educate people in cultural awareness, international business, and international team building.

It takes leadership to help improve communication. If top management lets employees give poor presentations, write sloppy reports, offend people from different cultures, or behave poorly at meetings, the employees will not want to improve their communication skills. Top management must set high expectations and lead by example. Some organizations send all information technology professionals to training that includes development of technical *and* communication skills. Successful organizations allocate time in project schedules for preparing drafts of important reports and presentations and incorporating feedback on the drafts. It is good practice to include time for informal meetings with customers to help develop relationships and provide staff to assist in relationship management. As with any other goal, improving communication can be achieved with proper planning, support, and leadership from top management.

## MEDIA SNAPSHOT

Communications technology, such as using e-mail and searching the Web, should help improve project communications, but it can also cause conflict. How? Most people have heard the term "slackers" before, referring to people who avoid work. But have you heard the term "cyberslackers?" Cyberslackers are people who should be working, but instead spend their time online doing non-work-related activities, such as annoying friends or co-workers by sending unimportant e-mails. A recent study by Websense suggested that employees are using the Web more and more for personal reasons, and it is costing U.S.

*continued*

companies $178 billion annually, or $5,000 per employee.[13] Websense determined this number by surveying managers, who estimated that each employee was using the Internet for personal use for 5.9 hours a week on average, and then multiplying these numbers by the average American hourly salary. Cyberslacking is not a new phenomenon, nor is it restricted to people in certain countries. In 2000, Internet security company Surfcontrol estimated that every employee in Australia was taking the equivalent of a two-week "cyber-holiday" each year, costing the nation $22.5 billion annually.[14]

A two-year research project by postgraduate psychology student Kerryann Wyatt found that cyberslacking could be using as much as a quarter of the time employees spend online, and even more if they have outgoing personalities. Of the many Internet distractions tempting employees away from their work, e-mailing friends was the most popular, followed by general Internet searches, according to Wyatt's research. One of the goals of this study was to attempt to match personality types with Internet misuse at work. Wyatt said her findings indicated that, contrary to previous studies, introverted participants were not the key offenders. The report said: "People scoring highly in extroversion were significantly more likely to send higher numbers of both work and non work-related e-mails."[15]

A 2008 survey found that more than a quarter of U.S. employers have fired workers for misusing e-mail and one-third have fired workers for misusing the Internet on the job. Of the managers surveyed, 84 percent said employees were fired for accessing pornography or other inappropriate content on the Internet. Among managers who fired workers for e-mail misuse, 64 percent did so because the employee violated company policy, and 62 percent said the e-mail contained inappropriate or offensive language. Most employees who were fired knew that their computer usage was being monitored.[16]



## Running Effective Meetings

A well-run meeting can be a vehicle for fostering team building and reinforcing expectations, roles, relationships, and commitment to the project. However, a poorly run meeting can have a detrimental effect on a project. For example, a terrible kick-off meeting may cause some important stakeholders to decide not to support the project further. Many people complain about the time they waste in unnecessary or poorly planned and poorly executed meetings. Following are some guidelines to help improve time spent at meetings:

- *Determine if a meeting can be avoided*. Do not have a meeting if there is a better way of achieving the objective at hand. For example, a project manager might know that he or she needs approval from a top manager to hire another person for the project team. It could take a week or longer to schedule even a ten-minute meeting on the top manager's calendar. Instead, an e-mail or phone call describing the situation and justifying the request is a faster, more effective approach than having a meeting. However, many times you do need a face-to-face meeting, and it would not be appropriate to try to use e-mail or a phone call. Consider which medium would be most effective, as described earlier.
- *Define the purpose and intended outcome of the meeting*. Be specific about what should happen as a result of the meeting. Is the purpose to brainstorm ideas, provide status information, or solve a problem? Make the purpose of a meeting very clear to all meeting planners and participants. For example, if a project manager calls a meeting of all project team members without knowing

the true purpose of the meeting, everyone will start focusing on their own agendas and very little will be accomplished. All meetings should have a purpose and intended outcome.

- *Determine who should attend the meeting*. Do certain stakeholders have to be at a meeting to make it effective? Should only the project team leaders attend a meeting, or should the entire project team be involved? Many meetings are most effective with the minimum number of participants possible, especially if decisions must be made. Other meetings require many attendees. It is important to determine who should attend a meeting based on the purpose and intended outcome of the meeting.
- *Provide an agenda to participants before the meeting*. Meetings are most effective when the participants come prepared. Did they read reports before the meeting? Did they collect necessary information? Some professionals refuse to attend meetings if they do not have an agenda ahead of time. Insisting on an agenda forces meeting organizers to plan the meeting and gives potential attendees the chance to decide whether they really need to attend the meeting.
- *Prepare handouts and visual aids, and make logistical arrangements ahead of time*. By creating handouts and visual aids, the meeting organizers must organize their thoughts and ideas. This usually helps the entire meeting run more effectively. It is also important to make logistical arrangements by booking an appropriate room, having necessary equipment available, and providing refreshments or entire meals, if appropriate. It takes time to plan for effective meetings. Project managers and their team members must take time to prepare for meetings, especially important ones with key stakeholders.
- *Run the meeting professionally*. Introduce people, restate the purpose of the meeting, and state any ground rules that attendees should follow. Have someone facilitate the meeting to make sure important items are discussed, watch the time, encourage participation, summarize key issues, and clarify decisions and action items. Designate someone to take minutes and send the minutes out soon after the meeting. Minutes should be short and focus on the crucial decisions and action items from the meeting.
- *Set the ground rules for the meeting*. State up front how the meeting will be run. For example, can people speak at will, or will the facilitator lead discussions? Can attendees use their laptops or other electronic devices during the meeting? Don't assume that all meetings are run in the same way. Do what works best in each specific case.
- *Build relationships*. Depending on the culture of the organization and project, it may help to build relationships by making meetings fun experiences. For example, it may be appropriate to use humor, refreshments, or prizes for good ideas to keep meeting participants actively involved. If used effectively, meetings are a good way to build relationships.



## Using E-Mail, Instant Messaging, and Collaborative Tools Effectively

Since most people use e-mail and other electronic communications tools now, communications should improve, right? Not necessarily. In fact, few people have received any training or guidelines on when or how to use e-mail, instant messaging, or other collaborative tools, such as Microsoft SharePoint portals, Google Docs, or wikis. (A **SharePoint portal** allows users to create custom Web sites to access documents and applications stored on shared devices. **Google Docs** allow users to create, share, and edit documents, spreadsheets, and presentations online. A **wiki** is a Web site designed to enable anyone who accesses it to contribute or modify Web page content.) As discussed earlier in this chapter, e-mail is not an appropriate medium for several types of communications. The media choice table (Table 10-3) suggests that e-mail is not appropriate for assessing commitment, building consensus, mediating a conflict, resolving a misunderstanding, making an ironic statement, conveying a reference document, reinforcing one's authority, or maintaining confidentiality. The same is true for instant messaging and other electronic communications tools.



Even if people do know when to use e-mail or other tools for project communications, they also need to know how to use it. New features are added to e-mail, instant messaging, and collaborative software programs with every new release, but often users are unaware of these features and do not receive any training on how to use them. Do you know how to organize and file your e-mail messages, or do you have hundreds of e-mail messages sitting in your Inbox? Do you know how to use your address book or how to create distribution lists? Have you ever used sorting features to find e-mail messages by date, author, or key words? Do you use filtering software to prevent spam? Do you know how to share your desktop with instant messaging to teach someone how to use software on your computer? Do you know how to track and incorporate changes in Google Docs to create reports and spreadsheets as a collaborative effort? Does everyone on your project team know how to use important features of your SharePoint portal or wiki?

Even if you know how to use all the features of these communications systems, you will likely need to learn how to put ideas into words clearly. For example, the subject line for any e-mail messages you write should clearly state the intention of the e-mail. Folder and file names for collaborative projects should be clear and follow file naming conventions, if provided. A business professional who is not a very good writer may prefer to talk to people than to send an e-mail or instant message. Poor writing often leads to misunderstandings and confusion.

Project managers should do whatever they can to help their project stakeholders use e-mail, instant messaging, collaborative tools, or any other communications technologies effectively and not waste time with poor or unclear electronic communications.

The following guidelines will help you use several of these tools more effectively:

- Information sent via e-mail, instant messaging, or a collaborative tool should be appropriate for that medium, versus other media. If you can communicate the information better with a phone call or meeting, for example, then do so.
- Be sure to send the e-mail or instant message to the right people. Do not automatically "reply to all" on an e-mail, for example, if you do not need to.
- Use meaningful subject lines in e-mails so readers can quickly see what information the message will contain. If the entire message can be put in the subject line, put it there. For example, if a meeting is cancelled, just type that in the subject line. Also, do not continue replying to e-mail messages without changing the subject. The subject should always relate to the latest correspondence.

- Limit the content of the e-mail to one main subject. Send a second or third e-mail if it relates to a different subject.
- The body of the e-mail should be as clear and concise as possible and you should always reread your e-mail before you send it. Also, be sure to check your spelling using the spell check function. If you have three questions you need answered, number them as question 1, 2, and 3.
- Limit the number and size of e-mail attachments. If you can include a link to an online version of a document instead of attaching a file, do so.
- Delete e-mail that you do not need to save or respond to. Do not even open e-mail that you know is not important, such as spam. Use the e-mail blocking feature of the software, if available, to block unwanted junk mail.
- Make sure your virus protection software is up to date. Never open e-mail attachments if you do not trust the source.
- Respond to e-mail quickly, if possible. It will take you longer to open and read it again later. In addition, if you send an e-mail that does not require a response, make that clear as well.
- If you need to keep e-mail, file each message appropriately. Create folders with meaningful names to file the e-mail messages you want to keep. File them as soon as possible.
- Learn how to use important features of your e-mail, instant messaging, and collaborative software.
- Most people are comfortable with using e-mail, but some may not be familiar with using instant messaging. Develop a strategy for getting users up to date, and discuss when it's best to use instant messaging versus e-mail.
- Collaborative tools continue to advance. Make sure your team is using a good tool. Many, like Google Docs and several wikis, are available for free.
- Be sure to authorize the right people to share your collaborative documents. Also ensure that other security is in place. Confidential project documents should probably not be stored on Google Docs. Use more secure tools when needed. For example, Alaska Airlines uses an internally controlled wiki to improve project communications, as described in the following Best Practice feature.
- Make sure the right person can authorize changes to shared documents and that you back up files.
- Develop a logical structure for organizing and filing shared documents. Use good file-naming conventions for folder and document names.



## BEST PRACTICE

At a 2008 conference, John Petroske shared successful practices Alaska Airlines follows in using wikis to facilitate project communications and collaboration. As Petroske explains, a wiki is a fully editable Web site. To prevent anyone on the Internet from seeing or

*continued*

modifying content, it is important to make the project wikis accessible only behind a protective corporate firewall. You can also enable tracking and authentication mechanisms to deter abuse and prevent non-project members from modifying a wiki.

The main benefits wikis bring to project management include:

- *Better documentation*: Data can be stored in different network locations, and a wiki can centrally organize this information. For example, after a meeting or hallway conversation, any authorized wiki user can document the information and make it available on the wiki. The wiki includes hyperlinks to related project documents, such as meeting notes, status reports, product specifications, use case definitions, lessons learned reports, and so on.
- *Improved trust and information sharing*: Project wikis encourage the project team to openly share information. By putting all information in one place and allowing users to edit information, team members develop a better working relationship. There can be problems with misuse if someone intentionally deletes or changes information in a negative way, but that has not occurred at Alaska Airlines because project teams are dedicated to project success and self-police content.
- *Sustained growth*: In order to ensure the sustainability of a wiki, it is important to train project users, keep the wiki organized, understand how people are using the wiki, lead by example by using the wiki often and appropriately, protect the wiki behind a corporate firewall and back it up often, and provide style guides. The Alaska Airlines IT department even created a "Mother of All Wikis" to serve as an index for all the known project wikis. Site visitors are encouraged to help maintain the site by adding related information or links.[17]



## Using Templates for Project Communications

Many intelligent people have a hard time writing a performance report or preparing a ten-minute technical presentation for a customer review. Some people in these situations are too embarrassed to ask for help. To make preparing project communications easier, project managers need to provide examples and templates for common project communications items such as project descriptions, project charters, monthly performance reports, issue logs, and so on. Good documentation from past projects can be an ample source of examples. Samples and templates of both written and oral reports are particularly helpful for people who have never before had to write project documents or give project presentations. Finding, developing, and sharing relevant templates and sample documents are important tasks for many project managers. Several examples of project documentation such as a business case, project charter, scope statement, stakeholder analysis, WBS, Gantt chart, cost estimate, and so on are provided throughout this text. The companion Web site for this text includes the actual files used in creating the templates for these sample documents. A few of these templates and guidelines for preparing them are provided in this section.

Figure 10-3 shows a sample template for a one-page project description. This form could be used to show a "snapshot" of an entire project on one page. For example, top managers might require that all project managers provide a brief project description as part of a quarterly management review meeting. Peter Gumpert, the program manager in the opening case, might request this type of document from all of the project managers working for him to get an overall picture of what each project involves. According to Figure 10-3, a project description should include the project objective, scope, assumptions, cost information, and schedule information. This template suggests including information from the project's Gantt chart to highlight key deliverables and other milestones.



## Project X Descripton

**Objective:** Describe the objective of the project in one or two sentences. Focus on the business benefits of doing the project.

**Scope:** Briefly describe the scope of the project. What business functions are involved, and what are the main products the project will produce?

**Assumptions:** Summarize the most critical assumptions for the project.

**Cost:** Provide the total estimated cost of the project. If desired, list the total cost each year.

**Schedule:** Provide summary information from the project's Gantt chart, as shown. Focus on summary tasks and milestones.

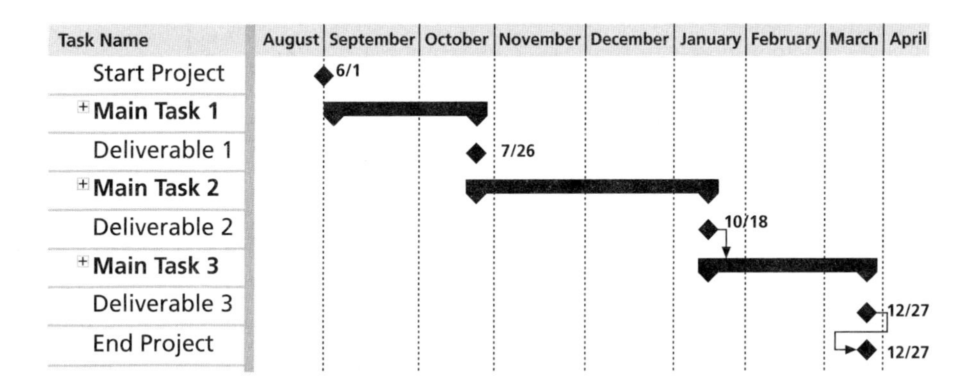


**FIGURE 10-3** Sample template for a project description

Table 10-5 shows a template for a monthly progress report. Sections of the progress report include accomplishments from the current period, plans for the next period, issues, and project changes. Recall that progress reports focus on accomplishments during a specific time period while status reports focus on where the project stands at a certain point in time. Because progress and status reports are important ways to communicate project information, it is important for project teams to tailor them to meet their specific needs. Some organizations, such as JWD Consulting from Chapter 3, combine both progress and status information on the same template.

**TABLE 10-5** Sample template for a monthly progress report

| | |
|---|---|
| **I.** | Accomplishments for January (or appropriate month):<br>• Describe most important accomplishments. Relate them to project's Gantt chart.<br>• Describe other important accomplishments, one bullet for each. If any issues were resolved from the previous month, list them as accomplishments. |
| **II.** | Plans for February (or following month):<br>• Describe most important items to accomplish in the next month. Again, relate them to project's Gantt chart.<br>• Describe other important items to accomplish, one bullet for each. |
| **III.** | **Issues:** Briefly list important issues that surfaced or are still important. Managers hate surprises and want to help the project succeed, so be sure to list issues. |
| **IV.** | **Project Changes (Dates and Description):** List any approved or requested changes to the project. Include the date of the change and a brief description. |



Table 10-6 provides an exhaustive list of all of the documentation that should be organized and filed at the end of a major project. From this list, you can see that a large project can generate a lot of documentation. In fact, some project professionals have observed that documentation for designing an airplane usually weighs more than the airplane itself. (Smaller projects usually generate much less documentation!)

**TABLE 10-6** Final project documentation items

| | |
|---|---|
| I. | Project description |
| II. | Project proposal and backup data (request for proposal, statement of work, proposal correspondence, and so on) |
| III. | Original and revised contract information and client acceptance documents |
| IV. | Original and revised project plans and schedules (WBS, Gantt charts and network diagrams, cost estimates, communications management plan, etc.) |
| V. | Design documents |
| VI. | Final project report |
| VII. | Deliverables, as appropriate |
| VIII. | Audit reports |
| IX. | Lessons-learned reports |
| X. | Copies of all status reports, meeting minutes, change notices, and other written and electronic communications |

The project manager and project team members should all prepare a **lessons-learned report**—a reflective statement documenting important things they have learned from working on the project. The project manager often combines information from all of the lessons-learned reports into a project summary report. See Chapter 3 for an example of this type of lessons-learned report. Some items discussed in lessons-learned reports include reflections on whether project goals were met, whether the project was successful or not, the causes of variances on the project, the reasoning behind corrective actions chosen, the use of different project management tools and techniques, and personal words of wisdom based on team members' experiences. On some projects, all project members are required to write a brief lessons-learned report; on other projects, just the team leads or project manager writes the report. These reports provide valuable reflections by people who know what really worked or did not work on the project. Everyone learns in different ways and has different insights into a project, so it is helpful to have more than one person provide inputs on the lessons-learned reports. These reports can be an excellent resource and help future projects run more smoothly. To reinforce the benefits of lessons-learned reports, some companies require new project managers to read several past project managers' lessons-learned reports and discuss how they will incorporate some of their ideas into their own projects. It is also important to organize and prepare project archives. **Project archives** are a complete set of organized project records that provide an accurate history of the project. These archives can provide valuable information for future projects as well.



In the past few years, more and more project teams have started putting all or part of their project information, including various templates and lessons-learned reports, on project Web sites. Project Web sites provide a centralized way of delivering project documents and other communications. Project teams can develop project Web sites using wikis or Web-authoring tools, such as Adobe Dreamweaver or Microsoft Expression Web. The home page for the project Web site should include summary information about the project, such as the background and objectives of the project. The home page should also include contact information, such as names and e-mail addresses for the project manager, other team members, or the Webmaster. Links should be provided to items such as project documents, a team member list, meeting minutes, a discussions area, if applicable, and other materials related to the project. If the project involves creating research reports, software, design documents, or other items that can also be accessed via the Web site, links can be provided to those files as well. The project team should also address other issues in creating and using a project Web site, such as security, access, and type of content that should be included on the site.

For more sophisticated Web sites, project teams can also use one of the many software products created specifically to assist in project communications through the Web. These products vary considerably in price and functionality, as described in Chapter 1. For example, Figure 10-4 provides a sample screen from the Microsoft Office Enterprise Project Management (EPM) Solution for 2007. Notice the categories of information on the left side of the screen where users can access their individual work information (i.e., tasks, timesheets, and issues and risks), information on projects, resources, reporting, approvals, personal settings, and shared documents.

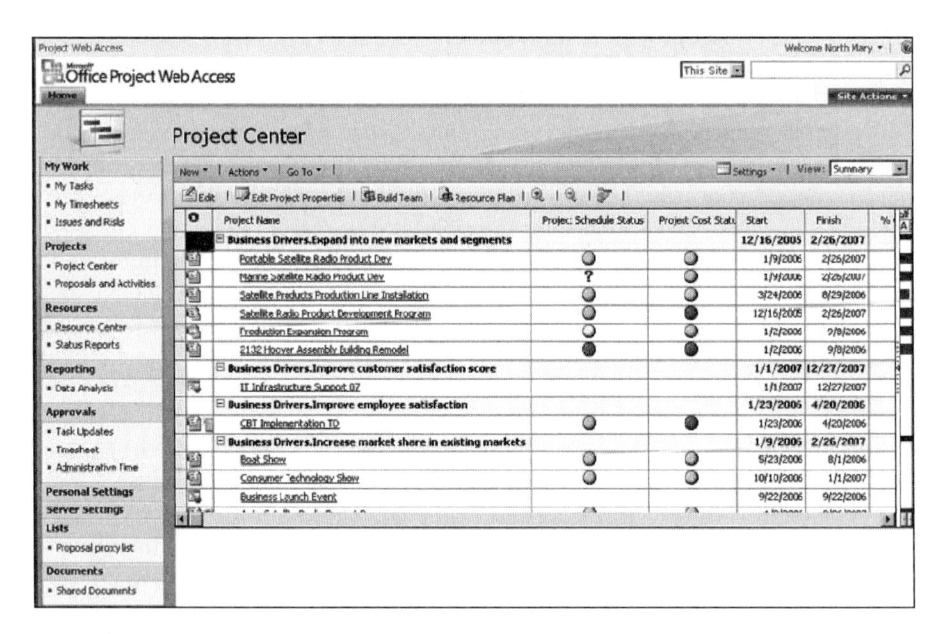


**FIGURE 10-4** Microsoft Office Enterprise Project Management (EPM) Solution

When the project team develops their project communications management plan, they should also determine what templates to use for key documentation. To make it even more convenient to use templates, the organization should make project templates readily available online for all projects. The project team should also understand top management's and customers' documentation expectations for each particular project. For example, if a project sponsor or customer wants a one-page monthly progress report for a specific project, but the project team delivers a 20-page report, there are communication problems. In addition, if particular customers or top managers want specific items in all final project reports, they should make sure the project team is aware of those expectations and modifies any templates for those reports to take these requirements into account.

## USING SOFTWARE TO ASSIST IN PROJECT COMMUNICATIONS

Many organizations are discovering how valuable project management software can be in communicating information about individual projects and multiple projects across the organization. Project management software can provide different views of information to help meet various communication needs. For example, senior managers might only need to see summary screens with colors indicating the overall health of all projects. Middle managers often want to see the status of milestones for all of the projects in their area. Project team members often need to see all project documentation. Often, one of the biggest communication problems on projects is providing the most recent project plans, Gantt charts, specifications, meeting information, change requests, and so on to all or selective stakeholders in a timely fashion. Most project management software allows users to insert hyperlinks to other project-related files. In Project 2007, you can insert a hyperlink from a task or milestone listed in a Gantt chart to another file that contains relevant information. For example, there might be a milestone that the project charter was signed. You can insert a hyperlink to the Word file containing the project charter from within the Gantt chart. You could also link appropriate tasks or milestones to Excel files that contain a staffing management plan or cost estimate or to Microsoft PowerPoint files with important presentations or other information. The Project 2007 file and all associated hyperlinked files could then be placed on a local area network server or Web server, allowing all project stakeholders easy access to important project information. (See Appendix A for more information on using Microsoft Project 2007 to assist in project communications management.)

Even though organizations routinely use many types of hardware and software to enhance communications, they need to take advantage of new technologies and adjust existing systems to serve the special communications needs of customers and project teams. In addition to diverse customer and project needs, they also have to address the changing expectations of consumers and the workforce. For example, several television shows have harnessed communications technology to engage their audiences by letting them vote online or via telephone for their favorite singers, e-mail celebrities, or access information on their Web sites. Many people, especially younger people, use instant messaging or cell phone text messages every day to communicate with friends. Some business and technical professionals also find instant messaging and text messaging to be a useful tool for quickly communicating with colleagues, customers, suppliers, and others. Web logs, or **blogs**, are easy-to-use journals on the Web that allow users to write entries, respond to another poster's comments, create links, upload pictures, and post comments to journal entries. Blogs have also become popular as a communication technology. If television shows and non-technical people can use advanced communications technologies, why can't project stakeholders?



Employers have made changes to meet changing expectations and needs in communications. The Telework Coalition reports that since 1990, the number of people telecommuting has grown from about 4 million to more than 45 million. At IBM, 40 percent of its 330,000 employees work from home, on the road, or at a client location on any given day.[18] On several IT projects, project managers have found that their team members can be more productive when they are allowed to work from home. Other project managers have no choice in the matter when some or all of their project team members work remotely. As described in Chapter 9, Project Human Resource Management, studies show that providing a quiet work environment and a dedicated workspace increases programmer productivity. Most people who work from home have well-equipped, comfortable home offices with fewer distractions and more space than corporate offices. Workers also appreciate the added bonus of avoiding traffic and having more flexible work schedules. However, it is important to make sure work is well defined and communications are in place to allow for remote workers to work effectively.

Several products are now available to assist individual consumers and organizations with communications. Many products were developed or enhanced in the early 2000s to address the issue of providing fast, convenient, consistent, and up-to-date project information. Webcasts are now a common tool for presenting video, graphics, sound, voice, and participant feedback live over the Web. Podcasts and YouTube videos have also become popular tools for providing various types of audio and video information, from exercise instructions to class lectures. Most working adults or college students have cell phones, and today it is common to see someone take and send a picture or send and receive text messages or e-mail with a cell phone. Many high school or college students are instant messaging or texting their friends to plan social activities or occasionally discuss academic topics.

These same technologies can enhance project communications. For even more powerful and integrated communications, enterprise project management software provides many workgroup functions that allow a team of people at different locations to work together on projects and share project information. Workgroup functions allow the exchange of messages through e-mail, an intranet, wireless devices, or the Web. For example, you can use Project 2007 to alert members about new or changed task assignments, and members can return status information and notify other workgroup members about changes in the schedule or other project parameters.



Microsoft Office Enterprise Project Management (EPM) Solution and similar products also provides the following tools to enhance communications:

- *Portfolio management*: By providing a centralized and consolidated view of programs and projects, the user can evaluate and prioritize activities across the organization. This feature makes it possible to maximize productivity, minimize costs, and keep activities aligned with strategic objectives.
- *Resource management*: Maximizing human resources is often the key to minimizing project costs. This feature enables the user to maximize resource use across the organization to help effectively plan and manage the workforce.
- *Project collaboration*: Sharing project information is often a haphazard endeavor. Project collaboration enables an organization to share knowledge immediately and consistently to improve communications and decision-making, eliminate redundancies, and take advantage of best practices for project management.

Communication is among the more important factors for success in project management. While technology can aid in the communications process and be the easiest aspect of the process to address, it is not the most important. Far more important is improving an organization's ability to communicate. Improving the ability to communicate often requires a cultural change in an organization that takes a lot of time, hard work, and patience. Information technology personnel, in particular, often need special coaching to improve their communications skills. The project manager's chief role in the communications process is that of facilitator. Project managers must educate all stakeholders—management, team members, and customers—on the importance of good project communications and ensure that the project has a communications management plan to help make good communication happen.

## CASE WRAP-UP

Christine Braun worked closely with Peter Gumpert and his project managers to develop a communications management plan for all of the Fiber-optic Undersea Telecommunications projects. Peter was very skilled at running effective meetings, so everyone focused on meeting specific objectives. Peter stressed the importance of keeping himself, the project managers, and other major stakeholders informed about the status of all projects. He emphasized that the project managers were in charge of their projects, and that he did not intend to tell them how to do their jobs. He just wanted to have accurate and consistent information to help coordinate all of the projects and make everyone's jobs easier. When some of the project managers balked at the additional work of providing more project information in different formats, Peter openly discussed the issues with them in more detail. He then authorized each project manager to use additional staff to help develop and follow standards for all project communications.

Christine used her strong technical and communications skills to create a Web site that included samples of important project documents, presentations, and templates for other people to download and use on their own projects. After determining the need for more remote communications and collaboration between projects, Christine and other staff members researched the latest hardware and software products. Peter authorized funds for a new project led by Christine to evaluate and then purchase several wireless handheld devices and enterprise project management software with wiki capability that could be accessed via the Web. All managers and technical staff received their own devices, and any project stakeholder could check out one of these handheld devices and get one-on-one training on how to use it with the new Web-based software. Even Peter learned how to use one and doesn't know how he got along without it.

## Chapter Summary

Failure to communicate is often the greatest threat to the success of any project, especially information technology projects. Communication is the oil that keeps a project running smoothly. Project communications management involves identifying stakeholders, planning communications, distributing information, managing stakeholder expectations, and reporting performance.

It is important to identify stakeholders and determine strategies for managing relationships with them in order to satisfy their needs and expectations. A stakeholder register and a stakeholder management strategy are key outputs of this process.

A communications management plan of some type should be created for all projects to help ensure good communications. Contents of this plan will vary based on the needs of the project.



The various methods for distributing project information include formal, informal, written, and verbal. It is important to determine the most appropriate means for distributing different types of project information. Project managers and their teams should focus on the importance of building relationships as they communicate project information. As the number of people that need to communicate increases, the number of communications channels also increases.

Reporting performance involves collecting and disseminating information about how well a project is moving toward meeting its goals. Project teams can use earned value charts and other forms of progress information to communicate and assess project performance. Status review meetings are an important part of communicating, monitoring, and controlling projects.

To improve project communications, project managers and their teams must develop good conflict management skills, as well as other communication skills. Conflict resolution is an important part of project communications management. The main causes of conflict during a project are schedules, priorities, staffing, technical opinions, procedures, cost, and personalities. A confrontational or problem-solving approach to managing conflict is often the best approach. Other suggestions for improving project communications include learning how to run more effective meetings, how to use e-mail, instant messaging, and collaborative software more effectively, and how to use templates for project communications.

New hardware and software continues to become available to help improve communications. As more people work remotely, it is important to make sure they have the necessary tools to be productive. Enterprise project management software provides many features to enhance communications across the organization.

## Quick Quiz

1. What do many experts agree is the greatest threat to the success of any project?
   a. lack of proper funding
   b. a failure to communicate
   c. poor listening skills
   d. inadequate staffing

2. Which communication skill is most important for information technology professionals for career advancement?
   a. writing
   b. listening
   c. speaking
   d. using communication technologies
3. Which of the following is not a process in project communications management?
   a. information planning
   b. information distribution
   c. performance reporting
   d. managing stakeholders
4. What popular book illustrates the concept that people are not interchangeable parts and uses the analogy that you cannot take nine women and produce a baby in one month?
   a. Covey's *7 Habits of Highly Effective People*
   b. Goldratt's *Critical Chain*
   c. Gates's *Business @ the Speed of Thought*
   d. Brooks's *The Mythical Man-Month*
5. If you add three more people to a project team of five, how many more communications channels will you add?
   a. 2
   b. 12
   c. 15
   d. 18
6. A ________________ report describes where the project stands at a specific point in time.
   a. status
   b. performance
   c. forecast
   d. earned value
7. What tool can you use to help manage stakeholders by ranking scope, time, and cost goals in order of importance and provide guidelines on balancing these constraints?
   a. triple constraint matrix
   b. expectations matrix
   c. issue log
   d. priority log

8. You have two project stakeholders who do not get along at all. You know they both enjoy traveling, so you discuss great travel destinations when they are both in the room together to distract them from arguing with each other. What conflict-handling mode are you using?
    a. confrontation
    b. compromise
    c. smoothing
    d. withdrawal
9. Which of the following is not a guideline to help improve time spent at meetings?
    a. Determine if a meeting can be avoided.
    b. Invite extra people who support your project to make it run more smoothly.
    c. Define the purpose and intended outcome of the meeting.
    d. Build relationships.
10. A ________________ report is a reflective statement documenting important things that people learned from working on the project.
    a. final project
    b. lessons-learned
    c. project archive
    d. progress

### Quick Quiz Answers

1. b; 2. c; 3. a; 4. d; 5. d; 6. a; 7. b; 8. c; 9. b; 10. b

## Discussion Questions

1. Think of examples in the media that poke fun at the communications skills of technical professionals, such as Dilbert® cartoons. How does poking fun at technical professionals' communications skills influence the industry and educational programs?
2. Discuss the use of a stakeholder register and a stakeholder management strategy to assist in managing stakeholders.
3. What items should a communications management plan address? How can a stakeholder analysis assist in preparing and implementing parts of this plan?
4. Discuss the advantages and disadvantages of different ways of distributing project information.
5. What are some of the ways to create and distribute project performance information?
6. How can an expectations management matrix help a project manager in making important decisions?

7. Explain why you agree or disagree with some of the suggestions provided in this chapter for improving project communications, such as creating a communications management plan, stakeholder analysis, or performance reports, for example. What other suggestions do you have?
8. How can software assist in project communications? How can it hurt project communications?

## Exercises



1. Create a stakeholder management strategy using Table 10-1 as a guide. Assume your organization has a project to determine employees' training needs and then provide in-house and external sources for courses in developing communications skills for employees. Stakeholders could be various levels and types of employees, suppliers, the Human Resources department in charge of the project, and so on. Determine at least five specific stakeholders for the project, and be creative in developing your potential management strategies
2. Create a stakeholder communications strategy using the information from Exercise 1. List at least one type of project communication for each stakeholder and the format for disseminating information to each. Use Table 10-2 as an example.
3. Review the following scenarios, and then write a paragraph for each one describing what media you think would be most appropriate to use and why. See Table 10-3 for suggestions.
    a. Many of the technical staff on the project come in from 9:30 a.m. to 10:00 a.m. while the business users always come in before 9:00 a.m. The business users have been making comments. The project manager wants to have the technical people come in by 9:00, although many of them leave late.
    b. Your company is bidding on a project for the entertainment industry. You know that you need new ideas on how to put together the proposal and communicate your approach in a way that will impress the customer.
    c. Your business has been growing successfully, but you are becoming inundated with phone calls and e-mails asking similar types of questions.
    d. You need to make a general announcement to a large group of people and want to make sure they get the information.
4. How many different communications channels does a project team with six people have? How many more communications channels would there be if the team grew to ten people?
5. Review the templates for various project documents provided in this chapter. Pick one of them and apply it to a project of your choice. Make suggestions for improving the template.
6. Write a lessons-learned report for a project of your choice using the template provided on the companion Web site and sample in Chapter 3 as guides. Do you think it is important for all project managers and team members to write lessons-learned reports? Would you take the time to read them if they were available in your organization? Why or why not?
7. Research new software products that assist in communications management for large projects. Write a two-page paper summarizing your findings. Include Web sites for software vendors and your opinion of some of the products.

## Running Case

Several issues have arisen on the Recreation and Wellness Intranet Project. The person from the HR department supporting the project left the company, and the team really needs more support from that group. One of the members of the user group supporting the project is extremely vocal and hard to work with, and other users can hardly get a word in at meetings. The project manager, Tony, is getting weekly status reports from all of his team members, but many of them do not address challenges people are obviously facing. The team is having difficulties deciding how to communicate various project reports and documents and where to store all of the information being generated. Recall that the team members include you, a programmer/analyst and aspiring project manager; Patrick, a network specialist; Nancy, a business analyst; and Bonnie, another programmer/analyst.



1. Create a stakeholder management strategy for the project. Include at least four stakeholders. Be creative in developing potential management strategies.
2. Prepare a partial communications management plan to address some of the challenges mentioned in #1.
3. Prepare a template and sample of a good weekly progress report that could be used for this project. Include a list of tips to help team members provide information on these reports.
4. Write a one-page paper describing two suggested approaches to managing the conflict presented by the hard-to-work-with user.

## Companion Web Site

Visit the companion Web site for this text (*www.cengage.com/mis/schwalbe*) to access:

- References cited in the text and additional suggested readings for each chapter
- Template files
- Lecture notes
- Interactive quizzes
- Podcasts
- Links to general project management Web sites
- And more

See the Preface of this text for additional information on accessing the companion Web site.

## Key Terms

**blogs** — easy to use journals on the Web that allow users to write entries, create links, and upload pictures, while readers can post comments to journal entries

**collaborating mode** — a conflict-handling mode where decision makers incorporate different viewpoints and insights to develop consensus and commitment

**communications management plan** — a document that guides project communications

**compromise mode** — using a give-and-take approach to resolving conflicts; bargaining and searching for solutions that bring some degree of satisfaction to all the parties in a dispute

**confrontation mode** — directly facing a conflict using a problem-solving approach that allows affected parties to work through their disagreements

**expectations management matrix** — a tool to help understand unique measures of success for a particular project

**forcing mode** — using a win-lose approach to conflict resolution to get one's way

**forecasts** — used to predict future project status and progress based on past information and trends

**Google Docs** — online applications offered by Google that allow users to create, share, and edit documents, spreadsheets, and presentations online

**groupthink** — conformance to the values or ethical standards of a group

**issue** — a matter under question or dispute that could impede project success

**issue log** — a tool to document and monitor the resolution of project issues

**lessons-learned report** — reflective statements written by project managers and their team members to document important things they have learned from working on the project

**progress reports** — reports that describe what the project team has accomplished during a certain period of time

**project archives** — a complete set of organized project records that provide an accurate history of the project

**SharePoint portal** — allows users to create custom Web sites to access documents and applications stored on shared devices

**smoothing mode** — deemphasizing or avoiding areas of differences and emphasizing areas of agreements

**stakeholder register** — a public document that includes details related to the identified project stakeholders

**stakeholder management strategy** — an approach to help increase the support of stakeholders throughout the project

**status reports** — reports that describe where the project stands at a specific point in time

**wiki** — a Web site that has a page or pages designed to enable anyone who accesses it to contribute or modify content

**withdrawal mode** — retreating or withdrawing from an actual or potential disagreement



## End Notes


[1] Marcos P. Sivitanides, James R. Cook, Roy B. Martin, and Beverly A. Chiodo, "Verbal Communication Skills Requirements for Information Systems Professionals," *Journal of Information Systems Education* 7, no. 1 (Spring 1995), pp. 38–43.

[2] Mark E. McMurtrey, James P. Downey, Steven M. Zeltmann, and William H. Friedman, "Critical Skill Sets of Entry-Level IT Professionals: An Empirical Examination of Perceptions from Field Personnel," *Journal of Information Technology Education* 7 (2008), p. 116.

[3] *Reuters*, "Asia Quakes Damage Cables; Internet, Banks Affected," (December 27, 2006).

[4] Ying Lou, "Initial repairs to quake-damaged cables pushed back by one week," *Bloomberg News* (December 31, 2006).

[5] Albert Mehrabian, *Silent messages: Implicit communication of emotions and attitudes*, 2nd ed. (Belmont, CA: Wadsworth, 1981).

[6] Marion Walker, "E-mail is out at this office, at least on Fridays," *Minneapolis Star Tribune* (November 10, 2004).

[7] Kirkpatrick, David, "How one CEO Facebooked his company," *Fortune* (June 13, 2008).

[8] Ibid.

[9] "New Research Reveals Collaboration Is a Key Driver of Business Performance Around the World: Verizon Business, Microsoft sponsor international study; create first-of-its-kind collaboration index to measure impact of communications culture, technologies," *Microsoft PressPass* (June 5, 2006).

[10] Frederick P. Brooks, *The Mythical Man-Month*, 2nd ed. (Boston: Addison-Wesley, 1995).



[11] Carol Hildenbrand, "Loud and Clear," *CIO Magazine* (April 15, 1996).

[12] "Constructive Team Conflict" *Wharton Leadership Digest* 1, no. 6 (March 1997).

[13] Peter Saalfield, "Internet misuse costs businesses $178 billion annually," *IDG News Service* (July 19, 2005).

[14] Louisa Hearn, "Study probes web habits of office slackers," *Sydney Morning Herald* (January 17, 2006).

[15] Ibid.

[16] Nancy Gohring, "Over 50% of companies have fired workers for e-mail, Net abuse," *ComputerWorld Security* (February 28, 2008).

[17] John Petroske, "Promoting Project Communication Using Wikis," ProMAC 2008 Conference Proceedings (September 2008).

[18] Michael Cooney, "Telecommute. Kill a Career?" *Network World* (January 17, 2007).

CHAPTER 11

# PROJECT RISK MANAGEMENT

## LEARNING OBJECTIVES

**After reading this chapter, you will be able to:**

- Understand what risk is and the importance of good project risk management
- Discuss the elements involved in risk management planning and the contents of a risk management plan
- List common sources of risks on information technology projects
- Describe the process of identifying risks and be able to create a risk register
- Discuss the qualitative risk analysis process and explain how to calculate risk factors, create probability/impact matrixes, and apply the Top Ten Risk Item Tracking technique to rank risks
- Explain the quantitative risk analysis process and how to apply decision trees, simulation, and sensitivity analysis to quantify risks
- Provide examples of using different risk response planning strategies to address both negative and positive risks
- Discuss what is involved in monitoring and controlling risks
- Describe how software can assist in project risk management

## OPENING CASE

Cliff Branch was the president of a small information technology consulting firm that specialized in developing Internet applications and providing full-service support. The staff consisted of programmers, business analysts, database specialists, Web designers, project managers, and so on. The firm had 50 full-time people and planned to hire at least ten more in the next year. It also planned to increase the number of part-time consultants they used. The company had done very well the past few years, but was recently having difficulty winning contracts. Spending time and resources to respond to various requests for proposals from prospective clients was becoming expensive. Many clients were starting to require presentations and even some prototype development before awarding a contract.

Cliff knew he had an aggressive approach to risk and liked to bid on the projects with the highest payoff. He did not use a systematic approach to evaluate the risks involved in various projects before bidding on them. He focused on the profit potentials and on how challenging the projects were. His strategy was now causing problems for the company because it was investing heavily in the preparation of proposals, yet winning few contracts. Several employees, who were not currently working on projects, were still on the payroll, and some of their part-time consultants were actively pursuing other opportunities since they were being underutilized. What could Cliff and his company do to get a better understanding of project risks? Should Cliff adjust his strategy for deciding what projects to pursue? How?



## THE IMPORTANCE OF PROJECT RISK MANAGEMENT

Project risk management is the art and science of identifying, analyzing, and responding to risk throughout the life of a project and in the best interests of meeting project objectives. A frequently overlooked aspect of project management, risk management can often result in significant improvements in the ultimate success of projects. Risk management can have a positive impact on selecting projects, determining the scope of projects, and developing realistic schedules and cost estimates. It helps project stakeholders understand the nature of the project, involves team members in defining strengths and weaknesses, and helps to integrate the other project management knowledge areas.

Good project risk management often goes unnoticed, unlike crisis management. With crisis management, there is an obvious danger to the success of a project. The crisis, in turn, receives the intense interest of the entire project team. Resolving a crisis has much greater visibility, often accompanied by rewards from management, than successful risk management. In contrast, when risk management is effective, it results in fewer problems, and for the few problems that exist, it results in more expeditious resolutions. It may be difficult for outside observers to tell whether risk management or luck was responsible for the smooth development of a new system, but project teams will always know that their projects worked out better because of good risk management. Managing project risks takes dedicated, talented professionals. In response to this need, PMI introduced the PMI Risk Management Professional (PMI-RMP)$^{SM}$ credential in 2008. (Consult PMI's Web site for further information.)

All industries, especially the software development industry, tend to underestimate the importance of project risk management. William Ibbs and Young H. Kwak performed a study to assess project management maturity. The 38 organizations participating in the study were divided into four industry groups: engineering and construction, telecommunications, information systems/software development, and high-tech manufacturing. Survey participants answered 148 multiple-choice questions to assess how mature their organization was in the project management knowledge areas of scope, time, cost, quality, human resources, communications, risk, and procurement. The rating scale ranged from 1 to 5, with 5 being the highest maturity rating. Table 11-1 shows the results of the survey. Notice that risk management was the only knowledge area for which all ratings were less than 3. This study shows that all organizations should put more effort into project risk management, especially the information systems/software development industry, which had the lowest rating of 2.75 (emphasized in bold in Table 11-1).[1]



**TABLE 11-1** Project management maturity by industry group and knowledge area

| KEY: 1 = Lowest Maturity Rating | | 5 = Highest Maturity Rating | | |
|---|---|---|---|---|
| Knowledge Area | Engineering/ Construction | Tele-communications | Information Systems | Hi-Tech Manufacturing |
| *Scope* | 3.52 | 3.45 | 3.25 | 3.37 |
| *Time* | 3.55 | 3.41 | 3.03 | 3.50 |
| *Cost* | 3.74 | 3.22 | 3.20 | 3.97 |
| *Quality* | 2.91 | 3.22 | 2.88 | 3.26 |
| *Human resources* | 3.18 | 3.20 | 2.93 | 3.18 |
| *Communications* | 3.53 | 3.53 | 3.21 | 3.48 |
| ***Risk*** | **2.93** | **2.87** | **2.75** | **2.76** |
| *Procurement* | 3.33 | 3.01 | 2.91 | 3.33 |

A similar survey was completed with software development companies in Mauritius, South Africa in 2003. The average maturity rating was only 2.29 for all knowledge areas, on a scale of 1-5, with 5 being the highest maturity rating. The lowest maturity rating in this study was also in the area of project risk management, with an average maturity rating of only 1.84. Cost management had the highest maturity rating of 2.5, and the authors of the survey noted that organizations in the study were often concerned with cost overruns and had metrics in place to help control costs. The authors also found that maturity rating was closely linked to the success rate of projects, and they noted the fact that the poor rating for risk management was a likely cause of project problems/failures.[2]

KLCI Research Group surveyed 260 software organizations worldwide in 2001 to study software risk management practices. Below are some of their findings:

- 97 percent of the participants said they had procedures in place to identify and assess risk.
- 80 percent identified anticipating and avoiding problems as the primary benefit of risk management.
- 70 percent of the organizations had defined software development processes.
- 64 percent had a Project Management Office.

Figure 11-1 shows the main benefits from software risk management practices cited by survey respondents. In addition to anticipating/avoiding problems, risk management practices helped software project managers prevent surprises, improve negotiations, meet customer commitments, and reduce schedule slips and cost overruns.[3]



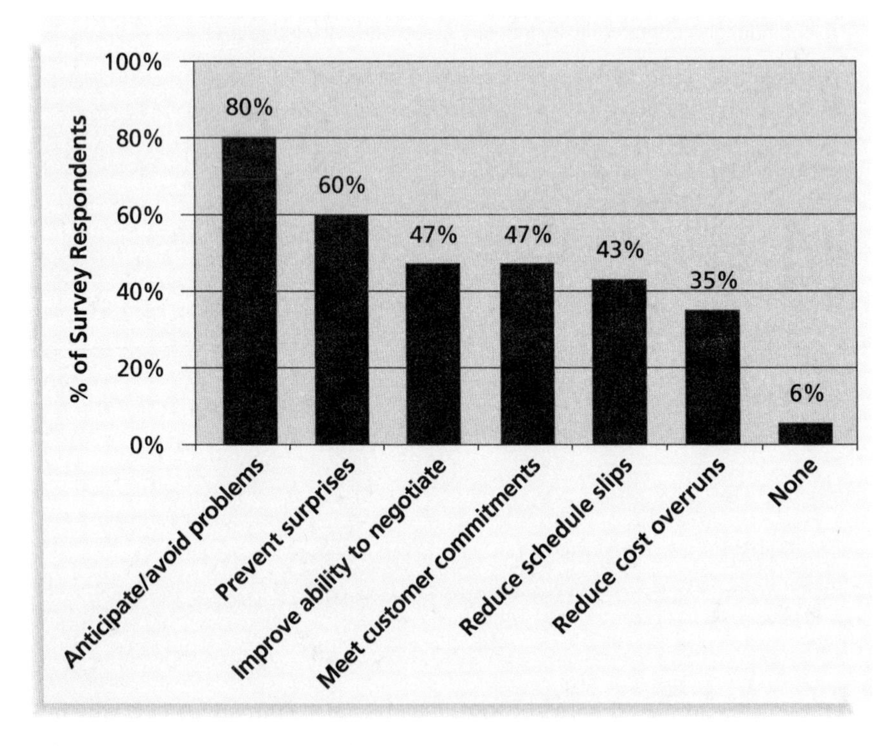


**FIGURE 11-1** Benefits from software risk management practices

Although many organizations know that they do not do a good job of managing project risk, little progress seems to have been made in improving risk management on a project level or an enterprise level. Several books and articles have been written on the topic in recent years. For example, Dr David Hilson, PMP, wrote an article about the importance of project risk management shortly after the stock market declines in the fall of 2008. Hillson says,

"There is no doubt that all sectors of industry and society are facing real challenges in coping with the current fallout from the credit crunch. But risk management should not be regarded as a nonessential cost to be cut in these difficult times. Instead, organisations should use the insights offered by the risk process to ensure that they can handle the inevitable uncertainties and emerge in the best possible position in [the] future. With high levels of volatility surrounding us on all sides, risk management is more needed now than ever, and cutting it would be a false economy. Rather than treating risk management as part of the problem, we should see it as a major part of the solution."[4]

## MEDIA SNAPSHOT

Many people around the world suffered from financial losses as various financial markets dropped in the fall of 2008, even after the $700 billion bailout bill was passed by the U.S. Congress. According to a global survey of 316 financial services executives conducted in July 2008, over 70 percent of respondents believed that the losses stemming from the credit crisis were largely due to failures to address risk management issues. They identified several challenges in implementing risk management, including data and company culture issues. For example, access to relevant, timely and consistent data continues to be a major obstacle in many organizations. Many respondents also said that fostering a culture of risk management was a major challenge.

Executives and lawmakers finally started paying attention to risk management. Fifty-nine percent of survey respondents said the credit crisis prompted them to scrutinize their risk management practices in greater detail, and many institutions are revisiting their risk management practices. The Financial Stability Forum (FSF) and the Institute for International Finance (IIF) are now calling for closer scrutiny of the risk management process.

Rodney Nelsestuen, an analyst for TowerGroup, agrees. "Enterprise risk management has taken on new importance as stockholders, boards of directors and regulators demand better, more timely analysis of risk and a deeper understanding of how the institution is impacted by the dynamic risk environment of a global financial community."[5]



Before you can improve project risk management, you must understand what risk is. A basic dictionary definition says that risk is "the possibility of loss or injury." This definition highlights the negativity often associated with risk and suggests that uncertainty is involved. Project risk management involves understanding potential problems that might occur on the project and how they might impede project success. The *PMBOK® Guide, Fourth Edition* refers to this type of risk as a negative risk or threat. However, there are also positive risks or opportunities, which can result in good things happening on a project. A general definition of a project **risk**, therefore, is an uncertainty that can have a negative or positive effect on meeting project objectives.

In many respects, negative risk management is like a form of insurance. It is an activity undertaken to lessen the impact of potentially adverse events on a project. Positive risk management is like investing in opportunities. It is important to note that risk management *is* an investment—there are costs associated with it. The investment an organization is willing to make in risk management activities depends on the nature of the project, the experience of the project team, and the constraints imposed on both. In any case, the cost for risk management should not exceed the potential benefits.

If there is so much risk in information technology projects, why do organizations pursue them? Many companies are in business today because they took risks that created great opportunities. Organizations survive over the long term when they pursue opportunities. Information technology is often a key part of a business's strategy; without it, many businesses might not survive. Given that all projects involve uncertainties that can have negative or positive outcomes, the question is how to decide which projects to pursue and how to identify and manage project risk throughout a project's life cycle.

## BEST PRACTICE

Some organizations make the mistake of only addressing tactical and negative risks when performing project risk management. David Hillson (*www.risk-doctor.com*) suggests overcoming this problem by widening the scope of risk management to encompass both *strategic risks* and *upside opportunities*, which he refers to as integrated risk management. Benefits of this approach include:

- Bridging the strategy and tactics gap to ensure that project delivery is tied to organizational needs and vision
- Focusing projects on the benefits they exist to support, rather than producing a set of deliverables
- Managing opportunities proactively as an integral part of business processes at both strategic and tactical levels
- Providing useful information to decision-makers at all levels when the environment is uncertain
- Allowing an appropriate level of risk to be taken intelligently with full awareness of the degree of uncertainty and its potential effects on objectives.[6]

Several risk experts suggest that organizations and individuals strive to find a balance between risks and opportunities in all aspects of projects and their personal lives. The idea of striving to balance risks and opportunities suggests that different organizations and people have different tolerances for risk. Some organizations or people have a neutral tolerance for risk, some have an aversion to risk, and others are risk-seeking. These three preferences for risk are part of the utility theory of risk.

**Risk utility** or **risk tolerance** is the amount of satisfaction or pleasure received from a potential payoff. Figure 11-2 shows the basic difference between risk-averse, risk-neutral, and risk-seeking preferences. The y-axis represents utility, or the amount of pleasure received from taking a risk. The x-axis shows the amount of potential payoff, opportunity, or dollar value of the opportunity at stake. Utility rises at a decreasing rate for a **risk-averse** person. In other words, when more payoff or money is at stake, a person or organization that is risk-averse gains less satisfaction from the risk, or has lower tolerance for the risk. Those who are **risk-seeking** have a higher tolerance for risk, and their satisfaction increases when more payoff is at stake. A risk-seeking person prefers outcomes that are more uncertain and is often willing to pay a penalty to take risks. A **risk-neutral** person achieves a balance between risk and payoff.

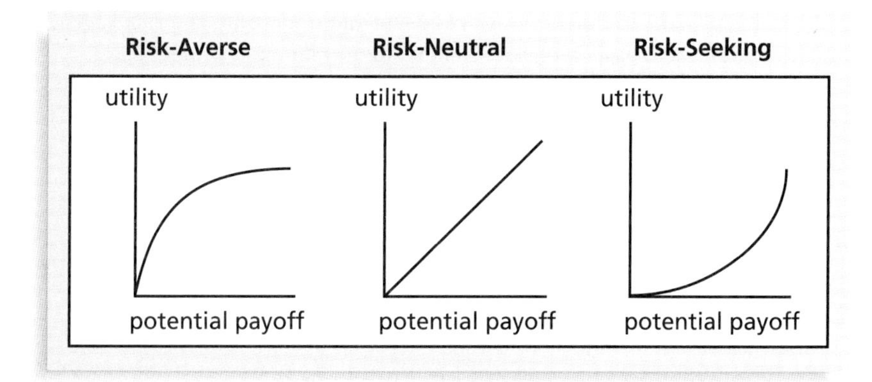




**FIGURE 11-2** Risk utility function and risk preference

The goal of project risk management can be viewed as minimizing potential negative risks while maximizing potential positive risks. The term **known risks** is sometimes used to describe risks that the project team have identified and analyzed. Known risks can be managed proactively. However, **unknown risks**, or risks that have not been identified and analyzed, cannot be managed. As you can imagine, good project managers know it is good practice to take the time to identify and manage project risks. There are six major processes involved in risk management:

1. *Planning risk management* involves deciding how to approach and plan the risk management activities for the project. By reviewing the project scope statement; cost, schedule, and communications management plans; enterprise environmental factors; and organizational process assets, project teams can discuss and analyze risk management activities for their particular projects. The main output of this process is a risk management plan.
2. *Identifying risks* involves determining which risks are likely to affect a project and documenting the characteristics of each. The main output of this process is the start of a risk register, as described in more detail later in this chapter.
3. *Performing qualitative risk analysis* involves prioritizing risks based on their probability and impact of occurrence. After identifying risks, project teams can use various tools and techniques to rank risks and update information in the risk register. The main output is risk register updates.
4. *Performing quantitative risk analysis* involves numerically estimating the effects of risks on project objectives. The main output of this process is also risk register updates.
5. *Planning risk responses* involves taking steps to enhance opportunities and reduce threats to meeting project objectives. Using outputs from the preceding risk management processes, project teams can develop risk response strategies that often result in updates to the risk register, project management plan, and other project documents as well as risk-related contract decisions.
6. *Monitoring and controlling risk* involves monitoring identified and residual risks, identifying new risks, carrying out risk response plans, and evaluating the effectiveness of risk strategies throughout the life of the project. The main

outputs of this process include change requests and updates to the risk register, organizational process assets, project management plan, and project documents.

Figure 11-3 summarizes these processes and outputs, showing when they occur in a typical project.

**Planning**
Process: **Plan risk management**
Output: Risk management plan
Process: **Identify risks**
Output: Risk register
Process: **Perform qualitative risk analysis**
Output: Risk register updates
Process: **Perform quantitative risk analysis**
Output: Risk register updates
Process: **Plan risk responses**
Outputs: Risk register updates, risk-related contract decisions, project management plan updates, project document updates

**Monitoring and Controlling**
Process: **Monitor and control risks**
Outputs: Risk register updates, organizational process assets updates, change requests, project management plan updates, project document updates

**Project Start** **Project Finish**



**FIGURE 11-3** Project risk management summary

The first step in project risk management is deciding how to address this knowledge area for a particular project by performing risk management planning.

## PLANNING RISK MANAGEMENT

Planning risk management is the process of deciding how to approach and plan for risk management activities for a project, and the main output of this process is a risk management plan. A **risk management plan** documents the procedures for managing risk throughout the project. Project teams should hold several planning meetings early in the project's life cycle to help develop the risk management plan. The project team should review project documents as well as corporate risk management policies, risk categories, lessons-learned reports from past projects, and templates for creating a risk management plan. It is also important to review the risk tolerances of various stakeholders. For example, if the project sponsor is risk-averse, the project might require a different approach to risk management than if the project sponsor were a risk seeker.

A risk management plan summarizes how risk management will be performed on a particular project. Like other specific knowledge area plans, it becomes a subset of the project management plan. Table 11-2 lists the general topics that a risk management plan should address. It is important to clarify roles and responsibilities, prepare budget and schedule estimates for risk-related work, and identify risk categories for consideration. It is also important to describe how risk management will be done, including assessment of risk probabilities and impacts as well as the creation of risk related documentation. The level of detail included in the risk management plan can vary with the needs of the project.

**TABLE 11-2** Topics addressed in a risk management plan

| Topic | Questions to Answer |
|---|---|
| Methodology | How will risk management be performed on this project? What tools and data sources are available and applicable? |
| Roles and responsibilities | Who are the individuals responsible for implementing specific tasks and providing deliverables related to risk management? |
| Budget and schedule | What are the estimated costs and schedules for performing risk-related activities? |
| Risk categories | What are the main categories of risks that should be addressed on this project? Is there a risk breakdown structure for the project? (See the information on risk breakdown structures later in this section.) |
| Risk probability and impact | How will the probabilities and impacts of risk items be assessed? What scoring and interpretation methods will be used for the qualitative and quantitative analysis of risks? How will the probability and impact matrix be developed? |
| Revised stakeholders' tolerances | Have stakeholders' tolerances for risk changed? How will those changes affect the project? |
| Tracking | How will the team track risk management activities? How will lessons learned be documented and shared? How will risk management processes be audited? |
| Risk documentation | What reporting formats and processes will be used for risk management activities? |



In addition to a risk management plan, many projects also include contingency plans, fallback plans, and contingency reserves.

- **Contingency plans** are predefined actions that the project team will take if an identified risk event occurs. For example, if the project team knows that a new release of a software package may not be available in time for them to use it for their project, they might have a contingency plan to use the existing, older version of the software.
- **Fallback plans** are developed for risks that have a high impact on meeting project objectives, and are put into effect if attempts to reduce the risk are not effective. For example, a new college graduate might have a main plan and

several contingency plans on where to live after graduation, but if none of those plans works out, a fallback plan might be to live at home for a while. Sometimes the terms contingency plan and fallback plan are used interchangeably.

- **Contingency reserves** or **contingency allowances** are provisions held by the project sponsor or organization to reduce the risk of cost or schedule overruns to an acceptable level. For example, if a project appears to be off course because the staff is inexperienced with some new technology and the team had not identified that as a risk, the project sponsor may provide additional funds from contingency reserves to hire an outside consultant to train and advise the project staff in using the new technology.

Before you can really understand and use the other project risk management processes on information technology projects, it is necessary to recognize and understand the common sources of risk.



## COMMON SOURCES OF RISK ON INFORMATION TECHNOLOGY PROJECTS

Several studies have shown that information technology projects share some common sources of risk. For example, the Standish Group did a follow-up study to the CHAOS research, which they called Unfinished Voyages. This study brought together 60 information technology professionals to elaborate on how to evaluate a project's overall likelihood of being successful. Table 11-3 shows the Standish Group's success potential scoring sheet and the relative importance of the project success criteria factors. If a potential project does not

**TABLE 11-3** Information technology success potential scoring sheet

| Success Criterion | Relative Importance |
| --- | --- |
| User involvement | 19 |
| Executive management support | 16 |
| Clear statement of requirements | 15 |
| Proper planning | 11 |
| Realistic expectations | 10 |
| Smaller project milestones | 9 |
| Competent staff | 8 |
| Ownership | 6 |
| Clear visions and objectives | 3 |
| Hardworking, focused staff | 3 |
| **Total** | **100** |

receive a minimum score, the organization might decide not to work on it or to take actions to reduce the risks before it invests too much time or money.[7]

The Standish Group provides specific questions for each success criterion to help decide the number of points to assign to a project. For example, the five questions related to user involvement include the following:

- Do I have the right user(s)?
- Did I involve the user(s) early and often?
- Do I have a quality relationship with the user(s)?
- Do I make involvement easy?
- Did I find out what the user(s) need(s)?

The number of questions corresponding to each success criterion determines the number of points each positive response is assigned. For example, in the case of user involvement there are five questions. For each positive reply, you would get 3.8 (19/5) points; 19 represents the weight of the criterion, and 5 represents the number of questions. Therefore, you would assign a value to the user involvement criterion by adding 3.8 points to the score for each question you can answer positively.



Many organizations develop their own risk questionnaires. Broad categories of risks described on these questionnaires might include:

- *Market risk*: If the information technology project is to produce a new product or service, will it be useful to the organization or marketable to others? Will users accept and use the product or service? Will someone else create a better product or service faster, making the project a waste of time and money?
- *Financial risk*: Can the organization afford to undertake the project? How confident are stakeholders in the financial projections? Will the project meet NPV, ROI, and payback estimates? If not, can the organization afford to continue the project? Is this project the best way to use the organization's financial resources?
- *Technology risk*: Is the project technically feasible? Will it use mature, leading edge, or bleeding edge technologies? When will decisions be made on which technology to use? Will hardware, software, and networks function properly? Will the technology be available in time to meet project objectives? Could the technology be obsolete before a useful product can be produced? You can also break down the technology risk category into hardware, software, and network technology, if desired.
- *People risk*: Does the organization have or can they find people with appropriate skills to complete the project successfully? Do people have the proper managerial and technical skills? Do they have enough experience? Does senior management support the project? Is there a project champion? Is the organization familiar with the sponsor/customer for the project? How good is the relationship with the sponsor/customer?
- *Structure/process risk*: What is the degree of change the new project will introduce into user areas and business procedures? How many distinct user groups does the project need to satisfy? With how many other systems does the new project/system need to interact? Does the organization have processes in place to complete the project successfully?

KPMG, a large consulting firm, published a study in 1995 that found that 55 percent of **runaway projects** (i.e., projects with significant cost or schedule overruns) did *no* risk management at all, that 38 percent did some (but half did not use their risk findings after the project was underway), and that 7 percent did not know whether they did risk management or not.[8] This study suggests that performing risk management is important to improving the likelihood of project success and preventing runaway projects.

The timing of risk management is also an important consideration. For example, Cincinnati, Ohio–based Comair is a regional airline that operates in 117 cities and carries about 30,000 passengers on 1,130 flights a day. Comair's IT managers knew in the late 1990s that they had to address the replacement of an aging legacy system used to manage flight crews. The application was one of the oldest in the company (11 years old at the time), written in Fortran (code that no one at Comair was fluent in) and the only system left that ran on the airline's old IBM AIX platform. Although managers and crew addressed possible options for replacing the system, they kept putting it off as other priorities emerged. A replacement system was finally approved in 2004, but the switch didn't happen soon enough. "Over the holidays, the legacy system failed, bringing down the entire airline, canceling or delaying 3,900 flights, and stranding nearly 200,000 passengers. The network crash cost Comair and its parent company, Delta Air Lines, $20 million, damaged the airline's reputation and prompted an investigation by the Department of Transportation." Had Comair or Delta acted sooner, they could have taken steps to mitigate that risk to avoid the disaster.[9]



Reviewing a proposed project in terms of the Standish Group's success criteria, a risk questionnaire, or any other similar tool is a good method for understanding common sources of risk on information technology projects. It is also useful to review the work breakdown structure (WBS) for a project to see if there might be specific risks by WBS categories. For example, if one item on the WBS involves preparing a press release and no one on the project team has ever done that, it could be a negative risk if it is not handled professionally.

A risk breakdown structure is a useful tool to help project managers consider potential risks in different categories. Similar in structure to a work breakdown structure, a **risk breakdown structure** is a hierarchy of potential risk categories for a project. Figure 11-4 shows a sample risk breakdown structure that might apply to many information technology projects. The highest-level categories are business, technical, organizational, and project management. Competitors, suppliers, and cash flow are categories that fall under business risks. Under technical risks are the categories of hardware, software, and network. Notice how the risk breakdown structure provides a simple, one-page chart to help ensure a project team is considering important risk categories related to all information technology projects. For example, Cliff and his managers in the opening case could have benefited from considering several of the categories listed under project management—estimates, communication, and resources. They could have discussed these and other types of risks related to the projects their company bid on and developed appropriate strategies for optimizing positive risks and minimizing negative ones.

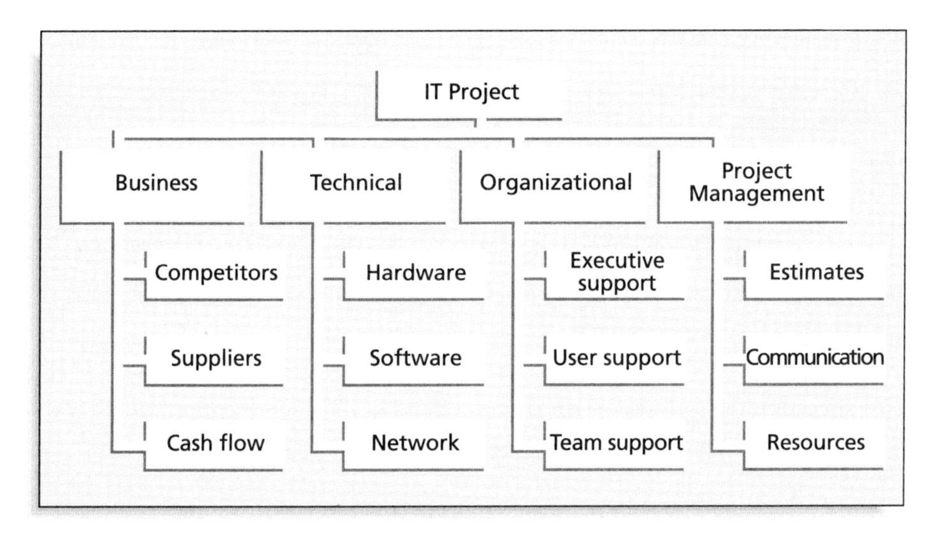


**FIGURE 11-4** Sample risk breakdown structure



In addition to identifying risk based on the nature of the project or products produced, it is also important to identify potential risks according to project management knowledge areas, such as scope, time, cost, and quality. Notice that one of the major categories in the risk breakdown structure in Figure 11-4 is project management. Table 11-4 lists potential negative risk conditions that can exist within each knowledge area.[10]

**TABLE 11-4** Potential negative risk conditions associated with each knowledge area

| Knowledge Area | Risk Conditions |
|---|---|
| ***Integration*** | Inadequate planning; poor resource allocation; poor integration management; lack of post-project review |
| ***Scope*** | Poor definition of scope or work packages; incomplete definition |
| ***Time*** | Errors in estimating time or resource availability; errors in determining the critical path; poor allocation and management of float; early release of competitive products |
| ***Cost*** | Estimating errors; inadequate productivity, cost, change, or contingency |
| ***Quality*** | Poor attitude toward quality; substandard design/materials/ workmanship; inadequate quality assurance program |
| ***Human resource*** | Poor conflict management; poor project organization and definition of responsibilities; absence of leadership |

TABLE 11-4 Potential negative risk conditions associated with each knowledge area (continued)

| Knowledge Area | Risk Conditions |
|---|---|
| *Communications* | Carelessness in planning or communicating; lack of consultation with key stakeholders |
| *Risk* | Ignoring risk; unclear analysis of risk; poor insurance management |
| *Procurement* | Unenforceable conditions or contract clauses; adversarial relations |



Understanding common sources of risk is very helpful in risk identification, which is the next step in project risk management.

# IDENTIFYING RISKS

Identifying risks is the process of understanding what potential events might hurt or enhance a particular project. It is important to identify potential risks early, but you must also continue to identify risks based on the changing project environment. Also remember that you cannot manage risks if you do not first identify them. By understanding common sources of risks and reviewing a project's planning documents (for risk, cost, schedule, and quality management), activity cost and duration estimates, the scope baseline, stakeholder register, other project documents, enterprise environmental factors, and organizational process assets, project managers and their teams can identify many potential risks.

## Suggestions for Identifying Risks

There are several tools and techniques for identifying risks. Project teams often begin the risk identification process by reviewing project documentation, recent and historical information related to the organization, and assumptions that might affect the project. Project team members and outside experts often hold meetings to discuss this information and ask important questions about them as they relate to risk. After identifying potential risks at this initial meeting, the project team might then use different information-gathering techniques to further identify risks. Five common information-gathering techniques include brainstorming, the Delphi technique, interviewing, root cause analysis, and SWOT analysis.

**Brainstorming** is a technique by which a group attempts to generate ideas or find a solution for a specific problem by amassing ideas spontaneously and without judgment. This approach can help the group create a comprehensive list of risks to address later in the qualitative and quantitative risk analysis processes. An experienced facilitator should run the brainstorming session and introduce new categories of potential risks to keep the ideas flowing. After the ideas are collected, the facilitator can group and categorize the ideas to make them more manageable. Care must be taken, however, not to overuse or misuse brainstorming. Although businesses use brainstorming widely to generate new ideas, the psychology literature shows that individuals, working alone, produce a greater number of ideas than the same individuals produce through brainstorming in small, face-to-face groups. Group effects, such as fear of social disapproval, the effects of authority hierarchy, and domination of the session by one or two very vocal people often inhibit idea generation for many participants.[11]

An approach to gathering information that helps prevent some of the negative group affects found in brainstorming is the Delphi technique. The basic concept of the **Delphi technique** is to derive a consensus among a panel of experts who make predictions about future developments. Developed by the Rand Corporation for the U.S. Air Force in the late 1960s, the Delphi technique is a systematic, interactive forecasting procedure based on independent and *anonymous* input regarding future events. The Delphi technique uses repeated rounds of questioning and written responses, including feedback to earlier-round responses, to take advantage of group input, while avoiding the biasing effects possible in oral panel deliberations. To use the Delphi technique, you must select a panel of experts for the particular area in question. For example, Cliff Branch from the opening case could use the Delphi technique to help him understand why his company is no longer winning many contracts. Cliff could assemble a panel of people with knowledge in his business area. Each expert would answer questions related to Cliff's scenario, and then Cliff or a facilitator would evaluate their responses, together with opinions and justifications, and provide that feedback to each expert in the next iteration. Cliff would continue this process until the group responses converge to a specific solution. If the responses diverge, the facilitator of the Delphi technique needs to determine if there is a problem with the process.

**Interviewing** is a fact-finding technique for collecting information in face-to-face, phone, e-mail, or instant-messaging discussions. Interviewing people with similar project experience is an important tool for identifying potential risks. For example, if a new project involves using a particular type of hardware or software, someone with recent experience with that hardware or software could describe problems he or she had on a past project. If someone has worked with a particular customer, he or she might provide insight into potential risks involved in working for that group again. It is important to be well-prepared for leading interviews; it often helps to create a list of questions to use as a guide during the interview.

It is not uncommon for people to identify problems or opportunities without really understanding them. Before suggesting courses of action, it is important to identify the root cause of a problem or opportunity. Root cause analysis (discussed earlier in Chapter 8, Project Quality Management) often results in identifying even more potential risks for a project.

Another technique (described in Chapter 4, Project Integration Management) is a SWOT analysis of strengths, weaknesses, opportunities, and threats, which is often used in strategic planning. SWOT analysis can also be used during risk identification by having project teams focus on the broad perspectives of potential risks for particular projects. For example, before writing a particular proposal, Cliff Branch could have a group of his employees discuss in detail what their company's strengths are, what their weaknesses are related to that project, and what opportunities and threats exist. Do they know that several competing firms are much more likely to win a certain contract? Do they know that winning a particular contract will likely lead to future contracts and help expand their business? Applying SWOT to specific potential projects can help identify the broad risks and opportunities that apply in that scenario.

Three other techniques for risk identification include the use of checklists, analysis of assumptions, and creation of diagrams:

- Checklists, based on risks that have been encountered in previous projects, provide a meaningful template for understanding risks in a current project. You can use checklists similar to those developed by the Standish Group and other IT research consultants to help identify risks on information technology projects.
- It is important to analyze project assumptions to make sure they are valid. Incomplete, inaccurate, or inconsistent assumptions might lead to identifying more risks.
- Diagramming techniques include using cause-and-effect diagrams or fishbone diagrams, flow charts, and influence diagrams. Recall from Chapter 8, Project Quality Management, that fishbone diagrams help you trace problems back to their root cause. System or process **flowcharts** are diagrams that show how different parts of a system interrelate. For example, many programmers create flowcharts to show programming logic. (A sample flowchart is provided in Chapter 8.) Another type of diagram, an **influence diagram**, represents decision problems by displaying essential elements, including decisions, uncertainties, causality, and objectives, and how they influence each other. (See other references, such as *www.lumina.com/software/influencediagrams.html*, for detailed information on influence diagrams.)



## The Risk Register

The main output of the risk identification process is a list of identified risks and other information needed to begin creating a risk register. A **risk register** is a document that contains results of various risk management processes, often displayed in a table or spreadsheet format. It is a tool for documenting potential risk events and related information. **Risk events** refer to specific, uncertain events that may occur to the detriment or enhancement of the project. For example, negative risk events might include the performance failure of a product produced as part of a project, delays in completing work as scheduled, increases in estimated costs, supply shortages, litigation against the company, strikes, and so on. Examples of positive risk events include completing work sooner or cheaper than planned, collaborating with suppliers to produce better products, good publicity resulting from the project, and so on.

Table 11-5 provides a sample of the format for a risk register that Cliff and his managers from the opening case might use on a new project. Actual data that might be entered for one of the risks is included below the table. Notice the main headings often included in the register. Many of these items are described in more detail later in this chapter.

- *An identification number for each risk event*: The project team may want to sort or quickly search for specific risk events, so they need to identify each risk with some type of unique descriptor, such as an identification number.
- *A rank for each risk event*: The rank is usually a number, with 1 being the highest ranked risk.
- *The name of the risk event*: For example, defective server, late completion of testing, reduced consulting costs, or good publicity.
- *A description of the risk event*: Because the name of a risk event is often abbreviated, it helps to provide a more detailed description. For example, reduced

consulting costs might be expanded in the description to say that the organization might be able to negotiate lower-than-average costs for a particular consultant because the consultant really enjoys working for that company in that particular location.

- *The category under which the risk event falls*: For example, defective server might fall under the broader category of technology or hardware technology.
- *The root cause of the risk*: The root cause of the defective server might be a defective power supply.
- *Triggers for each risk*: **Triggers** are indicators or symptoms of actual risk events. For example, cost overruns on early activities may be symptoms of poor cost estimates. Defective products may be symptoms of a low-quality supplier. Documenting potential risk symptoms for projects also helps the project team identify more potential risk events.
- *Potential responses to each risk*: A potential response to the risk event of a defective server might be the inclusion of a clause in a contract with the supplier to replace a defective server within a certain time period at a negotiated cost.
- *The* **risk owner** *or person who will own or take responsibility for the risk*: For example, a certain person might be in charge of any server-related risk events and managing response strategies.
- *The probability of the risk occurring*: There might be a high, medium, or low probability of a certain risk event occurring. For example, the risk might be low that the server would actually be defective.
- *The impact to the project if the risk occurs*: There might be a high, medium, or low impact to project success if the risk event actually occurs. A defective server might have a high impact on successfully completing a project on time.
- *The status of the risk*: Did the risk event occur? Was the response strategy completed? Is the risk no longer relevant to the project? For example, a contract clause may have been completed to address the risk of a defective server.

437

**TABLE 11-5** Sample risk register

| No. | Rank | Risk | Description | Category | Root Cause | Triggers | Potential Responses | Risk Owner | Probability | Impact | Status |
|---|---|---|---|---|---|---|---|---|---|---|---|
| R44 | 1 | | | | | | | | | | |
| R21 | 2 | | | | | | | | | | |
| R7 | 3 | | | | | | | | | | |

For example, the following data might be entered for the first risk in the register as follows. Notice that Cliff's team is taking a very proactive approach in managing this risk.

- *No.*: R44
- *Rank*: 1
- *Risk*: New customer
- *Description*: We have never done a project for this organization before and don't know too much about them. One of our company's strengths is building

good customer relationships, which often leads to further projects with that customer. We might have trouble working with this customer since they are new to us.

- *Category*: People risk
- *Root cause*: We won a contract to work on a project without really getting to know the customer.
- *Triggers*: The project manager and other senior managers realize that we don't know much about this customer and could easily misunderstand their needs or expectations.
- *Risk responses*: Make sure the project manager is sensitive to the fact that this is a new customer and takes the time to understand them. Have the PM set up a meeting to get to know the customer and clarify their expectations. Have Cliff attend the meeting, too.
- *Risk owner*: Our project manager
- *Probability*: Medium
- *Impact*: High
- *Status*: PM will set up the meeting within the week.



After identifying risks, the next step is to understand which risks are most important by performing qualitative risk analysis.

## PERFORMING QUALITATIVE RISK ANALYSIS

Qualitative risk analysis involves assessing the likelihood and impact of identified risks, to determine their magnitude and priority. This section describes examples of using a probability/impact matrix to produce a prioritized list of risks. It also provides examples of using the Top Ten Risk Item Tracking technique to produce an overall ranking for project risks and to track trends in qualitative risk analysis. Finally, it discusses the importance of expert judgment in performing risk analysis.

### Using Probability/Impact Matrixes to Calculate Risk Factors

People often describe a risk probability or consequence as being high, medium or moderate, or low. For example, a meteorologist might predict that there is a high probability, or likelihood, of severe rain showers on a certain day. If that day happens to be your wedding day and you are planning a large outdoor wedding, the consequences or impact of severe showers might also be high.

A project manager can chart the probability and impact of risks on a **probability/impact matrix or chart**. A probability/impact matrix or chart lists the relative probability of a risk occurring on one side of a matrix or axis on a chart and the relative impact of the risk occurring on the other. Many project teams would benefit from using this simple technique to help them identify risks that they need to pay attention to. To use this approach, project stakeholders list the risks they think might occur on their projects. They then label each risk as being high, medium, or low in terms of its probability of occurrence and its impact if it did occur.

The project manager then summarizes the results in a probability/impact matrix or chart, as shown in Figure 11-5. For example, Cliff Branch and some of his project managers in the opening case could each identify three negative and positive potential risks for a particular project. They could then label each risk as being high, medium, or low in terms of probability and impact. For example, one project manager might list a severe market downturn as a negative risk that's low in probability but high in impact. Cliff may have listed the same risk as being medium in both probability and impact. The team could then plot all of the risks on a matrix or chart, combine any common risks, and decide where those risks should be on the matrix or chart. The team should then focus on any risks that fall in the high sections of the probability/impact matrix or chart. For example, Risks 1 and 4 are listed as high in both categories of probability and impact. Risk 6 is high in the probability category but low in the impact category. Risk 9 is high in the probability category and medium in the impact category, and so on. The team should then discuss how they plan to respond to those risks if they occur, as discussed later in this chapter in the section on risk response planning.



| Probability \ Impact | Low | Medium | High |
|---|---|---|---|
| High | risk 6 | risk 9 | risk 1<br>risk 4 |
| Medium | risk 3<br>risk 7 | risk 2<br>risk 5<br>risk 11 | |
| Low | | risk 8<br>risk 10 | risk 12 |

**FIGURE 11-5** Sample probability/impact matrix

It may be useful to create a separate probability/impact matrix or chart for negative risks and positive risks to make sure both types of risks are adequately addressed. Some project teams also collect data based on the probability and impact of risks in terms of negatively or positively affecting scope, time, and cost goals. Qualitative risk analysis is normally done quickly, so the project team has to decide what type of approach makes the most sense for their project.

Some project teams develop a single number for a risk score by simply multiplying a numeric score for probability by a numeric score for impact. A more sophisticated approach to using probability/impact information is to calculate risk factors. To quantify risk probability and consequence, the U.S. Defense Systems Management College (DSMC) developed a technique for calculating **risk factors**—numbers that represent the overall risk of specific events, based on their probability of occurring and the consequences to the project if they do occur. The technique makes use of a probability/impact matrix that shows the probability of risks occurring and the impact or consequences of the risks.

Probabilities of a risk occurring can be estimated based on several factors, as determined by the unique nature of each project. For example, factors evaluated for potential hardware or software technology risks could include the technology not being mature, the technology being too complex, and an inadequate support base for developing the technology. The impact of a risk occurring could include factors such as the availability of fallback solutions or the consequences of not meeting performance, cost, and schedule estimates.



Figure 11-6 provides an example of how risk factors were used to graph the probability of failure, and consequence of failure, for proposed technologies in a research study on which the author worked to design more reliable aircraft. The figure classifies potential technologies (dots on the chart) as high-, medium-, or low-risk, based on the probability of failure and consequences of failure. The study researchers highly recommended that the U.S. Air Force invest in the low- to medium-risk technologies and suggested that it not pursue the high-risk technologies.[12] The rigor behind using the probability/impact matrix and risk factors can provide a much stronger argument than simply stating that risk probabilities or consequences are high, medium, or low.

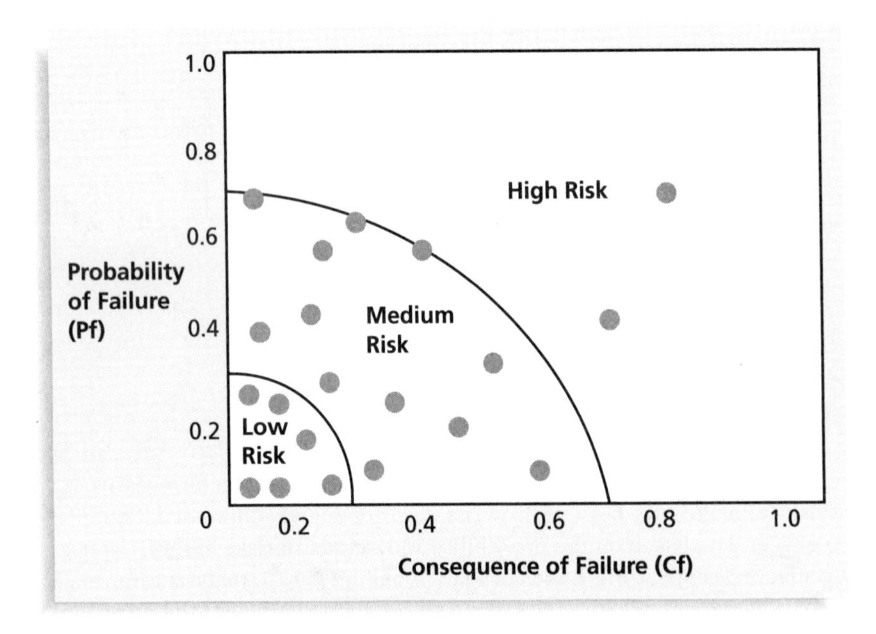


**FIGURE 11-6** Chart showing high-, medium-, and low-risk technologies

## Top Ten Risk Item Tracking

**Top Ten Risk Item Tracking** is a qualitative risk analysis tool, and in addition to identifying risks, it maintains an awareness of risks throughout the life of a project by also helping to monitor risks. It involves establishing a periodic review of the project's most significant risk items with management and, optionally, with the customer. The review begins with a summary of the status of the top-ten sources of risk on the project. The summary includes each item's current ranking, previous ranking, number of times it appears on the list over a period of time, and a summary of progress made in resolving the risk item since the previous review. The Microsoft Solution Framework (MSF) includes a risk management model that includes developing and monitoring a top-ten master list of risks. MSF is the methodology Microsoft uses for managing projects. It combines aspects of software design and development, and building and deploying infrastructure, into a single-project life cycle for guiding technology solutions of all kinds. (Consult Microsoft's Web site for more information on MSF.)



Table 11-6 provides an example of a Top Ten Risk Item Tracking chart that could be used at a management review meeting for a project. This particular example includes only the top five negative risk events. Notice that each risk event is ranked based on the current month, previous month, and how many months it has been in the top ten. The last column briefly describes the progress for resolving each particular risk item. You can have separate charts for negative and positive risks or combine them into one chart.

**TABLE 11-6** Example of Top Ten Risk Item Tracking

| | MONTHLY RANKING | | | |
|---|---|---|---|---|
| Risk Event | Rank This Month | Rank Last Month | Number of Months in Top Ten | Risk Resolution Progress |
| Inadequate planning | 1 | 2 | 4 | Working on revising the entire project management plan |
| Poor definition | 2 | 3 | 3 | Holding meetings with project customer and sponsor to clarify scope |
| Absence of leadership | 3 | 1 | 2 | After previous project manager quit, assigned a new one to lead the project |
| Poor cost estimates | 4 | 4 | 3 | Revising cost estimates |
| Poor time estimates | 5 | 5 | 3 | Revising schedule estimates |

A risk management review accomplishes several objectives. First, it keeps management and the customer (if included) aware of the major influences that could prevent or enhance the project's success. Second, by involving the customer, the project team may be able to consider alternative strategies for addressing the risks. Third, it is a means of promoting confidence in the project team by demonstrating to management and the customer that the team is aware of the significant risks, has a strategy in place, and is effectively carrying out that strategy.

The main output of qualitative risk analysis is updating the risk register. The ranking column of the risk register should be filled in, along with a numeric value or high/medium/low rating for the probability and impact of the risk event. Additional information is often added for risk events, such as identification of risks that need more attention in the near term or those that can be placed on a watch list. A **watch list** is a list of risks that are low priority, but are still identified as potential risks. Qualitative analysis can also identify risks that should be evaluated on a quantitative basis, as described in the next section.



## PERFORMING QUANTITATIVE RISK ANALYSIS

Quantitative risk analysis often follows qualitative risk analysis, yet both processes can be done together or separately. On some projects, the team may only perform qualitative risk analysis. The nature of the project and availability of time and money affect the type of risk analysis techniques used. Large, complex projects involving leading-edge technologies often require extensive quantitative risk analysis. The main techniques for quantitative risk analysis include data gathering, quantitative risk analysis and modeling techniques, and expert judgment. Data gathering often involves interviewing experts and collecting probability distribution information. This section focuses on using the quantitative risk analysis and modeling techniques of decision tree analysis, simulation, and sensitivity analysis.

### Decision Trees and Expected Monetary Value

A **decision tree** is a diagramming analysis technique used to help select the best course of action in situations in which future outcomes are uncertain. A common application of decision tree analysis involves calculating expected monetary value. **Expected monetary value (EMV)** is the product of a risk event probability and the risk event's monetary value. Figure 11-7 uses the decision of which project(s) an organization might pursue to illustrate this concept. Suppose Cliff Branch's firm was trying to decide if it should submit a proposal for Project 1, Project 2, both projects, or neither project. The team could draw a decision tree with two branches, one for Project 1 and one for Project 2. The firm could then calculate the expected monetary value to help make this decision.

To create a decision tree, and to calculate expected monetary value specifically, you must estimate the probabilities, or chances, of certain events occurring. For example, in Figure 11-7 there is a 20 percent probability or chance ($P$ = .20) that Cliff's firm will win the contract for Project 1, which is estimated to be worth $300,000 in profits—the outcome of the top branch in the figure. There is an 80 percent probability ($P$ = .80) that it will not win the contract for Project 1, and the outcome is estimated to be –$40,000, meaning that the firm will have to invest $40,000 into Project 1 with

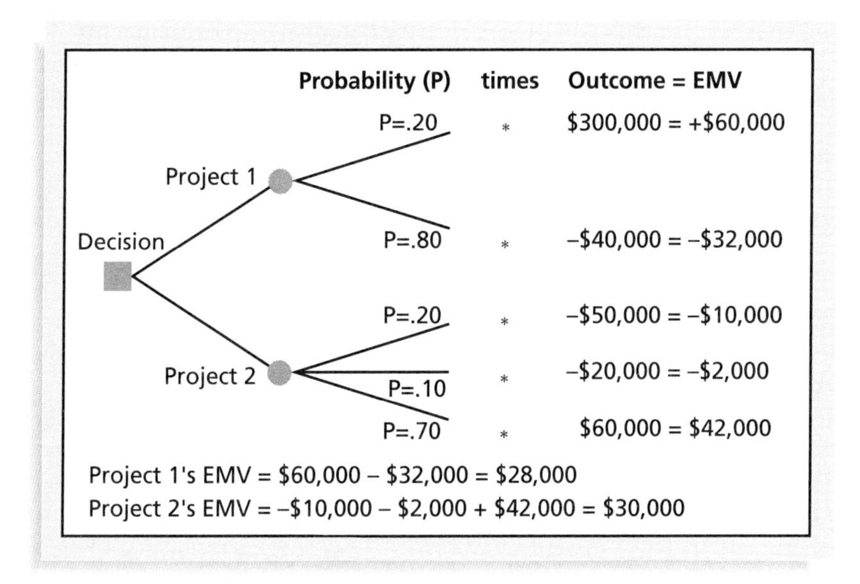


**FIGURE 11-7** Expected monetary value (EMV) example



no reimbursement if it is not awarded the contract. The sum of the probabilities for outcomes for each project must equal one (for Project 1, 20 percent plus 80 percent). Probabilities are normally determined based on expert judgment. Cliff or other people in his firm should have some sense of their likelihood of winning certain projects.

Figure 11-7 also shows probabilities and outcomes for Project 2. Suppose there is a 20 percent probability that Cliff's firm will lose $50,000 on Project 2, a 10 percent probability that it will lose $20,000, and a 70 percent probability that it will earn $60,000. Again, experts would need to estimate these dollar amounts and probabilities.

To calculate the expected monetary value (EMV) for each project, multiply the probability by the outcome value for each potential outcome for each project and sum the results. To calculate expected monetary value for Project 1, going from left to right, multiply the probability by the outcome for each branch and sum the results. In this example, the EMV for Project 1 is $28,000.

$$.2(\$300{,}000) + .8(-\$40{,}000) = \$60{,}000 - \$32{,}000 = \$28{,}000$$

The EMV for Project 2 is $30,000.

$$.2(-\$50{,}000) + .1(-\$20{,}000) + .7(\$60{,}000) = -\$10{,}000 - \$2{,}000 + \$42{,}000 = \$30{,}000$$

Because the EMV provides an estimate for the total dollar value of a decision, you want to have a positive number; the higher the EMV, the better. Since the EMV is positive for both Projects 1 and 2, Cliff's firm would expect a positive outcome from each and could bid on both projects. If it had to choose between the two projects, perhaps because of limited resources, Cliff's firm should bid on Project 2 because it has a higher EMV.

Also notice in Figure 11-7 that if you just looked at the potential outcome of the two projects, Project 1 looks more appealing. You could earn $300,000 in profits from Project 1, but you can only earn $60,000 for Project 2. If Cliff were a risk seeker, he would naturally want to bid on Project 1. However, there is only a 20 percent chance of getting that $300,000 on Project 1, and there is a 70 percent chance of earning $60,000 on Project 2. Using EMV helps account for all possible outcomes and their probabilities of occurrence, thereby reducing the tendency to pursue overly aggressive or conservative risk strategies.

## Simulation

A more sophisticated quantitative risk analysis technique is simulation. Simulation uses a representation or model of a system to analyze the expected behavior or performance of the system. Most simulations are based on some form of Monte Carlo analysis. **Monte Carlo analysis** simulates a model's outcome many times to provide a statistical distribution of the calculated results. Monte Carlo analysis can determine that a project will finish by a certain date only 10 percent of the time, and determine another date for which the project will finish 50 percent of the time. In another words, the Monte Carlo analysis can predict the probability of finishing by a certain date or the probability that the cost will be equal to or less than a certain value.



You can use several different types of distribution functions when performing a Monte Carlo analysis. The example below is a simplified approach.

The basic steps of a Monte Carlo analysis are:

1. Assess the range for the variables being considered. In other words, collect the most likely, optimistic, and pessimistic estimates for the variables in the model. For example, if you are trying to determine the likelihood of meeting project schedule goals, the project network diagram would be your model. You would collect the most likely, optimistic, and pessimistic time estimates for each task. Notice that this step is similar to collecting data for performing PERT estimates. However, instead of applying the same PERT weighted average formula, you go on to the following steps when performing a Monte Carlo simulation.
2. Determine the probability distribution of each variable. What is the likelihood of that variable falling between the optimistic and most likely estimates? For example, if an expert assigned to do a particular task provides a most likely estimate of ten weeks, an optimistic estimate of eight weeks, and a pessimistic estimate of 15 weeks, you then ask what the probability is of completing that task between 8 and 10 weeks. The expert might respond that there is a 20 percent probability.
3. For each variable, such as the time estimate for a task, select a random value based on the probability distribution for the occurrence of the variable. For example, using the above scenario, you would randomly pick a value between 8 weeks and 10 weeks 20 percent of the time and a value between 10 weeks and 15 weeks 80 percent of the time.
4. Run a deterministic analysis or one pass through the model using the combination of values selected for each one of the variables. For example, the one task described above might have a value of 12 on the first run. All of the other tasks would have one random value assigned to them on that first run, also, based on their estimates and probability distributions.

5. Repeat Steps 3 and 4 many times to obtain the probability distribution of the model's results. The number of iterations depends on the number of variables and the degree of confidence required in the results, but it typically lies between 100 and 1,000. Using the project schedule as an example, the final simulation results will show you the probability of completing the entire project within a certain time period.

Figure 11-8 illustrates the results from a Monte Carlo–based simulation of a project schedule. The simulation was done using Microsoft Project and Risk+ software. On the left side of Figure 11-8 is a chart displaying columns and an S-shaped curve. The height of each column, read by the scale on the left of the chart, indicates how many times the project was completed in a given time interval during the simulation run, which is the sample count. In this example, the time interval is two working days, and the simulation was run 250 times. The first column shows that the project was completed by 1/29 or January 29 (using month/day format) only two times during the simulation. The S-shaped curve, read from the scale on the right of the chart, shows the cumulative probability of completing the project on or before a given date. The right side of Figure 11-8 shows the information in tabular form. For example, there is a 10 percent probability that the project will be completed by 2/8, a 50 percent chance of completion by 2/17, and a 90 percent chance of completion by 2/25.

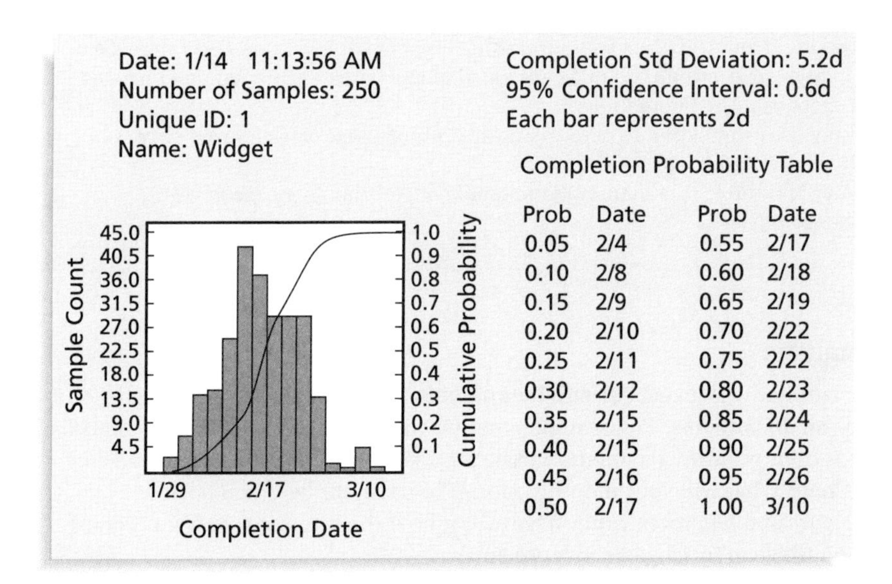


| Prob | Date | Prob | Date |
|---|---|---|---|
| 0.05 | 2/4 | 0.55 | 2/17 |
| 0.10 | 2/8 | 0.60 | 2/18 |
| 0.15 | 2/9 | 0.65 | 2/19 |
| 0.20 | 2/10 | 0.70 | 2/22 |
| 0.25 | 2/11 | 0.75 | 2/22 |
| 0.30 | 2/12 | 0.80 | 2/23 |
| 0.35 | 2/15 | 0.85 | 2/24 |
| 0.40 | 2/15 | 0.90 | 2/25 |
| 0.45 | 2/16 | 0.95 | 2/26 |
| 0.50 | 2/17 | 1.00 | 3/10 |

**FIGURE 11-8** Sample Monte Carlo–based simulation results for project schedule

As you can imagine, people use software to perform the steps required for a Monte Carlo analysis. Several PC-based software packages are available that perform Monte Carlo simulations. Many products will show you what the major risk drivers are, based on the simulation results. For example, a wide range for a certain task estimate might be causing most of the uncertainty in the project schedule. You will learn more about using simulation and other software related to project risk management later in this chapter.

## WHAT WENT RIGHT?

A large aerospace company used Monte Carlo analysis to help quantify risks on several advanced-design engineering projects. The U.S. National Aerospace Plane (NASP) project involved many risks. The purpose of this multibillion-dollar project was to design and develop a vehicle that could fly into space using a single-stage-to-orbit approach. A single-stage-to-orbit approach meant the vehicle would have to achieve a speed of Mach 25 (25 times the speed of sound) without a rocket booster. A team of engineers and business professionals worked together in the mid-1980s to develop a software model for estimating the time and cost of developing the NASP project. This model was then linked with simulation software to determine the sources of cost and schedule risk for the project. The company then used the results of the Monte Carlo analysis to determine how it would invest its internal research and development funds. Although the NASP project was terminated, the resulting research has helped develop more advanced materials and propulsion systems used on many modern aircraft.

Microsoft Excel is a common tool for performing quantitative risk analysis. Microsoft provides examples of how to use Excel to perform Monte Carlo simulation from its Web site, and explains how several companies use Monte Carlo simulation as an important tool for decision-making:

- General Motors uses simulation for forecasting net income for the corporation, predicting structural costs and purchasing costs of vehicles, and determining the company's susceptibility to different kinds of risk, such as interest rate changes and exchange rate fluctuations.
- Eli Lilly uses simulation to determine the optimal plant capacity that should be built for each drug.
- Procter & Gamble uses simulation to model and optimally hedge foreign exchange risk.[13]



### Sensitivity Analysis

Many people are familiar with using **sensitivity analysis** to see the effects of changing one or more variables on an outcome. For example, many people perform a sensitivity analysis to determine what their monthly payments will be for a loan given different interest rates or periods of the loan. What will your monthly mortgage payment be if you borrow $100,000 for 30 years at a 6 percent rate? What will it be if the interest rate is 7 percent? What if you pay off the loan in 15 years at 5 percent?

Many professionals use sensitivity analysis to help make several common business decisions, such as determining break-even points based on different assumptions. People often use spreadsheet software like Microsoft Excel to perform sensitivity analysis. Figure 11-9 shows an example Excel file created to quickly show the break-even point for a product based on various inputs: the sales price per unit, the manufacturing cost per unit, and fixed monthly expenses. The current inputs result in a break-even point of 6,250 units sold. Users of this spreadsheet can change inputs and see the effects on the break-even point in chart format. Project teams often create similar models to determine the sensitivity of various project variables. For example, Cliff's team could develop sensitivity analysis models to estimate their profits on jobs by varying how many hours it would take them to do the job, costs per hour, and so on.

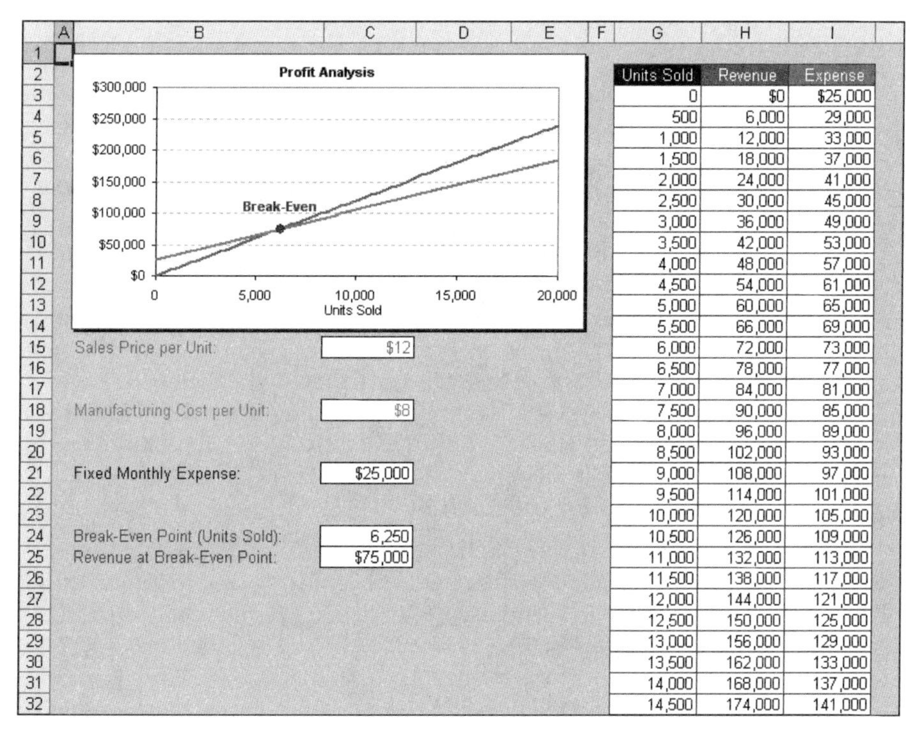


| Units Sold | Revenue | Expense |
|---|---|---|
| 0 | $0 | $25,000 |
| 500 | 6,000 | 29,000 |
| 1,000 | 12,000 | 33,000 |
| 1,500 | 18,000 | 37,000 |
| 2,000 | 24,000 | 41,000 |
| 2,500 | 30,000 | 45,000 |
| 3,000 | 36,000 | 49,000 |
| 3,500 | 42,000 | 53,000 |
| 4,000 | 48,000 | 57,000 |
| 4,500 | 54,000 | 61,000 |
| 5,000 | 60,000 | 65,000 |
| 5,500 | 66,000 | 69,000 |
| 6,000 | 72,000 | 73,000 |
| 6,500 | 78,000 | 77,000 |
| 7,000 | 84,000 | 81,000 |
| 7,500 | 90,000 | 85,000 |
| 8,000 | 96,000 | 89,000 |
| 8,500 | 102,000 | 93,000 |
| 9,000 | 108,000 | 97,000 |
| 9,500 | 114,000 | 101,000 |
| 10,000 | 120,000 | 105,000 |
| 10,500 | 126,000 | 109,000 |
| 11,000 | 132,000 | 113,000 |
| 11,500 | 138,000 | 117,000 |
| 12,000 | 144,000 | 121,000 |
| 12,500 | 150,000 | 125,000 |
| 13,000 | 156,000 | 129,000 |
| 13,500 | 162,000 | 133,000 |
| 14,000 | 168,000 | 137,000 |
| 14,500 | 174,000 | 141,000 |

**FIGURE 11-9** Sample sensitivity analysis for determining break-even point

The main outputs of quantitative risk analysis are updates to the risk register, such as revised risk rankings or detailed information behind those rankings. The quantitative analysis also provides high-level information in terms of the probabilities of achieving certain project objectives. This information might cause the project manager to suggest changes in contingency reserves. In some cases, projects may be redirected or canceled based on the quantitative analysis, or they might cause the initiation of new projects to help the current one succeed, such as the NASP project described in the What Went Right example.

## PLANNING RISK RESPONSES

After an organization identifies and quantifies risks, it must develop an appropriate response to them. Developing a response to risks involves developing options and defining strategies for reducing negative risks and enhancing positive risks.

The four basic response strategies for negative risks are:

1. **Risk avoidance** or eliminating a specific threat, usually by eliminating its causes. Of course, not all risks can be eliminated, but specific risk events can be. For example, a project team may decide to continue using a specific piece of hardware or software on a project because they know it works. Other products that could be used on the project may be available, but if the project

team is unfamiliar with them, they could cause significant risk. Using familiar hardware or software eliminates this risk.

2. **Risk acceptance** or accepting the consequences should a risk occur. For example, a project team planning a big project review meeting could take an active approach to risk by having a contingency or backup plan and contingency reserves if they cannot get approval for a specific site for the meeting. On the other hand, they could take a passive approach and accept whatever facility their organization provides them.
3. **Risk transference** or shifting the consequence of a risk and responsibility for its management to a third party. For example, risk transference is often used in dealing with financial risk exposure. A project team may purchase special insurance or warranty protection for specific hardware needed for a project. If the hardware fails, the insurer must replace it within an agreed-upon period of time.
4. **Risk mitigation** or reducing the impact of a risk event by reducing the probability of its occurrence. Suggestions for reducing common sources of risk on information technology projects were provided at the beginning of this chapter. Other examples of risk mitigation include using proven technology, having competent project personnel, using various analysis and validation techniques, and buying maintenance or service agreements from subcontractors.



Table 11-7 provides general mitigation strategies for technical, cost, and schedule risks on projects.[14] Note that increasing the frequency of project monitoring and using a work breakdown structure (WBS) and Critical Path Method (CPM) are strategies for all three areas. Increasing the project manager's authority is a strategy for mitigating technical and cost risks, and selecting the most experienced project manager is recommended for reducing schedule risks. Improving communication is also an effective strategy for mitigating risks.

**TABLE 11-7** General risk mitigation strategies for technical, cost, and schedule risks

| Technical Risks | Cost Risks | Schedule Risks |
|---|---|---|
| Emphasize team support and avoid stand-alone project structure | Increase the frequency of project monitoring | Increase the frequency of project monitoring |
| Increase project manager authority | Use WBS and CPM | Use WBS and CPM |
| Improve problem handling and communication | Improve communication, project goals understanding, and team support | Select the most experienced project manager |
| Increase the frequency of project monitoring | Increase project manager authority | |
| Use WBS and CPM | | |

The four basic response strategies for positive risks are:

- **Risk exploitation** or doing whatever you can to make sure the positive risk happens. For example, suppose Cliff's company funded a project to provide new computer classrooms for a nearby school in need. They might select one of their top project managers to organize news coverage of the project, write a press release, or hold some other public event to ensure the project produces good public relations for the company, which could lead to more business.
- **Risk sharing** or allocating ownership of the risk to another party. Using the same example of implementing new computer classrooms, the project manager could form a partnership with the school's principal, school board, or parent-teacher organization to share responsibility for achieving good public relations for the project. Or the company might partner with a local training firm that agrees to provide free training for all of the teachers on how to use the new computer classrooms.
- **Risk enhancement** or changing the size of the opportunity by identifying and maximizing key drivers of the positive risk. For example, an important driver of getting good public relations for the computer classrooms project might be getting the students, parents, and teachers aware of and excited about the project. They could then do their own formal or informal advertising of the project and Cliff's company, which in turn might interest other groups and could generate more business.
- **Risk acceptance** also applies to positive risks when the project team cannot or chooses not to take any actions toward a risk. For example, the computer classrooms project manager might just assume the project will result in good public relations for their company without doing anything extra.



The main outputs of risk response planning include risk-related contractual agreements, updates to the project management plan and other project documents, and updates to the risk register. For example, if Cliff's company decided to partner with a local training firm on the computer classrooms project to share the opportunity of achieving good public relations, it could write a contract with that firm. The project management plan and its related plans might need to be updated if the risk response strategies require additional tasks, resources, or time to accomplish them. Risk response strategies often result in changes to the WBS and project schedule, so plans with that information must be updated. The risk response strategies also provide updated information for the risk register by describing the risk responses, risk owners, and status information.

Risk response strategies often include identification of residual and secondary risks as well as contingency plans and reserves, as described earlier. **Residual risks** are risks that remain after all of the response strategies have been implemented. For example, even though a more stable hardware product may have been used on a project, there may still be some risk of it failing to function properly. **Secondary risks** are a direct result of implementing a risk response. For example, using the more stable hardware may have caused a risk of peripheral devices failing to function properly.

## MONITORING AND CONTROLLING RISKS

Monitoring and controlling risks involves executing the risk management processes to respond to risk events. Executing the risk management processes means ensuring that risk awareness is an ongoing activity performed by the entire project team throughout the entire project. Project risk management does not stop with the initial risk analysis. Identified risks may not materialize, or their probabilities of occurrence or loss may diminish. Previously identified risks may be determined to have a greater probability of occurrence or a higher estimated loss value. Similarly, new risks will be identified as the project progresses. Newly identified risks need to go through the same process as those identified during the initial risk assessment. A redistribution of resources devoted to risk management may be necessary because of relative changes in risk exposure.



Carrying out individual risk management plans involves monitoring risks based on defined milestones and making decisions regarding risks and their response strategies. It may be necessary to alter a strategy if it becomes ineffective, implement a planned contingency activity, or eliminate a risk from the list of potential risks when it no longer exists. Project teams sometimes use **workarounds**—unplanned responses to risk events—when they do not have contingency plans in place.

Risk reassessment, risk audits, variance and trend analysis, technical performance measurements, reserve analysis, and status meetings or periodic risk reviews such as the Top Ten Risk Item Tracking method are all tools and techniques for performing risk monitoring and control. Outputs of this process are risk register updates, organizational process assets updates (such as lessons-learned information that might help future projects), change requests, and updates to the project management plan and other project documents.

## USING SOFTWARE TO ASSIST IN PROJECT RISK MANAGEMENT

As demonstrated in several parts of this chapter, you can use a variety of software tools to enhance various risk management processes. Most organizations use software to create, update, and distribute information in their risk registers. The risk register is often a simple Microsoft Word or Excel file, but it can also be part of a more sophisticated database. Spreadsheets can aid in tracking and quantifying risks, preparing charts and graphs, and performing sensitivity analysis. Software can be used to create decision trees and estimate expected monetary value.

More sophisticated risk management software, such as Monte Carlo simulation software, can help you develop models and use simulations to analyze and respond to various risks. Several high-end project management tools include simulation capabilities. You can also purchase add-on software to perform Monte Carlo simulations using Excel (such as Decisioneering Crystal Ball or Palisade @Risk for Excel) or Project 2007 (such as CS Solutions Risk+ or Palisade @Risk for Project). Several software packages have also been specifically created for project risk management. Although it has become easier to do sophisticated risk analysis with new software tools, project teams must be careful not to over-rely on using software when performing project risk management. If a risk is not identified, it cannot be managed, and it takes intelligent, experienced people to do a good job of identifying risks. It also takes hard work to develop and implement good risk response strategies. Software should be used as a tool to help make good decisions in project risk management, not as a scapegoat for when things go wrong.

Well-run projects, like a master violinist's performance, an Olympic athlete's gold medal win, or a Pulitzer Prize–winning book, appear to be almost effortless. Those on the outside—whether audiences, customers, or managers—cannot observe the effort that goes into a superb performance. They cannot see the hours of practice, the edited drafts, or the planning, management, and foresight that create the appearance of ease. To improve information technology project management, project managers should strive to make their jobs look easy—it reflects the results of a well-run project.

## CASE WRAP-UP

Cliff Branch and two of his senior people attended a seminar on project risk management where the speaker discussed several techniques, such as estimating the expected monetary value of projects, Monte Carlo simulations, and so on. Cliff asked the speaker how these techniques could be used to help his company decide which projects to bid on, since bidding on projects often required up-front investments with the possibility of no payback. The speaker walked through an example of EMV and then ran a quick Monte Carlo simulation. Cliff did not have a strong math background and had a hard time understanding the EMV calculations. He thought the simulation was much too confusing to have any practical use for him. He believed in his gut instincts much more than any math calculation or computer output.

The speaker finally sensed that Cliff was not impressed, so she explained the importance of looking at the odds of winning project awards and not just at the potential profits. She suggested using a risk-neutral strategy by bidding on projects that the company had a good chance of winning (50 percent or so) and that had a good profit potential, instead of focusing on projects that they had a small chance of winning and that had a larger profit potential. Cliff disagreed with this advice, and he continued to bid on high-risk projects. The two other managers who attended the seminar now understood why the firm was having problems—their leader loved taking risks, even if it hurt the company. They soon found jobs with competing companies, as did several other employees.

## Chapter Summary

Risk is an uncertainty that can have a negative or positive effect on meeting project objectives. Projects, by virtue of their unique nature, involve risk. Many organizations do a poor job of project risk management, if they do any at all. Successful organizations realize the value of good project risk management.

Risk management is an investment; that is, there are costs associated with identifying risks, analyzing those risks, and establishing plans to address those risks. Those costs must be included in cost, schedule, and resource planning.

Risk utility or risk tolerance is the amount of satisfaction or pleasure received from a potential payoff. Risk seekers enjoy high risks, risk-averse people do not like to take risks, and risk-neutral people seek to balance risks and potential payoff.



Project risk management is a process in which the project team continually assesses what may negatively or positively impact the project, determines the probability of such events occurring, and determines the impact should such events occur. It also involves analyzing and determining alternate strategies to deal with risks. The six main processes involved in risk management are planning risk management, identifying risks, performing qualitative risk analysis, performing quantitative risk analysis, planning risk responses, and monitoring and controlling risks.

Planning risk management is the process of deciding how to approach and plan for risk management activities for a particular project. A risk management plan is a key output of the risk management planning process, and a risk register is a key output of the other risk management processes. Contingency plans are predefined actions that a project team will take if an identified risk event occurs. Fallback plans are developed for risks that have a high impact on meeting project objectives, and are implemented if attempts to reduce the risk are not effective. Contingency reserves or contingency allowances are provisions held by the project sponsor or organization to reduce the risk of cost or schedule overruns to an acceptable level.

Information technology projects often involve several risks: lack of user involvement, lack of executive management support, unclear requirements, poor planning, and so on. Lists developed by the Standish Group and other organizations can help you identify potential risks on information technology projects. A risk breakdown structure is a useful tool that can help project managers consider potential risks in different categories. Lists of common risk conditions in project management knowledge areas can also be helpful in identifying risks, as can information-gathering techniques such as brainstorming, the Delphi technique, interviewing, and SWOT analysis. A risk register is a document that contains results of various risk management processes, often displayed in a table or spreadsheet format. It is a tool for documenting potential risk events and related information. Risk events refer to specific, uncertain events that may occur to the detriment or enhancement of the project.

Risks can be assessed qualitatively and quantitatively. Tools for qualitative risk analysis include a probability/impact matrix and the Top Ten Risk Item Tracking technique. Tools for quantitative risk analysis include decision trees and Monte Carlo simulation. Expected monetary value (EMV) uses decision trees to evaluate potential projects based on their expected value. Simulations are a more sophisticated method for creating estimates to help you determine the likelihood of meeting specific project schedule or cost goals. Sensitivity analysis is used to show the effects of changing one or more variables on an outcome.

The four basic responses to risk are avoidance, acceptance, transference, and mitigation. Risk avoidance involves eliminating a specific threat or risk. Risk acceptance means accepting the consequences of a risk, should it occur. Risk transference is shifting the consequence of a risk and responsibility for its management to a third party. Risk mitigation is reducing the impact of a risk event by reducing the probability of its occurrence. The four basic response strategies for positive risks are risk exploitation, risk sharing, risk enhancement, and risk acceptance.

Monitoring and controlling risks involves executing the risk management processes and the risk management plan to respond to risks. Outputs of this process include risk register updates, organizational process assets updates, change requests, and updates to the project management plan and other project documents.

Several types of software can assist in project risk management. Monte Carlo–based simulation software is a particularly useful tool for helping get a better idea of project risks and top sources of risk or risk drivers.



## Quick Quiz

1. ______________ is an uncertainty that can have a negative or positive effect on meeting project objectives.
   a. Risk utility
   b. Risk tolerance
   c. Risk management
   d. Risk
2. A person who is risk-______________ receives greater satisfaction when more payoff is at stake and is willing to pay a penalty to take risks.
   a. averse
   b. seeking
   c. neutral
   d. aware
3. Which risk management process involves prioritizing risks based on their probability and impact of occurrence?
   a. planning risk management
   b. identifying risks
   c. performing qualitative risk analysis
   d. performing quantitative risk analysis
4. Your project involves using a new release of a common software application, but if that release is not available, your team has ______________ plans to use the current release.
   a. contingency
   b. fallback
   c. reserve
   d. mitigation

5. Which risk identification tool involves deriving a consensus among a panel of experts by using anonymous input regarding future events?
   a. risk breakdown structure
   b. brainstorming
   c. interviewing
   d. Delphi technique
6. A risk ____________ is a document that contains results of various risk management processes, often displayed in a table or spreadsheet format.
   a. management plan
   b. register
   c. breakdown structure
   d. probability/impact matrix
7. ____________ are indicators or symptoms of actual risk events, such as a cost overrun on early activities being a symptom of poor cost estimates.
   a. Probabilities
   b. Impacts
   c. Watch list items
   d. Triggers
8. Suppose there is a 30 percent chance that you will lose $10,000 and a 70 percent chance that you will earn $100,000 on a particular project. What is the project's estimated monetary value?
   a. –$30,000
   b. $70,000
   c. $67,000
   d. –$67,000
9. ____________ is a quantitative risk analysis tool that uses a model of a system to analyze the expected behavior or performance of the system.
   a. Simulation
   b. Sensitivity analysis
   c. Monte Carlo analysis
   d. EMV
10. Your project team has decided not to use an upcoming release of software because it might cause your schedule to slip. Which negative risk response strategy are you using?
   a. avoidance
   b. acceptance
   c. transference
   d. mitigation

### Quick Quiz Answers

1. d; 2. b; 3. c; 4. a; 5. d; 6. b; 7. d; 8. c; 9. a; 10. a

## Discussion Questions

1. Discuss the risk utility function and risk preference chart in Figure 11-2. Would you rate yourself as being risk-averse, risk-neutral, or risk-seeking? Give examples of each approach from different aspects of your life, such as your current job, your personal finances, romances, and eating habits.
2. What are some questions that should be addressed in a risk management plan?
3. Discuss the common sources of risk on information technology projects and suggestions for managing them. Which suggestions do you find most useful? Which do you feel would not work in your organization? Why?
4. What is the difference between using brainstorming and the Delphi technique for risk identification? What are some of the advantages and disadvantages of each approach? Describe the contents of a risk register and how the risk register is used in several risk management processes.
5. Describe how to use a probability/impact matrix and the Top Ten Risk Item Tracking approaches for performing qualitative risk analysis. How could you use each technique on a project?
6. Explain how to use decision trees and Monte Carlo analysis for quantifying risk. Give an example of how you could use each technique on an information technology project.
7. Provide realistic examples of each of the risk response strategies for both negative and positive risks.
8. List the tools and techniques for performing risk monitoring and control.
9. How can you use Excel to assist in project risk management? What other software can help project teams make better risk management decisions?



## Exercises

1. Suppose your college or organization is considering a new project that would involve developing an information system that would allow all employees and students/customers to access and maintain their own human resources–related information, such as address, marital status, tax information, and so on. The main benefits of the system would be a reduction in human resources personnel and more accurate information. For example, if an employee, student, or customer had a new telephone number or e-mail address, he or she would be responsible for entering the new data in the new system. The new system would also allow employees to change their tax withholdings or pension plan contributions. Identify five potential risks for this new project, being sure to list some negative and positive risks. Provide a detailed description of each risk and propose strategies for addressing each risk. Document your results in a two-page paper.

2. Review a document related to risk management, such as Microsoft's Security Risk Management Guide available from the companion Web site for this text. Does this guide address most of the topics related to risk management planning as described in this text? Document your analysis in a two-page paper.
3. Research risk management software. Are many products available? What are the main advantages of using them in managing projects? What are the main disadvantages? Write a two-page paper discussing your findings, and include at least three references.
4. Suppose your organization is deciding which of four projects to bid on. Information on each is in the table below. Assume that all up-front investments are not recovered, so they are shown as negative profits. Draw a diagram and calculate the EMV for each project. Write a few paragraphs explaining which projects you would bid on. Be sure to use the EMV information and your personal risk tolerance to justify your answer.



| Project | Chance of Outcome | Estimated Profits |
|---|---|---|
| Project 1 | 50 percent | $120,000 |
| | 50 percent | –$50,000 |
| Project 2 | 30 percent | $100,000 |
| | 40 percent | $ 50,000 |
| | 30 percent | –$60,000 |
| Project 3 | 70 percent | $ 20,000 |
| | 30 percent | –$ 5,000 |
| Project 4 | 30 percent | $ 40,000 |
| | 30 percent | $ 30,000 |
| | 20 percent | $ 20,000 |
| | 20 percent | –$50,000 |

5. Find an example of a company that took a big risk on an information technology project and succeeded. In addition, find an example of a company that took a big risk and failed. Summarize each project and situation in a two-page paper where you should also discuss whether you believe that anything besides luck makes a difference between success and failure.

## Running Case

Tony and his team identified some risks during the first month of the Recreation and Wellness Intranet Project. However, all they did was document them in a list. They never ranked them or developed any response strategies. Since several problems have been occurring on the project, such as key team members leaving the company, users being uncooperative, and team members not providing good status information, Tony has decided to be more proactive in managing risks. He also wants to address positive as well as negative risks.

1. Create a risk register for the project, using Table 11-5 and the data below it as a guide. Identify six potential risks, including risks related to the problems described above. Include negative and positive risks.

2. Plot the six risks on a probability/impact matrix, using Figure 11-7. Also assign a numeric value for the probability and impact of each risk on meeting the main project objective. Use a scale of 1 to 10 in assigning the values, with 1 being low and 10 being high. For a simple risk factor calculation, multiply these two values (the probability score and the impact score). Add a column to your risk register to the right of the impact column called Risk Score. Enter the new data in the risk register. Write your rationale for how you determined the scores for one of the negative risks and one of the positive risks.
3. Develop a response strategy for one of the negative risks and one of the positive risks. Enter the information in the risk register. Also write a separate paragraph describing what specific tasks would need to be done to implement the strategy. Include time and cost estimates for each strategy, as well.



## Companion Web Site

Visit the companion Web site for this text (*www.cengage.com/mis/schwalbe*) to access:

- References cited in the text and additional suggested readings for each chapter
- Template files
- Lecture notes
- Interactive quizzes
- Podcasts
- Links to general project management Web sites
- And more

See the Preface of this text for additional information on accessing the companion Web site.

## Key Terms

**brainstorming** — a technique by which a group attempts to generate ideas or find a solution for a specific problem by amassing ideas spontaneously and without judgment

**contingency allowances** — provisions held by the project sponsor or organization to reduce the risk of cost or schedule overruns to an acceptable level; also called *contingency reserves*

**contingency plans** — predefined actions that the project team will take if an identified risk event occurs

**contingency reserves** — provisions held by the project sponsor or organization to reduce the risk of cost or schedule overruns to an acceptable level; also called *contingency allowances*

**decision tree** — a diagramming analysis technique used to help select the best course of action in situations in which future outcomes are uncertain

**Delphi technique** — an approach used to derive a consensus among a panel of experts, to make predictions about future developments

**expected monetary value (EMV)** — the product of the risk event probability and the risk event's monetary value

**fallback plans** — plans developed for risks that have a high impact on meeting project objectives, to be implemented if attempts to reduce the risk are not effective

**flowcharts** — diagrams that show how various elements of a system relate to each other

**influence diagram** — diagram that represents decision problems by displaying essential elements, including decisions, uncertainties, and objectives, and how they influence each other

**interviewing** — a fact-finding technique that is normally done face-to-face, but can also occur through phone calls, e-mail, or instant messaging

**known risks** — risks that the project team have identified and analyzed and can be managed proactively

**Monte Carlo analysis** — a risk quantification technique that simulates a model's outcome many times, to provide a statistical distribution of the calculated results

**probability/impact matrix or chart** — a matrix or chart that lists the relative probability of a risk occurring on one side of a matrix or axis on a chart and the relative impact of the risk occurring on the other



**residual risks** — risks that remain after all of the response strategies have been implemented

**risk** — an uncertainty that can have a negative or positive effect on meeting project objectives

**risk acceptance** — accepting the consequences should a risk occur

**risk-averse** — having a low tolerance for risk

**risk avoidance** — eliminating a specific threat or risk, usually by eliminating its causes

**risk breakdown structure** — a hierarchy of potential risk categories for a project

**risk enhancement** — changing the size of an opportunity by identifying and maximizing key drivers of the positive risk

**risk events** — specific uncertain events that may occur to the detriment or enhancement of the project

**risk exploitation** — doing whatever you can to make sure the positive risk happens

**risk factors** — numbers that represent overall risk of specific events, given their probability of occurring and the consequence to the project if they do occur

**risk management plan** — a plan that documents the procedures for managing risk throughout a project

**risk mitigation** — reducing the impact of a risk event by reducing the probability of its occurrence

**risk-neutral** — a balance between risk and payoff

**risk owner** — the person who will take responsibility for a risk and its associated response strategies and tasks

**risk register** — a document that contains results of various risk management processes, often displayed in a table or spreadsheet format

**risk-seeking** — having a high tolerance for risk

**risk sharing** — allocating ownership of the risk to another party

**risk tolerance** — the amount of satisfaction or pleasure received from a potential payoff; also called *risk utility*

**risk transference** — shifting the consequence of a risk and responsibility for its management to a third party

**risk utility** — the amount of satisfaction or pleasure received from a potential payoff; also called *risk tolerance*

**runaway projects** — projects that have significant cost or schedule overruns

**secondary risks** — risks that are a direct result of implementing a risk response

**sensitivity analysis** — a technique used to show the effects of changing one or more variables on an outcome

**Top Ten Risk Item Tracking** — a qualitative risk analysis tool for identifying risks and maintaining an awareness of risks throughout the life of a project

**triggers** — indications for actual risk events

**unknown risks** — risks that have not been identified and analyzed so they cannot be managed proactively

**watch list** — a list of risks that are low priority, but are still identified as potential risks

**workarounds** — unplanned responses to risk events when there are no contingency plans in place

## End Notes


[1] C. William Ibbs and Young Hoon Kwak, "Assessing Project Management Maturity," *Project Management Journal* 31, no. 1 (March 2000), pp. 32–43.

[2] Aneerav Sukhoo, Andries Barnard, Mariki M. Eloff, and John A. Van der Poll, "An Assessment of Software Project Management Maturity in Mauritius," *Issues in Informing Science and Information Technology* 2 (May 2003), pp. 671–690.

[3] Peter Kulik and Catherine Weber, "Software Risk Management Practices—2001," KLCI Research Group (August 2001).

[4] David Hillson, "Boom, bust, and risk management," *Project Manager Today* (September 2008).

[5] SAS, "Survey: Better risk management would have lessened credit crisis," SAS Press Release (September 18, 2008).

[6] David Hillson, "Integrated Risk Management as a Framework for Organisational Success," PMI Global Congress Proceedings (2006).

[7] The Standish Group, "Unfinished Voyages" StandishGroup.com (1996).

[8] Andy Cole, "Runaway Projects—Cause and Effects," *Software World* 26, no. 3, (1995), pp. 3–5.

[9] Stephanie Overby, "Bound to Fail," CIO Magazine (May 1, 2005).

[10] R. Max Wideman, "Project and Program Risk Management: A Guide to Managing Project Risks and Opportunities," Upper Darby, PA, II–4 (1992).

[11] J. Daniel Couger, *Creative Problem Solving and Opportunity Finding*, Boyd & Fraser Publishing Company (1995).

[12] McDonnell Douglas Corporation, "Hi-Rel Fighter Concept," Report MDC B0642 (1988).

[13] Microsoft Corporation, "Introduction to Monte Carlo simulation" (*http://office.microsoft.com/en-us/excel/HA011118931033.aspx*) (accessed December 18, 2008).

[14] Jean Couillard, "The Role of Project Risk in Determining Project Management Approach," *Project Management Journal* 25, no. 4 (December 1995), pp. 3–15.

CHAPTER 12

# PROJECT PROCUREMENT MANAGEMENT

## LEARNING OBJECTIVES

**After reading this chapter, you will be able to:**

- Understand the importance of project procurement management and the increasing use of outsourcing for information technology projects
- Describe the work involved in planning procurements for projects, including determining the proper type of contract to use and preparing a procurement management plan, statement of work, source selection criteria, and make-or-buy analysis
- Discuss what is involved in conducting procurements and strategies for obtaining seller responses, selecting sellers, and awarding contracts
- Understand the process of administering procurements by managing procurement relationships and monitoring contract performance
- Describe the process of closing procurements
- Discuss types of software available to assist in project procurement management

## OPENING CASE

Marie McBride could not believe how much money her company was paying for outside consultants to help the company finish an important operating system conversion project. The consulting company's proposal said it would provide experienced professionals who had completed similar conversions, and that the job would be finished in six months or less with four consultants working full time. Nine months later her company was still paying high consulting fees, and half of the original consultants on the project had been replaced with new people. One new consultant had graduated from college only two months before and had extremely poor communications skills. Marie's internal staff complained that they were wasting time training some of these "experienced professionals." Marie talked to her company's purchasing manager about the contract, fees, and special clauses that might be relevant to the problems they were experiencing.

Marie was dismayed at how difficult it was to interpret the contract. It was very long and obviously written by someone with a legal background. When she asked what her company could do since the consulting firm was not following its proposal, the purchasing manager stated that the proposal was not part of the official contract. Marie's company was paying for time and materials, not specific deliverables. There was no clause stating the minimum experience level required for the consultants, nor were there penalty clauses for not completing the work on time. There was a termination clause, however, meaning the company could terminate the contract. Marie wondered why her company had signed such a poor contract. Was there a better way to deal with procuring services from outside the company?



## THE IMPORTANCE OF PROJECT PROCUREMENT MANAGEMENT

**Procurement** means acquiring goods and/or services from an outside source. The term procurement is widely used in government; many private companies use the terms *purchasing* and *outsourcing*. Organizations or individuals who provide procurement services are referred to as suppliers, vendors, contractors, subcontractors, or sellers, with suppliers being the most widely used term. Many information technology projects involve the use of goods and services from outside the organization.

As described in Chapter 2, outsourcing has become a hot topic for research and debate, especially the implication of outsourcing to other countries, referred to as offshoring. The outsourcing statistics below are from an Information Technology Association of America (ITAA)–sponsored report:

- Spending for global sources of computer software and services is expected to grow at a compound annual rate of about 20 percent, increasing from about $15 billion in 2005 to $38 billion in 2010.
- Total savings from offshore resources during the same time period are estimated to grow from $8.7 billion to $20.4 billion. The cost savings and use of offshore resources lower inflation, increase productivity, and lower interest rates, which boosts business and consumer spending and increases economic activity.

- Although global outsourcing displaces some IT workers, total employment in the United States increases, according to ITAA, as the benefits ripple through the economy. "The incremental economic activity that follows offshore IT outsourcing creates over 257,000 net new jobs in 2005 and is expected to create over 337,000 net new jobs by 2010."[1]

Politicians debate on whether offshore outsourcing helps their own country or not. Andy Bork, chief operating officer of a computer network support service provider, describes outsourcing as an essential part of a healthy business diet. He describes good vs. bad outsourcing as something like good vs. bad cholesterol. He says that most people view offshore outsourcing as being bad because it takes jobs away from domestic workers. However, many companies are realizing that they can use offshore outsourcing *and* create more jobs at home. For example, Atlanta-based Delta Air Lines created 1,000 call-center jobs in India in 2003, saving $25 million, which enabled it to add 1,200 job positions for reservations and sales agents in the United States.[2] Other companies, like Wal-Mart, successfully manage the majority of their information technology projects in-house with very little commercial software and no outsourcing at all. (See the Suggested Reading on the companion Web site on "Wal-Mart's Way.")



Deciding whether to outsource, what to outsource, and how to outsource are important topics for many organizations throughout the world. In a 2008 survey, 74 percent of 600 global procurement executives believed that procurement issues are a high priority for their companies. About half of respondents also said that their companies focus too much on cost reduction instead of value creation. They also believe they are missing opportunities by not focusing on using technology to improve procurement processes. For example, 72 percent of respondents "have less than 10 percent of their spend channeled through eProcurement and eSourcing applications."[3]

Most organizations use some form of outsourcing to meet their information technology needs, spending most money within their own country. A 2008 report on IT outsourcing trends in the U.S. and Canada revealed the following:

- Application development is the most popular form of IT outsourcing and was used by 53 percent of organizations surveyed. Of the surveyed organizations, 44 percent outsourced application maintenance, 40 percent outsourced Web site or e-commerce systems, and 37 percent outsourced disaster recovery services.
- The IT function with the largest percentage of work outsourced is disaster recovery services, accounting for 50 percent of total IT outsourcing. Many organizations see the benefit in having an outside party perform offsite storage or maintenance of a recovery facility. Desktop support is the second most outsourced IT function (48 percent), followed closely by data center operations and help desk (47 percent each) and Web site or e-commerce systems (46 percent). IT security is at the bottom of the list, with only 29 percent of the work being outsourced.
- Even though application development and maintenance are frequently outsourced, they are a low percentage of the amount of total IT work outsourced. Application development and maintenance are often outsourced selectively since most organizations choose to do many projects in-house.[4]

Because outsourcing is a growing area, it is important for project managers to understand project procurement management. Many organizations are turning to outsourcing to:

- *Reduce both fixed and recurrent costs*. Outsourcing suppliers are often able to use economies of scale that may not be available to the client alone, especially for hardware and software. It can also be less expensive to outsource some labor costs to other organizations in the same country or offshore. Companies can also use outsourcing to reduce labor costs on projects by avoiding the costs of hiring, firing, and reassigning people to projects or paying their salaries when they are between projects.
- *Allow the client organization to focus on its core business*. Most organizations are not in business to provide information technology services, yet many have spent valuable time and resources on information technology functions when they should have focused on core competencies such as marketing, customer service, and new product design. By outsourcing many information technology functions, employees can focus on jobs that are critical to the success of the organization.
- *Access skills and technologies*. Organizations can gain access to specific skills and technologies when they are required by using outside resources. For example, a project may require an expert in a particular field or require the use of expensive hardware or software for one particular month on a project. Planning for this procurement ensure that the needed skills or technology will be available for the project.
- *Provide flexibility*. Outsourcing to provide extra staff during periods of peak workloads can be much more economical than trying to staff entire projects with internal resources. Many companies cite quicker flexibility in staffing as a key reason for outsourcing.
- *Increase accountability*. A well-written **contract**—a mutually binding agreement that obligates the seller to provide the specified products or services and obligates the buyer to pay for them—can clarify responsibilities and sharpen focus on key deliverables of a project. Because contracts are legally binding, there is more accountability for delivering the work as stated in the contract.



Organizations must also consider reasons they might *not* want to outsource. When an organization outsources work, it often does not have as much control over those aspects of projects that suppliers carry out. In addition, an organization could become too dependent on particular suppliers. If those suppliers went out of business or lost key personnel, it could cause great damage to a project. Organizations must also be careful to protect strategic information that could become vulnerable in the hands of suppliers. According to Scott McNeally, CEO of Sun Microsystems, Inc., "What you want to handle in-house is the stuff that gives you an edge over your competition—your core competencies. I call it your 'secret sauce.' If you're on Wall Street and you have your own program for tracking and analyzing the market, you'll hang onto that. At Sun, we have a complex program for testing microprocessor designs, and we'll keep it."[5] Project teams must think carefully about procurement issues and make wise decisions based on the unique needs of their projects and organizations. They can also change their minds on outsourcing as business conditions change.

In December 2002, when the financial services company JPMorgan Chase announced a seven-year, $5 billion deal to outsource much of its data processing to IBM, both companies bragged that the contract was the largest of its kind. It seemed like a win-win situation—IBM would make money and reduce costs, and JPMorgan Chase could push for innovation. However, in September 2004, JPMorgan Chase revoked the contract less than two years into its existence because the procurement plan no longer fit with JPMorgan Chase's business strategy. According to Austin Adams, chief information officer at JPMorgan Chase, "We believe managing our own technology infrastructure is best for the long-term growth and success of our company as well as our shareholders." However, IBM said the canceled contract was simply a result of JPMorgan Chase's merger earlier that year with Bank One. It tried to shrug off the loss of a large business deal. "The combined firm found itself with an abundance of IT assets," IBM spokesperson James Sciales said. "This decision was like other business decisions related to the merger."[6]



Outsourcing can also cause problems in other areas for companies and nations as a whole. For example, many people in Australia are concerned about outsourcing software development. "The Australian Computer Society says sending work offshore may lower the number of students entering IT courses, deplete the number of skilled IT professionals, and diminish the nation's strategic technology capability. Another issue is security, which encompasses the protection of intellectual property, integrity of data, and the reliability of infrastructure in offshore locations."[7]

The success of many information technology projects that use outside resources is often due to good project procurement management. **Project procurement management** includes the processes required to acquire goods and services for a project from outside the performing organization. Organizations can be either the buyer or the seller of products or services under a contract.

There are four main processes in project procurement management:

1. *Planning procurements* involves determining what to procure, when, and how. In procurement planning, one must decide what to outsource, determine the type of contract, and describe the work for potential sellers. **Sellers** are contractors, suppliers, or providers who provide goods and services to other organizations. Outputs of this process include a procurement management plan, statements of work, make-or-buy decisions, procurement documents, source selection criteria, and change requests.
2. *Conducting procurements* involves obtaining seller responses, selecting sellers, and awarding contracts. Outputs include selected sellers, procurement contract awards, resource calendars, change requests, and updates to the project management plan and other project documents.
3. *Administering procurements* involves managing relationships with sellers, monitoring contract performance, and making changes as needed. The main outputs of this process include procurement documentation, organizational

process asset updates, change requests, and project management plan updates.

4. *Closing procurements* involves completion and settlement of each contract, including resolution of any open items. Outputs include closed procurements and organizational process asset updates.

Figure 12-1 summarizes these processes and outputs, showing when they occur in a typical project.



**Planning**
Process: **Plan procurements**
Outputs: Procurement management plan, procurement statements of work, make-or-buy decisions, procurement documents, source selection criteria, and change requests

**Executing**
Process: **Conduct procurements**
Outputs: Selected sellers, procurement contract award, resource calendars, change requests, project management plan updates, project document updates

**Monitoring and Controlling**
Process: **Administer procurements**
Outputs: Procurement documentation, organizational process asset updates, change requests, project management plan updates

**Closing**
Process: **Close procurements**
Outputs: Closed procurements, organizational process asset updates

**Project Start** **Project Finish**

**FIGURE 12-1** Project procurement management summary

## PLANNING PROCUREMENTS

Planning procurements involves identifying which project needs can best be met by using products or services outside the organization. It involves deciding whether to procure, how to procure, what to procure, how much to procure, and when to procure. An important output of this process is the make-or-buy decision. A **make-or-buy decision** is one in which an organization decides if it is in its best interests to make certain products or perform certain services inside the organization, or if it is better to buy them from an outside organization. If there is no need to buy any products or services from outside the organization, then there is no need to perform any of the other procurement management processes.

For many projects, properly outsourcing some information technology functions can be a great investment, as shown in the following examples of What Went Right.

## WHAT WENT RIGHT?

The Boots Company PLC, a pharmacy and health care company in Nottingham, England, outsourced its information technology systems to IBM in October 2002. The Boots Company signed a ten-year contract worth about $1.1 billion and expected to save $203.9 million over that period compared with the cost of running the systems itself. IBM managed and developed The Boots Company's systems infrastructure "from the mainframes to the tills in our 1,400 stores, to the computer on my desk," said spokesperson Francis Thomas. More than 400 Boots employees were transferred to IBM's payroll but continued to work at Boots' head office, with extra IBM staff brought in as needed. Thomas added, "The great thing about this is that if IBM has an expert in Singapore and [if] we need that expertise, we can tap into it for three months. It keeps our costs on an even keel."[8] It is not uncommon for long contracts to be renegotiated, becoming either shorter or longer in length. In May 2006, Boots and IBM began discussing amendments to their contract because much of Boots' IT infrastructure renewal program (including a new pharmacy system and an SAP rollout) was complete. A Boots spokesman said that the company achieved its goal much quicker than planned. In contrast, in 2005 Boots renegotiated its £90 million, seven-year IT contract it initially signed in 2002 with Xansa, extending it for another two years to 2011 in a £26 million deal.[9] It is also not uncommon to take advantage of competition and the changing marketplace for major procurements. In 2008, Boots announced that it would have up to six different suppliers competing to supply its IT products and services over the next year. "The company is keeping its business system management team and service management in-house, including helpdesk and project management."[10]

Properly planning purchases and acquisitions and writing good contracts can also save organizations millions of dollars. Many companies centralize purchasing for products, such as personal computers, software, and printers, to earn special pricing discounts. For example, in the mid-1980s the U.S. Air Force awarded a five-year, multimillion-dollar contract to automate 15 Air Force Systems Command bases. The project manager and contracting officer decided to allow for a unit pricing strategy for some items required in the contract, such as the workstations and printers. By not requiring everything to be negotiated at a fixed cost, the winning supplier lowered its final bid by more than $40 million.[11]



Inputs needed for planning procurements include the scope baseline, requirements documentation, teaming agreements, the risk register, risk-related contract decisions, activity resource requirements, the project schedule, activity cost estimates, the cost performance baseline, enterprise environmental factors, and organizational process assets. For example, a large clothing company might consider outsourcing the delivery of, maintenance of, and basic user training and support for laptops supplied to its international sales and marketing force. If there were suppliers who could provide this service well at a reasonable price, it would make sense to outsource, because this could reduce fixed and recurring costs for the clothing company and let them focus on their core business of selling clothes.

It is important to understand why a company would want to procure goods or services and what inputs are needed to plan purchases and acquisitions. In the opening case, Marie's company hired outside consultants to help complete an operating system conversion project because it needed people with specialized skills for a short period of time. This is a common occurrence in many information technology projects. It can be more effective to hire skilled consultants to perform specific tasks for a short period of time than to hire or keep employees on staff full time.

However, it is also important to define clearly the scope of the project, the products, services, or results required, market conditions, and constraints and assumptions. In Marie's case, the scope of the project and services required were relatively clear, but her company may not have adequately discussed or documented the market conditions or constraints and assumptions involved in using the outside consultants. Were there many companies that provided consultants to do operating conversion projects similar to theirs? Did the project team investigate the background of the company that provided the consultants? Did they list important constraints and assumptions for using the consultants, such as limiting the time that the consultants had to complete the conversion project or the minimum years of experience for any consultant assigned to the project? It is very important to answer these types of questions before going into an outsourcing agreement.



## Tools and Techniques for Planning Procurements

There are several tools and techniques to help project managers and their teams in planning procurements, including make-or-buy analysis, expert judgment, and contract types.

### Make-or-Buy Analysis

Make-or-buy analysis is a general management technique used to determine whether an organization should make or perform a particular product or service inside the organization or buy from someone else. This form of analysis involves estimating the internal costs of providing a product or service and comparing that estimate to the cost of outsourcing. Consider a company that has 1,000 international salespeople with laptops. Using make-or-buy analysis, the company would compare the cost of providing those services using internal resources to the cost of buying those services from an outside source. If supplier quotes were less than its internal estimates, the company should definitely consider outsourcing the training and user support services. Another common make-or-buy decision, though more complex, is whether a company should develop an application itself or purchase software from an outside source and customize it to the company's needs.

Many organizations also use make-or-buy analysis to decide if they should either purchase or lease items for a particular project. For example, suppose you need a piece of equipment for a project that has a purchase price of $12,000. Assume it also had a daily operational cost of $400. Suppose you could lease the same piece of equipment for $800 per day, including the operational costs. You can set up an equation in which the purchase cost equals the lease cost to determine when it makes sense financially to lease or buy the equipment. In this example, $d$ = the number of days you need the piece of equipment. The equation would then be:

$$\$800d = \$12{,}000 + \$400d$$

Subtracting $400*d* from both sides, you get:

$$\$400d = \$12{,}000$$

Dividing both sides by $400, you get:

$$d = 30$$

which means that the purchase cost equals the lease cost in 30 days. So, if you need the equipment for less than 30 days, it would be more economical to lease it. If you need the equipment for more than 30 days, you should purchase it. In general, leasing is often cheaper for meeting short-term needs, but more expensive for long-term needs.

### Expert Judgment

Experts inside an organization and outside an organization could provide excellent advice in planning purchases and acquisitions. Project teams often need to consult experts within their organization as part of good business practice. Internal experts might suggest that the company in the above example could not provide quality training and user support for the 1,000 laptop users since the service involves so many people with different skill levels in so many different locations. Experts in the company might also know that most of their competitors outsource this type of work and know who the qualified outside suppliers are. It is also important to consult legal experts since contracts for outsourced work are legal agreements.

Experts outside the company, including potential suppliers themselves, can also provide expert judgment. For example, suppliers might suggest an option for salespeople to purchase the laptops themselves at a reduced cost. This option would solve problems during employee turnover—exiting employees would own their laptops and new employees would purchase a laptop through the program. An internal expert might then suggest that employees receive a technology bonus to help offset what they might view as an added expense. Expert judgment, both internal and external, is an asset in making many procurement decisions.

### Types of Contracts

Contract type is an important consideration. Different types of contracts can be used in different situations. Three broad categories of contracts are fixed price or lump sum, cost reimbursable, and time and material. A single contract can actually include all three of these categories, if it makes sense for that particular procurement. For example, you could have a contract with a seller that includes purchasing specific hardware for a fixed price or lump sum, some services that are provided on a cost reimbursable basis, and other services that are provided on a time and material basis. Project managers and their teams must understand and decide which approaches to use to meet their particular project needs. It is also important to understand when and how you can take advantage of unit pricing in contracts.

**Fixed-price** or **lump-sum contracts** involve a fixed total price for a well-defined product or service. The buyer incurs little risk in this situation since the price is predetermined. The sellers often pad their estimate somewhat to reduce their risk, realizing their price must still be competitive. For example, a company could award a fixed-price contract to purchase 100 laser printers with a certain print resolution and print speed to be delivered to one location within two months. In this example, the product and delivery date are well defined. Several sellers could create fixed price estimates for completing the job. Fixed-price contracts may also include incentives for meeting or exceeding selected project objectives. For example, the contract could include an incentive fee paid if the laser printers are delivered within one month. A firm-fixed-price (FFP) contract has the least amount of risk for the buyer, followed by a fixed-price incentive (FPI) contract.

Contracts can also include incentives to prevent or reduce cost overruns. For example, according to the U.S. Federal Acquisition Regulation (FAR) 16.4, fixed-price incentive fee contracts can include a **Point of Total Assumption (PTA)**, which is the cost at which the contractor assumes total responsibility for each additional dollar of contract cost. Contractors do not want to reach the point of total assumption because it hurts them financially, so they have an incentive to prevent cost overruns. The PTA is calculated with the following formula:

$$\text{PTA} = (\text{ceiling price} - \text{target price})/\text{government share} + \text{target cost}$$

For example, given the following information, assuming all dollars are in millions:



$$\text{Ceiling price} = \$1,250$$
$$\text{Target price} = \$1,100$$
$$\text{Target cost} = \$1,000$$
$$\text{Share: } 75\%$$
$$\text{PTA} = (\$1,250 - \$1,100)/.75 + \$1,000 = \$1,200$$[12]

**Cost-reimbursable contracts** involve payment to the supplier for direct and indirect actual costs. Recall from Chapter 7 that direct costs are costs that can be directly related to producing the products and services of the project. They normally can be traced back to a project in a cost-effective way. Indirect costs are costs that are not directly related to the products or services of the project, but are indirectly related to performing the project. They normally cannot be traced back in a cost-effective way. For example, the salaries for people working directly on a project and hardware or software purchased for a specific project are direct costs, while the cost of providing a work space with electricity, a cafeteria, and so on are indirect costs. Indirect costs are often calculated as a percentage of direct costs. Cost-reimbursable contracts often include fees, such as a profit percentage or incentives for meeting or exceeding selected project objectives. These contracts are often used for projects that include providing goods and services that involve new technologies. The buyer absorbs more of the risk with cost-reimbursable contracts than they do with fixed-price contracts. Three types of cost-reimbursable contracts, in order of lowest to highest risk to the buyer, include cost plus incentive fee, cost plus fixed fee, and cost plus percentage of costs.

- With a **cost plus incentive fee (CPIF) contract**, the buyer pays the supplier for allowable performance costs along with a predetermined fee and an incentive bonus. See the Media Snapshot for an example of providing financial incentives to complete an important construction project ahead of schedule. Incentives are also often provided to suppliers for reducing contract costs. If the final cost is less than the expected cost, both the buyer and the supplier benefit from the cost savings, according to a pre-negotiated share formula. For example, suppose the expected cost of a project is $100,000, the fee to the supplier is $10,000, and the share formula is 85/15, meaning that the buyer absorbs 85 percent of the uncertainty and the supplier absorbs 15 percent. If the final price is $80,000, the cost savings are $20,000. The supplier would be paid the final cost and the fee plus an incentive of $3,000 (15 percent of $20,000), for a total reimbursement of $93,000.

Contract incentives can be extremely effective. On August 1, 2007, tragedy struck Minneapolis, Minnesota, when a bridge on I-35W crossing the Mississippi River suddenly collapsed, killing 13 motorists, injuring 150 people, and leaving a mass of concrete and steel in the river and on its banks. The Minnesota Department of Transportation (MnDOT) acted quickly to find a contractor to rebuild the bridge. They also provided a strong incentive to finish the bridge as quickly as possible, ensuring quality and safety along the way.

Peter Sanderson, project manager for the joint venture of Flatiron-Manson, hired to build the bridge, led his team in completing the project three months ahead of schedule, and the new bridge opened on September 18, 2008. The contractors earned $25 million in incentive fees on top of their $234 million contract for completing the bridge ahead of schedule.

Why did MnDOT offer such a large incentive fee for finishing the project early? "I-35W in Minneapolis is a major transportation artery for the Twin Cities and entire state. Each day this bridge has been closed, it has cost road users more than $400,000," MnDOT Commissioner Tom Sorel remarked. "Area residents, business owners, motorists, workers and others have been affected by this corridor's closure. The opening of this bridge reconnects our community."[13]



- With a **cost plus fixed fee (CPFF) contract**, the buyer pays the supplier for allowable performance costs plus a fixed fee payment usually based on a percentage of estimated costs. This fee does not vary, however, unless the scope of the contract changes. For example, suppose the expected cost of a project is $100,000 and the fixed fee is $10,000. If the actual cost of the contract rises to $120,000 and the scope of the contract remains the same, the contractor will still receive the fee of $10,000.
- With a **cost plus award fee (CPAF) contract**, the buyer pays the supplier for allowable performance costs plus an award fee based on the satisfaction of subjective performance criteria. For example, you could consider the tip or gratuity you would give a server in a restaurant as a simple example, as long as there is no set gratuity percentage. You still have to pay for the cost of your meal, but you can decide on the tip amount based on your satisfaction with the food, drinks, and services provided. This type of contract is not usually subject to appeals.
- With a **cost plus percentage of costs (CPPC) contract**, the buyer pays the supplier for allowable performance costs along with a predetermined percentage based on total costs. From the buyer's perspective, this is the least desirable type of contract because the supplier has no incentive to decrease costs. In fact, the supplier may be motivated to increase costs, since doing so will automatically increase profits based on the percentage of costs. This type of contract is prohibited for U.S. federal government use, but it is sometimes used in private industry, particularly in the construction industry. All of the risk is borne by the buyer.

**Time and material (T&M) contracts** are a hybrid of both fixed-price and cost-reimbursable contracts. For example, an independent computer consultant might have a contract with a company based on a fee of $80 per hour for his or her services plus a fixed price of $10,000 for providing specific materials for the project. The materials fee might also be based on approved receipts for purchasing items, with a ceiling of $10,000. The consultant would send an invoice to the company each week or month, listing the materials fee, the number of hours worked, and a description of the work produced. This type of contract is often used for services that are needed when the work cannot be clearly specified and total costs cannot be estimated in a contract. Many contract programmers and consultants, such as those Marie's company hired in the opening case, prefer time and material contracts.

**Unit pricing** can also be used in various types of contracts to require the buyer to pay the supplier a predetermined amount per unit of product or service. The total value of the contract is a function of the quantities needed to complete the work. Consider an information technology department that might have a unit price contract for purchasing computer hardware. If the company purchases only one unit, the cost might be $1,000. If it purchases 10 units, the cost would be $10,000. This type of pricing often involves volume discounts. For example, if the company purchases between 10 and 50 units, the contracted cost might be $900 per unit. If it purchases over 50 units, the cost might go down to $800 per unit. This flexible pricing strategy is often advantageous to both the buyer and the seller. (See the second example in the What Went Right earlier in this chapter.)



Any type of contract should include specific clauses that take into account issues unique to the project. For example, if a company uses a time and material contract for consulting services, the contract should stipulate different hourly rates based on the level of experience of each individual contractor. The services of a junior programmer with no Bachelor's degree and less than three years' experience might be billed at $40 per hour, whereas the services of a senior programmer with at least a Bachelor's degree and more than ten years of experience might be billed at $80 per hour.

Figure 12-2 summarizes the spectrum of risk to the buyer and supplier for different types of contracts. Buyers have the lowest risk with firm-fixed price contracts, because they know exactly what they will need to pay the supplier. Buyers have the most risk with cost plus percentage of costs (CPPC) contracts because they do not know what the supplier's costs will be in advance, and the suppliers may be motivated to keep increasing costs. From the supplier's perspective, there is the least risk with a CPPC contract and the most risk with the firm-fixed price contract.

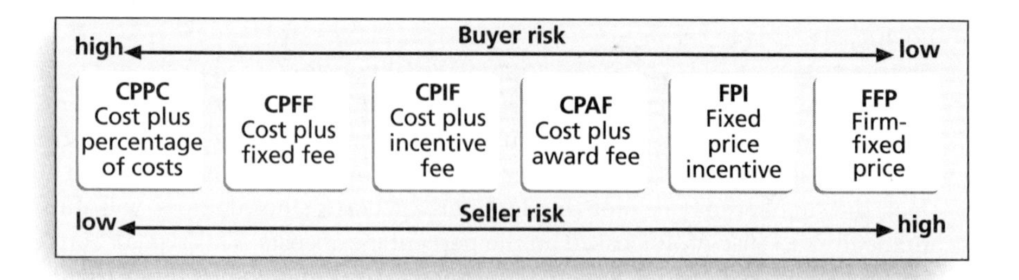


**FIGURE 12-2** Contract types versus risk

Time and material contracts and unit-price contracts can be high- or low-risk, depending on the nature of the project and other contract clauses. For example, if an organization is unclear on what work needs to be done, it cannot expect a supplier to sign a firm-fixed price contract. However, the buyer could find a consultant or group of consultants to work on specific tasks based on a predetermined hourly rate. The buying organization could evaluate the work produced every day or week to decide if it wants to continue using the consultants. In this case the contract would include a **termination clause**—a contract clause that allows the buyer or supplier to end the contract. Some termination clauses state that the buyer can terminate a contract for any reason and give the supplier only 24 hours' notice. Suppliers must often give a one-week notice to terminate a contract and must have sufficient reasons for the termination. The buyer could also include a contract clause specifying hourly rates that are based on education and experience of consultants. These contract clauses reduce the risk incurred by the buyer while providing flexibility for accomplishing the work.



In addition to make-or-buy decisions and change requests based on the procurement decision, important outputs of planning procurements are a procurement management plan, statement of work, procurement documents (i.e., requests for proposals or quotes), and source selection criteria.

## Procurement Management Plan

As stated earlier, every project management knowledge area includes some planning. The procurement management plan is a document that describes how the procurement processes will be managed, from developing documentation for making outside purchases or acquisitions to contract closure. Like other project plans, contents of the procurement management plan will vary with project needs. Some of the topics that can be included in a procurement management plan include:

- Guidelines on types of contracts to be used in different situations
- Standard procurement documents or templates to be used, if applicable
- Guidelines for creating contract work breakdown structures, statements of work, and other procurement documents
- Roles and responsibilities of the project team and related departments, such as the purchasing or legal department
- Guidelines on using independent estimates for evaluating sellers
- Suggestions on managing multiple providers
- Processes for coordinating procurement decisions, such as make-or-buy decisions, with other project areas, such as scheduling and performance reporting
- Constraints and assumptions related to purchases and acquisitions
- Lead times for purchases and acquisitions
- Risk mitigation strategies for purchases and acquisitions, such as insurance contracts and bonds
- Guidelines for identifying prequalified sellers and organizational lists of preferred sellers
- Procurement metrics to assist in evaluating sellers and managing contracts

## Statement of Work

The **statement of work (SOW)** is a description of the work required for the procurement. Some organizations use the term statement of work to describe a document for describing internal work, as well. If a SOW is used as part of a contract to describe only the work required for that particular contract, it is called a *contract statement of work*. The contract SOW is a type of scope statement that describes the work in sufficient detail to allow prospective suppliers to determine if they are capable of providing the goods and services required and to determine an appropriate price. A contract SOW should be clear, concise, and as complete as possible. It should describe all services required and include performance reporting. It is important to use appropriate words in a contract SOW, such as *must* instead of *may*. For example, *must* implies that something has to be done; *may* implies that there is a choice involved in doing something or not. The contract SOW should specify the products and services required for the project, use industry terms, and refer to industry standards.



Many organizations use samples and templates to generate SOWs. Figure 12-3 provides a basic outline or template for a contract SOW that Marie's organization could use when they hire outside consultants or purchase other goods or services. For example, for the operating

**Statement of Work (SOW)**

I. **Scope of Work:** Describe the work to be done in detail. Specify the hardware and software involved and the exact nature of the work.

II. **Location of Work:** Describe where the work must be performed. Specify the location of hardware and software and where the people must perform the work.

III. **Period of Performance:** Specify when the work is expected to start and end, working hours, number of hours that can be billed per week, where the work must be performed, and related schedule information.

IV. **Deliverables Schedule:** List specific deliverables, describe them in detail, and specify when they are due.

V. **Applicable Standards:** Specify any company or industry-specific standards that are relevant to performing the work.

VI. **Acceptance Criteria:** Describe how the buyer organization will determine if the work is acceptable.

VII. **Special Requirements:** Specify any special requirements such as hardware or software certifications, minimum degree or experience level of personnel, travel requirements, and so on.

**FIGURE 12-3** Statement of Work (SOW) template

system conversion project, Marie's company should specify the specific manufacturer and model number for the hardware involved, the former operating systems and new ones for the conversion, the number of pieces of each type of hardware involved (mainframes, midrange computers, or PCs), and so on. The contract SOW should also specify the location of the work, the expected period of performance, specific deliverables and when they are due, applicable standards, acceptance criteria, and special requirements. A good contract SOW gives bidders a better understanding of the buyer's expectations. A contract SOW should become part of the official contract to ensure that the buyer gets what the supplier bid on.

## Procurement Documents

Planning procurements also involves preparing the documents needed for potential sellers to prepare their responses and determining the evaluation criteria for the contract award. The project team often uses standard forms and expert judgment as tools to help them create relevant procurement documents and evaluation criteria.



Two common examples of procurement documents include a Request for Proposal (RFP) and a Request for Quote (RFQ). A **Request for Proposal (RFP)** is a document used to solicit proposals from prospective suppliers. A **proposal** is a document prepared by a seller when there are different approaches for meeting buyer needs. For example, if an organization wants to automate its work practices or find a solution to a business problem, it can write and issue an RFP so suppliers can respond with proposals. Suppliers might propose various hardware, software, and networking solutions to meet the organization's need. Selections of winning sellers are often made on a variety of criteria, not just the lowest price. Developing an RFP is often a very time-consuming process. Organizations must do proper planning to ensure they adequately describe what they want to procure, what sellers should include in their proposals, and how they will evaluate proposals.

A **Request for Quote (RFQ)** is a document used to solicit quotes or bids from prospective suppliers. A **bid**, also called a *tender* or *quote* (short for quotation), is a document prepared by sellers providing pricing for standard items that have been clearly defined by the buyer. Organizations often use an RFQ for solicitations that involve specific items. For example, if a company wanted to purchase 100 personal computers with specific features, it might issue an RFQ to potential suppliers. RFQs usually do not take nearly as long to prepare as RFPs, nor do responses to them. Selections are often made based on the lowest price bid.

Writing a good RFP is a critical part of project procurement management. Many people have never had to write or respond to an RFP. To generate a good RFP, expertise is invaluable. Many examples of RFPs are available within different companies, from potential contractors, and from government agencies. There are often legal requirements involved in issuing RFPs and reviewing proposals, especially for government projects. It is important to consult with experts familiar with the contract planning process for particular organizations. To make sure the RFP has enough information to provide the basis for a good proposal, the buying organization should try to put itself in the suppliers' shoes. Could you develop a good proposal based on the information in the RFP? Could you determine detailed pricing and schedule information based on the RFP? Developing a good RFP is difficult, as is writing a good proposal.

Figure 12-4 provides a basic outline or template for an RFP. The main sections of an RFP usually include a statement of the purpose of the RFP, background information on the organization issuing the RFP, the basic requirements for the products and services being proposed, the hardware and software environment (usually important information for information technology related proposals), a description of the RFP process, the statement of work and schedule information, and possible appendices. A simple RFP might be three to five pages long, while an RFP for a larger, more complicated procurement might be hundreds of pages long.



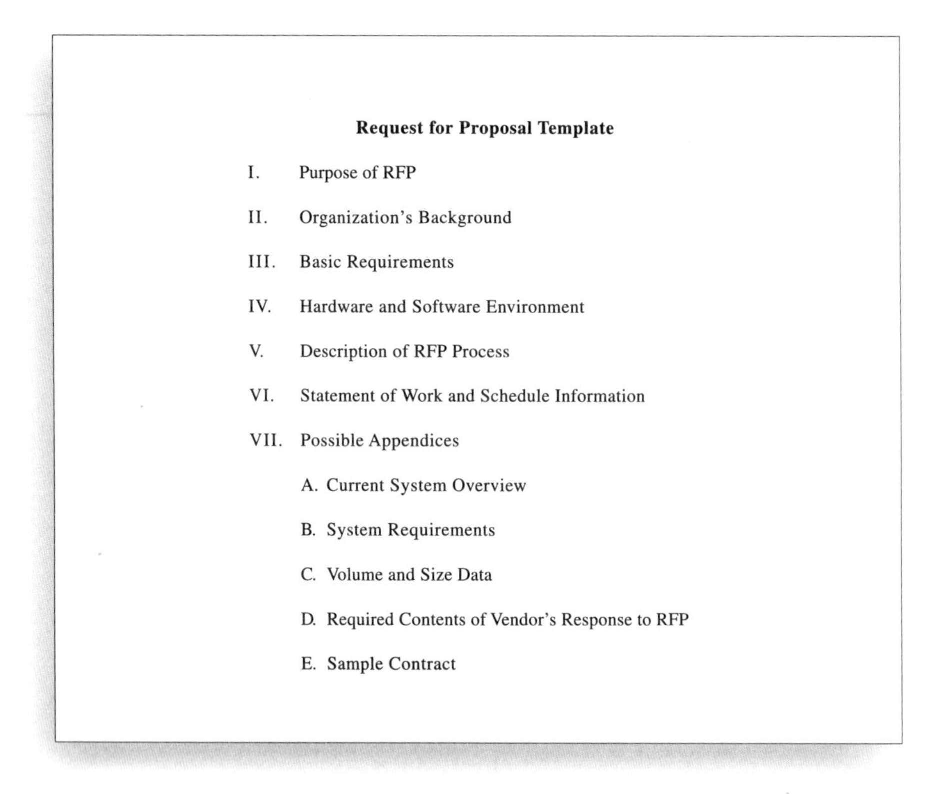

**Request for Proposal Template**

I. Purpose of RFP

II. Organization's Background

III. Basic Requirements

IV. Hardware and Software Environment

V. Description of RFP Process

VI. Statement of Work and Schedule Information

VII. Possible Appendices

A. Current System Overview

B. System Requirements

C. Volume and Size Data

D. Required Contents of Vendor's Response to RFP

E. Sample Contract

**FIGURE 12-4** Request for Proposal (RFP) template

Other terms used for RFQs and RFPs include *invitations for bid*, *invitations for negotiation*, and *initial contractor responses*. Regardless of what they are called, all procurement documents should be written to facilitate accurate and complete responses from prospective sellers. They should include background information on the organization and project, the relevant statement of work, a schedule, a description of the desired form of response, evaluation criteria, pricing forms, and any required contractual provisions. They should also be rigorous enough to ensure consistent, comparable responses, but flexible enough to allow consideration of seller suggestions for better ways to satisfy the requirements.

## Source Selection Criteria

It is very important for organizations to prepare some form of source selection evaluation criteria, preferably before they issue a formal RFP. Organizations use criteria to rate or score proposals, and they often assign a weight to each criterion to indicate how important it is. Some examples of criteria include the technical approach (30 percent weight), management approach (30 percent weight), past performance (20 percent weight), and price (20 percent weight). The criteria should be specific and objective. For example, if the buyer wants the supplier's project manager to be a certified Project Management Professional (PMP), the procurement documents should state that requirement clearly and follow it during the award process. Losing bidders may pursue legal recourse if the buyer does not follow a fair and consistent evaluation process.

Organizations should heed the saying, "Let the buyer beware." It is critical to evaluate proposals based on more than the appearance of the paperwork submitted. A key factor in evaluating bids, particularly for projects involving information technology, is the past performance record of the bidder. The RFP should require bidders to list other similar projects they have worked on and provide customer references for those projects. Reviewing performance records and references helps to reduce the risk of selecting a supplier with a poor track record. Suppliers should also demonstrate their understanding of the buyer's need, their technical and financial capabilities, their management approach to the project, and their price for delivering the desired goods and services. It is also crucial to write the contract to protect the buyer's interests.



Some information technology projects also require potential sellers to deliver a technical presentation as part of their proposal. The proposed project manager should lead the potential seller's presentation team. When the outside project manager leads the proposal presentation, the organization can begin building a relationship with the potential seller from the beginning. Visits to contractor sites can also help the buyer get a better feeling for the seller's capabilities and management style.

## CONDUCTING PROCUREMENTS

After planning for procurements, the next procurement management process involves deciding whom to ask to do the work, sending appropriate documentation to potential sellers, obtaining proposals or bids, selecting a seller, and awarding a contract. Prospective sellers do some of the work in this process, normally at no cost to the buyer or project. The buying organization is responsible for advertising the work, and for large procurements, they often hold some sort of bidders' conference to answer questions about the job. Two of the main outputs of this process are a selected seller and procurement contract award.

Organizations can advertise to procure outside goods and services in many different ways. Sometimes a specific supplier might be the number-one choice for the buyer. In this case, the buyer gives procurement information to just that company. If the preferred supplier responds favorably, both organizations proceed to work together. Many organizations have formed good working relationships with certain suppliers, so they want to continue working with them.

In many cases, however, there may be more than one supplier qualified to provide the goods and services. Providing information and receiving bids from multiple sources often takes advantage of the competitive business environment. Offshore outsourcing, as described earlier, has increased tremendously as organizations find suitable sellers around the globe. As a result of pursuing a competitive bidding strategy, the buyer can receive better goods and services than expected at a lower price.

A bidders' conference, also called a *supplier conference* or *pre-bid conference*, is a meeting with prospective sellers prior to preparation of their proposals or bids. These conferences help ensure that everyone has a clear, common understanding of the buyer's desired products or services. In some cases, the bidders' conference might be held online via a Webcast or using other communications technology. Buyers will also post procurement information on a Web site and post answers to frequently asked questions. Before, during, or after the bidders' conference, the buyer may incorporate responses to questions into the procurement documents as amendments.

Once buyers receive proposals or bids, they can select a supplier or decide to cancel the procurement. Selecting suppliers or sellers, often called *source selection*, involves evaluating proposals or bids from sellers, choosing the best one, negotiating the contract, and awarding the contract. It can be a long, tedious process, especially for large procurements. Several stakeholders in the procurement process should be involved in selecting the best supplier for the project. Often teams of people are responsible for evaluating various sections of the proposals. There might be a technical team, a management team, and a cost team to focus on each of those major areas. Buyers typically develop a short list of the top three to five suppliers to reduce the work involved in selecting a source.



Experts in source selection highly recommend that buyers use formal proposal evaluation sheets during source selection. Figure 12-5 provides a sample proposal evaluation sheet that the project team might use to help create a short list of the best three to five proposals. Notice that this example is a form of a weighted scoring model as described in Chapter 4, Project Integration Management. The score for a criterion would be calculated by multiplying the weight of that criterion by the rating for that proposal. Adding up the scores would provide the total weighted score for each proposal. The proposals with the highest weighted scores should be included in the short list of possible sellers. Experts also recommend that technical criteria should not be given more weight than management or cost criteria. Many organizations have suffered the consequences of paying too much attention to the technical aspects of proposals. For example, the project might cost much more than expected or take longer to complete because the source selection team focused only on technical aspects of proposals. Paying too much attention to technical aspects of proposals is especially likely to occur on information technology projects. However, it is often the supplier's management team—not the technical team—that makes procurement successful.

| | | Proposal 1 | | Proposal 2 | | Proposal 3, etc. | |
|---|---|---|---|---|---|---|---|
| Criteria | Weight | Rating | Score | Rating | Score | Rating | Score |
| Technical approach | 30% | | | | | | |
| Management approach | 30% | | | | | | |
| Past performance | 20% | | | | | | |
| Price | 20% | | | | | | |
| Total score | 100% | | | | | | |

**FIGURE 12-5** Sample proposal evaluation sheet

After developing a short list of possible sellers, organizations often follow a more detailed proposal evaluation process. For example, they might list more detailed criteria for important categories, such as the management approach. They might assign points for the potential project manager's educational background and PMP certification, his or her presentation (meaning the sellers had to give a formal presentation as part of the evaluation process), top management support for the project, and the organization's project management methodologies. If the criteria and evaluation are done well, the seller with the most points based on all of the criteria should be offered the contract.

It is customary to have contract negotiations during the source selection process. Sellers on the short list are often asked to prepare a best and final offer (BAFO). People who negotiate contracts for a living often conduct these negotiations for contracts that involve large amounts of money. In addition, top managers from both the buying and selling organizations usually meet before making final decisions. The final output is a contract that obligates the seller to provide the specified products or services and obligates the buyer to pay for them. It is also appropriate on some projects to prepare a contract management plan to describe details about how the contract will be managed.



## ADMINISTERING PROCUREMENTS

Administering procurements ensures that the seller's performance meets contractual requirements. The contractual relationship is a legal relationship and as such is subject to state and federal contract laws. It is very important that appropriate legal and contracting professionals be involved in writing and administering contracts.

Ideally, the project manager, a project team member, or an active user involved in the project should be actively involved in writing and administering the contract, so that everyone understands the importance of good procurement management. The project team should also seek expert advice in working with contractual issues. Project team members must be aware of potential legal problems they might cause by not understanding a contract. For example, most projects involve changes, and these changes must be handled properly for items under contract. Without understanding the provisions of the contract, a project manager may not realize he or she is authorizing the contractor to do additional work at additional costs. Therefore, change control is an important part of the contract administration process.

It is critical that project managers and team members watch for constructive change orders. **Constructive change orders** are oral or written acts or omissions by someone with actual or apparent authority that can be construed to have the same effect as a written change order. For example, if a member of the buyer's project team has met with the contractor on a weekly basis for three months to provide guidelines for performing work, he or she can be viewed as an apparent authority. If he or she tells the contractor to redo part of a report that has already been delivered and accepted by the project manager, that action can be viewed as a constructive change order and the contractor can legally bill the buyer for the additional work. Likewise, if this apparent authority tells the contractor to skip parts of a critical review meeting in the interests of time, the omission of that information is not the contractor's fault.

The following suggestions help ensure adequate change control and good contract administration:

- Changes to any part of the project need to be reviewed, approved, and documented by the same people in the same way that the original part of the plan was approved.
- Evaluation of any change should include an impact analysis. How will the change affect the scope, time, cost, and quality of the goods or services being provided? There must also be a baseline to understand and analyze changes.
- Changes must be documented in writing. Project team members should document all important meetings and telephone calls.
- When procuring complex information systems, project managers and their teams must stay closely involved to make sure the new system will meet business needs and work in an operational environment. Do not assume everything will go fine because you hired a reputable supplier. The buying organization needs to provide expertise as well.
- Have backup plans in case the new system does not work as planned when it is put into operation.
- Several tools and techniques can help in contract administration, such as a formal contract change control system, buyer-conducted procurement performance reviews, inspections and audits, performance reporting, payment systems, claims administration, and records management systems.



## BEST PRACTICE

Accenture, one of the leading IT outsourcing firms, summarized survey responses from 565 executives from several countries to develop a list of best practices from experienced outsourcers throughout the world. Its seven suggestions include the following:

1. *Build in Broad Business Outcomes Early and Often*: Incorporate business outcomes as a performance measure from the outset of the arrangement.
2. *Hire a Partner, Not Just a Provider*: Look for an outsourcing provider that brings a wide set of skills and strengths, and a long-term track record of delivering results, in addition to competitive pricing.
3. *It's More Than a Contract, It's a Business Relationship*: Give as much attention to performance measurement and the quality of your relationship with your provider as you do to the contract.
4. *Leverage Gain-Sharing*: Use risk/reward provisions as incentives for higher-performance outsourcing.
5. *Use Active Governance*: Use active governance to manage the outsourcing relationship for maximum performance.
6. *Assign a Dedicated Executive*: Task your talented executives with the mission of optimizing your outsourcing arrangements
7. *Focus Relentlessly on Primary Objectives*: Be clear about objectives—cost, process improvement and the ability to focus on the core business are the most common among outsourcing veterans.[14]

## CLOSING PROCUREMENTS

The final process in project procurement management is closing procurements, sometimes referred to as contract closure. Contract closure involves completion and settlement of contracts and resolution of any open items. The project team should determine if all work required in each contract was completed correctly and satisfactorily. They should also update records to reflect final results and archive information for future use.

Tools to assist in contract closure include procurement audits, negotiated settlements, and a records management system. Procurement audits are often done during contract closure to identify lessons learned in the entire procurement process. Organizations should strive to improve all of their business processes, including procurement management. Ideally, all procurements should end in a negotiated settlement between the buyer and seller. If negotiation is not possible, then some type of alternate disputes resolution such as mediation or arbitration can be used, and if all else fails, litigation in courts can be used to settle contracts. A records management system provides the ability to easily organize, find, and archive procurement-related documents. It is often an automated system, or at least partially automated, since there can be a large amount of information related to project procurement.



Outputs from contract closure include closed procurements and updates to organizational process assets. The buying organization often provides the seller with formal written notice that the contract has been completed. The contract itself should include requirements for formal acceptance and closure.

## USING SOFTWARE TO ASSIST IN PROJECT PROCUREMENT MANAGEMENT

Over the years, organizations have used various types of productivity software to assist in project procurement management. For example, most organizations use word-processing software to write proposals or contracts, spreadsheet software to create proposal evaluation worksheets, databases to track suppliers, and presentation software to present procurement-related information.

Many companies are now using more advanced software to assist in procurement management. In fact, the term "e-procurement" often describes various procurement functions that are now done electronically. A 2008 Wikipedia entry for e-procurement described seven types of e-procurement:[15]

- *Web-based ERP (Electronic Resource Planning)*: Creating and approving purchasing requisitions, placing purchase orders and receiving goods and services by using a software system based on Internet technology.
- *e-MRO (Maintenance, Repair and Overhaul)*: The same as web-based ERP except that the goods and services ordered are non-product related MRO supplies.
- *e-sourcing*: Identifying new suppliers for a specific category of purchasing requirements using Internet technology.
- *e-tendering*: Sending requests for information and prices to suppliers and receiving the responses of suppliers using Internet technology.

- *e-reverse auctioning*: Using Internet technology to buy goods and services from a number of known or unknown suppliers.
- *e-informing*: Gathering and distributing purchasing information both from and to internal and external parties using Internet technology.
- *e-marketsites*: Expands on Web-based ERP to open up value chains. Buying communities can access preferred suppliers' products and services, add to shopping carts, create requisition, seek approval, receipt purchase orders and process electronic invoices with integration to suppliers' supply chains and buyers' financial systems.

Many different Web sites and software tools can assist in procurement functions. For example, most business travelers use the Web to purchase airline tickets and to reserve rental cars and hotel rooms for business trips. With the rise of applications for smartphones such as the Apple iPhone and T-Mobile G1, shoppers can even take a picture of a barcode on all types of products and compare prices with competing stores in the area so that they know they are getting the best deal. Likewise, many organizations can purchase numerous items online, or they can buy specialized software to help streamline their procurement activities. Companies such as Perfect Commerce, Ariba, and others started providing corporate procurement services over the Internet. Other established companies, such as Oracle, SAS, and Baan, have developed new software products to assist in procurement management. Traditional procurement methods were very inefficient and costly, and e-procurement services have proved to be very effective in reducing the costs and burdens of procurement.



Organizations can also take advantage of information available on the Web, in industry publications, or in various discussion groups offering advice on selecting suppliers. For example, many organizations invest millions of dollars in enterprise project management software. Before deciding which seller's software to use, organizations use the Internet to find information describing specific products provided by various suppliers, prices, case studies, and current customer information to assist in making procurement decisions. Buyers can also use the Internet to hold bidders' conferences, as described earlier in this chapter, or to communicate procurement-related information.

As with any information or software tool, organizations must focus on using the information and tools to meet project and organizational needs. Many nontechnical issues are often involved in getting the most value out of new technologies, especially new e-procurement software. For example, organizations must often develop partnerships and strategic alliances with other organizations to take advantage of potential cost savings. Organizations should practice good procurement management in selecting new software tools and managing relationships with the chosen suppliers.

The processes involved in project procurement management follow a clear, logical sequence. However, many project managers are not familiar with the many issues involved in purchasing goods and services from other organizations. If projects will benefit by procuring goods or services, then project managers and their teams must follow good project procurement management. As outsourcing for information technology projects increases, it is important for all project managers to have a fundamental understanding of this knowledge area.

## CASE WRAP-UP

After reading the contract for her company's consultants carefully, Marie McBride found a clause giving her company the right to terminate the contract with a one-week notice. She met with her project team to get their suggestions. They still needed help completing the operating system conversion project. One team member had a friend who worked for a competing consulting firm. The competing consulting firm had experienced people available, and their fees were lower than the fees in the current contract. Marie asked this team member to help her research other consulting firms in the area that could work on the operating system conversion project. Marie then requested bids from these companies. She personally interviewed people from the top three suppliers' management teams and checked their references for similar projects.

Marie worked with the purchasing department to terminate the original contract and issue a new one with a new consulting firm that had a much better reputation and lower hourly rates. This time, she made certain the contract included a statement of work, specific deliverables, and requirements stating the minimum experience level of consultants provided. The contract also included incentive fees for completing the conversion work within a certain time period. Marie learned the importance of good project procurement management.

## Chapter Summary

Procurement, purchasing, or outsourcing is acquiring goods and/or services from an outside source. Information technology outsourcing continues to grow, both within an organization's own country and offshore. Organizations outsource to reduce costs, focus on their core business, access skills and technologies, provide flexibility, and increase accountability. It is becoming increasingly important for information technology professionals to understand project procurement management.

Project procurement management processes include planning procurements, conducting procurements, administering procurements, and closing procurements.

Planning procurements involves deciding what to procure or outsource, what type of contract to use, and how to describe the effort in a statement of work. Make-or-buy analysis helps an organization determine whether it can cost-effectively procure a product or service. Project managers should consult internal and external experts to assist them with procurement planning because many legal, organizational, and financial issues are often involved.



The basic types of contracts are fixed price, cost reimbursable, and time and material. Fixed-price contracts involve a fixed total price for a well-defined product and entail the least risk to buyers. Cost-reimbursable contracts involve payments to suppliers for direct and indirect actual costs and require buyers to absorb some of the risk. Time and material contracts are a hybrid of fixed-price and cost-reimbursable contracts and are commonly used by consultants. Unit pricing involves paying suppliers a predetermined amount per unit of service and imposes different levels of risk on buyers, depending on how the contract is written. It is important to decide which contract type is most appropriate for a particular procurement. All contracts should include specific clauses that address unique aspects of a project and that describe termination requirements.

A statement of work (SOW) describes the work required for the procurement in enough detail to allow prospective suppliers to determine if they are capable of providing the goods and services and to determine an appropriate price.

Conducting procurements involves obtaining seller responses, selecting sellers, and awarding contracts. Organizations should use a formal proposal evaluation form when evaluating suppliers. Technical criteria should not be given more weight than management or cost criteria during evaluation.

Administering procurements involves managing relationships with sellers, monitoring contract performance, and making changes as needed. The project manager and key team members should be involved in writing and administering the contract. Project managers must be aware of potential legal problems they might cause when they do not understand a contract. Project managers and teams should use change control procedures when working with outside contracts and should be especially careful about constructive change orders.

Closing procurements involves completion and settlement of each contract, including resolution of any open items. Procurement audits, negotiated settlements, and records management systems are tools and techniques for closing procurements.

Several types of software can assist in project procurement management. E-procurement software helps organizations save money in procuring various goods and services. Organizations can also use the Web, industry publications, and discussion groups to research and compare various suppliers.

## Quick Quiz

1. What IT function has the largest percentage of work outsourced?
   a. application development
   b. disaster recovery
   c. application maintenance
   d. help desk support
2. Your organization hired a specialist in a certain field to provide training for a short period of time. Which reason for outsourcing would this fall under?
   a. reducing costs
   b. allowing the client organization to focus on its core business
   c. accessing skills and technologies
   d. providing flexibility
3. In which project procurement management process is an RFP often written?
   a. planning procurements
   b. conducting procurements
   c. administering procurements
   d. selecting sellers
4. An item you need for a project has a daily lease cost of $200. To purchase the item, there is an investment cost of $6,000 and a daily cost of $100. Calculate the number of days when the lease cost would be the same as the purchase cost.
   a. 30
   b. 40
   c. 50
   d. 60
5. Which type of contract has the least amount of risk for the buyer?
   a. fixed-price
   b. cost plus incentive fee (CPIF)
   c. time and material
   d. cost plus fixed fee (CPFF)
6. The _____________ is the point at which the contractor assumes total responsibility for each additional dollar of contract cost.
   a. breakeven point
   b. Share Ratio Point
   c. Point of Reconciliation
   d. Point of Total Assumption

7. If your college or university wanted to get information from potential sellers for providing a new sports stadium, what type of document would they require of the potential sellers?
   a. RFP
   b. RFQ
   c. proposal
   d. quote
8. Buyers often prepare a ____________ list when selecting a seller to make this process more manageable.
   a. preferred
   b. short
   c. qualified suppliers
   d. BAFO
9. A proposal evaluation sheet is an example of a(n) ____________.
   a. RFP
   b. NPV analysis
   c. earned value analysis
   d. weighted scoring model
10. ____________ is a term used to describe various procurement functions that are now done electronically.
   a. E-procurement
   b. eBay
   c. E-commerce
   d. EMV



### Quick Quiz Answers

1. b; 2. c; 3. a; 4. d; 5. a; 6. d; 7. c; 8. b; 9. d; 10. a

## Discussion Questions

1. List five reasons why organizations outsource. Why is there a growing trend in outsourcing, especially offshore?
2. Explain the make-or-buy decision process and describe how to perform the financial calculations involved in the simple lease-or-buy example provided in this chapter. What are the main types of contracts if you decide to outsource? What are the advantages and disadvantages of each?
3. Do you think many information technology professionals have experience writing RFPs and evaluating proposals for information technology projects? What skills would be useful for these tasks?
4. How do organizations decide whom to send RFPs or RFQs?

5. How can organizations use a weighted decision matrix to evaluate proposals as part of seller selection?
6. List two suggestions for ensuring adequate change control on projects that involve outside contracts.
7. What is the main purpose of a procurement audit?
8. How can software assist in procuring goods and services? What is e-procurement software? Do you see any ethical issues with e-procurement? For example, should stores be able to block people with smartphones from taking pictures of barcodes in their stores to do comparison shopping?

## Exercises



1. Search the Internet for the term "IT outsourcing." Find at least two articles that discuss outsourcing, whether beneficial or controversial. Summarize the articles and answer the following questions in a two-page paper:
   - What are the main types of goods and services being outsourced?
   - Why are the organizations in the articles choosing to outsource?
   - Have the organizations in your articles benefited from outsourcing? Why or why not?
2. Interview someone who was involved in an information technology procurement process, such as a manager in your organization's IT department, and have him or her explain the process that was followed. Alternatively, find an article describing an IT procurement in an organization. Write a two-page paper describing the procurement and any lessons learned by the organization.
3. Suppose your company is trying to decide whether it should buy special equipment to prepare some of its high-quality publications itself or lease the equipment from another company. Suppose leasing the equipment costs $240 per day. If you decide to purchase the equipment, the initial investment is $6,800, and operations will cost $70 per day. After how many days will the lease cost be the same as the purchase cost for the equipment? Assume your company would only use this equipment for 30 days. Should your company buy the equipment or lease it?
4. Search online for samples of IT contracts. Use search phrases like "IT contract" or "sample contract." Analyze the key features of the contract. What type of contract was used and why? Review the language and clauses in the contract. What are some of the key clauses? List questions you have about the contract and try to get answers from someone familiar with contracts.
5. Review the SUNY Library Automation Migration RFP (available on the companion Web site under Chapter 12) or another RFP for an IT project. Write a two-page paper summarizing the purpose of the RFP and how well you think it describes the work required.
6. Draft the source selection criteria that might be used for evaluating proposals for providing laptops for all students, faculty, and staff at your college or university or all business professionals in your organization. Use Figure 12-5 as a guide. Include at least five criteria, and make the total weights add up to 100. Write a two-page paper explaining and justifying the criteria and their weights.

## Running Case

Senior management at Manage Your Health, Inc. (MYH) decided that it would be best to outsource the work involved in training employees on the soon to be rolled-out Recreation and Wellness system and provide incentives for employees to use the system and improve their health. MYH feels that the right outside company could get people excited about the system and provide a good incentive program. As part of the seller selection process, MYH will require interviews and samples of similar work to be physically presented to a review team. Recall that MYH has more than 20,000 full-time employees and more than 5,000 part-time employees. Assume the work would involve holding several instructor-led training sessions, developing a training video that could be viewed from the company's Intranet site, developing a training manual for the courses and for anyone to download from the Intranet site, developing an incentive program for using the system and improving health, creating surveys to assess the training and incentive programs, and developing monthly presentations and reports on the work completed. The initial contract would last one year, with annual renewal options.



1. Suppose that your team has discussed management's request. You agree that it makes sense to have another organization manage the incentive program for this new application, but you do not think it makes sense to outsource the training. Your company has a lot of experience doing internal training. You also know that your staff will have to support the system, so you want to develop the training to minimize future support calls. Write a one-page memo to senior management stating why you think the training should be done in-house.
2. Assume the source selection criteria for evaluating proposals is as follows:
    - Management approach, 15%
    - Technical approach, 15%
    - Past performance, 20%
    - Price, 20%
    - Interview results and samples, 30%

    Using Figure 12-5 as a guide and the weighted scoring model template, if desired, create a spreadsheet that could be used to enter ratings and calculate scores for each criterion and total weighted scores for three proposals. Enter scores for Proposal 1 as 80, 90, 70, 90, and 80, respectively. Enter scores for Proposal 2 as 90, 50, 95, 80, and 95. Enter scores for Proposal 3 as 60, 90, 90, 80, and 65. Add a paragraph summarizing the results and your recommendation on the spreadsheet. Print your results on one page.
3. Draft potential clauses you could include in the contract to provide incentives to the seller based on MYH achieving its main goal of improving employee health and lowering health care premiums as a result of this project. Be creative in your response, and document your ideas in a one-page paper.

## Companion Web Site

Visit the companion Web site for this text (*www.cengage.com/mis/schwalbe*) to access:

- References cited in the text and additional suggested readings for each chapter
- Template files

- Lecture notes
- Interactive quizzes
- Podcasts
- Links to general project management Web sites
- And more

See the Preface of this text for additional information on accessing the companion Web site.

## Key Terms

**bid** — also called a *tender* or *quote* (short for quotation), a document prepared by sellers providing pricing for standard items that have been clearly defined by the buyer

**constructive change orders** — oral or written acts or omissions by someone with actual or apparent authority that can be construed to have the same effect as a written change order



**contract** — a mutually binding agreement that obligates the seller to provide the specified products or services, and obligates the buyer to pay for them

**cost plus award fee (CPAF) contract** — a contract in which the buyer pays the supplier for allowable performance costs plus an award fee based on the satisfaction of subjective performance criteria

**cost plus fixed fee (CPFF) contract** — a contract in which the buyer pays the supplier for allowable performance costs plus a fixed fee payment usually based on a percentage of estimated costs

**cost plus incentive fee (CPIF) contract** — a contract in which the buyer pays the supplier for allowable performance costs along with a predetermined fee and an incentive bonus

**cost plus percentage of costs (CPPC) contract** — a contract in which the buyer pays the supplier for allowable performance costs along with a predetermined percentage based on total costs

**cost-reimbursable contracts** — contracts involving payment to the supplier for direct and indirect actual costs

**fixed-price contract** — contract with a fixed total price for a well-defined product or service; also called a *lump-sum contract*

**lump-sum contract** — contract with a fixed total price for a well-defined product or service; also called a *fixed-price contract*

**make-or-buy decision** — when an organization decides if it is in its best interests to make certain products or perform certain services inside the organization, or if it is better to buy them from an outside organization

**Point of Total Assumption (PTA)** — the cost at which the contractor assumes total responsibility for each additional dollar of contract cost in a fixed price incentive fee contract

**procurement** — acquiring goods and/or services from an outside source

**project procurement management** — the processes required to acquire goods and services for a project from outside the performing organization

**proposal** — a document prepared by sellers when there are different approaches for meeting buyer needs

**Request for Proposal (RFP)** — a document used to solicit proposals from prospective suppliers

**Request for Quote (RFQ)** — a document used to solicit quotes or bids from prospective suppliers

**sellers** — contractors, suppliers, or providers who provide goods and services to other organizations

**statement of work (SOW)** — a description of the work required for the procurement

**termination clause** — a contract clause that allows the buyer or supplier to end the contract

**time and material (T&M) contracts** — a hybrid of both fixed-price and cost-reimbursable contracts

**unit pricing** — an approach in which the buyer pays the supplier a predetermined amount per unit of service, and the total value of the contract is a function of the quantities needed to complete the work

## End Notes




[1] Global Insight, "Executive Summary: The Comprehensive Impact of Offshore IT Software and Services Outsourcing on the U.S. Economy and the IT Industry," Information Technology Association of America, *ITAA.org* (October 2005).

[2] Andy Bork, "Soft skills needed in a hard world," *Minneapolis Star Tribune* (May 24, 2004).

[3] KPMG, "KPMG Survey finds procurement a high priority at most companies, yet many are failing to implement cost, operational improvements," *PR Newswire* (September 10, 2008).

[4] Computer Economics, Inc. *IT Spending, Staffing, and Technology Trends: 2008/2009* (2008).

[5] Scott McNeally, "The Future of the Net: Why We Don't Want You to Buy Our Software," *Sun Executive Perspectives* (*www.sun.com/dot-com/perspectives/stop.html*) (Sun Microsystems, Inc., 1999): 1.

[6] Gretchen Morgenson, "IBM Shrugs Off Loss of Big Contract," *TechNewsWorld* (September 17, 2004).

[7] Stan Beer, "Is going offshore good for Australia?" *The Age* (September 21, 2004).

[8] Gillian Law, "IBM wins $1.1B outsourcing deal in England," *ComputerWorld* (October 1, 2002).

[9] Computer Business Review Online, "Boots scales back $1.3 billion IBM deal," CBRonline.com (May 15, 2006),

[10] Karl Finders, "Boots' multi-sourcing will use up to six suppliers," ComputerWeekly.com (June 2008).

[11] Kathy Schwalbe, Air Force Commendation Medal Citation (1986).

[12] Robert Antonio, "The Fixed-Price Incentive Firm Target Contract: Not As Firm As the Name Suggests," *WIFCON.com* (November 2003).

[13] Dick Rohland, "I-35W Bridge Completion Brings Closure to Minneapolis," ConstructionEquipmentGuide.com (October 4, 2008).

[14] Accenture, "Driving High-Performance Outsourcing: Best Practices from the Masters," (2004).

[15] Wikipedia, "E-procurement," *http://en.wikipedia.org/wiki/E-procurement* (accessed December 2008).

# GUIDE TO USING MICROSOFT PROJECT 2010

## INTRODUCTION

This appendix provides a concise guide to using Microsoft Office Project Professional 2010 (often referred to as Project 2010) to assist in performing project management functions. The Office Project 2010 family includes several different products:

- Project Standard 2010, a non-Web-based, stand-alone program for individuals who manage projects independently, similar to earlier versions of Project Standard.
- Project Professional 2010, which is basically Project Standard 2010 plus the ability to connect to Project Server 2010, if available. (Note: This guide does not include information on connecting to Project Server 2010.)
- Project Server 2010, built on SharePoint Server 2010, delivers flexible work management solutions across organizations, departments, and teams. Organizations should develop and apply many standards, templates, codes, and procedures before using Project Server 2010 to make the best use of its capabilities.

Each version of Project 2010 can help users manage different aspects of all nine project management knowledge areas. (Consult Microsoft's Web site for more details on all of the versions of Project 2010.) Most users, however, focus on using Project 2010 to assist with scope, time, cost, human resource, and communications management. This guide uses these project management knowledge areas as the context for learning how to use Project 2010. The basic order of steps in this appendix follows best practices in project management. That is, you should first determine the scope of a project, then the time, and then the resource and cost information. You can then set a baseline and enter actuals to track and communicate performance information.

Hundreds of project management software products are available today, but Microsoft Project is the clear market leader among midrange applications. Before you can use Project 2010 or any project management software effectively, you must understand the fundamental concepts of project management, such as creating work breakdown structures, linking tasks, entering duration estimates, and so on. See the Suggested Readings on the companion Web site (*www.cengage.com/mis/schwalbe*) for recommendations on other resources to help you gain an even deeper understanding of Project 2010.

## New Features of Project 2010*

Project 2010 is not just a run-of-the-mill update. Microsoft really listened to users and has revised Project to meet user needs. Learning some of the new features might seem like a chore, but it is well worth the effort.

If you are familiar with Project 2007, it may be helpful to review some of the new features in Project 2010.

- *Improved user interface:* Project 2010 now includes the "ribbon" interface instead of using the traditional menus and toolbars similar to Office 2003. Commands are organized in logical groups under tabs, such as File, Task, Resource, Project, View, and Format. The File tab takes you to the new Backstage feature, a one-stop graphical destination for opening, saving, and printing your files. You can also now right-click different items, like a table cell or chart, to bring up commonly used commands quickly.
- *New viewing options:* Project 2010 includes several new views. A timeline view shows you a concise overview of the entire project schedule. You can easily add tasks to the timeline, print it, or paste it into an e-mail. The new team planner view lets you quickly see what your team members are working on, and you can move tasks from one person to another using this view. For example, if a resource is overallocated, you can drag a task to another resource to remove the overallocation. You can also add new columns quickly and use the new Zoom Slider at the lower right of the screen to zoom your schedule in and out. Also, the tab for viewing and printing reports is easier to navigate with more options for visual reports.
- *Manual scheduling:* Unlike previous versions of Project where tasks were automatically scheduled, Project 2010 uses manual schedule as its default. In past versions of Project, summary tasks were automatically calculated based on their subtasks. Resources were also adjusted automatically. With Project

*From Schwalbe's *An Introduction to Project Management, Third Edition*, 2010.

2010, this is no longer the case. For example, you might want to enter durations for summary tasks and then fill in the detailed information for their subtasks later. When you open a new file, Project reminds you that new tasks are manually scheduled and lets you easily switch to automatic scheduling, if desired. You can also use the new compare versions to see Gantt bars to more clearly see how one version of a project differs from another version.

- *Improved collaboration:* Project 2010 can provide an interface with the most popular portals used in the industry. Project now uses SharePoint instead of Project Web Access for collaboration. Project Server 2010 also provides integration with Microsoft Exchange 2010 to enable team members to manage and report on tasks directly from Microsoft Outlook. Remember that Project Standard does not include these collaboration features. You must have Project Professional and Project Server to use the enterprise features of Project.

## Before You Begin

This appendix assumes you are using Project 2010 with Windows 7, Vista, or XP and are familiar with other Windows-based applications. You can download a free trial of Project 2010 from Microsoft's Web site. Additionally, students and faculty can purchase software at deep discounts from several sources, including *www.journeyed.com*. Check your work by reviewing the many screen shots included in the steps or by using the solution files that are available for download from the companion Web site for this text or from your instructor.

### HELP

You need to be running Windows 7, Vista, or XP as your operating system to use Project 2010. If you have any technical difficulties, contact Microsoft's Support Services at *www.microsoft.com*. To complete the hands-on activities in the appendix, you will need to copy a set of files from the companion Web site for this text (*www.cengage.com/mis/schwalbe*) to your computer. You can also download the files from the author's Web site (*www.kathyschwalbe.com*) under Book FAQs.

This appendix uses a fictitious information technology project—the Project Tracking Database project—to illustrate how to use the software. The goal of this project is to create a database to keep track of all the projects a company is working on. Each section of the appendix includes hands-on activities for you to perform. When you begin each set of steps, make sure you are using the correct file. Before you begin your work you should have the finance.mpp, resource.mpp, and kickoffmeeting.doc files.

In addition, you will create the following files from scratch as you work through the steps:

- scope.mpp
- time.mpp
- tracking.mpp
- baseline.mpp
- level.mpp

Now that you understand project management concepts and the basic project management terminology, you will learn how to start Project 2010, review the Help facility and a template file, and begin to plan the Project Tracking Database project.

## OVERVIEW OF PROJECT 2010

The first step to mastering Project 2010 is to become familiar with the Help facility, major screen elements, views, and filters. This section describes each of these features.

### Starting Project 2010 and Using the Help Feature

To start Project 2010:

1. *Open Project 2010.* Click the **Start** button on the taskbar, point to **All Programs** in Windows XP or **Programs** in Windows 7 or Vista, point to or click **Microsoft Office,** and then click **Microsoft Office Project 2010.** Alternatively, a shortcut or icon might be available on the desktop; in this case, double-click the icon to start the software.
2. *Maximize Project 2010.* If the Project 2010 window does not fill the entire screen as shown in Figure A-1, click the **Maximize** button in the upper-right corner of the window.

**HELP**

All screen shots were taken using Windows XP, so your screen may look slightly different. Focus on the Project 2010 main screen.

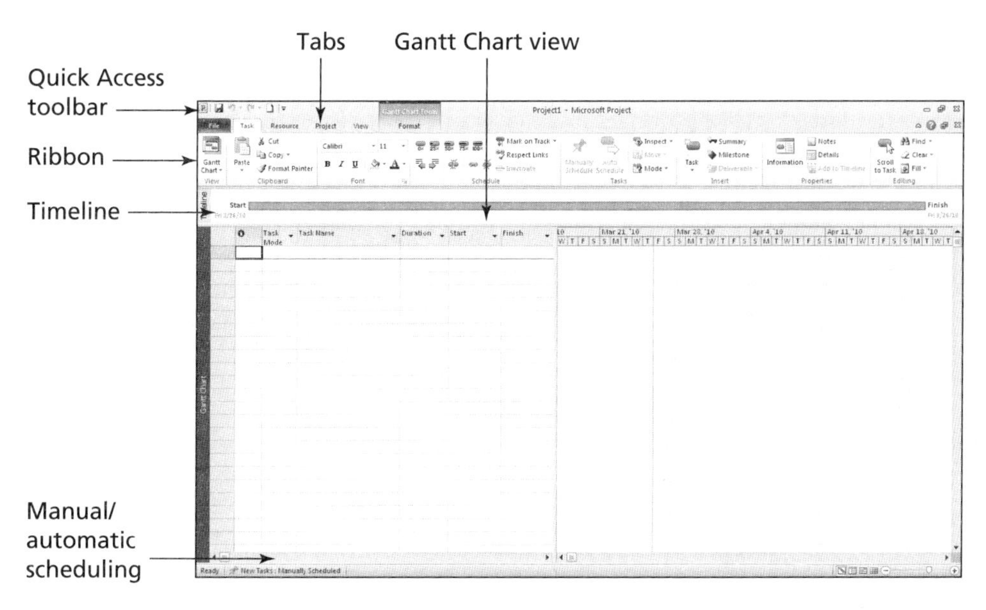


**FIGURE A-1** Project 2010 main screen

Project 2010 is now running and ready to use. Look at some of the elements of the Project 2010 screen. The default view is the Gantt Chart view under the Task tab, which shows tasks and other information in the Entry table as well as a calendar display. You can also use the Quick Access toolbar and tabs as in other Windows applications. Notice the new Timeline and Manual scheduling features on the main screen. You can access other views by clicking the View button on the far left of the Ribbon as shown in Figure A-2.

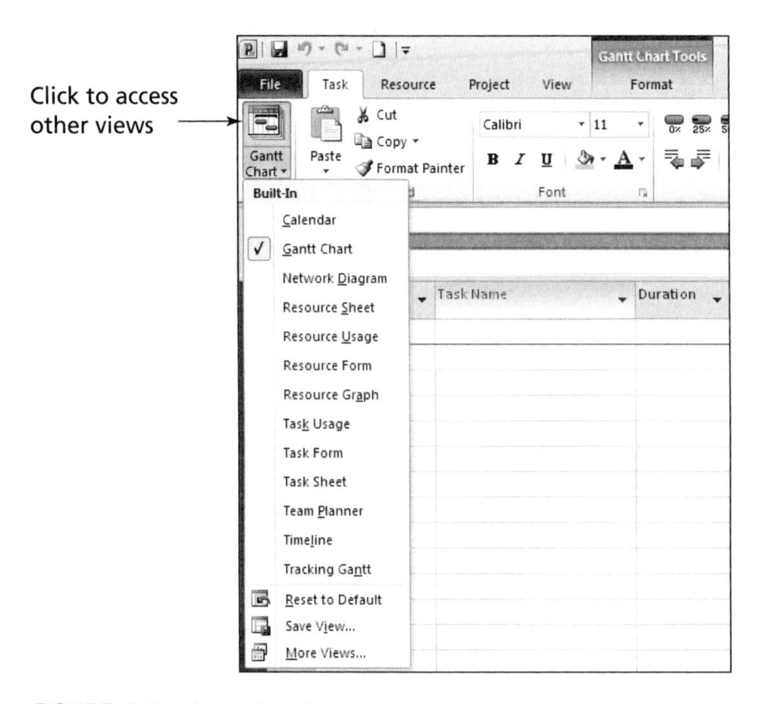


**FIGURE A-2** Accessing other views

Notice that when Project 2010 starts, it opens a new file named Project1, as shown in the title bar. If you open a second file, the name will be Project2, and so on, until you save and rename the file.

You can access other information to help you learn how to use Project 2010 under the Help menu. Figure A-3 shows the Help menu options available by clicking Help on the far right side of the Ribbon or by pressing F1.

**FIGURE A-3** Project 2010 Help menu options

Microsoft realizes that Project 2010 can take some time to learn and provides a number of resources on its Web site. Microsoft's Web site for Project 2010 (*www.microsoft.com/project*) provides files for users to download, case studies, articles, and other useful materials. Also see the Suggested Readings under Appendix A on the companion Web site for this text.

## Main Screen Elements

The Project 2010 default main screen is called the Gantt Chart view. At the top of the main screen, the Ribbon is similar to other *Microsoft Windows 2007* and *2010* programs. Project 2010 does have Ribbon interface, while Project 2007 did not. The order and appearance of buttons on the Ribbon may vary, depending on the features you are using and how the Ribbon is customized.

Figure A-4 shows the main screen elements of Project 2010. On the left side of the screen below the Ribbon is the Entry table. The Gantt chart calendar display appears on the right of the split bar, which separates the Entry table and the Gantt chart. The column to the left of the Task Name column is the Indicators column. The Indicators column displays indicators or symbols related to items associated with each task, such as task notes or hyperlinks to other files.

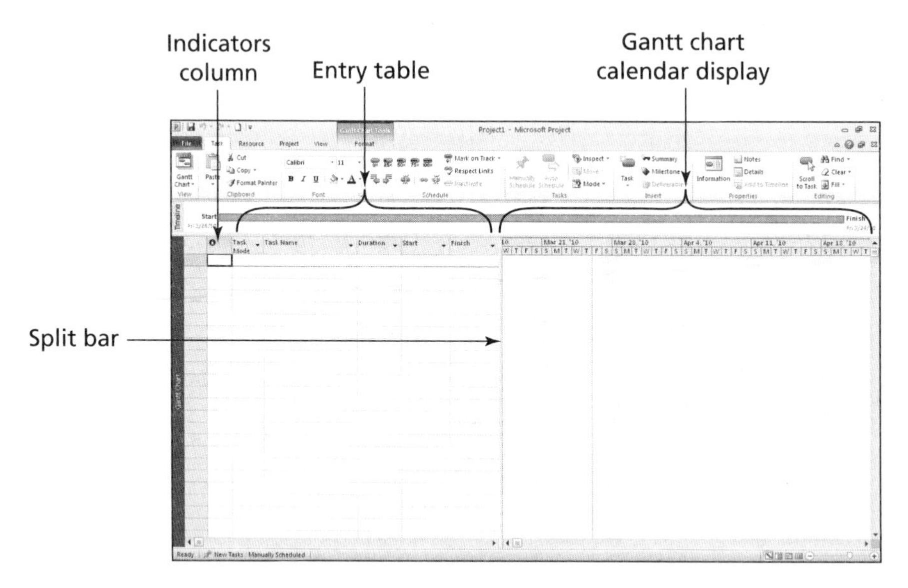


**FIGURE A-4** Project 2010 main screen elements

## TIP

Many features in Project 2010 are similar to ones in other Windows programs. For example, to collapse or expand tasks, click the appropriate symbols to the left of the task name. To access shortcut items, right-click in either the Entry table area or the Gantt chart. Many of the Entry table operations in Project 2010 are very similar to operations in Excel. For example, to adjust a column width, click and drag or double-click between the column heading titles.

If you select another view and want to return to the Gantt Chart view, click the Gantt Chart button on the Ribbon under the Tasks tab or the appropriate button under the Task views section under the View tab. If the Entry table on the left appears to be different, select the Tables button under the Data section from the View tab.

Notice the split bar that separates the Entry table from the Gantt chart. When you move the mouse over the split bar, your cursor changes to the resize pointer. Clicking and dragging the split bar to the right reveals other task information in the Entry table, including the Duration, Start date, Finish date, Predecessors, and Resource Names columns.

Next, you will open a template file to explore more screen elements. Project 2010 comes with several template files, and you can also access templates from Microsoft Office Online or other Web sites. To open template files on your computer, you normally click New under the File tab. A list of available templates displays, as shown in Figure A-5. For the following steps you will open a template file you downloaded from the companion Web site. It will not display under the list of available templates because it is a template from an older version of Project.

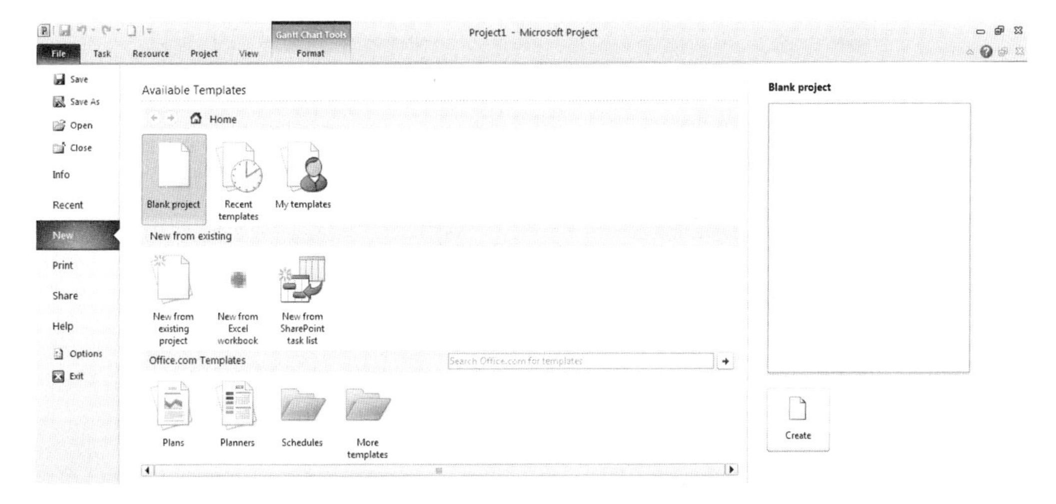


**FIGURE A-5** Opening a template file

## HELP

To access template files on your computer, you must do a full or custom install of Project 2010.

To open a template file and adjust Project 2010 screen elements:

1. *Open a template file*. Click the **Open** button under the File tab, browse to find the file named **finance.mpp** that you copied from the companion Web site, or author's Web site for this text, and then double-click the filename to open the file. This file provides a template for a project to implement a new finance and accounting system in a corporate environment.

## TIP

The finance.mpp file is identical to the template named "Finance and Accounting System Implementation" that comes with Project 2007.

2. *View the Note*. Move your mouse over the **Notes icon** in the Indicators column and read its contents. It is a good idea to provide a short note describing the purpose of project files. (You'll learn how to add notes in the Communications section of this guide.) Your screen should resemble Figure A-6.

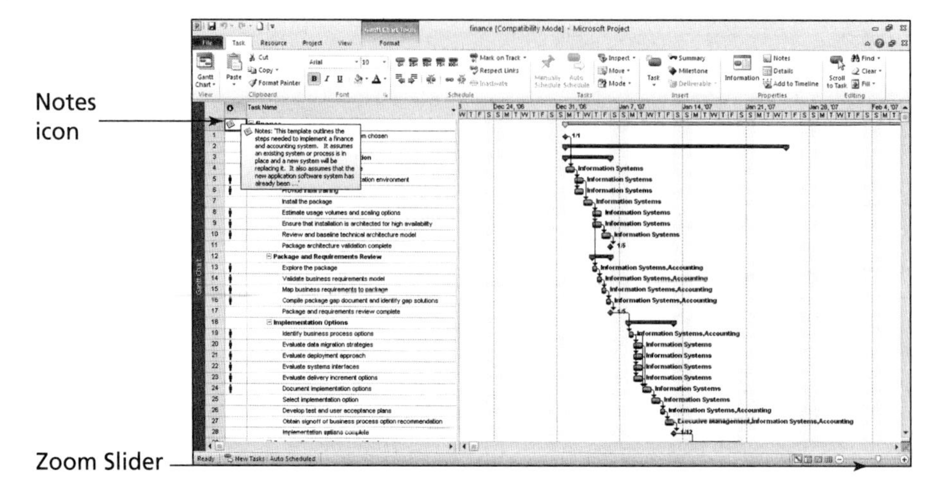


**FIGURE A-6** The finance.mpp file

3. *Adjust the timescale*. Click the **Zoom Out** icon on the **Zoom Slider bar** on the lower right of your screen twice to adjust the timescale so that you see the entire bar at the top of the Gantt chart. Notice that this project started on January 1, 2007, and ended in early April 2007. The first line in the Entry table displays the filename, and the bar in the Gantt chart next to that line shows the time line for the entire project.
4. *Select Outline Level 1 to display WBS level 2 tasks*. Under the View tab, click the **Outline** button's list arrow, and then click **Outline Level 1,** as shown in Figure A-7. Notice that only the outline level 1 items or WBS level 2 items display in the Entry table after you select Outline Level 1. The black bars on the Gantt chart represent the summary tasks. Note that according to the Project Management Institute (PMI), the entire project is normally referred to as WBS level 1, and the next highest level is called level 2. This view of the file also shows two milestone tasks in rows 1 and 137 indicating when a new Accounting and Finance system was chosen and when the project was completed. Recall that the black diamond symbol on a Gantt chart shows milestones.

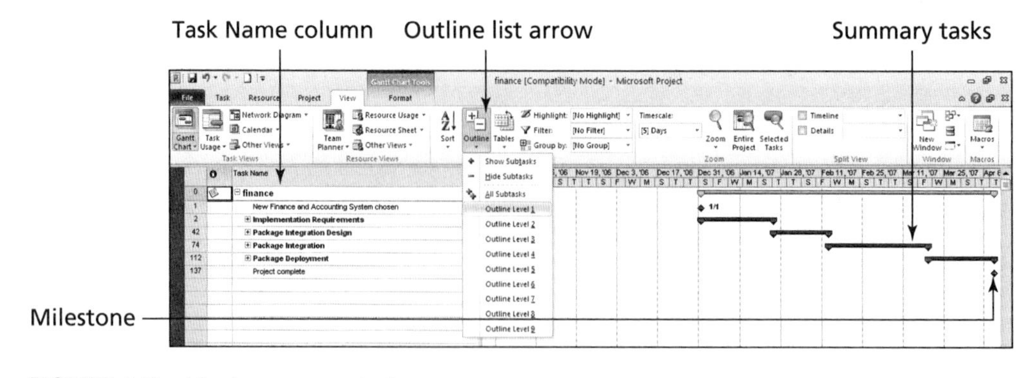


**FIGURE A-7** Viewing summary tasks

5. *Adjust the Task Name column*. Move the split bar to the right, if needed, to reveal the entire Task Name column. Move your cursor over the right-column gridline in the Task Name column heading until you see the resize pointer, and then double-click the **left mouse** button to resize the column width automatically.
6. *Move the split bar to reveal more Entry table columns*. Move the split bar to the right to reveal the Resource Names column. Your screen should resemble Figure A-8.

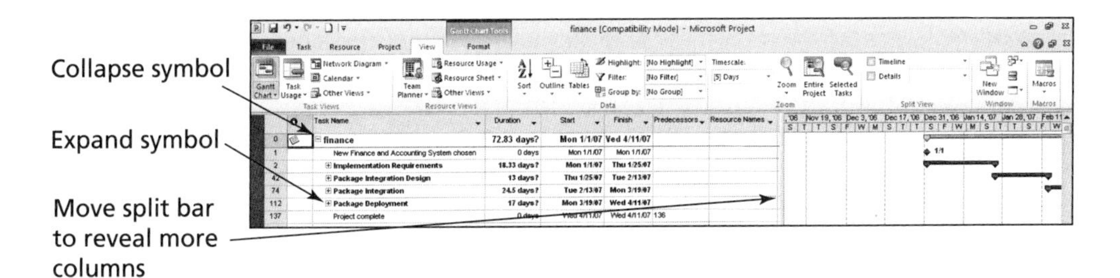


**FIGURE A-8** Adjusting screen elements

7. *Expand a task*. Click the **expand symbol** ⊞ to the left of Task 112, Package Deployment, to see its subtasks. Click the **collapse symbol** ⊟ to hide its subtasks.

## Project 2010 Views

Project 2010 provides many ways to display project information. These displays are called *views*. The views under the View tab include: Gantt Chart, Task Usage, Network Diagram, Calendar, Resource Usage, and Resource Sheet. The Resource Form and Resource Graph views are available under Other Views, and even more views are available under More Views, such as the Tracking Gantt and Leveling Gantt. These different views allow you to examine project information in different ways, which helps you analyze and understand what is happening on your project.

The View tab also provides access to different tables that display information in a variety of ways. Some tables that you can access from the View tab include Schedule, Cost, Tracking, and Earned Value. Some Project 2010 views, such as the Gantt Chart view, present a broad look at the entire project, whereas others, such as the Form views, focus on specific pieces of information about each task. Three main categories of views are available:

- *Graphical*: A chart or graphical representation of data using bars, boxes, lines, and images.
- *Task Sheet or Table*: A spreadsheet-like representation of data in which each task appears as a new row and each piece of information about the task is represented by a column. Different tables are applied to a task sheet to display different kinds of information.
- *Form*: A specific view of information for one task. Forms are used to focus on the details of one task.

Table A-1 describes some of the predesigned views within each category that Project 2010 provides to help you display the project or task information that you need.

**TABLE A-1** Common Project 2010 views

| Category of View | View Name | Description of View |
|---|---|---|
| Graphical | Gantt Chart | Standard format for displaying project schedule information that lists project tasks and their corresponding start and finish dates in a calendar format. Shows each task as a horizontal bar with the length and position corresponding to the timescale at the top of the Gantt chart. |
| | Network Diagram | Schematic display of the logical relationships or sequencing of project activities. Presents each task as a box with linking lines between tasks to show sequencing. Critical tasks appear in red. |
| Task Sheet or Table | Entry Table | Default table view showing columns for Task Name and Duration. By revealing more of the Entry table, you can enter start and end dates, predecessors, and resource names. |
| | Schedule Table | Displays columns for Task Name, Start, Finish, Late Start, Late Finish, Free Slack, and Total Slack. |
| | Cost Table | Displays columns for Task Name, Fixed Cost, Fixed Cost Accrual, Total Cost, Baseline, Variance, Actual, and Remaining. |
| | Tracking Table | Displays columns for Task Name, Actual Start, Actual Finish, % Complete, Physical % Complete, Actual Duration, Remaining Duration, Actual Cost, and Actual Work. |
| | Earned Value | Displays columns for Task Name, PV, EV, AC, SV, CV, EAC, BAC, and VAC. See the earned value section of this text for descriptions of these acronyms. |
| Form | Task Form | Displays detailed information about a single task in one window. |
| | Resource Form | Displays detailed information about a single resource in one window. |

Next, you will use the same file (finance.mpp) to access and explore some of the views available in Project 2010.

To access and explore different views:

1. *Show all subtasks*. Click the **Outline** button on the Ribbon under the View tab, and then click **All Subtasks.** Under Task Name click **Task 0, finance.**

2. *Explore the Network Diagram view*. Click the **Network Diagram** button on the Ribbon, and then move the **Zoom Slider** all the way to the left to see more of the diagram. Your screen should resemble Figure A-9.

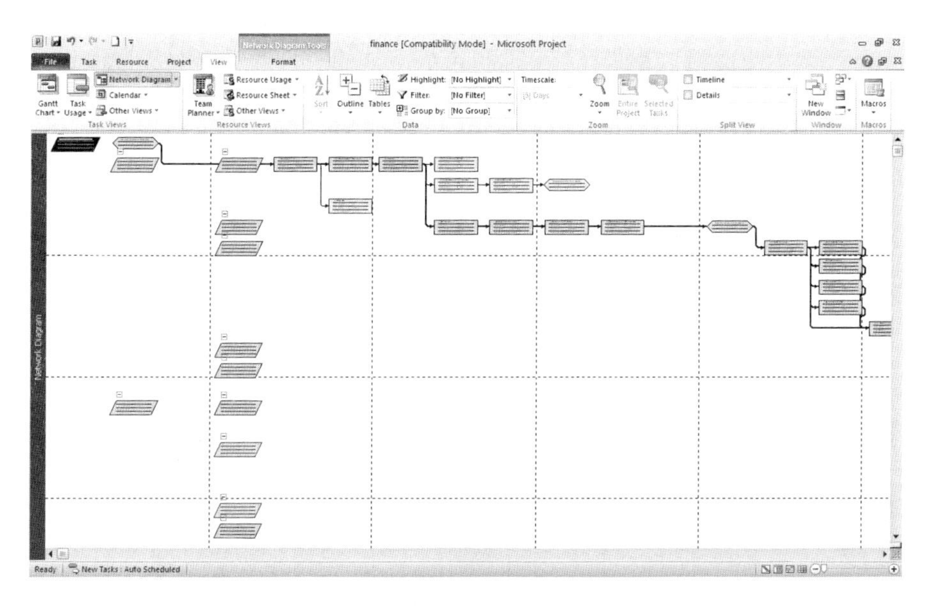


**FIGURE A-9** Network Diagram view of finance file

3. *Explore the Calendar view*. Click the **Calendar** button on the Ribbon. Your screen should resemble Figure A-10.

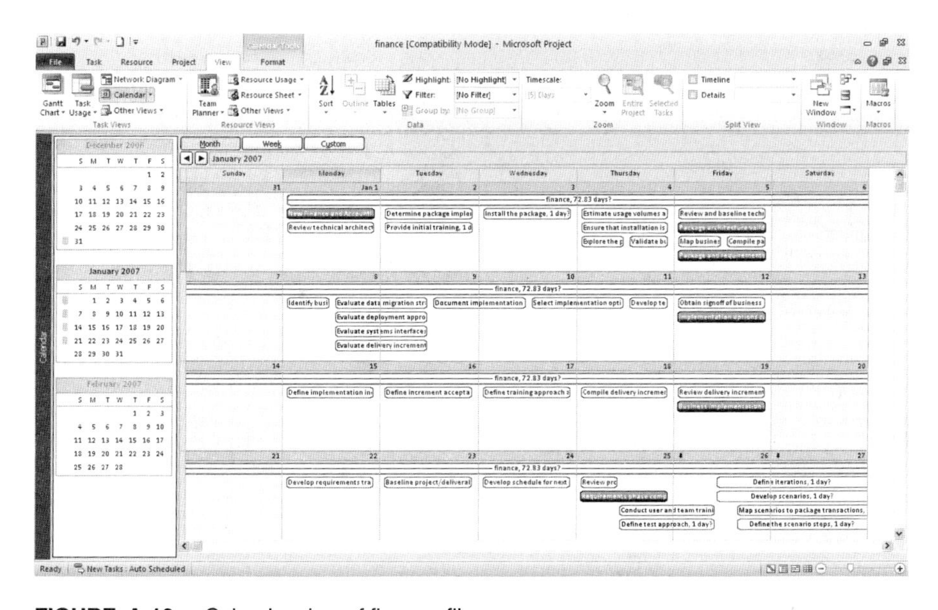


**FIGURE A-10** Calendar view of finance file

4. *Examine columns in the Entry table*. Click the **Gantt Chart** button on the Ribbon, and observe the information provided in each column of the Entry table.
5. *Examine the Table: Schedule view*. Click the **Tables** button under the Data section of the View tab, and then click **Schedule**. Notice that the columns to the left of the Gantt chart now display more detailed schedule information. Also notice that all of the text in the Task Name column is not visible. You can widen the column by moving the mouse to the right of the Task Name column and double-clicking the resize pointer. You can also move the split bar to the right to reveal more columns.
6. *Right-click the Select All button to access different table views*. Move your mouse to the **Select All** button to the left of the Task Name column symbol, and then right-click with your mouse. A shortcut menu to different table views displays, as shown in Figure A-11. Experiment with other table views and then return to the Table: Entry view.

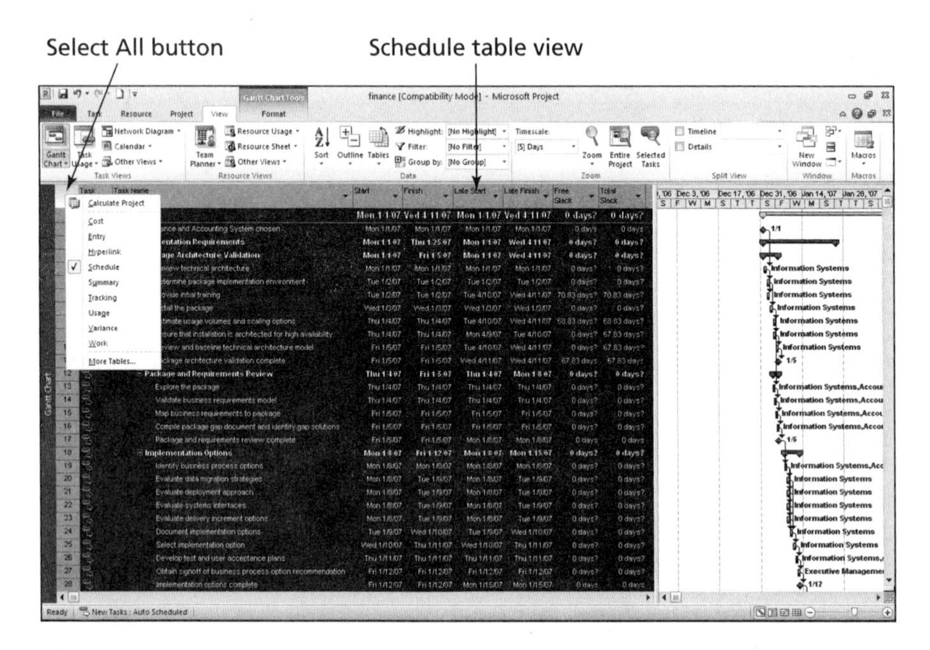


**FIGURE A-11** Changing table views

7. *Explore the Reports feature*. Click the **Project** tab, and then click the **Reports** button. The Reports dialog box displays, as shown in Figure A-12.

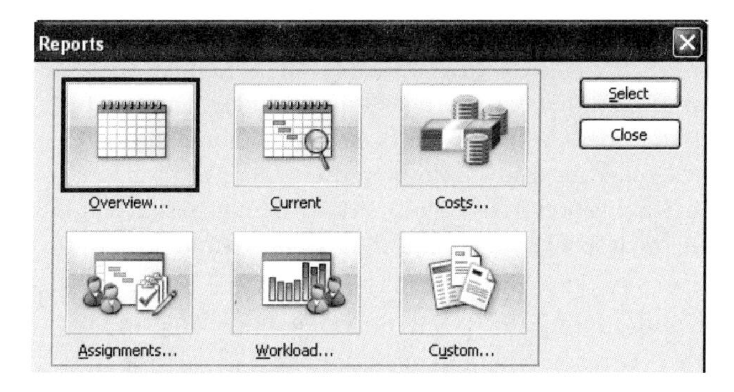


**FIGURE A-12** Reports dialog box

8. *View the Project Summary report*. Double-click **Overview** from the Reports dialog box, and then double-click **Project Summary** in the Overview Reports dialog box. Notice that Project 2010 switches to the Backstage (File tab) to make it easy for you to print or share your report, as shown in Figure A-13.

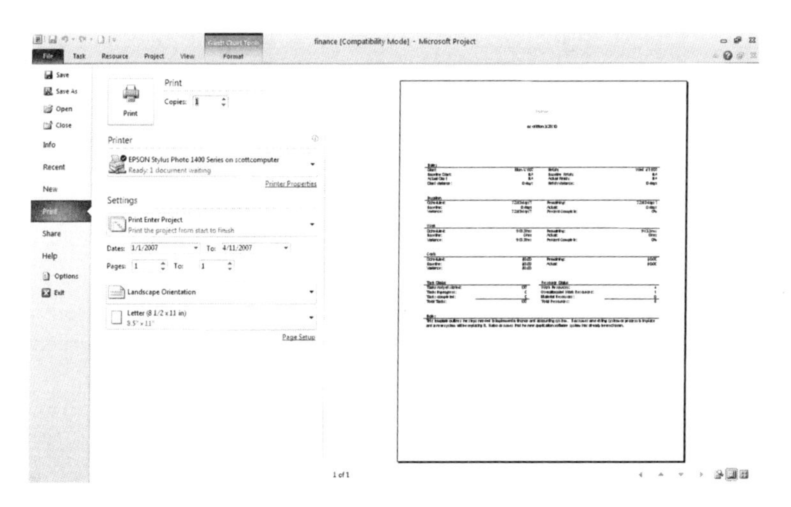


**FIGURE A-13** Previewing a report in the Backstage

9. *Examine reports and return to the Task tab*. Move the mouse to the right side of the screen to examine the report more closely. Notice that the insertion point now resembles a magnifying glass. Click inside the report to zoom in or zoom out. Click the **Project** tab again, and then experiment with viewing other reports. You will use several reports and other views later. Click the **Task** tab when you are finished.

## Project 2010 Filters

Project 2010 uses an underlying relational database to filter, sort, store, and display information. Filtering project information is very useful. For example, if a project includes hundreds of tasks, you might want to view only summary or milestone tasks to get an overview of the project. To get this type of overview of a project, select the Milestones or Summary Tasks filter from the Filter list. If you are concerned about the schedule, you can select a filter that shows only tasks on the critical path. Other filters include Completed Tasks, Late/Overbudget Tasks, and Date Range, which displays tasks based on dates you provide. As shown earlier, you can also click the Show button on the toolbar to display different levels in the WBS quickly. For example, Outline Level 1 shows the highest-level items in the WBS, Outline Level 2 shows the next level of detail in the WBS, and so on.

To explore Project 2010 filters:

1. *Apply a filter to see only milestone tasks*. From the Table: Entry view in the finance.mpp file, click the **View** tab, and then click the **Filter list arrow**, as shown in Figure A-14.

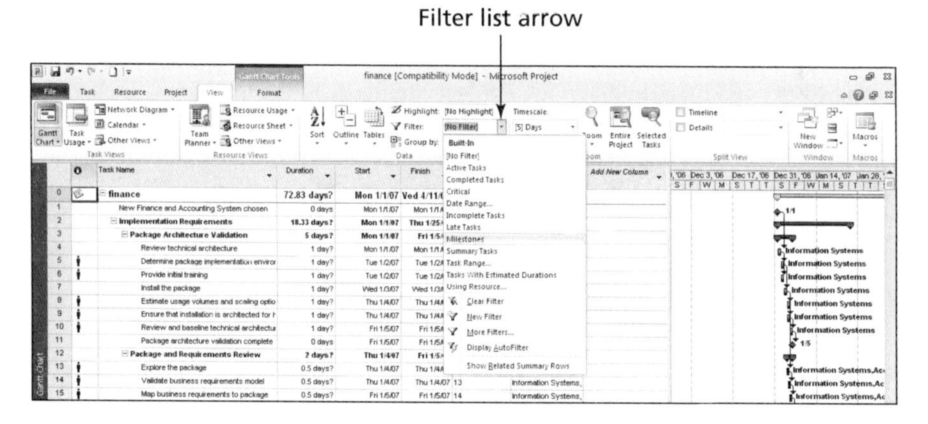


**FIGURE A-14** Using a filter

2. *Filter to show specific tasks*. Click **Milestones** in the list of filters, and move the split bar to the left to see all the milestones on the Gantt chart. Your screen should resemble Figure A-15. The black diamond symbol represents a milestone, a significant event on a project.

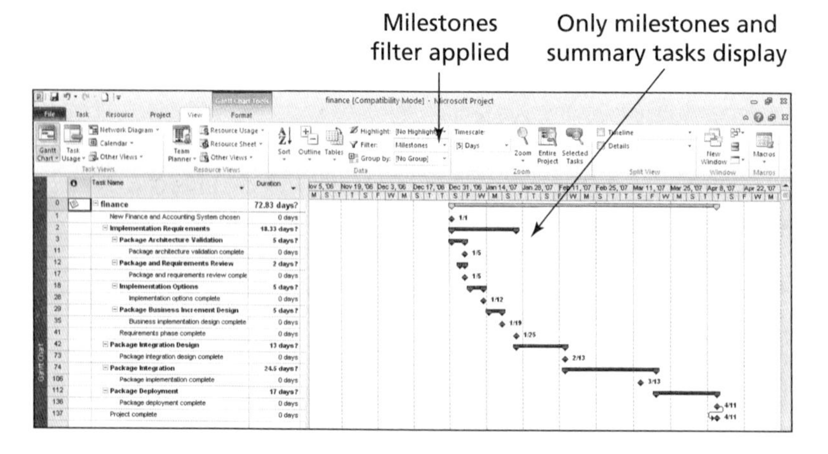


**FIGURE A-15** Milestone tasks filter for Project 2010 finance file

3. *Show summary tasks*. Click the **Filter** list arrow, scroll down until you see Summary Tasks, and then click **Summary Tasks.** Now only the summary tasks appear in the WBS. Experiment with other outline levels and filters.
4. *Close the file*. When you are finished reviewing the finance.mpp file, click **Close** from the File menu or click the **Close** button. A dialog box appears asking if you want to save changes. Click **No.**
5. *Exit Project 2010*. Select **Exit** from the File menu or click the **Close** button for Project 2010.

Now that you are familiar with the Project 2010 main screen elements, views, and filters, you will learn how to use Project 2010 to assist in project scope management by creating a new project file, developing a WBS, and setting a baseline.

## PROJECT SCOPE MANAGEMENT

Project scope management involves defining the work to perform to carry out the project. To use Project 2010, you must first determine the scope of the project. To begin determining the project's scope, create a new file with the project name and start date. Develop a list of tasks that need to be done to carry out the project. This list of tasks becomes the work breakdown structure (WBS). If you intend to track actual project information against the initial plan, you must set a baseline. In this section, you will learn how to create a new project file, develop a WBS, and set a baseline to help plan and manage the Project Tracking Database project. To start, you will enter the scope-related information.

**TIP**

In this section, you will go through several steps to create the scope.mpp Project 2010 file. If you want to download the completed file to check your work or continue to the next section, a copy of scope.mpp is available on the companion Web site for this text, from the author's Web site, or from your instructor. Try to complete an entire section of this appendix (project scope management, project time management, and so on) in one sitting to create the complete file. Also, be sure to save the scope.mpp file you create in a different folder than the one you download, and then compare the two files.

### Creating a New Project File

To create a new project file:

1. *Create a blank project*. Open Project 2010. A blank project file automatically opens when you start Project 2010. The default filenames are Project1, Project2, and so on. If Project 2010 is already open and you want to open a new file, click the **File** tab, click **New**, and select **Blank Project.**
2. *Open the Project Information dialog box*. Click the **Project** tab, and then click **Project Information** to display the Project Information dialog box, as shown in Figure A-16. The Project Information dialog box enables you to set dates for the project, select the calendar to use, and view project statistics. The project start

date will default to today's date. Note that in Figure A-16 the file was created on 3/31/10 and a Start date of 2/1/11 was entered. (See the following Help box for information about changing date formats).

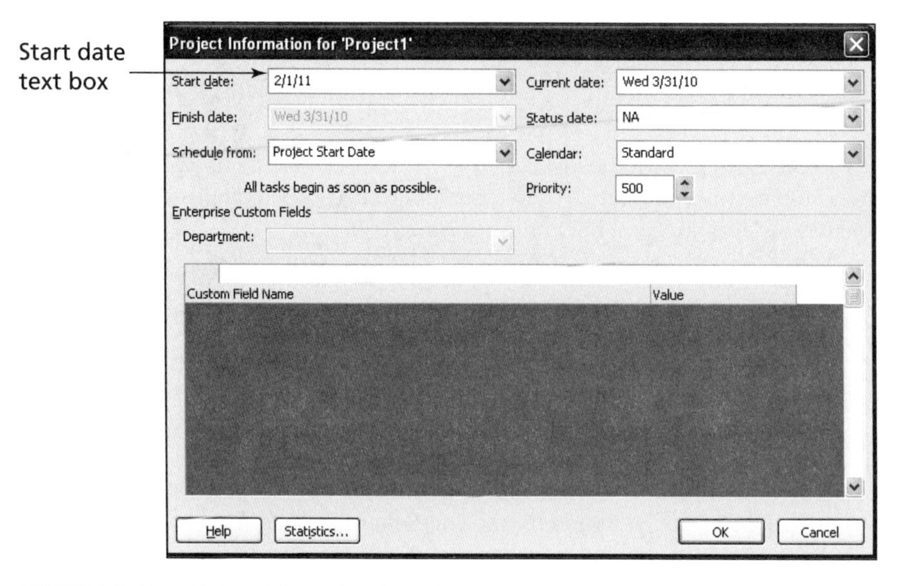


**FIGURE A-16** Project Information dialog box

3. *Enter the project start date*. In the Start date text box, enter **2/1/11.** Setting your project start date to 2/1/11 will ensure that your work matches the results that appear in this appendix. Leave the Finish date, Current date, and other information at the default settings. Click **OK.**

## HELP

This appendix uses American date formats. For example, 2/1/11 represents February 1, 2011. Be sure to enter dates in this format for these steps. However, you can change the date format by selecting Options from the File tab. Click the date format you want to use in the Date Format box under the General settings. You can also customize the Ribbon, change default currencies in the display, and so on under Project Options.

4. *Access advanced project properties*. Click the **File** tab, and then click **Info.** Click **Project Information** on the right side of the screen to access Advanced Properties, as shown in Figure A-17.

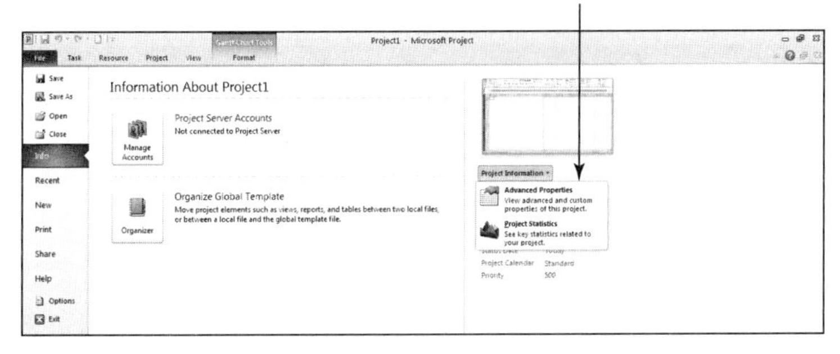


**FIGURE A-17** Accessing advanced project properties

5. *Enter advanced project information*. Type **Project Tracking Database** in the Title text box, type **Terry Dunlay** in the Author text box, as shown in Figure A-18, and then click **OK.** You may have some default information entered in the Project Properties dialog box, such as your company's name. Click the **Task** tab so you can see the Entry table and Gantt Chart view. Keep this file open for the next set of steps.

**FIGURE A-18** Project Properties dialog box

## Developing a Work Breakdown Structure

Before using Project 2010, you must develop a work breakdown structure (WBS) for your project. Developing a good WBS takes time, but it will make entering tasks into the Entry table easier if you develop the WBS first. It is also a good idea to establish milestones before entering tasks in Project 2010. You will use the information in Table A-2 to enter tasks for the Project Tracking Database project. Be aware that this example is much shorter and simpler than most WBSs.

To develop a WBS and enter milestones for the Project Tracking Database project:

1. *Enter task names*. Enter the 20 tasks in Table A-2 into the Task Name column in the order shown. Do not worry about durations or any other information at this time. Type the name of each task into the Task Name column of the Entry table, beginning with the first row. Press **Enter** or the **down arrow** key on your keyboard to move to the next row.

### HELP

If you accidentally skip a task, highlight the task row, right-click, and select Insert Task. To edit a task entry, click the text for that task, and either type over the old text or edit the existing text.

**TABLE A-2** Project Tracking Database tasks

| Order | Task Name |
|---|---|
| 1 | Initiating |
| 2 | Kickoff meeting |
| 3 | Develop project charter |
| 4 | Charter signed |
| 5 | Planning |
| 6 | Develop project plans |
| 7 | Review project plans |
| 8 | Project plans approved |
| 9 | Executing |
| 10 | Analysis |
| 11 | Design |
| 12 | Implementation |
| 13 | System implemented |

*(continued on next page)*

**TABLE A-2** Project Tracking Database tasks (*continued*)



| Order | Task Name |
|---|---|
| 14 | Controlling |
| 15 | Report performance |
| 16 | Control changes |
| 17 | Closing |
| 18 | Prepare final project report |
| 19 | Present final project |
| 20 | Project completed |

**TIP**

Entering tasks into Project 2010 and editing the information is similar to entering and editing data in an Excel spreadsheet. You can also easily copy and paste text from Excel or Word into Project, such as the list of tasks.

2. *Move the split bar to reveal more columns.* If necessary, move the split bar to the right to reveal the entire Task Name and Duration columns.
3. *Adjust the Task Name column width as needed.* To make all the text display in the Task Name column, move the mouse over the right-column gridline in the Task Name column heading until you see the resize pointer, and then click the **left mouse** button and drag the line to the right to make the column wider, or double-click to adjust the column width automatically.

This WBS separates tasks according to the project management process groups of initiating, planning, executing, controlling, and closing. These tasks will be the level 2 items in the WBS for this project. Remember that the whole project is considered level 1. It is a good idea to include all of these process groups because there are important tasks that must be done under each of them. Recall that the WBS should include *all* of the work required for the project. In the Project Tracking Database WBS, the tasks will be purposefully left at a high WBS level (level 3). You will create these levels, or the WBS hierarchy, next when you create summary tasks. For a real project, you would usually break the WBS into even more levels to provide more details to describe all the work involved in the project. For example, analysis tasks for a database project might be broken down further to include preparing entity relationship diagrams for the database and developing guidelines for the user interface. Design tasks might be broken down to include preparing prototypes, incorporating user feedback, entering data, and testing the database. Implementation tasks might include more levels, such as installing new hardware or software, training the users, fully documenting the system, and so on.

## Creating Summary Tasks

After entering the WBS tasks listed in Table A-2 into the Entry table, the next step is to show the WBS levels by creating summary tasks. The summary tasks in this example are Tasks 1 (initiating), 5 (planning), 9 (executing), 14 (controlling), and 17 (closing). You create summary tasks by highlighting and indenting their respective subtasks.

To create the summary tasks:

1. *Select lower level or subtasks*. Highlight **Tasks 2** through **4** by clicking the cell for Task 2 and dragging the mouse through the cells to Task 4.
2. *Indent subtasks*. Click the **Indent task** button on the Ribbon under the Schedule group of the Task tab (or press Alt + Shift + right arrow) so your screen resembles Figure A-19. After the subtasks (Tasks 2 through 4) are indented, notice that Task 1 automatically becomes boldface, which indicates that it is a summary task. A collapse symbol appears to the left of the new summary task name. Clicking the **collapse** symbol ⊟ will collapse the summary task and hide the subtasks beneath it. When subtasks are hidden, an **expand** symbol ⊞ appears to the left of the summary task name. Clicking the expand symbol will expand the summary task. Also, notice that the symbol for the summary task on the Gantt chart appears as a black line with arrows indicating the start and end dates.

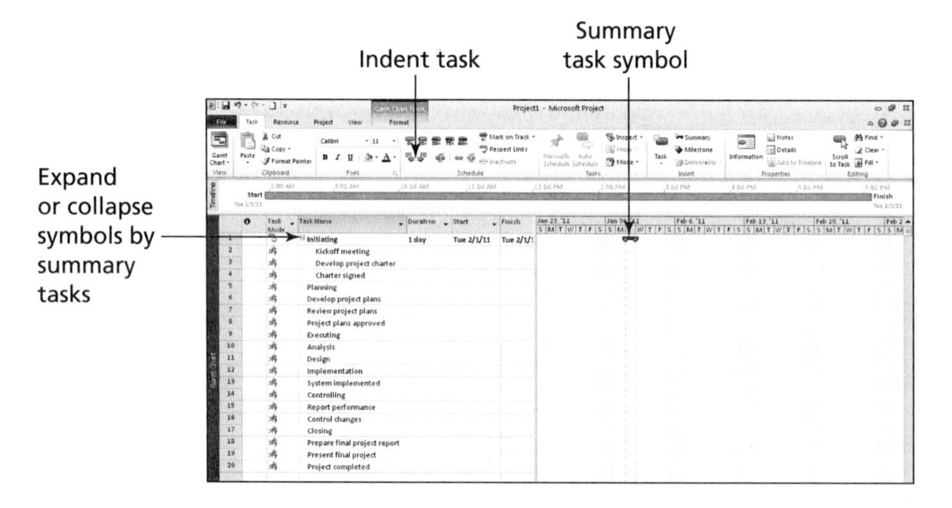


**FIGURE A-19** Indenting tasks to create the WBS hierarchy

3. *Create other summary tasks and subtasks*. Create subtasks and summary tasks for Tasks 5, 9, 14, and 17 by following the same steps. Indent **Tasks 6** through **8** to make Task 5 a summary task. Indent **Tasks 10** through **13** to make Task 9 a summary task. Indent **Tasks 15** through **16** to make Task 14 a summary task. Indent **Tasks 18** through **20** to make Task 17 a summary task. Widen the Task Name column to see all of your text, as needed.

> **TIP**
>
> To change a task from a summary task to a subtask or to change its level up one in the WBS, you can "outdent" the task instead of indenting it. To outdent the task, click the cell of the task or tasks you want to change, and then click the **Outdent Task** button on the Ribbon (the button just to the left of the Indent Task button). You can also press Alt + Shift + Right Arrow to indent tasks and Alt Shift Left Arrow to outdent tasks. Remember, the tasks in Project 2010 should be entered in an appropriate WBS format with several levels in the hierarchy.

### Numbering Tasks

Depending on how Project 2010 is set up on your computer, you may or may not see numbers associated with tasks as you enter and indent them.

To display automatic numbering of tasks using the standard tabular numbering system for a WBS:

1. *Show outline numbers*. Click the **Format** tab, and then click the **Outline Number** checkbox under the Show/Hide group. Project 2010 adds the appropriate WBS numbering to the task names.
2. *Show project summary task*. Click the **Project Summary** checkbox just below the Outline Number checkbox. Notice that a new task has been added under row 0.
3. *Adjust the file*. Widen the Task Name column and move the split bar so only that column displays. Your file should resemble Figure A-20.

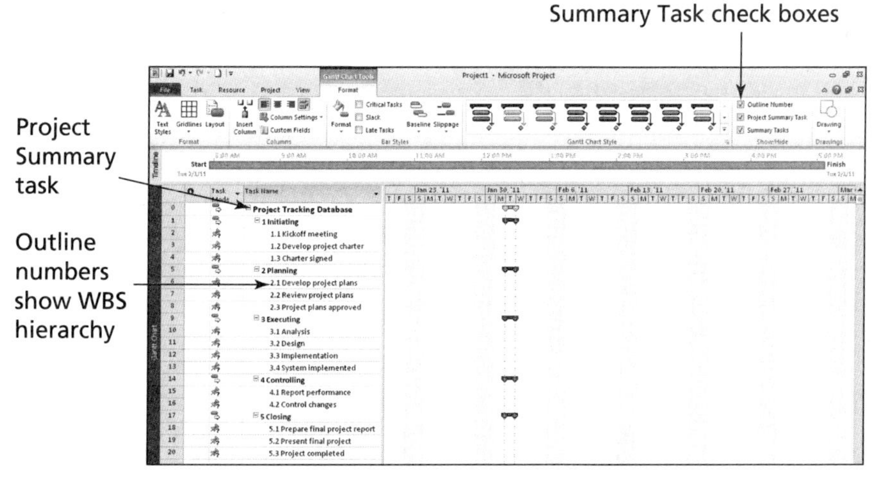


**FIGURE A-20** Adding Outline Numbers and Project Summary Task

## Saving Project Files with or without a Baseline

An important part of project management is tracking performance against a baseline, or approved plan. Project 2010 does not prompt you to save a file with or without a baseline each time you save it, as some previous versions did. The default is to save without a baseline. It is important to wait until you are ready to save your file with a baseline because Project 2010 will show changes against a baseline. Because you are still developing your project file for the Project Tracking Database project, you want to save the file without a baseline. Later in this appendix, you will save the file with a baseline by selecting Tools, Tracking, and then Set Baseline. You will then enter actual information to compare planned and actual performance data.

To save a file without a baseline:

1. *Save your file*. Click the **File** tab and then click **Save**, or click the **Save** button on the Quick Access toolbar.
2. *Enter a filename*. In the Save As dialog box, type **scope** in the File name text box. Browse to the location in which you want to save the file, and then click **Save**. Your Project 2010 file should look like Figure A-20, but with the file name showing as scope instead of Project1.

**HELP**

If you want to download the Project 2010 file scope.mpp to check your work or continue to the next section, a copy is available on the companion Web site for this text, the author's Web site, and from your instructor. If you downloaded the scope.mpp file that comes with this text, you can save your file with a different name or in a different location to avoid overwriting that file. Keep this in mind for other files you save as well.

Now that you have finished entering all 20 tasks, created the summary tasks and subtasks, set the options to show the standard WBS tabular numbering system, and saved your file, you will learn how to use the Project 2010 time management features.

# PROJECT TIME MANAGEMENT

Many people use Project 2010 for its time management features. The first step in using these features, after inputting the WBS for the project, is to enter durations for tasks or specific dates when tasks will occur. Inserting durations or specific dates will automatically update the Gantt chart. To use Project 2010 to adjust schedules automatically and to do critical path analysis, you must also enter task dependencies. After entering durations and task dependencies, you can view the network diagram and critical path information. This section describes how to use each of these time management features. It also explains the difference between the new manual and automatic scheduling features of Project 2010.

## Manual and Automatic Scheduling

If you have used earlier versions of Project, you probably noticed that when you entered a task, it was automatically assigned a duration of one day, and Start and Finish dates were also automatically entered. This is still the case in Project 2010 if you use automatic scheduling for a task. If you use manual scheduling, no durations or dates are automatically entered. The other big change with manual scheduling is that summary task durations are not automatically calculated based on their subtasks when they are set up as manually scheduled tasks. Figure A-21 illustrates these differences. Notice that the Manual subtask 1 had no information entered for its duration, start, or finish dates. Also note that the duration for Manual summary task 1 is not dependent on the durations of its subtasks. For the Automatic summary task, its duration is dependent on its summary tasks, and information is entered for all of the durations, start, and end dates. You can switch between automatic and manual scheduling for tasks in the same file, as desired, by changing the Task Mode.

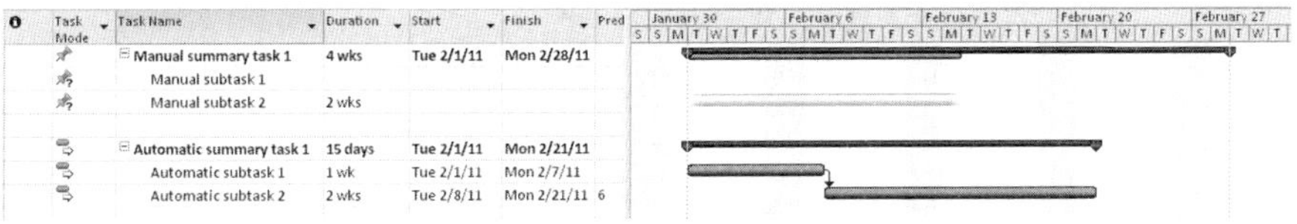

| Task Mode | Task Name | Duration | Start | Finish | Pred |
|---|---|---|---|---|---|
| | Manual summary task 1 | 4 wks | Tue 2/1/11 | Mon 2/28/11 | |
| | Manual subtask 1 | | | | |
| | Manual subtask 2 | 2 wks | | | |
| | | | | | |
| | Automatic summary task 1 | 15 days | Tue 2/1/11 | Mon 2/21/11 | |
| | Automatic subtask 1 | 1 wk | Tue 2/1/11 | Mon 2/7/11 | |
| | Automatic subtask 2 | 2 wks | Tue 2/8/11 | Mon 2/21/11 | 6 |

**FIGURE A-21** Manual versus automatic scheduling

When you move your mouse over the Task Mode column (shown in the far left in Figure A-21) Project 2010 displays the following information:

- A task can be Manually Scheduled or Automatically Scheduled.
- Manually Scheduled tasks have user-defined Start, Finish, and Duration values. Project will never change their dates, but may warn you if there are potential issues with the entered values.
- Automatically Scheduled tasks have Start, Finish, and Duration values calculated by Project based on dependencies, constraints, calendars, and other factors.

Project Help provides the following example of using both manual and automatic scheduling. You set up a preliminary project plan that's still in the proposal stage. You have a vague idea of major milestone dates but not much detail on other dates in various phases of the project. You build tasks and milestones using the Manually Scheduled task mode. The proposal is accepted and the tasks and deliverable dates become more defined. You continue to manually schedule those tasks and dates for a while, but as certain phases become well-defined, you decide to switch the tasks in those phases to the Automatically Scheduled task mode. By letting Project 2010 handle the complexities of scheduling, you can focus your attention on those phases that are still under development.

## Entering Task Durations

When you enter a task, Project 2010 assumes it is manually scheduled, so no duration is automatically entered. If you switch to the automatic scheduling mode, Project 2010 assigns those tasks a default duration of one day, followed by a question mark. To change the default duration, type a task's estimated duration in the Duration column. If you are unsure of an estimate and want to review it again later, enter a question mark after it. For example, you could enter 5d? for a task with an estimated duration of five days that you want to review later. You can then use the Tasks With Estimated Durations filter to see quickly the tasks for which you need to review duration estimates.

To indicate the length of a task's duration, you must type both a number and an appropriate duration symbol. If you type only a number, Project 2010 automatically enters days as the duration unit. Duration unit symbols include:

- d = days (default)
- w = weeks
- m = minutes
- h = hours
- mo or mon = months
- ed = elapsed days
- ew = elapsed weeks

For example, to enter one week for a task duration, type 1w in the Duration column. (You can also type wk, wks, week, or weeks, instead of just w.) To enter two days for a task duration, type 2d in the Duration column. The default unit is days, so if you enter 2 for the duration, it will be entered as 2 days. You can also enter elapsed times in the Duration column. For example, 2ed means two elapsed days, and 2ew means two elapsed weeks. You would use an elapsed duration for a task like "Allow paint to dry." The paint will dry in exactly the same amount of time regardless of whether it is a workday, a weekend, or a holiday.

**TIP**

If the Duration column is not visible, drag the split bar to the right until the Duration column is in view.

Entering time estimates or durations might seem like a straightforward process. However, you must follow a few important procedures:

- To mark a task as a milestone, enter 0 for the duration. You can also mark tasks that have a non-zero duration as milestones by checking the "Mark task as milestone" option in the Task Information dialog box on the Advanced tab. You simply double-click a task to access this dialog box. The milestone symbol for those tasks will appear at their start date.
- You can enter the exact start and finish dates for activities instead of entering durations in the automatic scheduling mode. To enter start and finish dates, move the split bar to the right to reveal the Start and Finish columns. You normally only enter start and finish dates in this mode when those dates are certain.
- If you want task dates to adjust according to any other task dates, do not enter exact start and finish dates. Instead, enter durations and then establish dependencies to related tasks.
- To enter recurring tasks, such as weekly meetings, select Recurring Task from the Task button under the Task tab, Insert group. Enter the task name, the duration, and when the task occurs. Project 2010 will automatically insert appropriate subtasks based on the length of the project and the number of tasks required for the recurring task.
- Project 2010 uses a default calendar with standard workdays and hours. Remember to change the default calendar if needed, as described later in this appendix.

Next, you will set task durations in the Project Tracking Database file (scope.mpp) that you created and saved in the previous section. If you did not create the file named scope.mpp, you can download it from the companion Web site for this text. You will create a new recurring task and enter its duration, and then you will enter other task durations. First, create a new recurring task called Status Reports above Task 15, Report performance.

To create a new recurring task:

1. *Insert a recurring task above Task 15, Report performance*. Open scope.mpp, if necessary, and then click **Report performance** (Task 15) in the Task Name column to select that task. Click the **Tasks** tab, if needed, and click the **Tasks** button drop-down box under Insert group, and then click **Recurring Task.** The Recurring Task Information dialog box opens.
2. *Enter task and duration information for the recurring task*. Type **Status Reports** as the task title in the Task Name text box. Type **1h** in the Duration text box. Select the **Weekly** radio button under Recurrence pattern. Make sure that 1 is entered in the **Recur every** list box. Select the **Wednesday** check box. In the Range of recurrence section, type **2/1/11** in the Start text box, click the **End by** radio button, and then type **6/15/11** in the End by text box. Click the **End by** list arrow to see the calendar, as shown in Figure A-22. You can use the calendar to enter the Start and End by dates. The new recurring task will appear above Task 15, Report performance, when you are finished.

## TIP

You can also enter a number of occurrences instead of an End by Date for a recurring task. You might need to adjust the End by date after you enter all of your task durations and dependencies. Remember, the date on your computer determines the date listed as Today in the calendar.

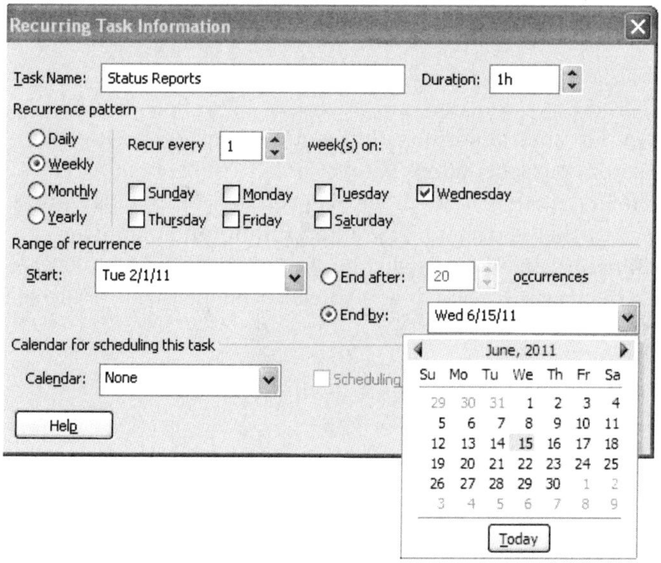

**FIGURE A-22** Recurring Task Information dialog box

3. *View the new summary task and its subtasks*. Click **OK.** Project 2010 inserts a new Status Reports subtask in the Task Name column. Expand the new subtask by clicking the **expand symbol** ⊞ to the left of Status Reports. To collapse the recurring task, click the **collapse symbol** ⊟.
4. *Adjust the columns shown and timescale*. Move the split bar so that the entire Finish column displays. Click the **Zoom Out** button (minus sign on the Zoom Slider in the lower right of your screen) three times (or as many as needed) to display the entire Gantt chart on your screen. Notice that the recurring task appears on the appropriate dates on the Gantt chart, as shown in Figure A-23.

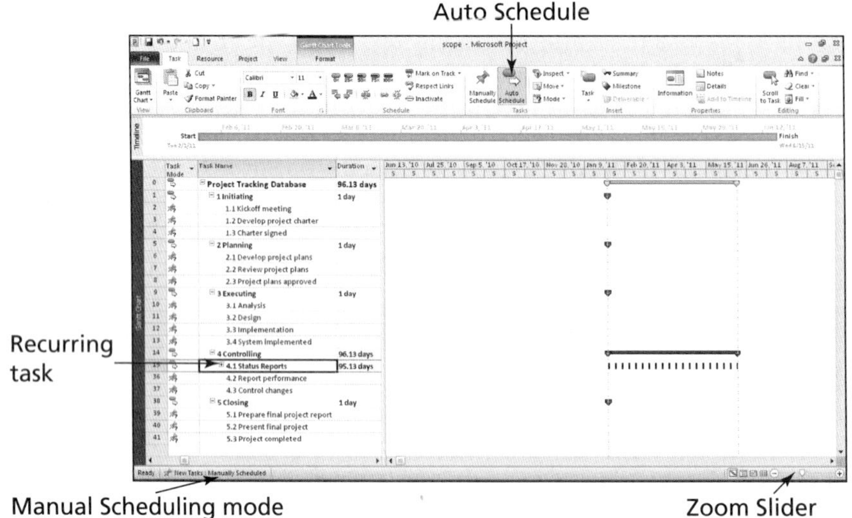


**FIGURE A-23** A recurring task

Use the information in Table A-3 to enter durations for the other tasks for the Project Tracking Database project using automatic scheduling. The Project 2010 row number is shown to the left of each task name in the table. Remember, you already entered a duration for the recurring task. Also, remember that you should *not* enter durations for summary tasks. When entering tasks in automatic scheduling mode, durations for summary tasks are automatically calculated to match the durations and dependencies of subtasks, as described further in the next section, Establishing Task Dependencies.

**TABLE A-3** Durations for Project Tracking Database tasks

| Task Number/Row | Task Name | Duration |
|---|---|---|
| 2 | Kickoff meeting | 2h |
| 3 | Develop project charter | 10d |
| 4 | Charter signed | 0 |
| 6 | Develop project plans | 3w |
| 7 | Review project plans | 4mo |
| 8 | Project plans approved | 0 |
| 10 | Analysis | 1mo |
| 11 | Design | 2mo |
| 12 | Implementation | 1mo |
| 13 | System implemented | 0 |
| 36 | Report performance | 6mo |
| 37 | Control changes | 6mo |
| 39 | Prepare final project report | 2w |
| 40 | Present final project | 1w |
| 41 | Project completed | 0 |

To enter task durations for the other tasks:

1. *Turn on automatic scheduling and enter the duration for Task 2.* Click row 2 and drag down through row 41 to select all of those tasks. Click the **Auto Schedule** button on the Ribbon under the Task group. Click the **Duration** column for row 2, Kickoff meeting, type **2h,** and then press **Enter.**

**TIP**

You can also change the scheduling mode for individual tasks by changing that icon in the Indicator column. For this file, we want all of the tasks to be automatically scheduled. If you know that at the start, you can click the icon on the lower left of the screen to make all new tasks automatically scheduled.

2. *Enter the duration for Task 3.* In the **Duration** column for row 3, Develop project charter, type **10d,** and then press **Enter.** You can also just type 10, since d or days is the default duration.
3. *Enter remaining task durations.* Continue to enter the durations using the information in Table A-3. Be careful not to make entries in the wrong row. Remember that the summary tasks are automatically entered, as is the recurring task.
4. Save your file and name it. Click the **File** tab, and then click **Save As.** Enter **time** as the filename, and then save the file to the desired location on your computer or network. Your file should resemble Figure A-24. Remember that you can use the Zoom Slider to adjust the timescale as needed. Notice that all of the tasks still begin on February 1. This will change in the next section when we add task dependencies. Keep this file open for the next set of steps.

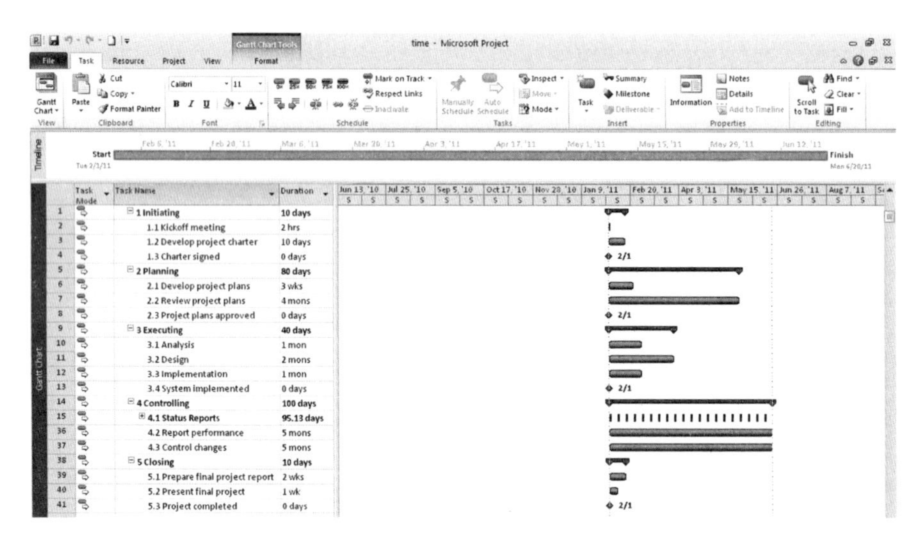


FIGURE A-24 Time file with durations entered

## Establishing Task Dependencies

To use Project 2010 to adjust schedules automatically and to do critical path analysis, you *must* determine the dependencies or relationships among tasks. Project 2010 provides three methods for creating task dependencies: using the Link Tasks button, using the Predecessors column of the Entry table or the Predecessors tab in the Task Information dialog box, or clicking and dragging the Gantt chart symbols for tasks with dependencies.

To create dependencies using the Link Tasks button, highlight tasks that are related and then click the Link Tasks button on the toolbar. For example, to create a finish-to-start dependency between Task 1 and Task 2, click any cell in row 1, drag down to row 2, and then click the Link Tasks button. The default type of link is finish-to-start. In the Project Tracking Database example, all the tasks use this default relationship. You will learn about other types of dependencies later in this appendix.

## TIP

Selecting tasks is similar to selecting cells in Excel. To select adjacent tasks, click and drag the mouse. You can also click the first task, hold down the Shift key, and then click the last task. To select nonadjacent tasks, hold down the Control (Ctrl) key as you click tasks in order of their dependencies.

When you use the Predecessors column of the Entry table to create dependencies, you must manually enter the information. To create dependencies manually, type the task row number of the preceding task in the Predecessors column of the Entry table. For example, Task 3 in Table A-3 has Task 2 as a predecessor, which can be entered in the Predecessors column, meaning that Task 3 cannot start until Task 2 is finished. To see the Predecessors column of the Entry table, move the split bar to the right. You can also double-click the task, click the Predecessors tab in the Task Information dialog box, and enter the predecessors there. (This is more work than just typing it in the Predecessor column.)

You can also create task dependencies by clicking the Gantt chart symbol for a task and then dragging to the Gantt chart symbol for a task that succeeds it. For example, you could click the Milestone symbol ◆ for Task 4, hold down the left mouse button, and drag to the Task Bar symbol for Task 6 to create a dependency, as shown in the Gantt chart in Figure A-25. After entering a dependency, it shows up in the predecessors column and as a line with an arrow connecting the dependent tasks on the Gantt chart.

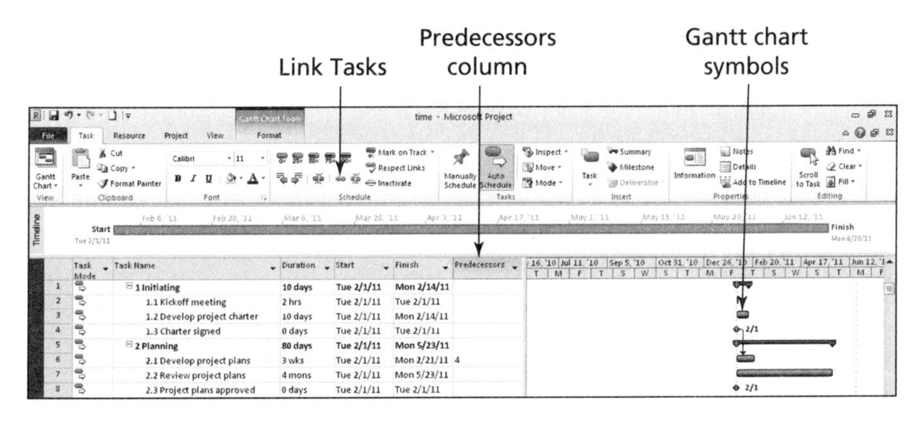


FIGURE A-25 Creating task dependencies

Next, you will use information from Figure A-26 to enter the predecessors for tasks as indicated. You will create some dependencies by manually typing the predecessors in the Predecessors column, some by using the Link Tasks button, some by using the Gantt chart symbols, and the remaining dependencies by using whichever method you prefer.

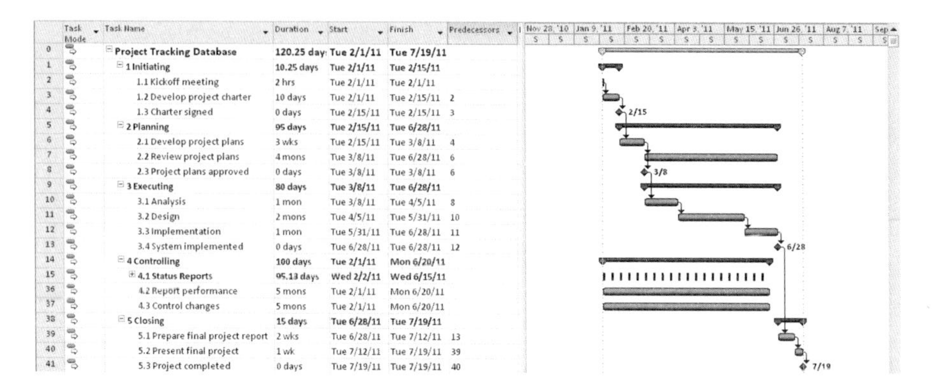

| | Task Mode | Task Name | Duration | Start | Finish | Predecessors |
|---|---|---|---|---|---|---|
| 0 | | **Project Tracking Database** | **120.25 day** | **Tue 2/1/11** | **Tue 7/19/11** | |
| 1 | | **1 Initiating** | **10.25 days** | **Tue 2/1/11** | **Tue 2/15/11** | |
| 2 | | 1.1 Kickoff meeting | 2 hrs | Tue 2/1/11 | Tue 2/1/11 | |
| 3 | | 1.2 Develop project charter | 10 days | Tue 2/1/11 | Tue 2/15/11 | 2 |
| 4 | | 1.3 Charter signed | 0 days | Tue 2/15/11 | Tue 2/15/11 | 3 |
| 5 | | **2 Planning** | **95 days** | **Tue 2/15/11** | **Tue 6/28/11** | |
| 6 | | 2.1 Develop project plans | 3 wks | Tue 2/15/11 | Tue 3/8/11 | 4 |
| 7 | | 2.2 Review project plans | 4 mons | Tue 3/8/11 | Tue 6/28/11 | 6 |
| 8 | | 2.3 Project plans approved | 0 days | Tue 3/8/11 | Tue 3/8/11 | 6 |
| 9 | | **3 Executing** | **80 days** | **Tue 3/8/11** | **Tue 6/28/11** | |
| 10 | | 3.1 Analysis | 1 mon | Tue 3/8/11 | Tue 4/5/11 | 8 |
| 11 | | 3.2 Design | 2 mons | Tue 4/5/11 | Tue 5/31/11 | 10 |
| 12 | | 3.3 Implementation | 1 mon | Tue 5/31/11 | Tue 6/28/11 | 11 |
| 13 | | 3.4 System implemented | 0 days | Tue 6/28/11 | Tue 6/28/11 | 12 |
| 14 | | **4 Controlling** | **100 days** | **Tue 2/1/11** | **Mon 6/20/11** | |
| 15 | | **4.1 Status Reports** | **95.13 days** | **Wed 2/2/11** | **Wed 6/15/11** | |
| 36 | | 4.2 Report performance | 5 mons | Tue 2/1/11 | Mon 6/20/11 | |
| 37 | | 4.3 Control changes | 5 mons | Tue 2/1/11 | Mon 6/20/11 | |
| 38 | | **5 Closing** | **15 days** | **Tue 6/28/11** | **Tue 7/19/11** | |
| 39 | | 5.1 Prepare final project report | 2 wks | Tue 6/28/11 | Tue 7/12/11 | 13 |
| 40 | | 5.2 Present final project | 1 wk | Tue 7/12/11 | Tue 7/19/11 | 39 |
| 41 | | 5.3 Project completed | 0 days | Tue 7/19/11 | Tue 7/19/11 | 40 |



**FIGURE A-26** Project Tracking Database file with durations and dependencies entered

To link tasks or establish dependencies for the Project Tracking Database project:

1. *Display the Predecessors column in the Entry table*. Move the split bar to the right to reveal the full Predecessors column in the time.mpp file. Widen the Task Name or other columns, if needed.
2. *Highlight the cell where you want to enter a predecessor, and then type the task number for the preceding task*. Click the **Predecessors cell** for Task 3, type **2**, and press **Enter.** Notice that as you enter task dependencies, the Gantt chart changes to reflect the new schedule.
3. *Enter predecessors for Task 4*. Click the **Predecessors cell** for Task 4, type **3**, and press **Enter.**
4. *Establish dependencies using the Link Tasks button*. To link Tasks 10 through 13, click the task name for Task 10, **Analysis,** and drag down through Task 13, **System Implemented.** Then, click the **Link Tasks** button on the Ribbon under the Schedule group of the Task tab.
5. *Create a dependency using Gantt chart symbols*. Click the **Milestone** symbol for Task 4 on the Gantt chart, hold down the **left mouse** button, and drag to the **Task Bar** symbol for Task 6.
6. *Enter remaining dependencies*. Link the other tasks by either manually entering the predecessors into the Predecessors column, by using the Link Tasks button, or by clicking and dragging the Gantt chart symbols. Note the visual change highlighting feature that displays as you enter dependencies. You can view the dependencies in Figure A-26. For example, Task 8 has Task 6 as a predecessor, Task 10 has Task 8 as a predecessor, Task 39 has 13, Task 40 has 39, and Task 41 has 40 as a predecessor. If you have entered all data correctly, the project should end on 7/19/11, or July 19, 2011.
7. *Adjust screen elements*. Move the position of the timescale and adjust columns as needed so your screen resembles Figure A-26. Double-check your screen to make sure you entered the dependencies correctly.
8. *Preview and save your file*. Click the **File** tab, and then select **Print** to preview your file in the Backstage, as shown in Figure A-27. If you need to make changes to the file before printing, click the **Task** tab to make changes, and then save your file again by clicking the **Save** button on the Quick Access

toolbar. If you desire, print your file by clicking the **Print** button in the Backstage (or File tab). Keep the file open for the next set of steps.

File tab shows the Backstage view of files

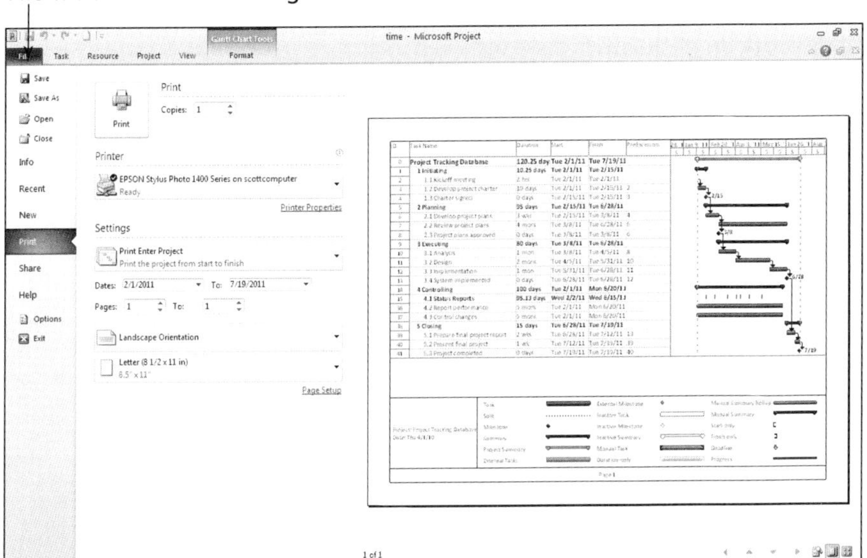


**FIGURE A-27** Project Tracking Database file in the Backstage

## Changing Task Dependency Types and Adding Lead or Lag Time

A task dependency or relationship describes how a task is related to the start or finish of another task. Project 2010 allows for four task dependencies: finish-to-start (FS), start-to-start (SS), finish-to-finish (FF), and start-to-finish (SF). By using these dependencies effectively, you can modify the critical path and shorten your project schedule. The most common type of dependency is finish-to-start (FS). All of the dependencies in the Project Tracking Database example are FS dependencies. However, sometimes you need to establish other types of dependencies. This section describes how to change task dependency types. It also explains how to add lead or lag times between tasks. You will shorten the duration of the Project Tracking Database project by adding lead time between some tasks.

To change a dependency type, open the Task Information dialog box for that task by double-clicking the task name. On the Predecessors tab of the Task Information dialog box, select a new dependency type from the Type column list arrow.

The Predecessors tab also allows you to add lead or lag time to a dependency. You can enter both lead and lag time using the Lag column on the Predecessors tab. Lead time reflects an overlap between tasks that have a dependency. For example, if Task B can start when its predecessor, Task A, is half-finished, you can specify a finish-to-start dependency with a lead time of 50% for the successor task. Enter lead times as negative numbers. In this example, enter –50% in the first cell of the Lag column. Adding lead times is also called *fast tracking* and is one way to compress a project's schedule.

Lag time is the opposite of lead time; it is a time gap or delay between tasks that have a dependency. If you need a two-day delay between the finish of Task C and the start of Task D, establish a finish-to-start dependency between Tasks C and D and specify a two-day lag time. Enter lag time as a positive value. In this example, type 2d in the Lag column.

In the Project Tracking Database example, notice that work on design tasks does not begin until all the work on the analysis tasks has been completed (see Rows 10 and 11), and work on implementation tasks does not begin until all the work on the design tasks has been completed (see Rows 11 and 12). In reality, it is rare to wait until all of the analysis work is complete before starting any design work, or to wait until all of the design work is finished before starting any implementation work. It is also a good idea to add some additional time, or a buffer, before crucial milestones, such as a system being implemented. To create a more realistic schedule, add lead times to the design and implementation tasks and lag time before the system implemented milestone.

To add lead and lag times:

1. *Open the Task Information dialog box for Task 11, Design*. In the Task Name column, double-click the text for Task 11, **Design.** The Task Information dialog box opens. Click the **Predecessors** tab.
2. *Enter lead time for Task 11*. Type **-10%** in the Lag column, as shown in Figure A-28. Click **OK.** You could also type a value such as -5d to indicate a five-day overlap. In the resulting Gantt chart, notice that the bar for this task has moved slightly to the left. Also, notice that the project completion date has moved.

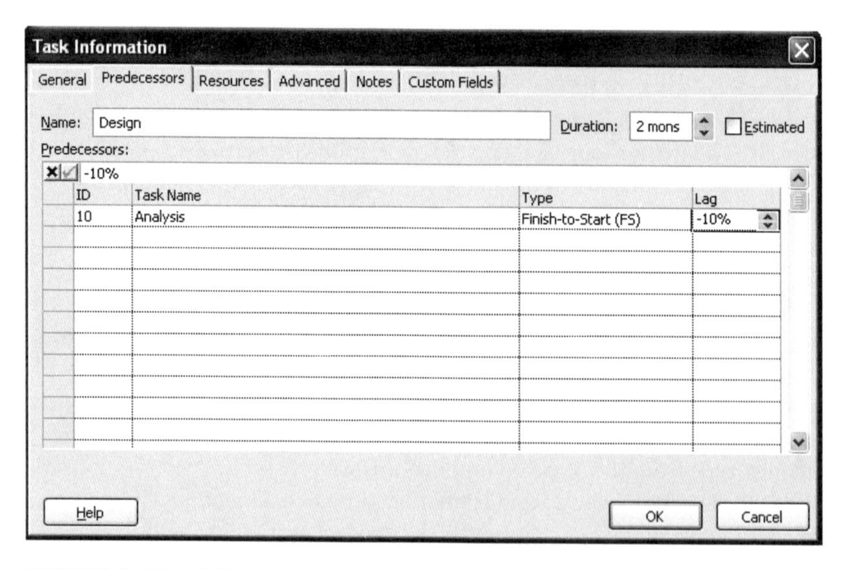


**FIGURE A-28** Adding lead or lag time to task dependencies

3. *Enter lead time for Task 12*. Double-click the text for Task 12, **Implementation,** type **-3d** in the Lag column, and click **OK.** Notice that the project end date has again changed.

4. *Enter lag time for Task 13*. Double-click the text for Task 13, **System Implemented,** type **5d** in the Lag column for this task, and click **OK.** Move the split bar to the right, if necessary, to reveal the Predecessors column. Also adjust the timescale using the Zoom Slider, if needed, to see all of the Gantt chart symbols. When you are finished, your screen should resemble Figure A-29. Notice the slight overlap in the taskbars for Task 10 and the short gap between the taskbars for Tasks 11, 12, and 13. Also notice the changes in the Predecessors column for tasks 11 and 12. The project completion date should be 7/19 after these changes are made.

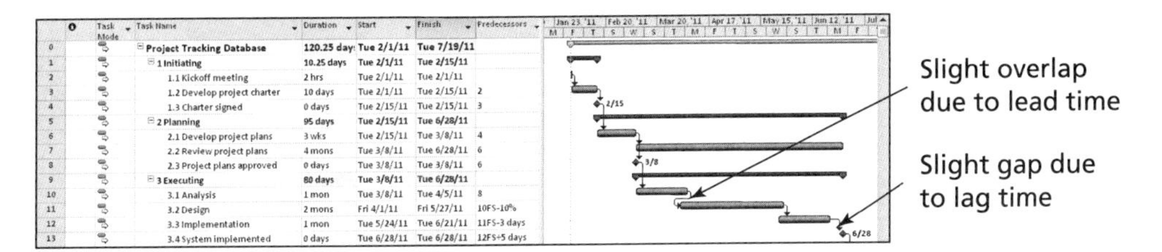


**FIGURE A-29** Schedule for Project Tracking Database file with lead and lag times

## TIP

You can enter or modify lead or lag times directly in the Predecessors column of the Entry table. Notice how the predecessors appear for Tasks 11, 12, and 13. For example, the Predecessors column for Task 11 shows 10FS-10%. This notation means that Task 11 has a finish-to-start (FS) dependency with Task 10 and a 10% lead. You can enter lead or lag times to task dependencies directly into the Predecessors column of the Entry table by using this same format: task row number, followed by type of dependency, followed by the amount of lead or lag.

5. *Add tasks to the Timeline*. Under Task Name, click **Charter Signed,** the task name for Task 4. Click the **Add to Timeline** button under the Properties group of the Task tab, as shown in Figure A-30. Notice that this task is now added to the Timeline. Add the other milestones, **Tasks 8, 13,** and **41** to the Timeline. The new Timeline feature makes it display high level schedule information. For example, you might want to just show the timeline to senior management as part of a presentation instead of a full Gantt chart. You can right-click the Timeline and select **Copy Timeline** (or click the **Copy Timeline** button under the Format tab) to easily copy it into an e-mail or presentation.

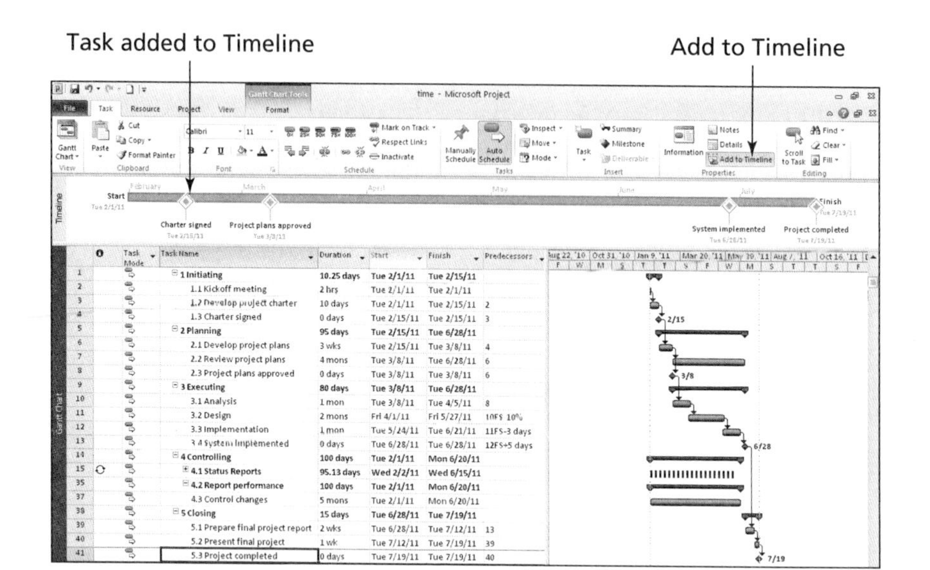


**FIGURE A-30** Adding tasks to the Timeline

6. *Save your file.* Click the **Save** button on the Quick Access toolbar. Keep this file open for the next set of steps.

## Gantt Charts

Project 2010 shows a Gantt chart as the default view to the right of the Entry table. Gantt charts show the timescale for a project and all of its activities. In Project 2010, dependencies between tasks are shown on the Gantt chart by the arrows between tasks. Many Gantt charts, however, do not show any dependencies. Instead, as you might recall, network diagrams or PERT charts are used to show task dependencies. This section explains important information about Gantt charts and describes how to make critical path information more visible in the Gantt Chart view.

You must follow a few important procedures when working with Gantt charts:

- To adjust the timescale, click the Zoom Out button or the Zoom In button on the Zoom Slider. Clicking these buttons automatically makes the dates on the Gantt chart show more or less information. For example, if the timescale for the Gantt chart is showing months and you click the Zoom Out button, the timescale will adjust to show quarters. Clicking Zoom Out again will display the timescale in years. Similarly, each time you click the Zoom In button, the timescale changes to display more detailed time information—from years to quarters, quarters to months, and months to weeks.
- You can also adjust the timescale and access more formatting options by clicking on the Gantt chart and selecting the Format tab. For example, by checking the Critical Task checkbox, critical tasks will display in red on the Gantt chart.
- You can view a tracking Gantt chart by setting a baseline for your entire project or for selected tasks and then entering actual durations for tasks. The Tracking

Gantt view displays two taskbars, one above the other, for each task. One taskbar shows planned or baseline start and finish dates, and the other taskbar shows actual start and finish dates. You will find a sample Tracking Gantt chart later in this appendix, after you enter actual information for the Project Tracking Database project.

Because you have already created task dependencies, you can now find the critical path for the Project Tracking Database project. You can view the critical tasks by changing the color of those items on the Gantt chart. Tasks on the critical path will automatically be red in the Network Diagram view, as described in the following section.

To make the Gantt chart symbols for the critical path tasks appear in red:

1. *Format the Gantt chart for critical tasks*. Click the **Format** tab, and then click the **Critical Tasks check box** under the Bar Styles group. The symbols for critical tasks display in red on the Gantt chart. Your screen should resemble Figure A-31.

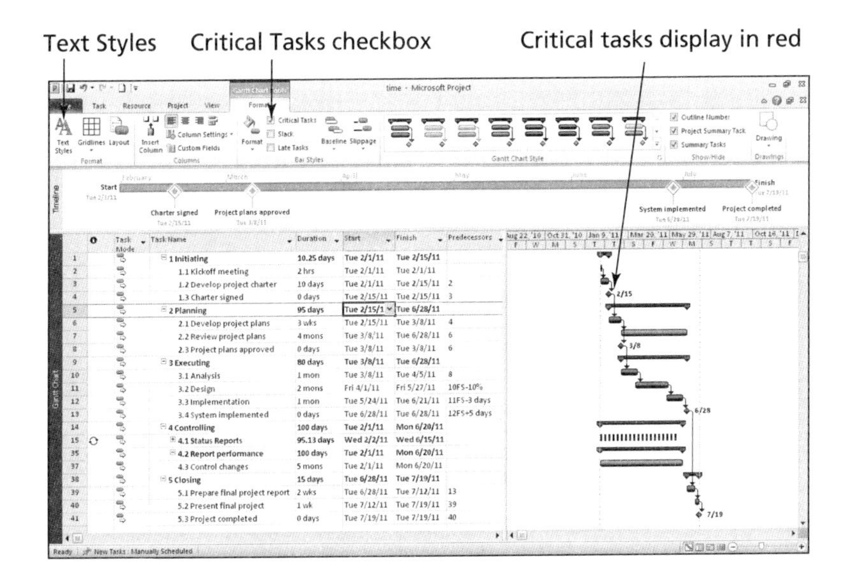


**FIGURE A-31** Formatting the Gantt chart

2. *Format the text color for critical tasks*. Click the **Text Styles** button under the Format group. Click the **Item to Change** list arrow, and then select Critical Tasks. Click the **Color** list arrow, and then select a different color, such as **red**, as shown in Figure A-32. Click **OK** to accept the changes.

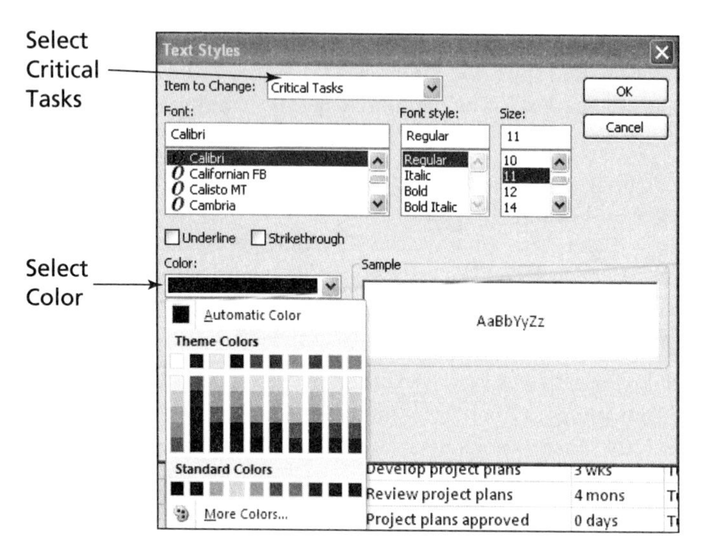


**FIGURE A-32** Formatting text for critical tasks

3. *Save your file*. Click **Save** on the Quick Access toolbar to save your file. It should resemble Figure A-33. Keep this file open for the next set of steps.

Text and Gantt chart symbols for critical tasks, such as Tasks 2 and 3, should display in red on your computer.

| | Task Mode | Task Name | Duration | Start | Finish | Predecessors |
|---|---|---|---|---|---|---|
| 0 | | **Project Tracking Database** | **120.25 day:** | **Tue 2/1/11** | **Tue 7/19/11** | |
| 1 | | **1 Initiating** | **10.25 days** | **Tue 2/1/11** | **Tue 2/15/11** | |
| 2 | | 1.1 Kickoff meeting | 2 hrs | Tue 2/1/11 | Tue 2/1/11 | |
| 3 | | 1.2 Develop project charter | 10 days | Tue 2/1/11 | Tue 2/15/11 | 2 |
| 4 | | 1.3 Charter signed | 0 days | Tue 2/15/11 | Tue 2/15/11 | 3 |
| 5 | | **2 Planning** | **95 days** | **Tue 2/15/11** | **Tue 6/28/11** | |
| 6 | | 2.1 Develop project plans | 3 wks | Tue 2/15/11 | Tue 3/8/11 | 4 |
| 7 | | 2.2 Review project plans | 4 mons | Tue 3/8/11 | Tue 6/28/11 | 6 |
| 8 | | 2.3 Project plans approved | 0 days | Tue 3/8/11 | Tue 3/8/11 | 6 |
| 9 | | **3 Executing** | **80 days** | **Tue 3/8/11** | **Tue 6/28/11** | |
| 10 | | 3.1 Analysis | 1 mon | Tue 3/8/11 | Tue 4/5/11 | 8 |
| 11 | | 3.2 Design | 2 mons | Fri 4/1/11 | Fri 5/27/11 | 10FS-10% |
| 12 | | 3.3 Implementation | 1 mon | Tue 5/24/11 | Tue 6/21/11 | 11FS-3 days |
| 13 | | 3.4 System implemented | 0 days | Tue 6/28/11 | Tue 6/28/11 | 12FS+5 days |
| 14 | | **4 Controlling** | **100 days** | **Tue 2/1/11** | **Mon 6/20/11** | |
| 15 | | **4.1 Status Reports** | **95.13 days** | **Wed 2/2/11** | **Wed 6/15/11** | |
| 35 | | **4.2 Report performance** | **100 days** | **Tue 2/1/11** | **Mon 6/20/11** | |
| 37 | | 4.3 Control changes | 5 mons | Tue 2/1/11 | Mon 6/20/11 | |
| 38 | | **5 Closing** | **15 days** | **Tue 6/28/11** | **Tue 7/19/11** | |
| 39 | | 5.1 Prepare final project report | 2 wks | Tue 6/28/11 | Tue 7/12/11 | 13 |
| 40 | | 5.2 Present final project | 1 wk | Tue 7/12/11 | Tue 7/19/11 | 39 |
| 41 | | 5.3 Project completed | 0 days | Tue 7/19/11 | Tue 7/19/11 | 40 |

**FIGURE A-33** Formatted Gantt chart

## Network Diagrams

The network diagrams in Project 2010 use the precedence diagramming method, with tasks displayed in rectangular boxes and relationships shown by lines connecting the boxes. In the Network Diagram view, tasks on the critical path are automatically shown in red.

To view the network diagram for the Project Tracking Database project:

1. *View the network diagram*. Click the **View** tab, and then click the **Network Diagram** button under the Task Views group.

2. *Adjust the Network Diagram view*. To see more tasks in the Network Diagram view, move the **Zoom Out** button on the Zoom Slider. Figure A-34 shows several of the tasks in the Project Tracking Database network diagram. Note that milestone tasks, such as Charter signed, appear as pointed rectangular boxes, while other tasks appear as rectangles. Tasks on the critical path automatically appear in red, while noncritical tasks appear in blue. Each task in the network diagram also shows information such as the start and finish dates, task ID, and duration. Move your mouse over the Charter signed box to see it in a larger view. A dashed line on a network diagram represents a page break. You often need to change some of the default settings for the Network Diagram view before printing it.

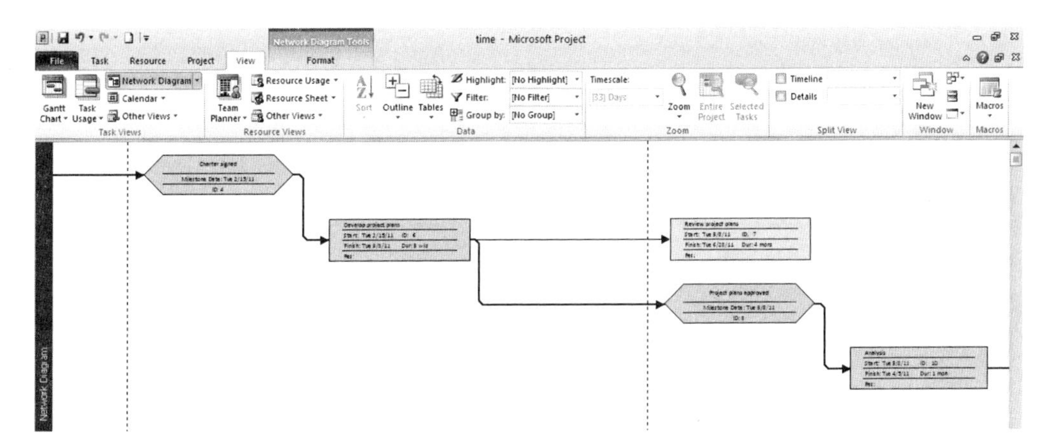


**FIGURE A-34** Network Diagram view

3. *Change the format/layout of the network diagram*. Click the **Format** tab, and then click the **Layout** button under the Format group. The Layout dialog box displays, as shown in Figure A-35. Note that you can change several layout options for the network diagram, such as manually positioning boxes, changing the link style, changing colors, and so on. Click **OK** to close the Layout dialog box.

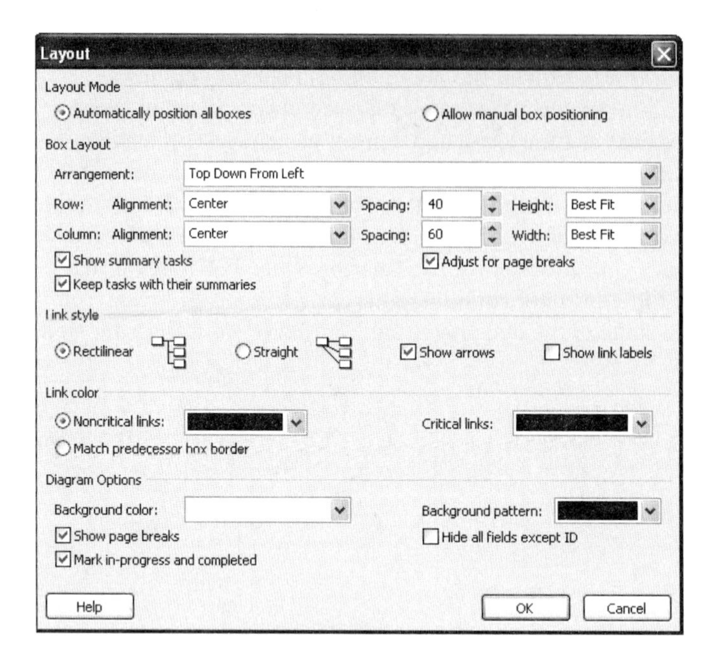


**FIGURE A-35** Changing the network diagram layout

4. *Return to Gantt Chart view*. Click the **View** tab, and then click the **Gantt Chart** button under the Task Views group. Keep this file open for the next set of steps.

**TIP**

Some users prefer to create or modify files in Network Diagram view instead of Gantt Chart view. To add a new task or node in Network Diagram view, select the Task tab and then click the Task button under the Insert group. Double-click the new node to add a task name and other information. Create finish-to-start dependencies between tasks in Network Diagram view by clicking the preceding node and dragging to the succeeding node. To modify the type of dependency and add lead or lag time, double-click the arrow between the dependent nodes.

## Critical Path Analysis

The critical path is the path through the network diagram with the least amount of slack; it represents the shortest possible time to complete the project. If a task on the critical path takes longer than planned, the project schedule will slip unless time is reduced on a task later on the critical path. Sometimes you can shift resources between tasks to help keep a project on schedule. Project 2010 has several views and reports to help analyze critical path information.

Two particularly useful features are the Schedule table view and the Critical Tasks report. The Schedule table view shows the early and late start dates for each task, the early and late finish dates for each task, and the free and total slack for each task. This information shows how flexible the schedule is and helps in making schedule compression decisions. The Critical Tasks report lists only tasks that are on the critical path for the project. If meeting schedule deadlines is essential for a project, project managers will want to monitor tasks on the critical path closely.

To access the Schedule table view and view the Critical Tasks report for a file:

1. *View the Schedule table*. Right-click the **Select All** button to the left of the Task Name column heading and select **Schedule.** The **Schedule** table replaces the Entry table to the left of the Gantt chart.
2. *Reveal all columns in the Schedule table*. Move the split bar to the right until you see the entire Schedule table. Your screen should resemble Figure A-36. This view shows the start and finish (meaning the early start and early finish) and late start and late finish dates for each task, as well as free and total slack. Right-click the **Select All** button and select **Entry** to return to the Entry table view.

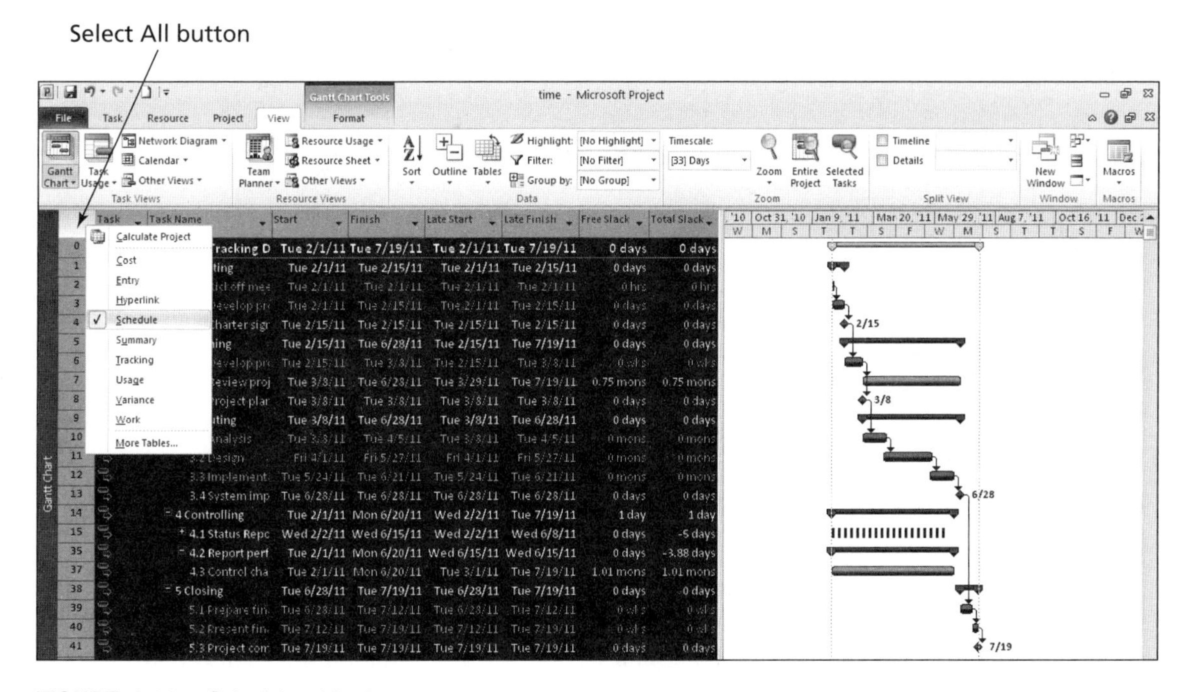


FIGURE A-36 Schedule table view

3. *Open the Reports dialog box*. Click the **Project** tab, and then click the **Reports** button under the Reports group. Double-click Overview to open the **Overview Reports** dialog box. Your screen should resemble Figure A-37.

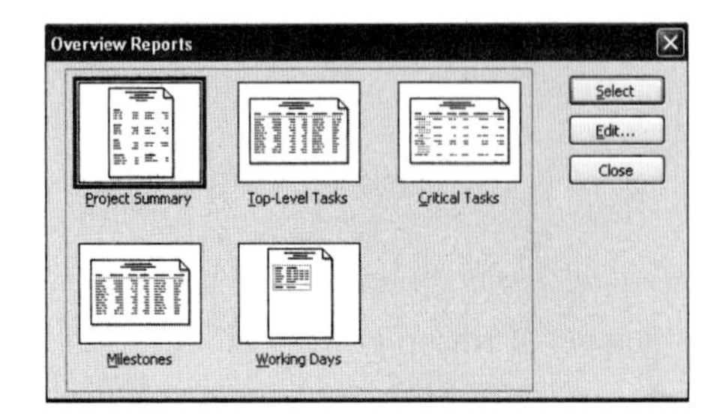


**FIGURE A-37** Accessing the Critical Tasks report

4. *Display the Critical Tasks report.* Double-click **Critical Tasks.** A Critical Tasks report as of today's date is displayed. You can click **Page Setup** in the Backstage to adjust settings before printing.
5. *Close the report and save your file.* When you are finished examining the Critical Tasks report, click the **Task tab.** Save your file again by clicking **Save** on the Quick Access toolbar. Close Project 2010 if you are not continuing to the next section.

**HELP**

If you want to download the Project 2010 file time.mpp to check your work or continue to the next section, a copy is available on the companion Web site for this text, the author's Web site, or from your instructor.

Now that you have entered task durations, established task dependencies, and reviewed the network diagram and critical path information, you are ready to explore some of the Project 2010 cost management features.

## PROJECT COST MANAGEMENT

Many people do not use Project 2010 for cost management. Most organizations have more established cost management software products and procedures in place, and many people simply do not know how to use the cost features of Project 2010. However, the cost features of Project 2010 make it possible to integrate total project information more easily. This section offers brief instructions for entering fixed and variable cost estimates and actual cost and time information after establishing a baseline plan. It also explains how to use Project 2010 for earned value management. More details on these features are available in Project 2010 Help, online tutorials, or other texts.

**TIP**

To complete the steps in this section, you need to use the Project 2010 file resource.mpp mentioned earlier. You can download it from the companion Web site for this text or from the author's Web site under Book FAQs.

## Fixed and Variable Cost Estimates

The first step to using the cost features of Project 2010 is entering cost-related information. You enter costs as fixed or variable based on per-use material costs or variable based on the type and amount of resources used. Costs related to personnel are often a significant part of project costs.

### Entering Fixed Costs in the Cost Table

The Cost table allows you to enter fixed costs related to each task. To access the Cost table, right-click the Select All button in the Entry table and select Cost. You can also assign a per-use cost to a resource that represents materials or supplies and use it as a base for calculating the total materials or supplies cost of a task. See Project 2010 Help for details on this feature.

### Entering Human Resource Costs

Human resources represent a major part of the costs on many projects. By defining and then assigning human resources and their related costs to tasks in Project 2010, you can calculate human resource costs, track how people are used, identify potential resource shortages that could force you to miss deadlines, and identify underutilized resources. It is often possible to shorten a project's schedule by reassigning underutilized resources. This section focuses on entering human resource costs and assigning resources to tasks. The following section describes other features of Project 2010 related to human resource management.

Several methods are available for entering resource information in Project 2010. One of the easiest is to enter basic resource information in the Resource Sheet, accessible from the View tab. The Resource Sheet allows you to enter the resource name, initials, resource group, maximum units, standard rate, overtime rate, cost/use, accrual method, base calendar, and code. Entering data into the Resource Sheet is similar to entering data into an Excel spreadsheet, and you can easily sort items by selecting Sort from the Project menu. In addition, you can use the Filter list on the Formatting toolbar to filter resources. Once you have established resources in the Resource Sheet, you can assign those resources to tasks in the Entry table with the list arrow that appears when you click a cell in the Resource Names column. The Resource Names column is the last column of the Entry table. You can also use other methods for assigning resources, as described in the following steps.

Next, you will use the Project 2010 file time.mpp, which you saved in the preceding section, to assign resources to tasks. (If you did not save your file, download it from the companion Web site or from the author's Web site.) Assume that there are four people working on the Project Tracking database project and that the only costs for this project are for these human resources. Kathy is the project manager; John is the business analyst; Mary is the database analyst; and Chris is an intern or trainee.

To enter basic information about each person into the Resource Sheet:

1. *Display the Resource Sheet view*. Open your Project 2010 file time.mpp, if necessary. Click the **View** tab, and then click the **Resource Sheet** button under the Resource Views group.
2. *Enter resource information*. Enter the information from Table A-4 into the Resource Sheet. Type the information as shown and press the **Tab** key to move to the next field. Notice that you are putting abbreviated job titles in the Initials column: PM stands for project manager; BA, business analyst; DA, database

analyst; and IN, intern. When you type the standard and overtime rates, you can just type 50, and Project 2010 will automatically enter $50.00/hr. The standard and overtime rates entered are based on hourly rates. You can also enter annual salaries by typing the annual salary number followed by /y for "per year." Leave the default values for the other columns in the Resource Sheet as they are. Your screen should resemble Figure A-38 when you are finished entering the resource data.

**TABLE A-4** Project Tracking Database resource data

| Resource Name | Initials | Group | Stand. Rate | Ovt. Rate |
|---|---|---|---|---|
| Kathy | PM | 1 | $50.00/h | $60.00/h |
| John | BA | 1 | $40.00/h | $50.00/h |
| Mary | DA | 1 | $40.00/h | $50.00/h |
| Chris | IN | 1 | $20.00/h | $25.00/h |

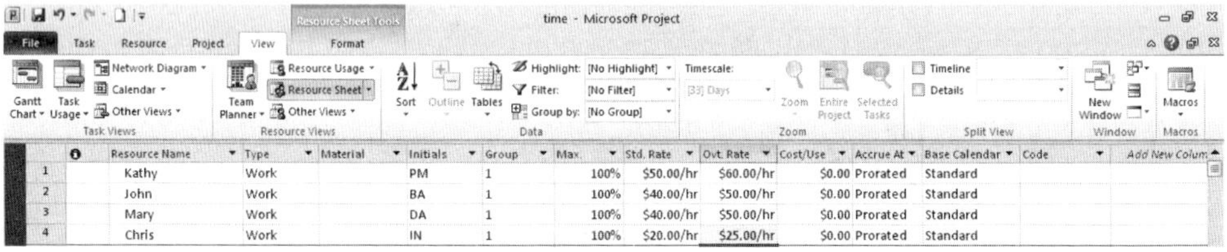

**FIGURE A-38** Resource Sheet view with resource data entered

## TIP

If you know that some people will be available for a project only part time, enter their percentage of availability in the Max Units column of the Resource Sheet. Project 2010 will then automatically assign those people based on their maximum units. For example, if someone can work only 25% of their time on a project throughout most of the project, enter 25% in the Max Units column for that person. When you enter that person as a resource for a task, his or her default number of hours will be 25% of a standard eight-hour workday, or two hours per day.

### Adjusting Resource Costs

To make a resource cost adjustment, such as a raise, double-click the person's name in the Resource Name column, select the Costs tab in the Resource Information dialog box, and then enter the effective date and raise percentage. You can also adjust other resource cost information, such as standard and overtime rates.

To give the project manager a 10% raise starting 4/1/11:

1. *Open the Resource Information dialog box*. In Resource Sheet view, double-click **Kathy** in the Resource Name column. The Resource Information dialog box opens.
2. *Enter an effective date for a raise*. Select the **Costs** tab, and then select **tab A**, if needed. Type **4/1/11** in the second cell in the Effective Date column and press **Enter.** Alternately, click the **list arrow** in the second cell and use the calendar that appears to enter the effective date, April 1, 2011.
3. *Enter the raise percentage*. Type **10%** in the second cell for the Standard Rate column, and then press **Enter.** The Resource Information screen should resemble Figure A-39. Notice that Project 2010 calculated the 10% raise to be $55.00 per hour. Click **OK.**

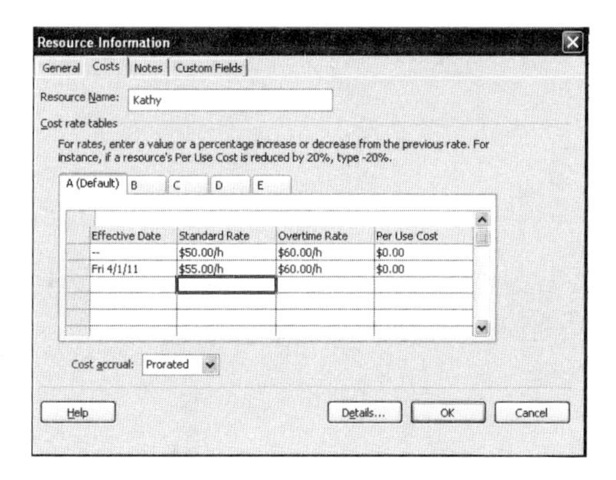


**FIGURE A-39** Adjusting resource costs

## Assigning Resources to Tasks

For Project 2010 to calculate resource costs, you must assign the appropriate resources to tasks in your WBS. There are several methods for assigning resources. The Resources column in the Entry table allows you to select resources using a list. You can use other methods to assign resources, such as the Assign Resources button or the split window, which is the recommended approach to have the most control over how resources are assigned because Project 2010 makes several assumptions about resource assignments that might change your schedule or costs. Next, you will use these three methods for assigning resources to the Project Tracking Database project.

### Assigning Resources Using the Entry Table

To assign resources using the Entry table:

1. *Select the task to which you want to assign resources*. Click the **Gantt Chart** button on the View tab, and right-click the **Select All** button and click **Entry** to return to the Entry table, if needed.
2. *Reveal the Resource Names column of the Entry table*. Move the split bar to the right to reveal the entire Resource Names column in the Entry table.

3. *Select a resource from the Resource Names column*. In the Resource Names column, click the cell associated with Task 2, **Kickoff meeting.** Click the cell's **list arrow,** and then click the checkboxes next to **John** and **Kathy** to assign them to Task 2, as shown in Figure A-40. Notice that the resource choices are based on information that you entered in the Resource Sheet. If you had not entered any resources, you would not have a list arrow or any choices to select. Click in another cell to see the entered resources in the Resource column and on the Task 2 Gantt chart symbol.

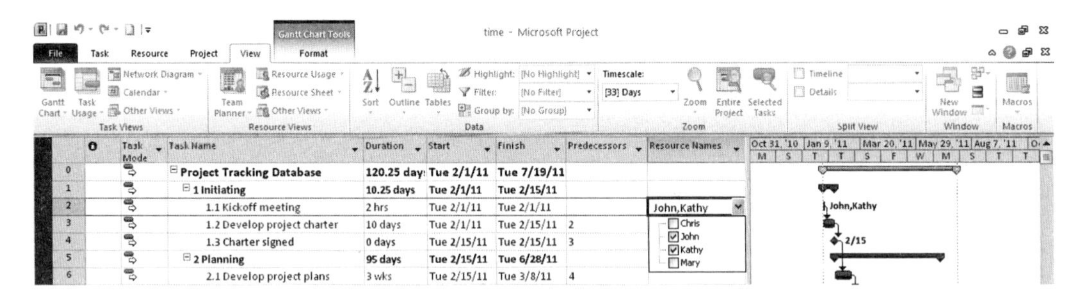


**FIGURE A-40** Assigning resources using the Resource column

4. *Clear the resource assignment*. Click the **Undo** button on the Quick Access toolbar to remove the resource assignments.

## Assigning Resources Using the Resource Tab

To assign resources using the Resource tab:

1. *Select the task to which you want to assign resources*. In the second row of the Task Name column, click **Kickoff meeting,** the task name for Task 2.
2. *Open the Assign Resources dialog box*. Click the **Resource** tab, and then click the **Assign Resources** button in the Assignments group. The Assign Resources dialog box, as shown in Figure A-41, displays.

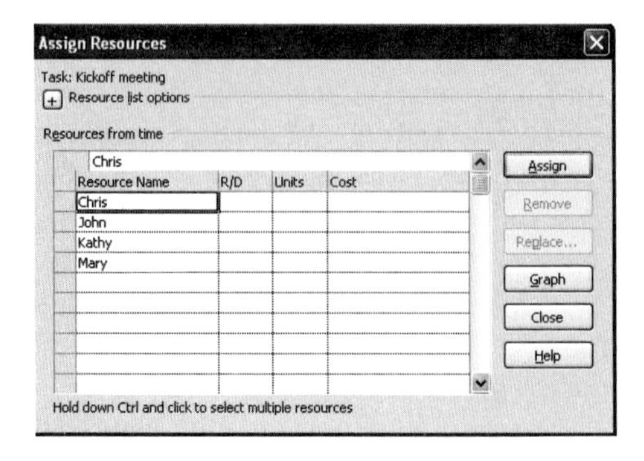


**FIGURE A-41** Assign Resources dialog box

3. *Assign Kathy to Task 2*. Click **Kathy** in the Resource Name column of the Assign Resources dialog box, and then click **Assign.** Notice that the duration estimate for Task 2 remains at 2 hours, and Kathy's name appears on the Gantt chart by the bar for Task 2.
4. *Assign John to Task 2*. Click **John** in the Resource Name column of the Assign Resources dialog box, and then click **Assign.** Click **Close** in the Assign Resources dialog box. Notice that the duration for Task 2 remains at 2 hours, but an Exclamation point appears in the Indicator column for Task 2, as shown in Figure A-42. The default selection is to "Increase total work because the task requires more person-hours. Keep duration constant." That is what you want, so **click that option,** and the indicator symbol disappears. *This feature is a big change from previous versions of Microsoft Project, which automatically reduced the duration when more resources were added.*

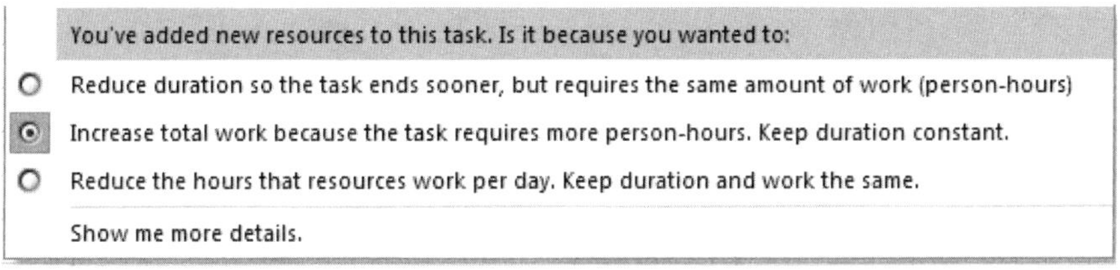

You've added new resources to this task. Is it because you wanted to:

- Reduce duration so the task ends sooner, but requires the same amount of work (person-hours)
- Increase total work because the task requires more person-hours. Keep duration constant.
- Reduce the hours that resources work per day. Keep duration and work the same.

Show me more details.

**FIGURE A-42** Indicator column options

**TIP**

Project 2010 includes a multiple-level undo feature, so you can click the Undo button on the Quick Access toolbar several times to undo several steps in your file.

### Assigning Resources Using the Split Window

Even though using the Assign Resources button seems simple, it is often better to use a split when assigning resources. When you assign resources using the split view, you have more control over how you enter information and can visually see the cost table and Gantt chart changes in the top window as you assign each resource in the bottom window.

To assign both Kathy and John to attend the two-hour kickoff meeting using the split window:

1. *Open the Cost table view*. Right-click the **Select All** button, and then click **Cost** to reveal the Cost table. Notice that Task 2 shows a cost of $180 after assigning Kathy and John to it.
2. *Split the window to reveal more information*. Click the **Details** button under the Properties group of the Resource tab. The Gantt Chart view is displayed at the top of the screen and a resource form is displayed at the bottom of the screen.
3. *Assign Kathy to Task 3*. Select Task 3, **Develop project charter** in the top window, and then click the **Assign Resources** button, select **Kathy**, and **Close** the

dialog box. Notice the changes in both windows, as shown in Figure A-43. By default, resources are assigned to work full-time on tasks, so 80 hours, or two weeks of full-time work, are entered for Kathy, and a cost of $4,000 is entered in the cost table. However, you only want to enter 40 hours for this task but leave its duration at two weeks.

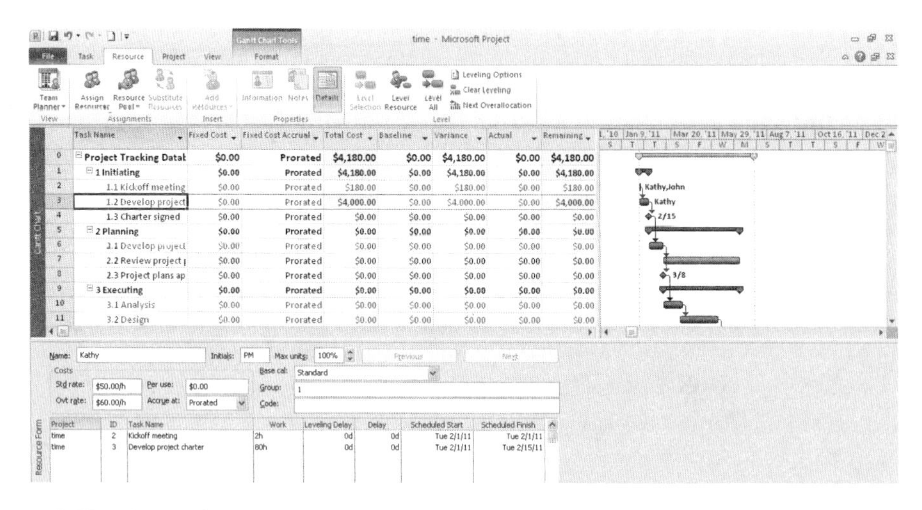


**FIGURE A-43** Split screen for entering resource information

4. *Make Task 3 a manually scheduled task and change Kathy's hours to 80*. With Task 3 selected, click the **Task** tab, and then click the **Manually Schedule** button under the Task group. In the bottom window, click the cell for the Work column in the bottom window for Task 2, and type **40h** instead of 80h. Press **OK.** Notice that the cost for Task 2 changed from $4,000 to $2,000, but the duration for the task remained the same. If you left the task as automatically scheduled, its duration would have been shortened. You could assign all of the resources and then change the hours for one person at a time using the split screen.
5. *Close the file and do not save it*. Close the file, but do not save the changes you made. Other resource information has been entered for you in the Project 2010 file named resource.mpp.

## HELP

A copy of the Project 2010 file resource.mpp is available on the companion Web site for this text, from the author's Web site under Book FAQs, or from your instructor. You must use this file to continue the steps in the next section.

As you can see, you must be careful when assigning resources and scheduling tasks. Project 2010 assumes that the durations of automatically scheduled tasks are not fixed, but are effort driven, and this assumption can create some problems when you are assigning resources.

## Viewing Project Cost Information

Once you enter resource information, Project 2010 automatically calculates resource costs for the project. There are several ways to view project cost information. You can view the Cost table to see cost information, or you can run various cost reports. Next, you will view cost information for the Project Tracking Database project.

To view cost information:

1. *Open resource.mpp*. Download **resource.mpp** from the companion Web site or the author's site. Open the file.
2. *Open the Cost table*. Right-click the **Select All** button, and then click **Cost.** The Cost table displays various cost information. Your screen should resemble Figure A-44. Note that by assigning resources, costs have been automatically calculated for tasks. You could also enter fixed costs for tasks by simply typing them into the appropriate cell in the cost table.

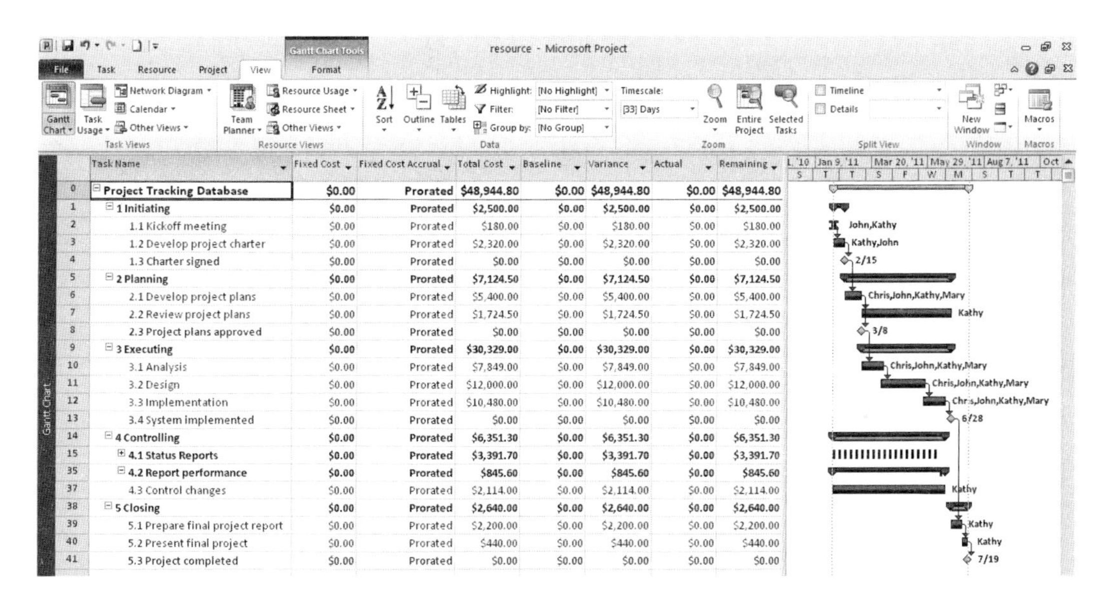

| | Task Name | Fixed Cost | Fixed Cost Accrual | Total Cost | Baseline | Variance | Actual | Remaining |
|---|---|---|---|---|---|---|---|---|
| 0 | Project Tracking Database | $0.00 | Prorated | $48,944.80 | $0.00 | $48,944.80 | $0.00 | $48,944.80 |
| 1 | 1 Initiating | $0.00 | Prorated | $2,500.00 | $0.00 | $2,500.00 | $0.00 | $2,500.00 |
| 2 | 1.1 Kickoff meeting | $0.00 | Prorated | $180.00 | $0.00 | $180.00 | $0.00 | $180.00 |
| 3 | 1.2 Develop project charter | $0.00 | Prorated | $2,320.00 | $0.00 | $2,320.00 | $0.00 | $2,320.00 |
| 4 | 1.3 Charter signed | $0.00 | Prorated | $0.00 | $0.00 | $0.00 | $0.00 | $0.00 |
| 5 | 2 Planning | $0.00 | Prorated | $7,124.50 | $0.00 | $7,124.50 | $0.00 | $7,124.50 |
| 6 | 2.1 Develop project plans | $0.00 | Prorated | $5,400.00 | $0.00 | $5,400.00 | $0.00 | $5,400.00 |
| 7 | 2.2 Review project plans | $0.00 | Prorated | $1,724.50 | $0.00 | $1,724.50 | $0.00 | $1,724.50 |
| 8 | 2.3 Project plans approved | $0.00 | Prorated | $0.00 | $0.00 | $0.00 | $0.00 | $0.00 |
| 9 | 3 Executing | $0.00 | Prorated | $30,329.00 | $0.00 | $30,329.00 | $0.00 | $30,329.00 |
| 10 | 3.1 Analysis | $0.00 | Prorated | $7,849.00 | $0.00 | $7,849.00 | $0.00 | $7,849.00 |
| 11 | 3.2 Design | $0.00 | Prorated | $12,000.00 | $0.00 | $12,000.00 | $0.00 | $12,000.00 |
| 12 | 3.3 Implementation | $0.00 | Prorated | $10,480.00 | $0.00 | $10,480.00 | $0.00 | $10,480.00 |
| 13 | 3.4 System implemented | $0.00 | Prorated | $0.00 | $0.00 | $0.00 | $0.00 | $0.00 |
| 14 | 4 Controlling | $0.00 | Prorated | $6,351.30 | $0.00 | $6,351.30 | $0.00 | $6,351.30 |
| 15 | 4.1 Status Reports | $0.00 | Prorated | $3,391.70 | $0.00 | $3,391.70 | $0.00 | $3,391.70 |
| 35 | 4.2 Report performance | $0.00 | Prorated | $845.60 | $0.00 | $845.60 | $0.00 | $845.60 |
| 37 | 4.3 Control changes | $0.00 | Prorated | $2,114.00 | $0.00 | $2,114.00 | $0.00 | $2,114.00 |
| 38 | 5 Closing | $0.00 | Prorated | $2,640.00 | $0.00 | $2,640.00 | $0.00 | $2,640.00 |
| 39 | 5.1 Prepare final project report | $0.00 | Prorated | $2,200.00 | $0.00 | $2,200.00 | $0.00 | $2,200.00 |
| 40 | 5.2 Present final project | $0.00 | Prorated | $440.00 | $0.00 | $440.00 | $0.00 | $440.00 |
| 41 | 5.3 Project completed | $0.00 | Prorated | $0.00 | $0.00 | $0.00 | $0.00 | $0.00 |

**FIGURE A-44** Cost table for the resource file

3. *Open the Cost Reports dialog box*. Click the **Project** tab, and then click **Reports.** Double-click **Costs** to display the Cost Reports dialog box.
4. *Set the time units for the report*. Click **Cash Flow,** if necessary, and then click **Edit.** The Crosstab Report dialog box is displayed. Click the **Column** list arrow to see other options, but leave it at months, as shown in Figure A-45. Sometimes you need to edit default settings to make reports meet user needs. Click **OK.**

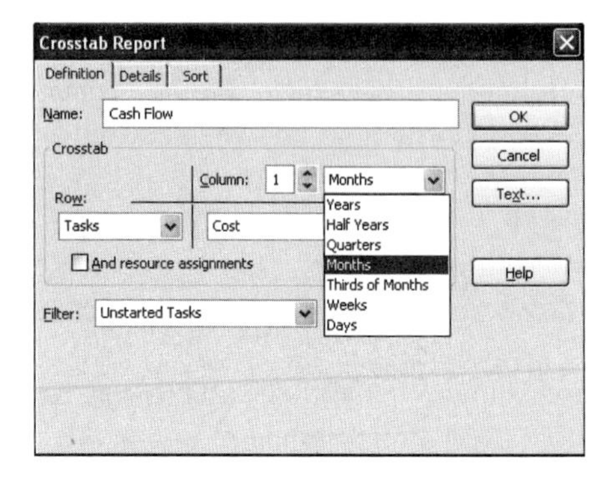


**FIGURE A-45** Crosstab Report dialog box

5. *View the Cash Flow report.* Click **Select** in the Cost Reports dialog box. A Cash Flow report for the Project Tracking Database project appears in the Backstage, as shown in Figure A-46. Notice how it displays by WBS by month. Click the **Project** tab to close the report and return to the Gantt Chart view.

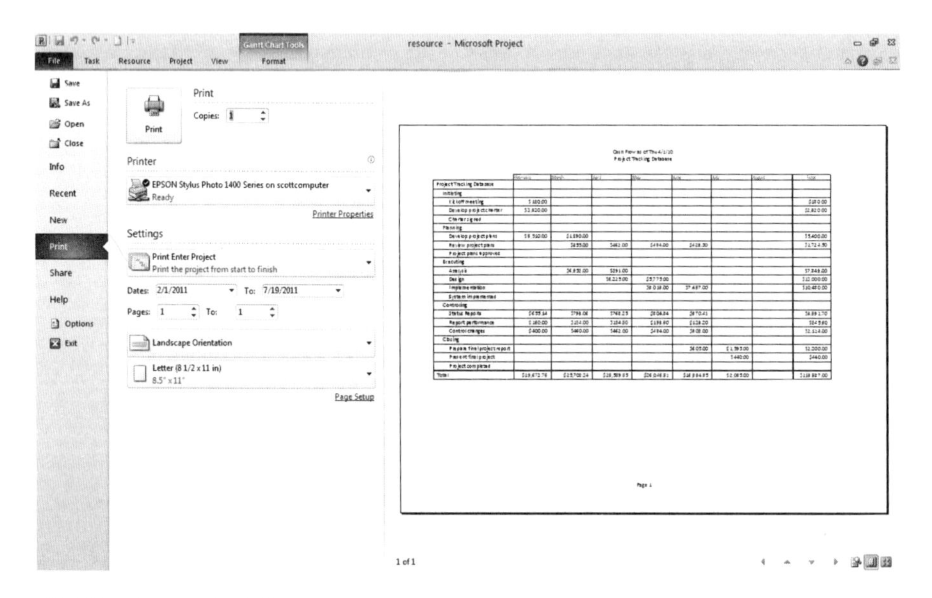


**FIGURE A-46** Cash Flow report

6. *View the Project Summary report.* Click the **Reports** button again, double-click **Overview,** and then double-click **Project Summary.** A Project Summary report for the Project Tracking Database project is displayed, listing information such as project baseline start and finish dates; actual start and finish dates; summaries of duration, work hours, and costs; as well as variance information. Because you have not yet saved the file as a baseline, much of this information is blank in the report. The Project Summary report provides a high-level

overview of a project. Click the **Project** tab to close the report after reviewing it. Experiment with other reports, as desired. Keep this file open for the next set of steps.

**TIP**

You can edit many of the report formats in Project 2010. Instead of double-clicking the report, select the desired report, and then click Edit.

The total estimated cost for this project, based on the information entered, should be $48,944.80, as shown in the Cash Flow and Project Summary reports. In the next section, you will save this data as a baseline plan and enter actual information.

## Baseline Plan, Actual Costs, and Actual Times

Once you complete the initial process of creating a plan—entering tasks, establishing dependencies, assigning costs, and so on—you are ready to set a baseline plan. By comparing the information in your baseline plan to an updated plan during the course of the project, you can identify and solve problems. After the project ends, you can use the baseline and actual information to plan similar, future projects more accurately. To use Project 2010 to help control projects, you must establish a baseline plan, enter actual costs, and enter actual durations.

### Establishing a Baseline Plan

An important part of project management is setting a baseline plan. If you plan to compare actual information such as durations and costs, you must first save the Project 2010 file as a baseline. Before setting a baseline, you must complete the baseline plan by entering time, cost, and human resources information. Be careful not to set a baseline until you have completed the baseline plan. If you do save a baseline before completing the baseline plan, Project 2010 allows you to save up to ten baselines. You can then clear unwanted baseline plans.

Even though you can clear a baseline plan or save multiple baselines, it is a good idea to save a separate, backup file of the original baseline for your project. Enter actuals and save that information in the main file, but always keep a backup baseline file without actuals. Keeping separate baseline and actual files will allow you to return to the original file in case you ever need to use it again.

To rename resource.mpp and then save it as a baseline plan in Project 2010:

1. *Save resource.mpp as a new file named baseline.mpp*. Click the **File** tab, and then click **Save As.** Type **baseline** as the filename, and then click **Save.** You can also download the baseline.mpp file from the Web site.
2. *Open the Save Baseline dialog box and save as a baseline*. Click the **Project tab**, click the **Set Baseline** button under the Schedule group, and click Set Baseline. Click the **Baseline** list arrow to reveal multiple baselines, as shown in Figure A-47. Notice that you can save up to ten baselines, and you can change the options if you do not want to save the Entire project. Also notice that after you set the baseline, the cost numbers in the Baseline column are updated. Keep the default settings, and click **OK.**

**TIP**

There are several options in the Save Baseline dialog box, as shown in Figure A-47. You can save the file as an interim plan if you anticipate several versions of the plan. You can also select the entire project or selected tasks when saving a baseline or interim plan. As mentioned above, you can save up to ten baselines with Project 2010.

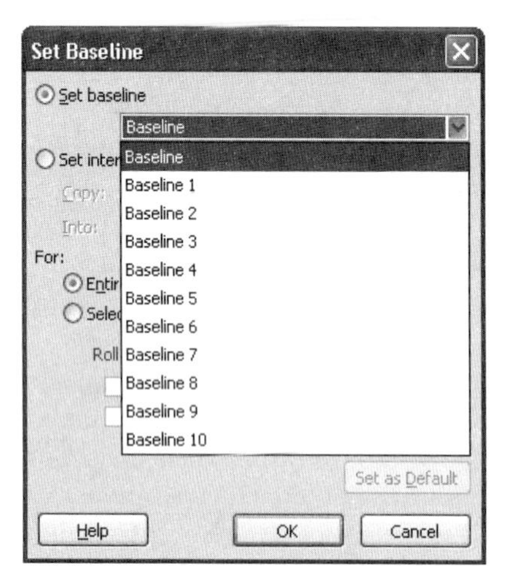


**FIGURE A-47** Set Baseline dialog box

### Entering Actual Costs and Times

After you set the baseline plan, you can track information on each of the tasks as the project progresses. You can also adjust planned task information for tasks still in the future. The Tracking table displays tracking information, and the Ribbon provides icons to help you enter this information. Figure A-48 describes each Tracking button.

| Button | Name | Description |
|---|---|---|
| 0% | 0% Complete | Marks the selected tasks as 0% complete. |
| 25% | 25% Complete | Marks the selected tasks as 25% complete. |
| 50% | 50% Complete | Marks the selected tasks as 50% complete. |
| 75% | 75% Complete | Marks the selected tasks as 75% complete. |
| 100% | 100% Complete | Marks the selected tasks as 100% complete. |
| | Mark on Track | Marks the selected tasks so that they are on schedule. Set the status date on the Project tab to control what date is used in this calculation. |
| | Update Tasks | Shows the Update Tasks dialog box. You can update information for the selected tasks, such as mark percent complete, set actual or remaining duration, modify actual start and finish dates, and create notes. |

**FIGURE A-48** Tracking buttons

To practice entering actual information, enter just a few changes to the baseline. Assume that Tasks 1 through 8 were completed as planned, but that Task 10 took longer than planned. First you need to change the tasks to be automatically scheduled so the actual information, such as a critical task taking longer than planned, automatically adjusts any dependent tasks.

To enter actual information for tasks:

1. *Return to the Task tab and make the tasks automatically scheduled.* Click the **Task** tab, where the tracking buttons are located. Click the task name for Task 2, drag down through the task name for Task 41, and then click the **Auto Schedule** button under the Tasks group.
2. *Display the Tracking table.* Right-click the **Select All** button, and then click **Tracking** to see more tracking-related information as you enter actual data. Widen the Task Name column to see all of the text, and then move the split bar to reveal all the columns in the Tracking table. Make other adjustments as needed so your screen resembles Figure A-49.

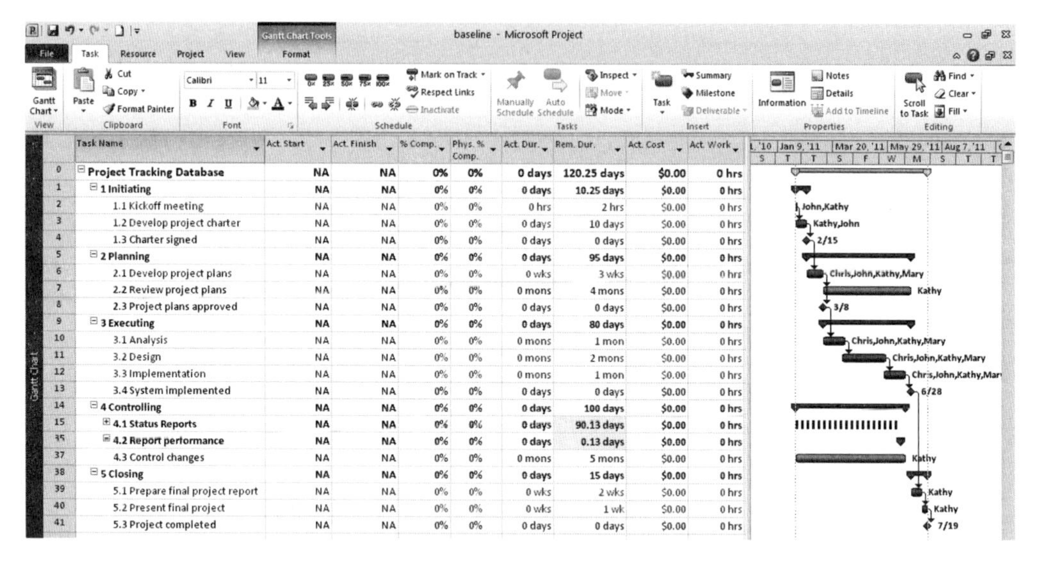

| | Task Name | Act. Start | Act. Finish | % Comp. | Phys. % Comp. | Act. Dur. | Rem. Dur. | Act. Cost | Act. Work |
|---|---|---|---|---|---|---|---|---|---|
| 0 | Project Tracking Database | NA | NA | 0% | 0% | 0 days | 120.25 days | $0.00 | 0 hrs |
| 1 | 1 Initiating | NA | NA | 0% | 0% | 0 days | 10.25 days | $0.00 | 0 hrs |
| 2 | 1.1 Kickoff meeting | NA | NA | 0% | 0% | 0 hrs | 2 hrs | $0.00 | 0 hrs |
| 3 | 1.2 Develop project charter | NA | NA | 0% | 0% | 0 days | 10 days | $0.00 | 0 hrs |
| 4 | 1.3 Charter signed | NA | NA | 0% | 0% | 0 days | 0 days | $0.00 | 0 hrs |
| 5 | 2 Planning | NA | NA | 0% | 0% | 0 days | 95 days | $0.00 | 0 hrs |
| 6 | 2.1 Develop project plans | NA | NA | 0% | 0% | 0 wks | 3 wks | $0.00 | 0 hrs |
| 7 | 2.2 Review project plans | NA | NA | 0% | 0% | 0 mons | 4 mons | $0.00 | 0 hrs |
| 8 | 2.3 Project plans approved | NA | NA | 0% | 0% | 0 days | 0 days | $0.00 | 0 hrs |
| 9 | 3 Executing | NA | NA | 0% | 0% | 0 days | 80 days | $0.00 | 0 hrs |
| 10 | 3.1 Analysis | NA | NA | 0% | 0% | 0 mons | 1 mon | $0.00 | 0 hrs |
| 11 | 3.2 Design | NA | NA | 0% | 0% | 0 mons | 2 mons | $0.00 | 0 hrs |
| 12 | 3.3 Implementation | NA | NA | 0% | 0% | 0 mons | 1 mon | $0.00 | 0 hrs |
| 13 | 3.4 System implemented | NA | NA | 0% | 0% | 0 days | 0 days | $0.00 | 0 hrs |
| 14 | 4 Controlling | NA | NA | 0% | 0% | 0 days | 100 days | $0.00 | 0 hrs |
| 15 | 4.1 Status Reports | NA | NA | 0% | 0% | 0 days | 90.13 days | $0.00 | 0 hrs |
| 35 | 4.2 Report performance | NA | NA | 0% | 0% | 0 days | 0.13 days | $0.00 | 0 hrs |
| 37 | 4.3 Control changes | NA | NA | 0% | 0% | 0 mons | 5 mons | $0.00 | 0 hrs |
| 38 | 5 Closing | NA | NA | 0% | 0% | 0 days | 15 days | $0.00 | 0 hrs |
| 39 | 5.1 Prepare final project report | NA | NA | 0% | 0% | 0 wks | 2 wks | $0.00 | 0 hrs |
| 40 | 5.2 Present final project | NA | NA | 0% | 0% | 0 wks | 1 wk | $0.00 | 0 hrs |
| 41 | 5.3 Project completed | NA | NA | 0% | 0% | 0 days | 0 days | $0.00 | 0 hrs |

**FIGURE A-49** Tracking table

3. *Set the Status Date*. Click the **Project** tab, and then click the **Status Date** button under the Status Group. Type **4/15/11**, and then click **OK.** It is important to set the status date to make the tracking features work properly.
4. *Mark Tasks 1 though 8 as 100% complete*. Click the **Task** tab. Click the Task Name for Task 1, **Initiating,** and drag down through Task 8 to highlight the first eight tasks. Click the **100% Complete** button under the Schedule group on the Ribbon. The columns with dates, durations, and cost information should now contain data instead of the default values, such as NA or 0. The % Comp column should display 100%. Adjust column widths if needed. Your screen should resemble Figure A-50. Note that you could also use the **Mark on Track** button to get the same results.

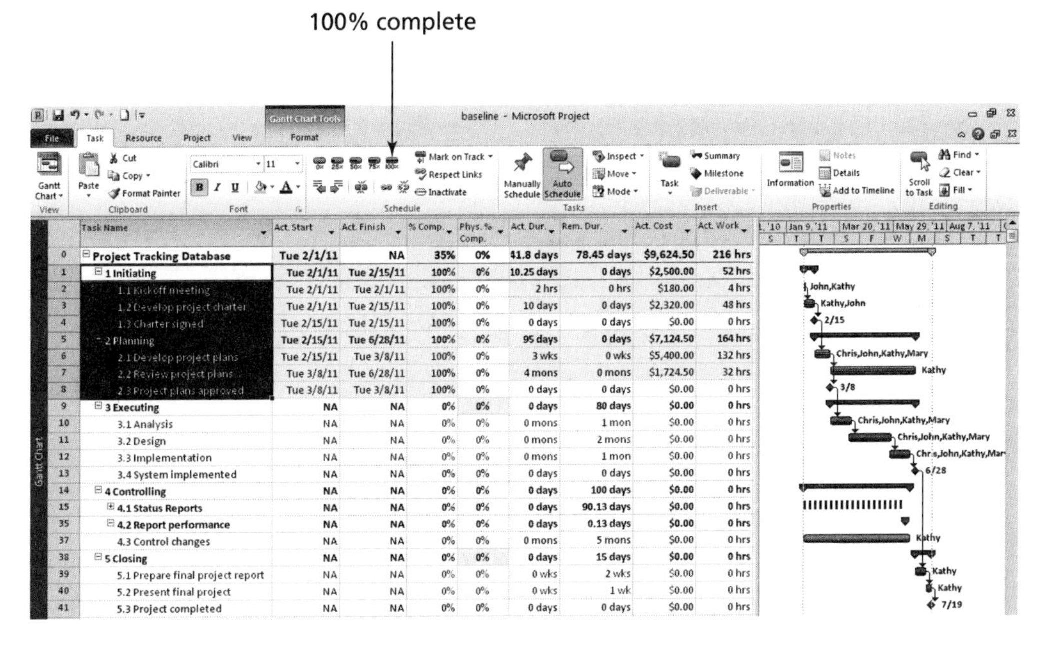


FIGURE A-50 Tracking table information

5. *Enter actual completion dates for Task 10*. Click the Task Name for Task 10, **Analysis**, and then click the **Mark on Track** drop-down, and select **Update Tasks.** The Update Tasks dialog box opens. For Task 10, enter the Actual Start date as **3/10/11** and the Actual Finish date as **4/15/11**, as shown in Figure A-51. Click **OK.** Notice how the information in the tracking sheet has changed and that the project completion date has moved from 7/19 to 7/29 because Task 10 was a critical task.

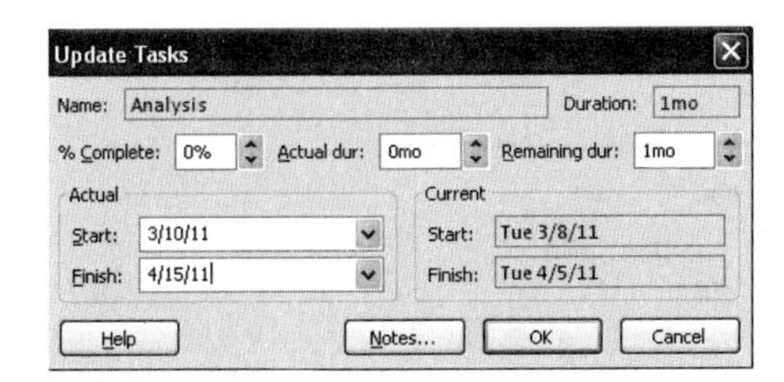


FIGURE A-51 Update Tasks dialog box

6. *View the Variance table*. Right-click the **Select All** button, and then select **Variance** to display the Variance table. Adjust column widths, move the split bar, and adjust the Zoom Slider as needed so your screen resembles Figure A-52. Notice that there is now a finish variance of 8.05 days.

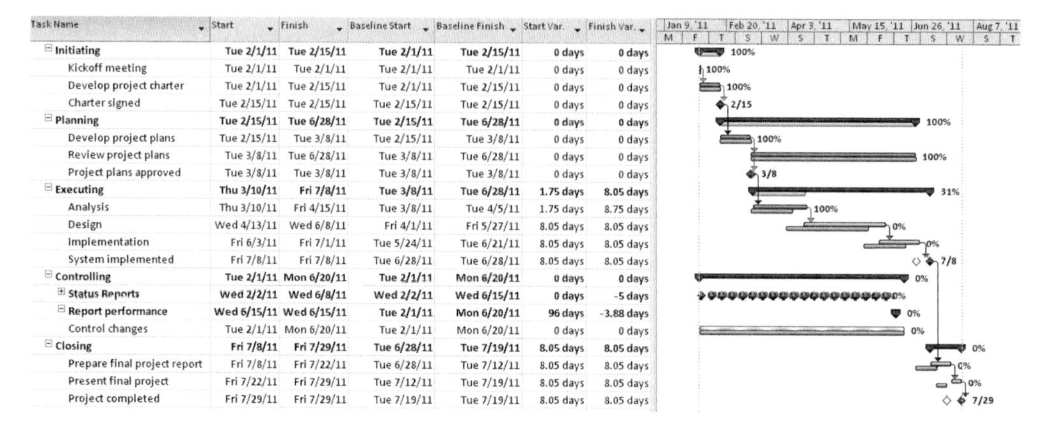

| Task Name | Start | Finish | Baseline Start | Baseline Finish | Start Var. | Finish Var. |
|---|---|---|---|---|---|---|
| Initiating | Tue 2/1/11 | Tue 2/15/11 | Tue 2/1/11 | Tue 2/15/11 | 0 days | 0 days |
| Kickoff meeting | Tue 2/1/11 | Tue 2/1/11 | Tue 2/1/11 | Tue 2/1/11 | 0 days | 0 days |
| Develop project charter | Tue 2/1/11 | Tue 2/15/11 | Tue 2/1/11 | Tue 2/15/11 | 0 days | 0 days |
| Charter signed | Tue 2/15/11 | Tue 2/15/11 | Tue 2/15/11 | Tue 2/15/11 | 0 days | 0 days |
| Planning | Tue 2/15/11 | Tue 6/28/11 | Tue 2/15/11 | Tue 6/28/11 | 0 days | 0 days |
| Develop project plans | Tue 2/15/11 | Tue 3/8/11 | Tue 2/15/11 | Tue 3/8/11 | 0 days | 0 days |
| Review project plans | Tue 3/8/11 | Tue 6/28/11 | Tue 3/8/11 | Tue 6/28/11 | 0 days | 0 days |
| Project plans approved | Tue 3/8/11 | Tue 3/8/11 | Tue 3/8/11 | Tue 3/8/11 | 0 days | 0 days |
| Executing | Thu 3/10/11 | Fri 7/8/11 | Tue 3/8/11 | Tue 6/28/11 | 1.75 days | 8.05 days |
| Analysis | Thu 3/10/11 | Fri 4/15/11 | Tue 3/8/11 | Tue 4/5/11 | 1.75 days | 8.75 days |
| Design | Wed 4/13/11 | Wed 6/8/11 | Fri 4/1/11 | Fri 5/27/11 | 8.05 days | 8.05 days |
| Implementation | Fri 6/3/11 | Fri 7/1/11 | Tue 5/24/11 | Tue 6/21/11 | 8.05 days | 8.05 days |
| System implemented | Fri 7/8/11 | Fri 7/8/11 | Tue 6/28/11 | Tue 6/28/11 | 8.05 days | 8.05 days |
| Controlling | Tue 2/1/11 | Mon 6/20/11 | Tue 2/1/11 | Mon 6/20/11 | 0 days | 0 days |
| Status Reports | Wed 2/2/11 | Wed 6/8/11 | Wed 2/2/11 | Wed 6/15/11 | 0 days | -5 days |
| Report performance | Wed 6/15/11 | Wed 6/15/11 | Tue 2/1/11 | Mon 6/20/11 | 96 days | -3.88 days |
| Control changes | Tue 2/1/11 | Mon 6/20/11 | Tue 2/1/11 | Mon 6/20/11 | 0 days | 0 days |
| Closing | Fri 7/8/11 | Fri 7/29/11 | Tue 6/28/11 | Tue 7/19/11 | 8.05 days | 8.05 days |
| Prepare final project report | Fri 7/8/11 | Fri 7/22/11 | Tue 6/28/11 | Tue 7/12/11 | 8.05 days | 8.05 days |
| Present final project | Fri 7/22/11 | Fri 7/29/11 | Tue 7/12/11 | Tue 7/19/11 | 8.05 days | 8.05 days |
| Project completed | Fri 7/29/11 | Fri 7/29/11 | Tue 7/19/11 | Tue 7/19/11 | 8.05 days | 8.05 days |

**FIGURE A-52** Variance table

7. *Review changes using the Tracking Gantt chart.* Click the **Gantt chart drop-down** under the View group of the Task tab, and then click **Tracking Gantt.** Move the split bar to the left to reveal more of the Gantt chart, and adjust the Zoom slider and other screen elements until your screen resembles Figure A-53. Notice that the Gantt chart bars for completed tasks have changed. There are also a few new symbols, such as white diamonds, representing slipped milestones, for Tasks 13 and 41. Because the Analysis task preceded others on the critical path, such as Design, Implementation, and so on, they were delayed in starting. A good project manager would take corrective action to make up for this lost time or renegotiate the completion date.

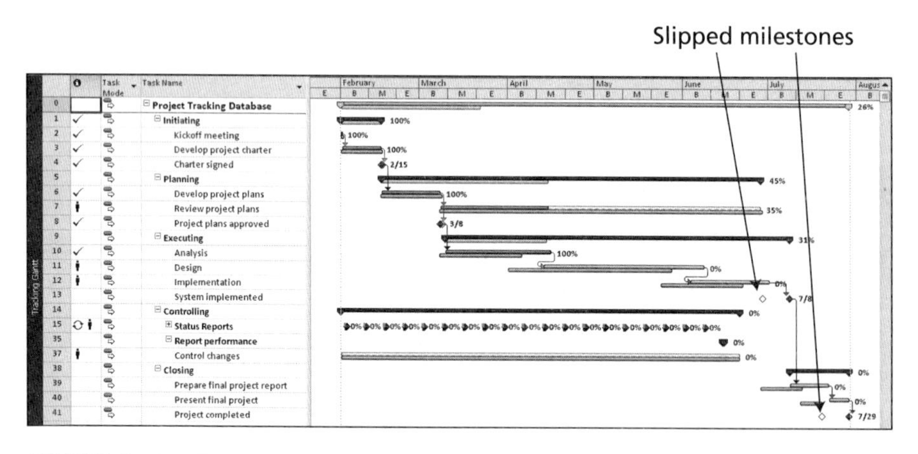


**FIGURE A-53** Tracking Gantt chart

8. *Review the cost table and enter actual costs.* Right-click the **Select All** button, and then select **Cost** to display the Cost table. Move the split bar to the right, if needed, to display all of the columns. Notice that the Actual costs for tasks 1 through 8 are the same as the Baseline costs because they were entered as

100% complete. Also notice that the Actual cost for task 10, which took longer than planned, seems very high. Project 2010 makes several assumptions in calculating costs. In this case, we want to enter our own Actual cost. First click **Task 10** and make it manually scheduled by clicking the **Task** tab and then clicking the **Manually Schedule** button under the Schedule group. Type **10,000** in the Actual column for task 10, Analysis. Notice the changes in the Cost table, as shown in Figure A-54.

| | Task Name | Fixed Cost | Fixed Cost Accrual | Total Cost | Baseline | Variance | Actual | Remaining |
|---|---|---|---|---|---|---|---|---|
| 0 | **Project Tracking Database** | **$0.00** | **Prorated** | **$50,068.38** | **$48,944.80** | **$1,123.58** | **$18,475.00** | **$31,593.38** |
| 1 | **Initiating** | **$0.00** | **Prorated** | **$2,500.00** | **$2,500.00** | **$0.00** | **$2,500.00** | **$0.00** |
| 2 | Kickoff meeting | $0.00 | Prorated | $180.00 | $180.00 | $0.00 | $180.00 | $0.00 |
| 3 | Develop project charter | $0.00 | Prorated | $2,320.00 | $2,320.00 | $0.00 | $2,320.00 | $0.00 |
| 4 | Charter signed | $0.00 | Prorated | $0.00 | $0.00 | $0.00 | $0.00 | $0.00 |
| 5 | **Planning** | **$0.00** | **Prorated** | **$7,124.50** | **$7,124.50** | **$0.00** | **$5,975.00** | **$1,149.50** |
| 6 | Develop project plans | $0.00 | Prorated | $5,400.00 | $5,400.00 | $0.00 | $5,400.00 | $0.00 |
| 7 | Review project plans | $0.00 | Prorated | $1,724.50 | $1,724.50 | $0.00 | $575.00 | $1,149.50 |
| 8 | Project plans approved | $0.00 | Prorated | $0.00 | $0.00 | $0.00 | $0.00 | $0.00 |
| 9 | **Executing** | **$0.00** | **Prorated** | **$32,480.00** | **$30,329.00** | **$2,151.00** | **$10,000.00** | **$22,480.00** |
| 10 | Analysis | ($22,840.00) | Prorated | $10,000.00 | $7,849.00 | $2,151.00 | $10,000.00 | $0.00 |
| 11 | Design | $0.00 | Prorated | $12,000.00 | $12,000.00 | $0.00 | $0.00 | $12,000.00 |
| 12 | Implementation | $0.00 | Prorated | $10,480.00 | $10,480.00 | $0.00 | $0.00 | $10,480.00 |
| 13 | System implemented | $0.00 | Prorated | $0.00 | $0.00 | $0.00 | $0.00 | $0.00 |
| 14 | **Controlling** | **$0.00** | **Prorated** | **$5,323.88** | **$6,351.30** | **($1,027.42)** | **$0.00** | **$5,323.88** |
| 15 | **Status Reports** | **$0.00** | **Prorated** | **$3,208.78** | **$3,391.70** | **($182.92)** | **$0.00** | **$3,208.78** |
| 35 | **Report performance** | **$0.00** | **Prorated** | **$1.10** | **$845.60** | **($844.50)** | **$0.00** | **$1.10** |
| 37 | Control changes | $0.00 | Prorated | $2,114.00 | $2,114.00 | $0.00 | $0.00 | $2,114.00 |
| 38 | **Closing** | **$0.00** | **Prorated** | **$2,640.00** | **$2,640.00** | **$0.00** | **$0.00** | **$2,640.00** |
| 39 | Prepare final project report | $0.00 | Prorated | $2,200.00 | $2,200.00 | $0.00 | $0.00 | $2,200.00 |
| 40 | Present final project | $0.00 | Prorated | $440.00 | $440.00 | $0.00 | $0.00 | $440.00 |
| 41 | Project completed | $0.00 | Prorated | $0.00 | $0.00 | $0.00 | $0.00 | $0.00 |

**FIGURE A-54** Updated Cost table

9. *Save your file as a new file named tracking.mpp*. Click the **File** tab, and then click **Save As.** Name the file **tracking**, and then click **Save.**

After you have entered some actuals, you can also review earned value information, as described in the next section.

## Earned Value Management

Earned value management is an important project management technique for measuring project performance. Because you have entered actual information for some of the tasks in the Project Tracking Database project, you can now view earned value information in Project 2010. You can also view an earned value report using the visual reports feature.

To view earned value information:

1. *View the Earned Value table*. Using the tracking.mpp file, right-click the **Select All** button, and then select **More Tables.** The More Tables dialog box opens. Double-click **Earned Value.**
2. *Display all the Earned Value table columns*. Move the split bar to the right to reveal all of the columns, as shown in Figure A-55. Note that the Earned Value table includes columns for each earned value acronym, such as SV, CV, and so on, as explained in this text. Also note that the EAC (Estimate at Completion) is higher than the BAC (Budget at Completion) starting with Task 9,

where the task took longer than planned to complete. Task 0, which represents the entire project, shows a CV of $2,050.30 and a SV of $6,544.40. Remember that not all of the actual information has been entered yet.

| | Task Name | Planned Value - PV (BCWS) | Earned Value - EV (BCWP) | AC (ACWP) | SV | CV | EAC | BAC | VAC |
|---|---|---|---|---|---|---|---|---|---|
| 0 | **Project Tracking Database** | **$22,575.03** | **$16,030.63** | **$18,080.93** | **($6,544.40)** | **($2,050.30)** | **$55,204.77** | **$48,944.80** | **($6,259.97)** |
| 1 | **Initiating** | **$2,500.00** | **$2,500.00** | **$2,500.00** | **$0.00** | **$0.00** | **$2,500.00** | **$2,500.00** | **$0.00** |
| 2 | Kickoff meeting | $180.00 | $180.00 | $180.00 | $0.00 | $0.00 | $180.00 | $180.00 | $0.00 |
| 3 | Develop project charter | $2,320.00 | $2,320.00 | $2,320.00 | $0.00 | $0.00 | $2,320.00 | $2,320.00 | $0.00 |
| 4 | Charter signed | $0.00 | $0.00 | $0.00 | $0.00 | $0.00 | $0.00 | $0.00 | $0.00 |
| 5 | **Planning** | **$5,975.00** | **$5,975.00** | **$5,975.00** | **$0.00** | **$0.00** | **$7,124.50** | **$7,124.50** | **$0.00** |
| 6 | Develop project plans | $5,400.00 | $5,400.00 | $5,400.00 | $0.00 | $0.00 | $5,400.00 | $5,400.00 | $0.00 |
| 7 | Review project plans | $575.00 | $575.00 | $575.00 | $0.00 | $0.00 | $1,724.50 | $1,724.50 | $0.00 |
| 8 | Project plans approved | $0.00 | $0.00 | $0.00 | $0.00 | $0.00 | $0.00 | $0.00 | $0.00 |
| 9 | **Executing** | **$10,774.00** | **$7,555.63** | **$9,605.93** | **($3,218.37)** | **($2,050.30)** | **$38,559.08** | **$30,329.00** | **($8,230.08)** |
| 10 | Analysis | $7,849.00 | $7,555.63 | $9,605.93 | ($293.37) | ($2,050.30) | $9,978.90 | $7,849.00 | ($2,129.90) |
| 11 | Design | $2,925.00 | $0.00 | $0.00 | ($2,925.00) | $0.00 | $12,000.00 | $12,000.00 | $0.00 |
| 12 | Implementation | $0.00 | $0.00 | $0.00 | $0.00 | $0.00 | $10,480.00 | $10,480.00 | $0.00 |
| 13 | System implemented | $0.00 | $0.00 | $0.00 | $0.00 | $0.00 | $0.00 | $0.00 | $0.00 |
| 14 | **Controlling** | **$3,326.03** | **$0.00** | **$0.00** | **($3,326.03)** | **$0.00** | **$5,323.88** | **$6,351.30** | **$1,027.42** |
| 15 | **Status Reports** | **$1,814.03** | **$0.00** | **$0.00** | **($1,814.03)** | **$0.00** | **$3,208.78** | **$3,391.70** | **$182.92** |
| 35 | **Report performance** | **$432.00** | **$0.00** | **$0.00** | **($432.00)** | **$0.00** | **$1.10** | **$845.60** | **$844.50** |
| 37 | Control changes | $1,080.00 | $0.00 | $0.00 | ($1,080.00) | $0.00 | $2,114.00 | $2,114.00 | $0.00 |
| 38 | **Closing** | **$0.00** | **$0.00** | **$0.00** | **$0.00** | **$0.00** | **$2,640.00** | **$2,640.00** | **$0.00** |
| 39 | Prepare final project report | $0.00 | $0.00 | $0.00 | $0.00 | $0.00 | $2,200.00 | $2,200.00 | $0.00 |
| 40 | Present final project | $0.00 | $0.00 | $0.00 | $0.00 | $0.00 | $440.00 | $440.00 | $0.00 |
| 41 | Project completed | $0.00 | $0.00 | $0.00 | $0.00 | $0.00 | $0.00 | $0.00 | $0.00 |

**FIGURE A-55** Earned Value table

3. *View the earned value chart.* Click the **Project** tab, and then click the **Visual Reports** button under the Reports group to open the Visual Reports dialog box. Click **Earned Value Over Time Report,** as shown in Figure A-56. Notice the sample of the selected report on the right side of the dialog box. If you have Excel, click **View** to see the resulting report as Project 2010 automatically creates Excel data and a chart based on your current file. Close Excel without saving the file, and then click the **Close** button of the Visual Reports dialog box.

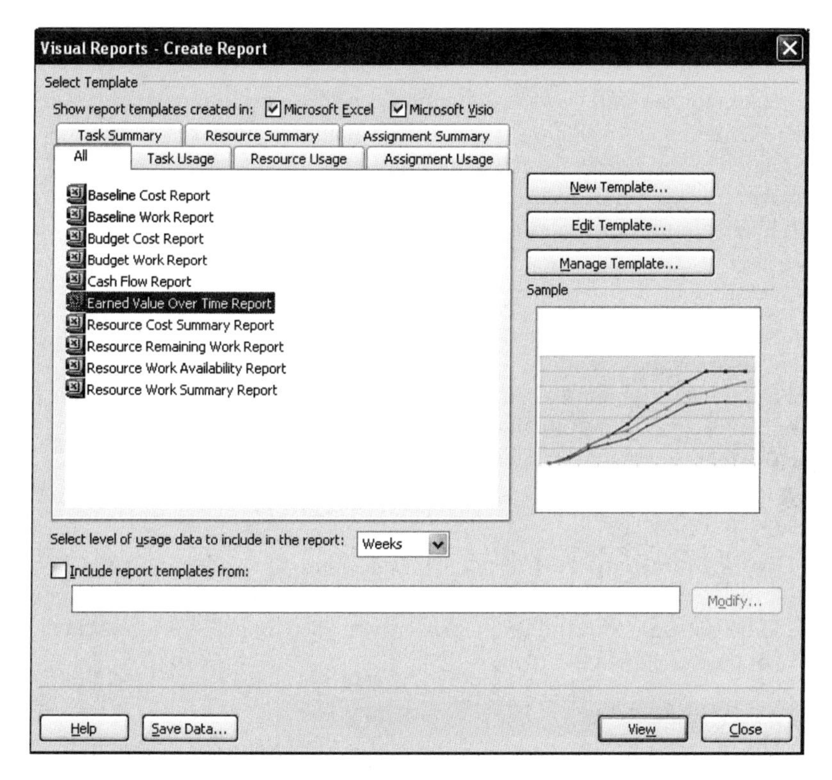


**FIGURE A-56** Visual Reports dialog box

## HELP

To take full advantage of the visual reports features, you must have specific software installed on your computer. You may also need to adjust data, such as displaying weeks or months versus quarters, to get the reports to display as desired.

4. *Save and close the file*. Click the **Save** button on the Quick Access toolbar, and then close the tracking.mpp file. You can also exit Project 2010 and take a break, if desired.

## HELP

If you want to download the Project 2010 files baseline.mpp and tracking.mpp to check your work or continue to the next section, a copy is available on the companion Web site, the author's Web site, or from your instructor.

Now that you have entered and analyzed various project cost information, you will examine some of the human resource management features of Project 2010.

In the project cost management section, you learned how to enter resource information into Project 2010 and how to assign resources to tasks. Two other helpful human resource features include resource calendars and histograms. In addition, it is important to know how to use Project 2010 to assist in resource leveling. You will also explore the new Team Planner feature, which can also help to level resources.

## Resource Calendars

When you created the Project Tracking Database file, you used the standard Project 2010 calendar. This calendar assumes that standard working hours are Monday through Friday, from 8:00 a.m. to 5:00 p.m., with an hour for lunch starting at noon. Rather than using this standard calendar, you can create a different calendar that takes into account each project's unique requirements. You can create a new calendar by using the Tasks pane or by changing the working time under the Tools menu.

To create a new base calendar:

1. *Open a new file and access the Change Working Time dialog box*. After opening a blank file in Project 2010, click the **Project** tab, and then click the **Change Working Time** button under the Properties group. The Change Working Time dialog box opens, as shown in Figure A-57.

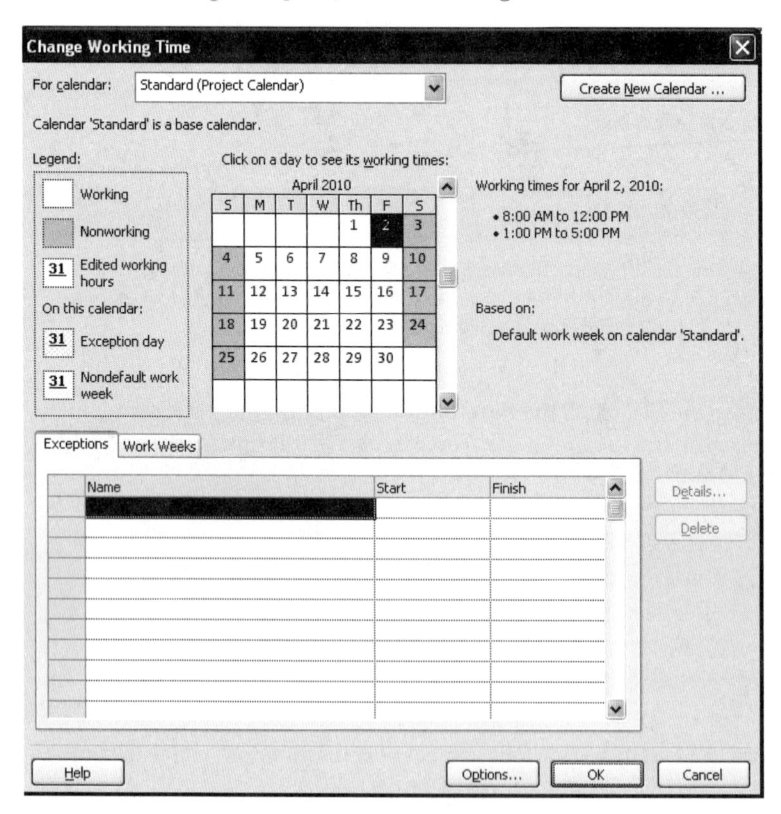


FIGURE A-57 Change Working Time dialog box

2. *Name the new base calendar*. In the Change Working Time dialog box, click **Create New Calendar.** The Create New Base Calendar dialog box opens. Click the **Create new base calendar** radio button, type **Mine** as the name of the new calendar in the **Name** text box, and then click **OK.** Click the **Options** button, and then make desired changes, such as changing the month when the fiscal year starts. Click **OK** to close the Change Working Time dialog box.

You can use this new calendar for the whole project, or you can assign it to specific resources on the project.

To assign the new calendar to the whole project:

1. *Open the Project Information dialog box*. Click the **Project** tab, and then click the **Change Working Time** button. The Project Information dialog box opens.
2. *Select a new calendar*. Click the **Calendar** list arrow to display a list of available calendars. Select your new calendar named **Mine** from this list, and then click **OK.**

To assign a specific calendar to a specific resource:

1. *Assign a new calendar*. Click the **View** tab, and then click the **Resource Sheet** button under the Resource Views group. Type **Me** in the Resource Name column, and press **Enter.**
2. *Select the calendar*. Click the **Base Calendar** cell for that resource name. If the Base Calendar column is not visible, click the horizontal scroll bar to view more columns. Click the **Base Calendar** list arrow to display the options, and then select **Mine,** if needed.
3. *Block off vacation time*. Double-click the resource name **Me** to display the Resource Information dialog box, and then click the **Change Working Time** button, located on the General tab in the Resource Information dialog box. You can block off vacation time for people by selecting the appropriate days on the calendar and marking them as nonworking days. Click **OK** to accept your changes, and then click **OK** to close the Resource Information dialog box.
4. *Close the file without saving it*. Click the **Close** icon, and click **No** when you are prompted to save the file.

## Resource Histograms

A resource histogram is a type of chart that shows the number of resources assigned to a project over time. A histogram by individual shows whether a person is over- or underallocated during certain periods. The Resource Graph helps you see which resources are overallocated, by how much, and when. It also shows you the percentage of capacity each resource is scheduled to work, so you can reallocate resources, if necessary, to meet the needs of the project.

To view resource histograms for the Project Tracking Database project:

1. *View the Resource Graph*. Open the **baseline.mpp** file, and click the **View** tab, if necessary. Click the **Other Views** button under the Resource Views group, and then click the **Resource Graph** button. A histogram for Chris appears, as shown in Figure A-58. The screen is divided into two sections: the left pane displays a person's name and the right pane displays a resource histogram for

that person. You may need to click the right scroll bar to get the histogram to appear on your screen. Notice that Kathy is overallocated slightly in the month of June, because the column for that month goes above the 100% line.

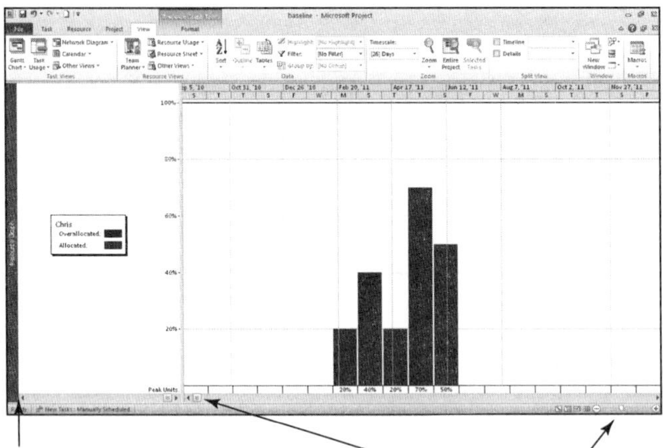


**FIGURE A-58** Resource Graph view

2. *View the other resources' histograms*. Click the **left scroll arrow** at the bottom of the resource name pane on the left side of the screen. The resource histogram for the next person appears. View the resource histograms for all four people, and then go back Kathy's. Click the **Zoom In** button several times until the information is displayed in days. Move the scroll bar until you see the information for the first day of the project, as shown in Figure A-59.

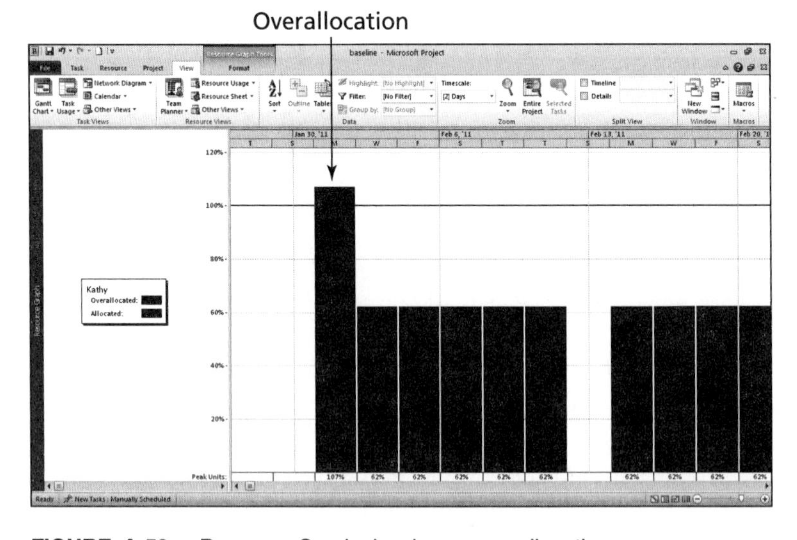


**FIGURE A-59** Resource Graph showing an overallocation

Notice that Kathy's histogram has a partially red bar on the first day of the project. This red portion of the bar means that she has been overallocated that day. The percentages at the bottom of each bar show the percentage each resource is assigned to work. For example, Kathy is scheduled to work 107% of her available time the first day, and 62% of her available time the next several days. You can use the Resource Usage view to see more details on the overallocation.

To see more details about an overallocated resource using the Resource Usage view:

1. *Display the Resource Usage view.* Click the **Resource Usage** button under the Resource Views group on the Ribbon.
2. *Adjust the information displayed.* On the right side of the screen, click the **scroll arrows** to display the hours Kathy is assigned to work each day. Widen the Resource Name column and make other adjustments as needed so your screen resembles Figure A-60.

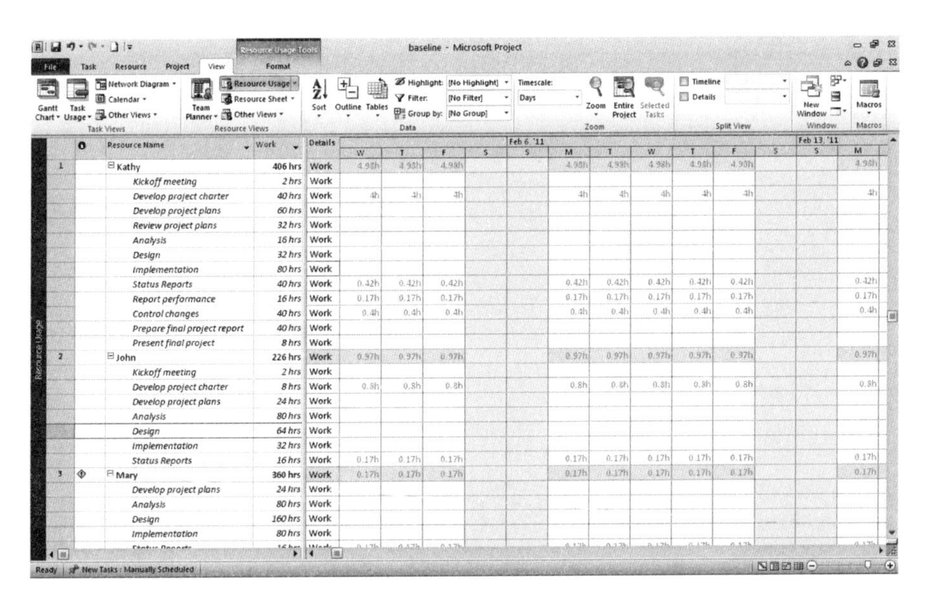


**FIGURE A-60** Resource Usage view

3. *Examine overallocation information.* Notice in the Resource Usage view (Figure A-60) that Kathy is only scheduled to work 5.57h on the first day of the project. Although Kathy is not scheduled to work more than eight hours on that day, some of the tasks were entered as hours, and Project 2010 assumes those hours start as soon as possible, thus causing the scheduling conflict. Also notice that Mary is overallocated. Next, you will remove the overallocations by using resource leveling.

## Resource Leveling

Resource leveling is a technique for resolving resource conflicts by delaying tasks. Resource leveling also creates a smoother distribution of resource usage. You can find detailed information on resource leveling in the Help facility.

To use resource leveling:

1. *View Kathy's Resource Graph again*. Click the **Resource Graph** button under the View tab to view Kathy's histogram again. Make adjustments as needed to see the graph for the first day of the project. Remember that just Kathy and Mary showed overallocations.
2. *Open the Resource Leveling dialog box*. Click the **Resource** tab, and then click **Leveling Options.** The Level Resources dialog box opens, as shown in Figure A-61.

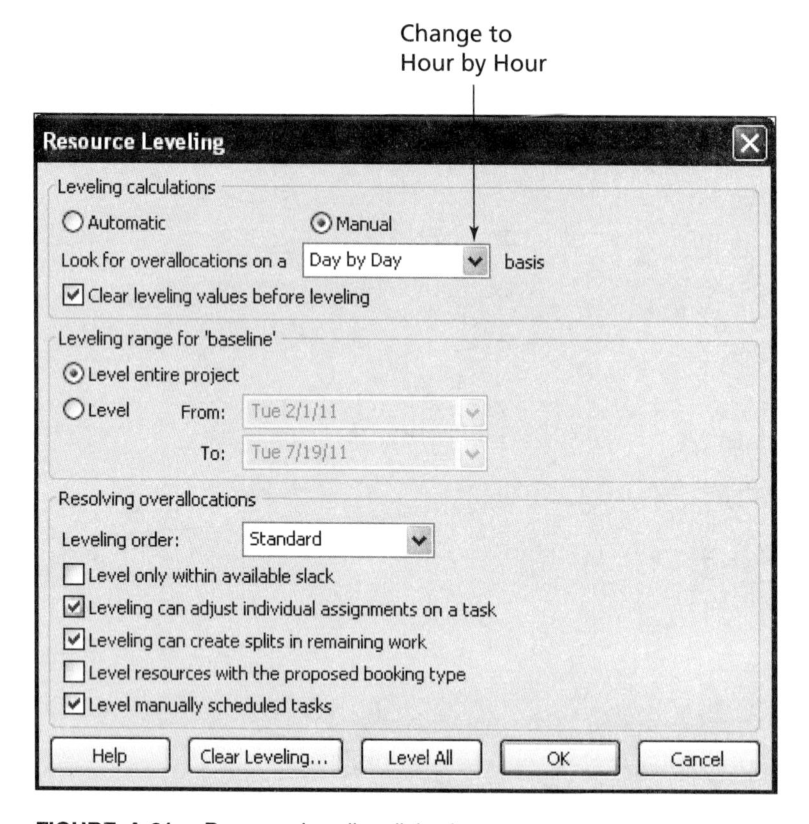


**FIGURE A-61** Resource Leveling dialog box

3. *Level using the Hour by Hour option*. Change the basis for leveling from Day by Day to **Hour by Hour,** click the **Level All** button, keep the other default options in the Level Now dialog box. Notice that Kathy's overallocation disappears. Click the **scroll arrow** in the resource name pane to reveal information about other resources to see that their overallocations are gone as well.

## TIP

If you want to undo the leveling, immediately click the Undo button on the Standard toolbar. Alternatively, you can return to the Resource Leveling dialog box, and click the Clear Leveling button.

4. *Save the file as level.mpp.* Click the **File** tab, click **Save As,** type **level** as the filename, and then click **Save.** Close the file and Project 2010.

## HELP

If you want to download the Project 2010 file level.mpp to check your work or continue to the next section, a copy is available on the companion Web site, the author's Web site, or from your instructor.

Consult the Project 2010 Help feature and use the keyword "level" for more information on resource leveling. Also, when setting options for this feature, be careful that the software adjusts resources only when it should. For example, the end date for a project might be pushed back if you set the leveling options not to level only within available slack. In this case, the project manager might prefer to ask his or her team to work a little overtime to remain on schedule.

## Using the New Team Planner Feature*

Another way to assign resources and reduce overallocations is by using the new Team Planner feature. Assume you have two people assigned to work on a project, Brian and Cindy, as shown in Figure A-62. Notice that Brian is assigned to work on both Task 1 and Task 2 full-time the first week. Therefore, Brian is overallocated. Cindy is scheduled to work on Task 3 full-time the second week, and Task 4, also scheduled for the second week, is not assigned yet.

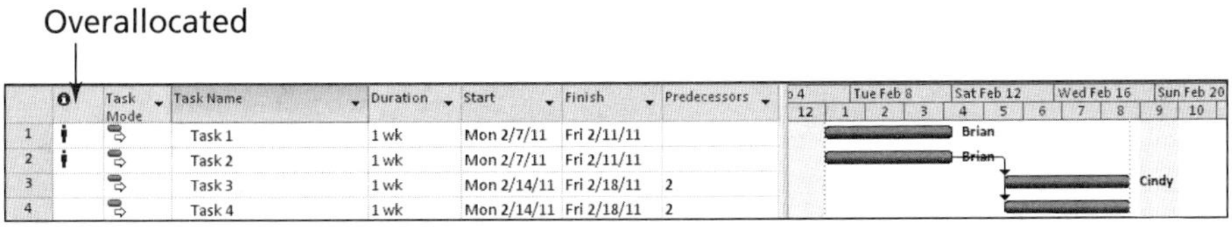

Overallocated

| | | Task Mode | Task Name | Duration | Start | Finish | Predecessors |
|---|---|---|---|---|---|---|---|
| 1 | | | Task 1 | 1 wk | Mon 2/7/11 | Fri 2/11/11 | |
| 2 | | | Task 2 | 1 wk | Mon 2/7/11 | Fri 2/11/11 | |
| 3 | | | Task 3 | 1 wk | Mon 2/14/11 | Fri 2/18/11 | 2 |
| 4 | | | Task 4 | 1 wk | Mon 2/14/11 | Fri 2/18/11 | 2 |

Tue Feb 8
Sat Feb 12
Wed Feb 16
Sun Feb 20
Brian
Brian
Cindy

**FIGURE A-62** Overallocated resource

You can click the Team Planner view under the View tab to see a screen similar to the top section of Figure A-63. Notice that Brian has both Tasks 1 and 2 assigned to him at the same time. These tasks and Brian's name display in red to show the overallocation. Cindy is assigned Task 3 the following week, and Task 4 is unassigned. By simply clicking and dragging Task 4 straight up so it is under Brian in Week 2 and dragging Task 2 straight down so it is under Cindy in Week 1, you can reassign those tasks and remove Brian's overallocation, as shown in the bottom section of Figure A-63. Many people will appreciate the simplicity of this new feature!

*From Schwalbe's *An Introduction to Project Management, Third Edition,* 2010.

Before moving tasks in the Team Planner View:

| Resource Name | Unscheduled Tasks | Jan 30, '11 | Feb 6, '11 | Feb 13, '11 |
|---|---|---|---|---|
| Brian | | | Task 1<br>Task 2 | |
| Cindy | | | | Task 3 |
| Unassigned Tasks: 1 | | | | |
| | | | | Task 4 |

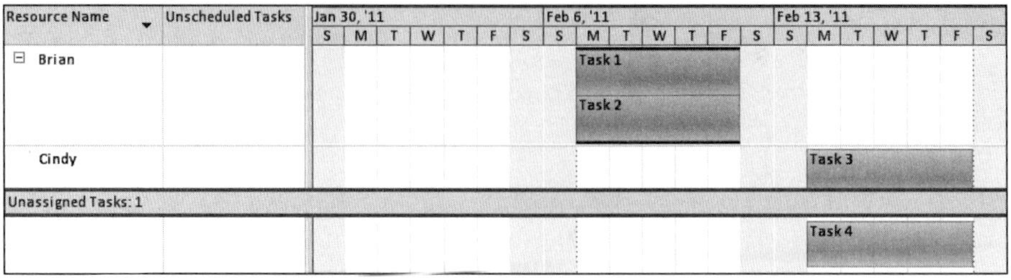

After moving tasks in the Team Planner View:

| Resource Name | Unscheduled Tasks | Jan 30, '11 | Feb 6, '11 | Feb 13, '11 |
|---|---|---|---|---|
| Brian | | | Task 1 | Task 4 |
| Cindy | | | Task 2 | Task 3 |

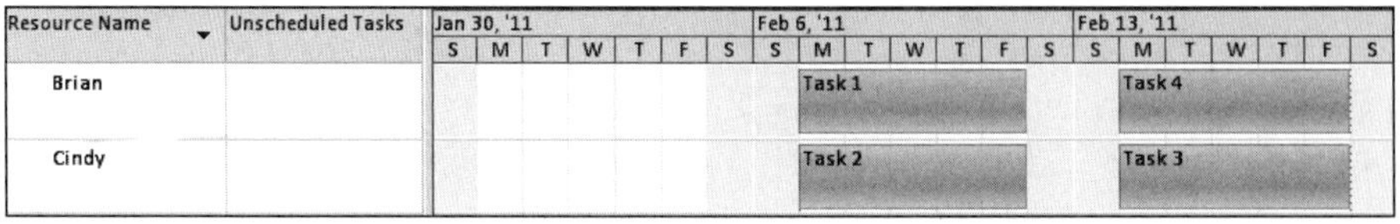

**FIGURE A-63** Adjusting resource assignments using the Team Planner feature

Now that you have learned how to change resource calendars, view resource histograms, and level resources, you are ready to learn how to use Project 2010 to assist in project communications management.

## PROJECT COMMUNICATIONS MANAGEMENT

Project 2010 can help you generate, collect, disseminate, store, and report project information. There are many different tables, views, reports, and formatting features to aid in project communications, as you have seen in the previous sections. This section highlights some common reports and views. It also describes how to use templates and insert hyperlinks from Project 2010 into other project documents and how to use Project 2010 in a workgroup setting.

### Common Reports and Views

To use Project 2010 to enhance project communications, it is important to know when to use the many different ways to collect, view, and display project information. Table A-5 provides a brief summary of Project 2010 features and their functions, which will help you understand when to use which feature. Examples of most of these features are provided as figures in this appendix.

You can see from Table A-5 that many different reports are available in Project 2010. The overview reports provide summary information that top management might want to see, such as a project summary or a report of milestone tasks. Current activities reports help project managers stay abreast of and control project activities. The reports of unstarted tasks and slipping tasks alert project managers to problem areas. Cost reports provide information related to the cash flow for the project, budget information, overbudget items, and earned value management.

**TABLE A-5** Functions of Project 2010 Features

| Feature | Function |
|---|---|
| Gantt Chart view, Entry table | Enter basic task information |
| Network Diagram view | View task dependencies and critical path graphically |
| Schedule table | View schedule information in a tabular form |
| Cost table | Enter fixed costs or view cost information |
| Resource Sheet view | Enter resource information |
| Resource Information and Gantt Chart split view | Assign resources to tasks |
| Team Planner | Help remove resource overallocations |
| Set Baseline | Save project baseline plan |
| Earned Value table | View earned value information |
| Resource Graph | View resource allocation |
| Resource Usage | View detailed resource usage |
| Resource Leveling | Level resources |
| Overview Reports | View project summary, top-level tasks, critical tasks, milestones, working days |
| Current Activities Reports | View unstarted tasks, tasks starting soon, tasks in progress, completed tasks, should-have-started tasks, slipping tasks |
| Cost Reports | View cash flow, budget, overbudget tasks, overbudget resources, earned value |
| Assignment Reports | View who does what, who does what when, to-do list, overallocated resources |
| Workload Reports | View task usage, resource usage |
| Custom Reports | Allow customization of each type of report |
| Insert Hyperlink | Insert hyperlinks to other files or Web sites |

Assignment reports help the entire project team by providing different views of who is doing what on a project. You can see who is overallocated by running the Overallocated Resources report or by viewing the two workload reports. You can also create custom reports based on any project information you have entered into Project 2010.

## Using Templates and Inserting Hyperlinks and Comments

This text provides examples of many templates that you can use to improve project communications. Because it is often difficult to create good project files, many organizations keep a repository of template or sample files. As shown earlier in this appendix, Project 2010 includes several template files, and you can also access several templates online. You must load the template files when installing Project 2010 or access templates via the Internet.

To access the Project 2010 templates:

1. *Access Project 2010 templates*. Start Project 2010, if necessary, click the **File** tab, and then click **New** to display the Available Templates, as shown in Figure A-64.

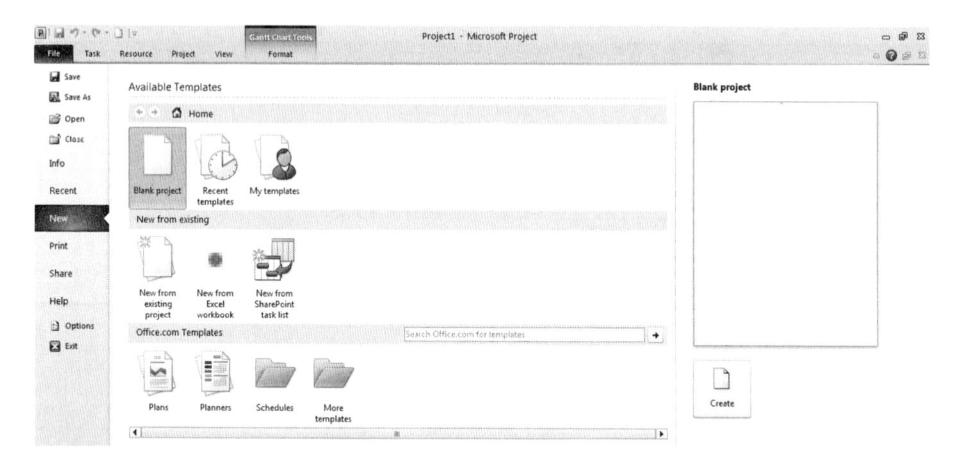


**FIGURE A-64** Opening templates

2. *Access templates from Office.com*. Click the **Schedules** icon in the Office.com Templates section at the bottom of the screen, as shown in Figure A-65. A list of templates available at the time appears. Experiment with finding and using various template files in Project 2010.

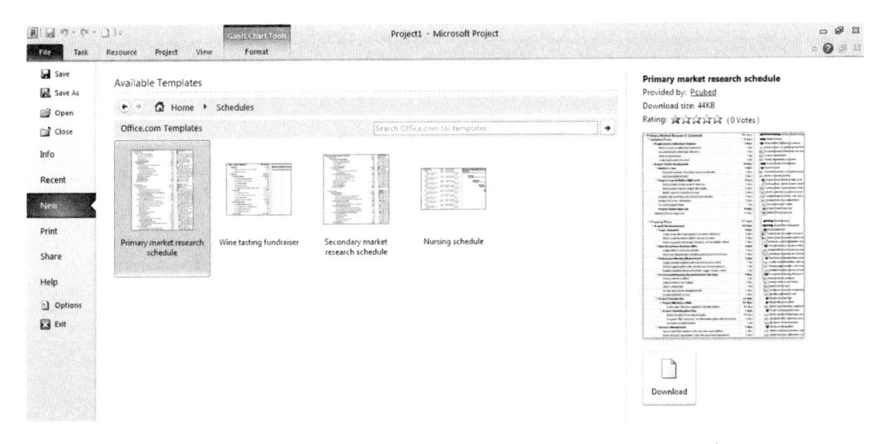


**FIGURE A-65** Schedules templates from Office.com

Using templates can help you prepare your project files, but be careful to address the unique needs of your project and organization. For example, even though the Home Move template file can provide guidance in preparing a Project 2010 file for a specific home move project, you need to tailor the file for your particular situation and schedule. You can also create your own Project 2010 templates by saving your file as a template using the Save As option from the File menu. Simply change the Save As Type option to Template. Project 2010 then gives the file an .mpt extension (Microsoft Project Template) to identify the file as a template.

In addition to using templates for Project 2010 files, it is helpful to use templates for other project documents and insert hyperlinks to them from Project 2010. For example, your organization might have templates for meeting agendas, project charters, status reports, and project management plans. See the companion Web site for examples of other template files. Next, you will create hyperlinks to some template files created in other applications.

To insert a hyperlink within a Project 2010 file:

1. *Display the Entry table*. Open the **baseline.mpp** file. The Entry table and Gantt Chart view should display.
2. *Select the task in which you want to insert a hyperlink*. Under Task Name, click **Kickoff meeting**, the task name for Task 2.
3. *Open the Insert Hyperlink dialog box*. Right-click the task name, and then click **Hyperlink.** The Insert Hyperlink dialog box opens, as shown in Figure A-66.

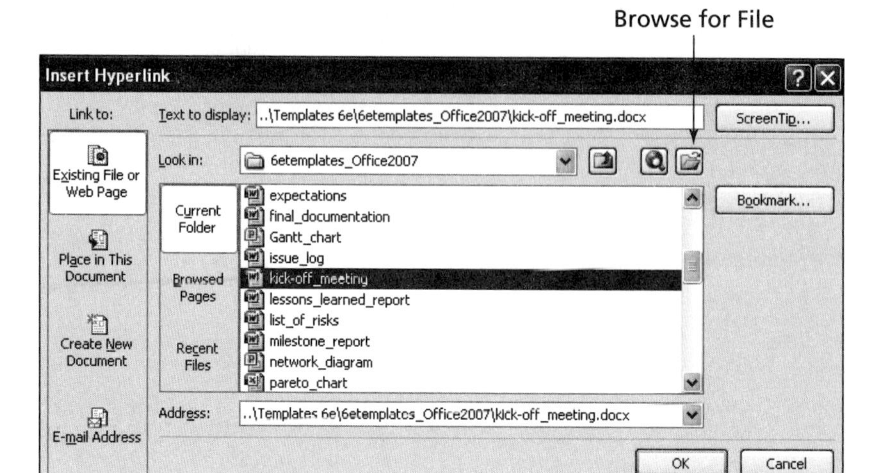


**FIGURE A-66** Insert Hyperlink dialog box

4. *Enter the filename of the hyperlink file*. Click the **Browse for File** button to browse for the kickoffmeeting.doc file you downloaded earlier. Double-click **kickoffmeeting.doc,** and then click **OK.**
5. *Review the Indicators column*. A **Hyperlink** icon appears in the Indicators column to the left of the Task Name for Task 2, next to the Overallocation icon. Move your mouse over the hyperlink icon until the mouse pointer changes to the Hand symbol to reveal the name of the hyperlinked file.

Clicking the Hyperlink button in the Indicators column or right-clicking, pointing to Hyperlink, and selecting Open will automatically open the hyperlinked file. Using hyperlinks is a good way to keep all project documents organized.

It is also a good idea to insert notes or comments into Project 2010 files to provide more information on specific tasks.

To insert a note for Task 4:

1. *Open the Task Information dialog box for Task 4*. Under Task Name, double-click **Charter signed,** the task name for Task 4. Click the **Notes** tab.
2. *Enter your note text*. Type **The charter was developed as a joint effort.** in the Notes text box, as shown in Figure A-67.

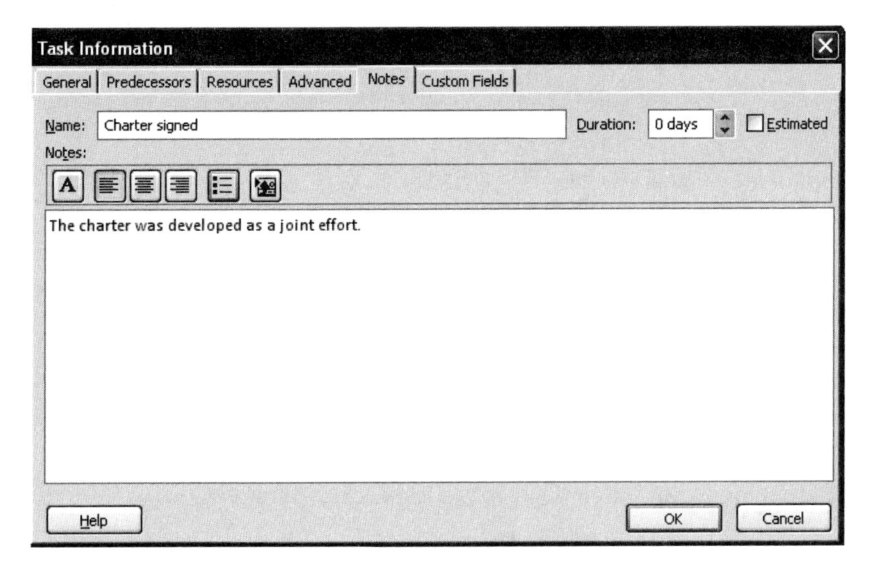


**FIGURE A-67** Task Information dialog box Notes tab

3. *See the resulting Notes icon*. Click **OK** to enter the note. The Notes icon appears in the Indicators column next to Task 4.
4. *Open the note*. Double-click the **Notes** icon in the Indicators column for Task 4 to view the note, and then click **OK.**
5. *Close the file without saving it*. Click **Close,** and then click **No** when you are prompted to save the file.

As you can see, Project 2010 is a very powerful tool. If used properly, it can greatly help users successfully manage projects.

# DISCUSSION QUESTIONS

1. What are some new features of Project 2010, and how do they differ from previous versions of Microsoft Project?
2. How do you use Project 2010 to create a WBS hierarchy?
3. Summarize how you use Project 2010 to assist in time management. How do you enter durations, link tasks, and view critical path information?
4. How can Project 2010 assist you in project cost management? What table view would you use to enter fixed costs? How do you enter resources and assign them to tasks? How can you view earned value information?
5. Briefly describe how to change resource calendars, view resource histograms, and level resources. How can you use the new Team Planner feature?
6. Summarize ways to communicate information with Project 2010. How do you link to other files from within your Project 2010 file? How can you find and use templates?

# EXERCISES

The more you practice using the features of Project 2010, the more quickly you will master the application and use it to manage projects. This section includes an exercise that you can do as a series of two homework assignments to make sure you understand the steps and concepts in this appendix. There are also exercises based on three examples of information technology projects that could benefit from using Project 2010 to assist in project scope, time, cost, human resource, and communications management. This section also includes an exercise you can use to apply Project 2010 on a real project. Simpler exercises that use Project 2010 are at the end of several chapters in this text.

## Exercise A-1: Homework Assignments

You can go through this appendix as a series of homework assignments. The following two assignments are used by the author of this text in her classes. Instructors can easily change which items to print out or modify instructions to make this assignment more unique.

### HW1: Project 2010, Part 1 (100 points, 25 points for each item)

See points for each item listed below. Note that you can download a free trial of Project 2010 from *www.microsoft.com* or use the PCs at school. Remember that you must be running Windows 7, Vista, or XP. Read and follow the hands-on instructions in this Guide to Using Project 2010. Read the first several pages of the appendix, do all the steps starting under the Overview of Project 2010 heading, and stop after you finish the steps right before the heading called Baseline Plan, Actual Costs, and Actual Times. The files mentioned are on the companion Web site and the author's site (*www.kathyschwalbe.com* under Book FAQs on the left). In the section on Project Scope Management, Table A-2 lists the first task as Initiating. After the word Initiating, enter your first and last name (e.g., Initiating – Kathy Schwalbe). Continue doing all of the steps, and print out only the items described below.

1. Your formatted Gantt chart, Figure A-33. Be sure it fits on one page before printing and that all the columns shown are visible.
2. The Schedule Table view, Figure A-36.
3. The Split screen view for entering resource information, Figure A-43. *Note:* Press the print screen key (PrtScrn [or PrtScrn and Fn] on your keyboard) and paste it into Word to print because it does not print out as shown from within Project 2010.
4. Create a new Project file that shows the WBS for a generic project. Make the main categories project management, concept, development, implementation, and close-out. Include at least four tasks and one milestone under each of these main categories. Enter 0 for the duration of the milestones, but do not enter any durations for the other tasks. Be sure to indent tasks and show the outline numbers before printing out the Gantt Chart view on one page.

HW2: Project 2010, Part 2 (100 points, 25 points for each item)

Do all the steps starting under the Baseline Plan, Actual Costs, and Actual Times heading through the end of the appendix. Remember to type your first and last name after the first task, Initiating.

1. Print out the Tracking Gantt Chart view, Figure A-53.
2. Print out the Resource Graph view showing an overallocation, Figure A-59.
3. Create your own file using the information in Figures A-62 and 63 to experiment with the Team Planner. Print out screen shots of both figures by copying and pasting them into a word processing software.
4. Write a one- to two-page, single-spaced paper summarizing what you think about Microsoft Project. What do you like and dislike about it? Do you think it would be useful for managing all projects, or just some? Which ones?

## Exercise A-2: Web Site Development

A nonprofit organization would like you to lead a Web site development project. The organization has Internet access that includes space on a Web server, but no experience developing Web sites. In addition to creating its Web site, the organization would like you to train two people on its staff to do simple Web page updates. The Web site should include the following information, as a minimum: description of the organization (mission, history, and recent events), list of services, and contact information. The organization wants the Web site to include graphics (photographs and other images) and have an attractive, easy-to-use layout.

1. *Project Scope Management*
   Create a WBS for this project and enter the tasks in Project 2010. Create milestones and summary tasks. Assume that the main WBS categories and some of the project management tasks are similar to tasks from the Project Tracking Database project. Some of the specific analysis, design, and implementation tasks will be to:
   a. Collect information on the organization in hardcopy and digital form (brochures, reports, organization charts, photographs, and so on).
   b. Research Web sites of similar organizations.
   c. Collect detailed information about the customer's design preferences and access to space on a Web server.

   d. Develop a template for the customer to review (background color for all pages, position of navigation buttons, layout of text and images, typography, including basic text font and display type, and so on).
   e. Create a site map or hierarchy chart showing the flow of Web pages.
   f. Digitize the photographs and find other images for the Web pages; digitize hard-copy text.
   g. Create the individual Web pages for the site.
   h. Test the pages and the site.
   i. Implement the Web site on the customer's Web server.
   j. Get customer feedback.
   k. Incorporate changes.
   l. Create training materials for the customer on how to update the Web pages.
   m. Train the customer's staff on updating the Web pages.
2. *Project Time Management*
   a. Enter realistic durations for each task, and then link the tasks as appropriate. Be sure that all tasks are linked (in some fashion) to the start and end of the project. Assume that you have four months to complete the entire project. *Hint:* Use the Project Tracking Database as an example.
   b. Print the Gantt Chart view and Network Diagram view for the project.
   c. Print the Schedule table to see key dates and slack times for each task.
3. *Project Cost Management*
   a. Assume that you have three people working on the project and each of them would charge $20 per hour. Enter this information in the Resource Sheet.
   b. Estimate that each person will spend an average of about five hours per week for the four-month period. Assign resources to the tasks, and try to make the final cost in line with this estimate.
   c. Print the budget report for your project.
4. *Project Human Resource Management*
   a. Assume that one project team member will be unavailable (due to vacation) for two weeks in the middle of the project. Make adjustments to accommodate this vacation so that the schedule does not slip and the costs do not change. Document the changes from the original plan and the new plan.
   b. Use the Resource Usage view to see each person's work each month. Print a copy of the Resource Usage view.
5. *Project Communications Management*
   a. Adjust the timescale on your Gantt chart to enable the chart to fit on one page. Then paste a copy of the Gantt chart in PowerPoint. You can use your Print Screen button to copy the image and paste it into PowerPoint. Also add key milestones to the Timeline and copy it into a second PowerPoint slide. Copy and print out both slides in PowerPoint on the page.
   b. Print a "To-do List" report for each team member.
   c. Create a "Who Does What Report" and print it out.

## Exercise A-3: Software Training Program

ABC Company has 50,000 employees and wants to increase employee productivity by setting up an internal software applications training program. The training program will teach employees how to use Microsoft software programs such as Vista, Word 2010, Excel 2010, PowerPoint 2010, Access 2010, and Project 2010. Courses will be offered in the evenings and on Saturdays and taught by qualified volunteer employees. Instructors will be paid $40 per hour. In the past, employees were sent to courses offered by local vendors during company time. In contrast, this internal training program should save the company money on training as well as make people more productive. The Human Resources department will manage the program, and any employee can take the courses. Employees will receive a certificate for completing courses, and a copy will be put in their personnel files. The company is not sure which vendor's off-the-shelf training materials to use. The company needs to set up a training classroom, survey employees on desired courses, find qualified volunteer instructors, and start offering courses. The company wants to offer the first courses within six months. One person from Human Resources is assigned full time to manage this project, and top management has pledged its support.

1. *Project Scope Management*
   Create a WBS for this project and enter the tasks in Project 2010. Create milestones and summary tasks. Assume that some of the project management tasks you need to do are similar to tasks from the Project Tracking Database example. Some of the tasks specific to this project will be to:
   a. Review off-the-shelf training materials from three major vendors and decide which materials to use.
   b. Negotiate a contract with the selected vendor for its materials.
   c. Develop communications information about this new training program. Disseminate the information via department meetings, e-mail, the company's intranet, and flyers to all employees.
   d. Create a survey to determine the number and type of courses needed and employees' preferred times for taking courses.
   e. Administer the survey.
   f. Solicit qualified volunteers to teach the courses.
   g. Review resumes, interview candidates for teaching the courses, and develop a list of preferred instructors.
   h. Coordinate with the Facilities department to build two classrooms with 20 personal computers each, a teacher station, and an overhead projection system (assume that Facilities will manage this part of the project).
   i. Schedule courses.
   j. Develop a fair system for signing up for classes.
   k. Develop a course evaluation form to assess the usefulness of each course and the instructor's teaching ability.
   l. Offer classes.
2. *Project Time Management*
   a. Enter realistic durations for each task and then link appropriate tasks. Be sure that all tasks are linked in some fashion to the start and end of the project. Use the Project Tracking Database as an example. Assume that you have six months to complete the entire project.
   b. Print the Gantt Chart view and Network Diagram view for the project.
   c. Print the Schedule table to see key dates and slack times for each task.

3. *Project Cost Management*
   a. Assume that you have four people from various departments available part time to support the full-time Human Resources person, Terry, on the project. Assume that Terry's hourly rate is $40. Two people from the Information Technology department will each spend up to 25% of their time supporting the project. Their hourly rate is $50. One person from the Marketing department is available 25% of the time at $40 per hour, and one person from Corporate is available 30% of the time at $35 per hour. Enter this information about time and hourly wages into the Resource Sheet. Assume that the cost to build the two classrooms will be $100,000, and enter it as a fixed cost.
   b. Using your best judgment, assign resources to the tasks.
   c. View the Resource Graphs for each person. If anyone is overallocated, make adjustments.
   d. Print the budget report for the project.
4. *Project Human Resource Management.*
   a. Assume that the Marketing person will be unavailable for one week, two months into the project, and for another week, four months into the project. Make adjustments to accommodate this unavailability so the schedule does not slip and costs do not change. Document the changes from the original plan and the new plan.
   b. Add to each resource a 5% raise that starts three months into the project. Print a new budget report.
   c. Use the Resource Usage view to see each person's work each month. Print a copy.
5. *Project Communications Management*
   a. Adjust the timescale on your Gantt chart to enable the chart to fit on one page. Then paste a copy of the Gantt chart in PowerPoint. You can use your Print Screen button to copy the image and paste it into PowerPoint. Also add key milestones to the Timeline and copy it into a second PowerPoint slide. Copy and print out both slides in PowerPoint on one page.
   b. Print a "To-do List" report for each team member.
   c. Review some of the other reports, and print out one that you think would help in managing the project.

## Exercise A-4: Project Tracking Database

Expand the Project Tracking Database example. Assume that XYZ Company wants to create a history of project information, and the Project Tracking Database example is the best format for this type of history. The company wants to track information on 20 past and current projects and wants the database to be able to handle 100 projects total. The company wants to track the following project information:

- Project name
- Sponsor name
- Sponsor department
- Type of project

- Project description
- Project manager
- Team members
- Date project was proposed
- Date project was approved or denied
- Initial cost estimate
- Initial time estimate
- Dates of milestones (e.g., project approval, funding approval, project completion)
- Actual cost
- Actual time
- Location of project files

1. *Project Scope Management*
   Open scope.mpp and add more detail to Executing tasks, using the information in Table A-6.

**TABLE A-6** Company XYZ Project Tracking Database executing tasks

| Analysis Tasks | Design Tasks | Implementation Tasks |
|---|---|---|
| Collect list of 20 projects | Gather detailed requirements for desired outputs from the database | Enter project data |
| Gather information on projects | Create fully attributed, normalized data model | Test database |
| Define draft requirements | Create list of edit rules for fields, determine default values, develop queries, and determine formats for reports | Make adjustments, as needed |
| Create entity relationship diagram for database | Develop list of queries required for database | Conduct user testing |
| Create sample entry screen | Review design information with customer | Make adjustments based on user testing |
| Create sample report | Make adjustments to design based on customer feedback | Create online Help, user manual, and other documentation |
| Develop simple prototype of database | Create full table structure in database prototype | Train users on the system |

*(continued)*

TABLE A-6 Company XYZ Project Tracking Database executing tasks (*continued*)

| Analysis Tasks | Design Tasks | Implementation Tasks |
|---|---|---|
| Review prototype with customer | Create data input screens | |
| Make adjustments based on customer feedback | Write queries | |
| | Create reports | |
| | Create main screen | |
| | Review new prototype with customer | |

2. *Project Time Management*
   a. Enter realistic durations for each task, and then link appropriate tasks. Make the additional tasks fit the time estimate: 20 days for analysis tasks, 30 days for design tasks, and 20 days for implementation tasks. Assume that all of the executing tasks have a total duration of 70 days. Do not overlap analysis, design, and implementation tasks.
   b. Print the Gantt Chart view and Network Diagram view for the project.
   c. Print the Schedule table to see key dates and slack times for each task.
3. *Project Cost Management*
   a. Use the resource and cost information provided in resource.mpp.
   b. Assign resources to the new tasks. Try to make the final cost about the same as that shown in the Project Tracking Database example: about $50,000.
   c. Print the budget report for the project.
4. *Project Human Resources Management*
   a. Two months after the project begins, give everyone on the team a 10% raise. Document the increase in costs that these raises cause.
   b. Use the Resource Usage view to see each person's work each month. Print a copy of the Resource Usage view.
5. *Project Communications Management*
   a. Adjust the timescale on your Gantt chart to enable the chart to fit on one page. Then paste a copy of the Gantt chart in PowerPoint. You can use your Print Screen button to copy the image and paste it into PowerPoint. Also add key milestones to the Timeline and copy it into a second PowerPoint slide. Copy and print out both slides in PowerPoint on one page.
   b. Print a "Top-Level Tasks" report.
   c. Review some of the other reports, and print out one that you think would help in managing the project.

## Exercise A-5: Real Project Application

If you are doing a group project as part of your class or for a project at work, use Project 2010 to create a detailed file describing the work you plan to do. Enter a sufficient WBS, estimate task durations, link tasks, enter milestones, enter resources and costs, assign resources, and so on. Save your file as a baseline and track your progress. View earned value information when you are halfway through the project or course. Continue tracking your progress until the project or course is finished. Print your Gantt chart, Resource Sheet, Project Summary report, and relevant information. Write a two- to three-page report describing your experience. What did you learn about Project 2010 from this exercise? What did you learn about managing a project? How do you think Project 2010 helps in managing a project? You may also want to interview people who use Project 2010 for their experiences and suggestions.

# ADVICE FOR THE PROJECT MANAGEMENT PROFESSIONAL (PMP) EXAM AND RELATED CERTIFICATIONS

## INTRODUCTION TO PROJECT MANAGEMENT CERTIFICATION PROGRAMS

This appendix provides information on project management certification programs and offers advice for earning certifications. It briefly describes various certification programs and provides detailed information on PMI's PMP and CompTIA's Project+ certifications, the structure and content of these exams, suggestions on preparing for the exams, tips for taking the exams, sample questions, and information on related certifications.

## WHAT IS PMP CERTIFICATION?

The Project Management Institute (PMI) offers certification as a Project Management Professional (PMP). As mentioned in Chapter 1, the number of people earning PMP certification grew rapidly in the past ten years. There are PMPs in more than 120 countries throughout the world. Detailed information about PMP certification, the PMP Certification Handbook, and an online application is available from PMI's Web site (*www.pmi.org*) under Career Development. The following information is quoted from PMI's Web site:

> "The Project Management Institute (PMI®) is the world's leading association for the project management profession. It administers a globally recognized, rigorous, education, and/or professional experience and examination-based professional credentialing program that maintains an ISO 9001 certification in Quality Management Systems. To get the latest information, please visit the breaking news section.

Earning a professional credential through PMI means that one has:

- Demonstrated the appropriate education and/or professional experience;
- Passed a rigorous examination;
- Agreed to abide by a professional code of conduct;
- Committed to maintaining their active credential through meeting continuing certification requirements.

PMI professional credentials—available to members of the Institute and non-members alike—are widely recognized and accepted throughout the world as evidence of a proven level of education, knowledge and experience in project management."[1]

Many companies and organizations are recommending or even requiring PMP certification for their project managers. A February 2003 newsletter reported that Microsoft chose PMI's PMP Certification Program as the certification of choice for its Microsoft Services Operation. Microsoft chose the PMP certification because of its global recognition and proven record in professional development for project managers.[2] *CertCities.com* ranked the PMP as No. 4 in their ten hottest certifications for 2006. It "made a strong showing this year, rising from its debut stop at No. 10 last year, thanks in part to even stronger buzz for this industry-neutral title within the IT community."[3]

*Certification Magazine* published its annual review of how certification affects salaries of information technology professionals. This industry-wide study uses real-world numbers to show how education and experience affect a person's bottom-line salary. During a down market, people might ask why they should seek additional technical certifications. According to Gary Gabelhouse of *Certification Magazine*, "Perhaps it is best expressed in two words: job security. In boom times, one constantly reviews the rate of growth in salary as a key personal-success measurement. However, in down times, job security is paramount."[4]

The October 2004 issue of *PM Network* included an article entitled "The Rise of PMP," which provided several examples of companies and countries that have made concerted efforts to increase their number of PMPs. Hewlett-Packard had only six registered PMPs in 1997, but by August 2004, it had more than 1,500 and was adding 500 per year. Although most PMPs are in the United States (51,498) and Canada (7,444), the PMP credential is growing in popularity in several countries, such as Japan (6,001), China (4,472), and India (2,281). Thomas Walenta, PMP, a senior project manager for IBM Germany, said, "The PMP credential lends portability to a career plan. It has significantly shaped careers in the IT industry, with global companies creating career models for project managers based on PMI certification requirements."[5]

As of January 2009, PMI also offers certifications as a Certified Associate in Project Management (CAPM), a PMI-Scheduling Professional (PMI-SP), a PMI-Risk Management Professional (PMI-RMP), and a Program Management Professional (PgMP). PMI developed the CAPM certification as a stepping-stone to PMP certification. Candidates for the CAPM certification must also meet specific education and experience requirements and pass an exam. As you can imagine, the requirements are not as rigorous as they are for the PMP exam. You might want to consider getting the CAPM, or just wait until you have enough experience to earn the more popular PMP certification. PMI reported only 6,729 active CAPMs by the end of December 2008 and 318,289 active PMPs.[6] As described on the following page, you need 4,500 hours of project experience or a minimum of three years if you have a bachelor's degree to qualify for PMP certification.

## What Are the Requirements for Earning and Maintaining PMP Certification?



You can now apply to take the PMP exam online. You can also fill out the application forms and mail them in with a check, if desired. *Before* you apply to take the exam, you must meet the following four requirements:

1. Have experience working in the field of project management. When you apply to take the exam, you enter the role you played in leading and directing project tasks on one or several projects and how many hours you worked in each of the five project management process groups for each project. Roles include:
   - Project contributor
   - Supervisor
   - Manager
   - Project leader
   - Project manager
   - Educator
   - Consultant
   - Administrator
   - Other

   *Note that you do not have to have experience as a project manager to take the PMP exam*—any of these roles will suffice. PMP certification requires all applicants with a bachelor's degree to have a minimum of three years of unique, nonoverlapping project management experience where at least 4,500 hours are spent leading and directing project tasks. Applicants without a bachelor's degree are required to have a minimum of five years of unique, non-overlapping project management experience where at least 7,500 hours were spent leading and directing project tasks. In both cases, the experience must be accrued within eight years of the date of application. You must fill out a simple form online (or by paper, if you choose not to apply online) listing the title of a project or projects you worked on; the start and end date for when you worked on the project(s); your role on the project; the number of hours you spent in leading and directing tasks related to the initiating, planning, executing, controlling, and closing processes; and a summary of the project tasks that you led and directed on the project. You must list some hours in each of the five process groups, but not for each project, if you worked on multiple projects. PMI staff will review your qualifications and let you know if you are qualified to take the PMP exam. *You cannot take the exam without this experience qualification.*
2. Document at least 35 contact hours of project management education. A contact hour is defined as one hour of participation in an educational activity. There is no time frame for this requirement. A university or college, a training company or independent consultant, a PMI chapter, a PMI Registered Education Provider (REP), a company-sponsored program, or a distance-learning company can provide the education. The hours must include content on project quality, scope, time, cost, human resources, communications, risk, procurement, and integration management. You must list the course title, institution, date, and number of hours for each course. For example, if you took

a "principles of management" course at a university 20 years ago, you could list that on the education form along with a one- or two-day PMP exam preparation class.

3. Agree to the PMP certificant and candidate agreement and release statement. This form certifies that application information is accurate and complete and that candidates will conduct themselves in accordance with the Code of Ethics and Professional Conduct, professional development requirements, and other PMI certification program policies and procedures. You can simply check a box saying you agree to this information when applying online. Note that PMI randomly audits 10–15 percent of applications, so be prepared to provide more detailed information as requested if audited, such as college transcripts, signatures of supervisors or managers to verify experience, etc.
4. Pay the appropriate exam fee. As of January 2009, the PMP certification fee was $405 for PMI members and $555 for non-members. The re-examination fee (if you don't pass the exam) is $275 for PMI members and $375 for non-members. The annual individual PMI membership fee is $129 (including the $10 application fee). Note that students, or anyone enrolled in a degree-granting program at an accredited, or globally equivalent, college or university, can join PMI at the student member rate of only $40 (including the $10 application fee). If you want to earn your PMP certification, it makes sense to join PMI, not only for the financial savings, but also for other benefits. Consult PMI's Web site for membership information.

The last step in earning PMP certification is passing the exam! After PMI sends you an eligibility letter to take the PMP exam, you can sign up to take it at several different testing sites. You must take the exam within one year of receiving your eligibility letter. The eligibility letter will include complete details for scheduling your exam. As of January 2009, the PMP exam consisted of 200 four-option multiple-choice questions, and 25 of those questions are considered pretest questions that do not affect your score. The pretest questions are randomly placed throughout the exam and are used to test the validity of future examination questions. Although you cannot use any study aids during the exam, you can bring a nonprogrammable calculator to assist in performing calculations required to answer some of the questions. You are also given two blank pieces of paper, so you can write down formulas or other information when you enter the exam room, but you cannot bring in any notes or other materials. The questions on each test are randomly selected from a large test bank, so each person taking the exam receives different questions. The exam is preceded by a 15-minute computer tutorial to familiarize you with the mechanics of taking the exam. Test takers have four hours to take this computerized exam, and a passing score is 60.5 percent, or at least 106 correct answers out of the 175 scored questions, as of January 2009. PMI reviews and revises the exam annually. *Be sure you consult PMI's Web site for any notices about changes to the PMP exam*. For example, the passing percentage was changed several times to achieve PMI's goal of having around 75 percent of people pass the exam. PMI uses the Modified Angoff Technique, a certification industry practice standard, to determine the passing score.

PMI offers a professional development program for maintaining the PMP certification. To maintain your PMP status, you must earn at least 60 Professional Development Units within three years, pay a recertification fee every three years when you renew your certification ($60 as of January 2009), and agree to continue to adhere to PMI's Code of Ethics and Professional Conduct. The Continuing Certification Requirements Handbook, available from PMI's Web site, provides more details on maintaining your PMP status.



## What Is the Structure and Content of the PMP Exam?

The PMP exam is based on information from the entire project management body of knowledge as well as the area of professional responsibility. Essentially, the exam reviews concepts and terminology in PMI's *PMBOK® Guide*, and texts such as this one will help to reinforce your understanding of key topics in project management. Table B-1 shows the approximate breakdown of questions on the PMP exam by process group as of January 2009. Candidates should review updated exam information on PMI's Web site to make sure they are using the latest information. Remember that this text is based on the *PMBOK® Guide, Fourth Edition*. The PMP exam will be based on the fourth edition starting July 1, 2009. There were not that many changes from the Third Edition, but make sure you study the correct edition for your exam. PMI also provides sample exam questions from their site as well. See a link to this site and other sites with free sample questions on the companion Web site.

**TABLE B-1** Breakdown of questions on the PMP exam by process groups

| Process Group | Percent of Questions on PMP Exam | Number of Questions on PMP Exam (out of 200) |
|---|---|---|
| Initiating | 11 | 22 |
| Planning | 23 | 46 |
| Executing | 27 | 54 |
| Monitoring and Controlling | 21 | 42 |
| Closing | 9 | 18 |
| Professional Responsibility | 9 | 18 |

Study Table 3-1 from Chapter 3 of this text to understand the relationships among project management process groups, activities, and knowledge areas. Table 3-1 briefly outlines which activities are performed during each of the project management process groups and what is involved in each of the knowledge areas. It is also important to understand what each of the project management activities includes. Several questions on the certification exam require an understanding of this framework for project management, and many questions require an understanding of the various tools and techniques described in the *PMBOK® Guide, Fourth Edition* and this text.

The PMP exam includes three basic types of questions:

1. *Conceptual questions* test your understanding of key terms and concepts in project management. For example, you should know basic definitions such as what a project is, what project management is, and what key activities are included in project scope management.

2. *Application questions* test your ability to apply techniques to specific problems. For example, a question might provide information for constructing a network diagram and ask you to find the critical path or determine how much slack is available on another path. You might be given cost and schedule information and be asked to find the schedule or cost performance index by applying earned value formulas.
3. *Evaluative questions* provide situations or scenarios for you to analyze, and your response will indicate how you would handle them. For example, a project might have many problems. A question might ask what you would do in that situation, given the information provided. Remember that all questions are multiple-choice, so you must select the best answer from the options provided.

## How Should You Prepare for the PMP Exam?

To prepare for the PMP exam, it is important to understand *your* learning and testing style, and to use whatever resources and study techniques work best for *you*. Below are some important questions to consider:

- *Are you a good test taker?* Some people are very good at studying and do well on multiple-choice exams, but others are not. If you have not taken a long multiple-choice test in a while, it may take you longer to prepare for the exam than others.
- *How confident do you need to be before taking the exam?* To pass the exam, you need to answer only 60.5 percent of the questions correctly. Your PMP certificate will not display your final score, so it does not matter if you get 61, 70, 80, 90, or even 100 percent correct.
- *How much information do you need to review before taking the exam?* The *PMBOK® Guide, Fourth Edition* and this text should be enough content information, but many people want to review even more information before taking the PMP exam. Several companies sell books, sample tests, CD-ROMs, and audiotapes, or provide courses designed to help people pass the PMP exam. There's even a *PMP Certification for Dummies* book available with a CD-ROM of sample questions. See the Suggested Readings section of the companion Web site for a list of suggested resources, many at no cost at all, such as *www.pmstudy.com* and *www.bestsamplequestions.com*.
- *How much time and money do you want to spend studying for the PMP exam?* Some people with little free time or money to spend on extra courses or materials would rather just take the PMP exam with little preparation; this will tell them what they need to study further if they don't pass on their first attempt. Even though there is a reexamination fee of $275 (for PMI members), this cost is generally less than what you would have to pay for most exam preparation courses. Spend the time and money you need to feel confident enough to pass the exam, but don't overextend yourself.
- *Do you know PMI's language?* Even if you think you know about project management, studying the material in the *PMBOK® Guide, Fourth Edition* before taking the exam will help you. Volunteer PMPs created the exam, and they often refer to information from the *PMBOK® Guide, Fourth Edition* when

writing questions. Many outstanding project managers might fail the exam if they don't use PMI's terminology or processes in their jobs.

- *Do you really understand the triple constraint of project management?* Many questions on the PMP exam are based on the scope, time, and cost knowledge areas. You should be familiar with project charters, WBSs, network diagrams, critical path analysis, cost estimates, earned value, and so on before you take the exam.
- *Do you want to meet other people in the field of project management as you study for your PMP exam?* Several chapters of PMI offer PMP exam review courses. These courses are often a good way to network with other local project managers or soon to be project managers. Many other organizations provide online and instructor-led courses in PMP preparation where you can also meet people in the field. Several organizations also provide their own in-house study groups as a means to network as well as pass the PMP exam.
- *Do you need extra support, peer pressure, or incentives to pass the PMP exam?* Having some support and positive peer pressure might help to ensure that you actually take and pass the exam in a timely manner. If you don't want to be part of a study group, even just telling a friend, colleague, or loved one that you have set a goal to pass the PMP exam by a certain date might provide motivation to actually do it. You could also reward yourself after you pass the exam.
- *How much are you willing to invest in getting PMP certification?* If you have the time and money, you could take one of the immersion courses several companies offer, like Cheetah Learning, Velociteach, Project Management Training Institute, or mScholar. These courses generally last four to five days and cost from $2,000 to $3,000. The company offering the course will have you come to class with your qualification to take the PMP exam already completed, and you actually take the PMP exam on the fourth or fifth day of the class. If you don't pass, many training companies will let you take the class again at no additional cost. If you don't have any extra money to spend, you can find several free resources (see the companion Web site) and join or form your own study group, or just go take the exam and see how you do. However, try not to over study for the exam; if you know the material well and are passing practice exams, trust yourself and your skills. If you need to take it a second time to pass, just do so.

## Ten Tips for Taking the PMP Exam

1. The PMP exam is computer based and begins with a short tutorial on how to use the testing software. The software makes it easy to mark questions you want to review later, so learning how to mark questions is helpful. Using this feature can give you a feel for how well you are doing on the test. It is a good idea to go through every question somewhat quickly and mark those questions on which you want to spend more time. If you mark 79 questions or less (the total number you can miss to get 60.5 percent on all 200 questions), you should

pass the exam. Remember that 25 of the questions are not scored, and you need 60.5 percent correct on the 175 scored questions.

2. The time allotted for the exam is four hours, and each multiple-choice question has four answer choices. You should have plenty of time to complete the exam. Use the first two to three hours to go through all of the questions. Do not spend too much time on any one question. As you work, mark each question that you would like to return to for further consideration. Then use the remaining time to check the questions you are not sure of. If you're a morning person, schedule your exam in the morning. If you work better after lunch, schedule an afternoon exam. Make sure you are alert and well rested when you go in to take the exam.
3. Some people believe it is better to change answers you are unsure about. If you think that a different answer is better, after reading the question again, then change your answer. Don't get hung up on any questions. Move on and focus on answering the questions you can answer correctly.
4. Do not try to read more into the questions than what is stated. There are no trick questions, but some may be poorly worded or just bad questions. Remember that they were written by volunteers and are part of a huge test bank. Most of the questions are relatively short, and there are only four options from which to choose the answer.
5. To increase your chances of getting the right answer, first eliminate obviously wrong options, and then choose among the remaining options. Take the time to read all of the options before selecting an answer. Remember, you have to pick the *best* answer available.
6. Some questions require doing calculations such as earned value management. It is worthwhile to memorize the earned value formulas to make answering these questions easier. You may use a nonprogrammable calculator while taking the exam, so be sure to bring one to make performing calculations easier.
7. You should be given two pieces of blank paper to use during the exam. You might want to bring the paper yourself to make sure it is available. Before starting the test, you should write down important equations so that you do not have to rely on your memory. When you come to a question involving calculations, write the calculations down so you can check your work for errors. See Table B-2 for a summary of formulas you should know for the PMP exam.
8. Read all questions carefully. A few sections of the test require that you answer three to four questions about a scenario. These questions can be difficult; it can seem as if two of the choices could be correct, although you can choose only one. Read the directions for these types of questions several times to be sure you know exactly what you are supposed to do. Also, remember important concepts such as the importance of using a WBS, emphasizing teamwork, and practicing professional integrity. You might want to skip the longer or more difficult questions and answer the shorter or easier ones first.
9. If you do not know an answer and need to guess, be wary of choices that include words such as always, never, only, must, and completely. These extreme words usually indicate incorrect answers because there are many exceptions to rules.

10. After an hour or two, take a short break to clear your mind. You might want to bring a snack to have during your break. You might also consider bringing earplugs if you're easily distracted by noises in the room.



**TABLE B-2** Formulas to know for the PMP exam

| **Time-Related Formulas** |
|---|
| Assume o = 6, m = 21, and p = 36 for the following examples, where o = optimistic, m = most likely, and p = pessimistic estimate |
| PERT weighted average = (o + 4m + p)/6<br>Example: PERT weighted average = (36 + 4(21) + 6)/6 = 21 |
| PERT standard deviation = (p – o)/6<br>Example: PERT standard deviation = (36 – 6)/6 = 5 |
| Range of outcomes using 1 std. dev. = 21 – 5 = 16 days and 21 + 5 = 26 days.<br>Range of outcomes using 2 std. dev. = 21 – 10 = 11 days and 21 + 10 = 31 days.<br>Range of outcomes using 3 std. dev. = 21 – 15 = 6 days and 21 + 15 = 36 days.<br>1 std dev. = **68.3%** of the population<br>2 std dev. = **95.5%** of the population<br>3 std dev. = **99.7%** of the population |
| **Cost/Earned Value Formulas** |
| Earned Value = EV |
| Actual Cost = AC |
| Planned Value = PV |
| Cost Variance = CV = EV – AC |
| Schedule Variance = SV = EV – PV |
| Cost Performance Index = CPI = EV/AC |
| Schedule Performance Index = SPI = EV/PV |
| BAC = Budget At Completion or the planned total budget for the project<br>EAC = Estimate At Completion = BAC/CPI |
| Estimated time at completion = estimated time/SPI |
| Estimate To Complete (ETC) = EAC – AC |

*(continued)*

TABLE B-2 Formulas to know for the PMP exam (*continued*)

| |
|---|
| Variance At Completion (VAC) = BAC – EAC<br>Remember, a negative value for a variance or equivalently an index less than 100 percent means over budget/behind schedule. |
| **Communications Formulas** |
| Number of communications channels = (n(n – 1))/2<br>Example: Assume n = 5, where n = number of people<br>Number of communications channels = (5(5 – 1))/2 = (5 * 4)/2 = 10 |
| **Procurement Formulas** |
| Point of Total Assumption (PTA) = (Ceiling Price – Target Price)/Government Share + Target Cost |
| Make or Buy Analysis: Create a formula so the "make" option equals the "buy" or "lease" option, and then solve for the number of days.<br>Example: Assume you can purchase equipment for $3,000 and it costs $100/day to operate OR you can lease the equipment for $400/day. In how many days is the lease price equal to the purchase price? Set up an equation where the cost to lease or buy the item is equal to the cost to purchase or make the item.<br>1. Let d = the number of days you'll use the equipment: $400d = $3,000 + $100d<br>2. Then solve for d. Subtract $100d from both sides to get $300d = $3,000<br>3. Then divide each side by $300 to get d = 10<br>Therefore, if you need the equipment for more than 10 days, it would be cheaper to buy it. |

## Sample PMP Exam Questions

A few sample questions similar to those you will find on the PMP exam are provided on the following pages. You can check your answers at the end of this appendix. If you miss seven or less out of these 20 questions, you are probably ready to take the PMP exam. You can find additional sample questions and their answers on the companion Web site, as well as links to other free sample tests from various Web sites.

1. A document that formally recognizes the existence of a project is a ________
    a. Gantt chart
    b. WBS
    c. project charter
    d. scope statement
2. Decomposition is used in developing ________.
    a. the management plan
    b. the communications plan
    c. the earned value
    d. the WBS

3. The critical path on a project represents ________.
   a. the shortest path through a network diagram
   b. the longest path through a network diagram
   c. the most important tasks on a project
   d. the highest-risk tasks on a project
4. If the earned value (EV) for a project is $30,000, the actual cost (AC) is $33,000, and the planned value (PV) is $25,000, what is the cost variance?
   a. $3,000
   b. –$3,000
   c. $5,000
   d. –$5,000
5. If the earned value (EV) for a project is $30,000, the actual cost (AC) is $33,000, and the planned value (PV) is $25,000, how is the project performing?
   a. The project is over budget and ahead of schedule.
   b. The project is over budget and behind schedule.
   c. The project is under budget and ahead of schedule.
   d. The project is under budget and behind schedule.
6. What is the target goal for defects per million opportunities using Six Sigma?
   a. 1
   b. 3.4
   c. 34
   d. 100
7. Project human resource management does not include which of the following processes?
   a. acquiring the project team
   b. developing the project team
   c. managing the project team
   d. estimating activity resources
8. If a project team goes from three to five people, how many more communications channels are there?
   a. 7
   b. 6
   c. 5
   d. 4
9. Your project team has identified several risks related to your project. You decide to take actions to reduce the impact of a particular risk event by reducing the probability of its occurrence. What risk response strategy are you using?
   a. risk avoidance
   b. risk acceptance
   c. risk mitigation
   d. contingency planning
10. Which type of contract provides the least amount of risk for the buyer?
    a. firm fixed price
    b. fixed price incentive
    c. cost plus incentive fee
    d. cost plus fixed fee

11. Suppose you have a project with four tasks as follows:
    - Task 1 can start immediately and has an estimated duration of 1.
    - Task 2 can start after Task 1 is completed and has an estimated duration of 4.
    - Task 3 can start after Task 2 is completed and has an estimated duration of 5.
    - Task 4 can start after Task 1 is completed and must be completed when Task 3 is completed. Its estimated duration is 8.

    What is the length of the critical path for this project?

    a. 9
    b. 10
    c. 11
    d. 12
12. In which project management process group is the most time and money typically spent?
    a. initiating
    b. planning
    c. executing
    d. monitoring and controlling
13. Creating a probability/impact matrix is part of which risk management process?
    a. plan risk management
    b. identify risks
    c. perform qualitative risk analysis
    d. perform quantitative risk analysis
14. It is crucial that your project team finish your project on time. Your team is using a technique to account for limited resources. You have also added a project buffer before the end date and feeding buffers before each critical task. What technique are you using?
    a. critical path analysis
    b. PERT
    c. critical chain scheduling
    d. earned value management
15. One of your senior technical specialists informs you that a major design flaw exists in a systems development project you are managing. You are already testing the system and planned to roll it out to more than 5,000 users in a month. You know that changing the design now will cause several cost and schedule overruns. As project manager, what should you do first?
    a. Issue a stop work order until you understand the extent of the flaw.
    b. Notify your project sponsor immediately to see if there are additional funds available to work on this problem.
    c. Notify your senior management and let them decide what to do.
    d. Hold a meeting as soon as possible with key members of your project team to discuss possible solutions to the problem.

16. You are a member of a large government project. You know that the contract insists that all equipment be manufactured in the United States. You see a senior member of your team replacing a company etching on a piece of equipment that was made in a foreign country. You confront this person, and he says he is following the project manager's orders. What should you do?
    a. Nothing; the project manager made the decision.
    b. Immediately report the violation to the government.
    c. Update your resume and look for another job.
    d. Talk to the project manager about the situation, and then decide what to do.
17. Which of the following is not an output of the integrated change control process?
    a. project management plan updates
    b. change request status updates
    c. forecasts
    d. project document updates
18. The ceiling price for a contract is $1.25 million, the target price is $1.1 million, the target cost is $1 million, and the government share is 75%. What is the point of total assumption?
    a. $1.2 million
    b. $1 million
    c. $1.1 million
    d. there is not enough information
19. Obtaining quotes, bids, offers, or proposals is part of which project procurement management process?
    a. plan procurements
    b. conduct procurements
    c. administer procurements
    d. close procurements
20. Your boss believes that all of your project team members avoid work as much as possible. He or she often uses threats and various control schemes to make sure people are doing their jobs. Which approach to managing people does your boss follow?
    a. Maslow's hierarchy of needs
    b. Theory X
    c. Theory Y
    d. Herzberg's motivation and hygiene factors

## WHAT IS PROJECT+ CERTIFICATION?

The Computing Technology Industry Association (CompTIA) is the world's largest developer of vendor-neutral IT certification exams. By January 2009, more than one million people worldwide have earned CompTIA certifications in topics such as PC service, networking, security, and Radio Frequency Identification (RFID). In April 2001, CompTIA started offering its IT Project+ certification, which was purchased from Prometric-Thomson Learning, and recognized as the Gartner Institute Certification Program. The certification was renamed Project+ in August 2004. According to CompTIA's manager of public relations, there are more than 11,000 people with CompTIA's Project+ certification by the end of 2008. Detailed information about Project+ certification is available from CompTIA's Web site (*www.comptia.org*) under certification. The following information is quoted from CompTIA's Web site in January 2009:

> CompTIA Project+ is a globally recognized project management certification that provides validation of fundamental project management skills. It covers the entire project life cycle from initiation and planning through execution, acceptance, support and closure. Unlike some project management certifications, CompTIA Project+ can be acquired in a quick and cost-effective manner. There are no prerequisites, and candidates are not required to submit an application or complete additional hours of continuing education. CompTIA Project+ gives project managers the skills necessary to complete projects on time and within budget, and creates a common project management language among project team members.[7]

## What Are the Requirements for Earning and Maintaining Project+ Certification?

You can register to take the Project+ exam online from CompTIA's Web site. Testing sites include Thomson Prometric and Pearson VUE. Unlike the PMP exam, there are very few requirements you need to meet to take the Project+ exam. CompTIA does not require that you have any work experience or formal education in project management before you can take the Project+ exam, but they do recommend 2,000 hours of work experience. The main requirements include paying a fee and passing the exam. Important information is summarized below:

1. As of January 2009, the cost for taking the Project+ exam is $239 for non-members in the United States.
2. To pass this 90-minute, 80-question exam, you must score at least 63 percent.
3. You do not need to renew your Project+ certification.
4. Project+ certification is one of the prerequisites or equivalents to Novell's Certified Novell Engineer (CNE), Certified Novell Administrator (CNA), and Certified Novell Instructor (CNI) certifications. CompTIA's Project+ certification is also a Continuing Certification Requirement (CCR) to maintaining a Master CNE certification.
5. You can earn college credit with the Project+ certification. For example, CompTIA's Web site says that Capella University will provide six credit hours to someone with the Project+ certification. Many colleges and universities will grant credit based on your experience or other certifications. There is usually some fee involved, but it's often much less than the cost of taking the courses.

## Additional Information on the Project+ Exam



Because there are no experience or education requirements for taking the Project+ exam, you might want to take it very early in your career. Once you have enough experience and education to take the PMP exam, you might want to earn and maintain PMP certification.

As stated already, the Project+ exam consists of 80 questions. Table B-3 shows the approximate breakdown of questions on the Project+ exam by four domain areas. Candidates should review updated exam information on CompTIA's Web site to make sure they are prepared for the exams. For example, CompTIA provides a detailed list of objectives you should understand before taking the Project+ exam. Studying information in this book will also help you prepare for the Project+ exam. You might also want to purchase an exam guide to get specific information and access to more sample questions.

**TABLE B-3** Breakdown of questions on the Project+ exam by domain areas

| Domain Area | Percent of Questions on Project+ Exam |
| --- | --- |
| Project Initiation and Scope Definition | 20 |
| Project Planning | 30 |
| Project Execution, Control, and, Coordination | 43 |
| Project Closure, Acceptance, and Support | 7 |

Much of the advice for taking the PMP exam also applies to taking the Project+ exam. Many people find the questions to be similar on both exams, although the PMP exam is longer and more comprehensive. Below are some of the main differences between the content and types of questions you will find on the Project+ exam:

- The Project+ exam includes some scenarios and content specific to the information technology industry. For example, you should understand the various systems development life cycles and issues that often occur on information technology projects.
- You should understand the various roles of people on information technology projects, such as business analysts, database analysts, programmers, and so on.
- Although many of the questions are multiple-choice like the PMP exam questions, several questions involve choosing two or more correct answers. Several questions also involve matching or putting items in order, called drag-and-drop questions.
- The Project+ exam is not based on the *PMBOK® Guide*, so you do not need to know the processes involved in the various knowledge areas. However, much of the terminology, concepts, tools, and techniques are the same on both exams.

## Sample Project+ Exam Questions

A few sample questions similar to those you will find on the Project+ exam are provided below. CompTIA also provides sample questions on their Web site at *http://certification.comptia.org/resources/practice_test.aspx*. You can check your answers at the end of this appendix.

1. Two software developers on your project disagree on how to design an important part of a system. There are several technologies and methodologies they could use. What should be the primary driver in deciding how to proceed?
   a. following corporate standards
   b. following industry standards
   c. meeting business needs
   d. using the lowest-cost approach
2. Match the following items to their descriptions:

   | | |
   |---|---|
   | Stakeholder | a. Acts as a liaison between the business area and developers |
   | Project manager | b. Writes software code |
   | Business analyst | c. Person involved in or affected by project activities |
   | Programmer | d. Responsible for managing project activities |

3. For a project to be successful, the project manager should strive to understand and meet certain goals. What are the three main project goals to meet? Select three answers.
   a. scope or performance goals
   b. time goals
   c. political goals
   d. cost goals
   e. stock price goals
4. You have received an incomplete project scope definition. Put the following actions in order of how you should proceed to complete them.
   a. Incorporate additional changes to the scope definition document.
   b. Review the draft scope definition document with your project team.
   c. Get signatures on the completed scope definition document.
   d. Rewrite the draft scope definition document with your users and project team.
5. What term is used to describe the process of reaching agreement on a collective decision?
   a. collaboration
   b. cooperation
   c. coordination
   d. consensus
6. What is a variance?
   a. a buffer in a duration estimate
   b. a small amount of money set aside for contingencies
   c. a form of risk management
   d. a deviation from the project plan

7. Which of the following would be legitimate reasons for a vendor to request a delay in delivering a product? Select two answers.
    a. The vendor may have underestimated the amount of time required to produce and deliver the product.
    b. The project contact from the vendor's organization may be going on vacation.
    c. The vendor might be able to provide a better product by delivering the product late.
    d. The vendor might lose money by delivering the product late.
8. When should you involve stakeholders in the change control process on information technology projects?
    a. before a change is submitted
    b. after a change is submitted
    c. when a change is submitted
    d. throughout the life of a project
9. Which of the following techniques can be used to help manage requirements? Select three answers.
    a. prototyping
    b. use case modeling
    c. JAD
    d. worst case modeling
10. Match the following items to their descriptions:

| | |
|---|---|
| Lessons learned | a. Leave a clear and complete history of a project |
| Project audits | b. Review project progress and results |
| Project archives | c. Document what went right or wrong on a project |

## WHAT OTHER EXAMS OR CERTIFICATIONS RELATED TO PROJECT MANAGEMENT ARE AVAILABLE?

In recent years, several organizations have been developing more certifications related to project management and information technology project management, in particular. Some involve taking exams while others require coursework or attendance at workshops. Below are brief descriptions of other existing exams or certifications related to project management:

- Microsoft provides certification as a Microsoft Office Specialist (MOS) to recognize proficiency in using its software products. In October 2006, Microsoft Learning launched a new Microsoft Office Project 2007 Certification program. You can become a Microsoft Certified Technology Specialist (MCTS) in either Managing Projects with Microsoft Office Project 2007 or Enterprise Project Management with Microsoft Office Project Server 2007. You can also earn a Microsoft Certified IT Professional (MCITP) in enterprise Project Management with Microsoft Office Project Server 2007. Its purpose is "to help advance project management as a profession and maximize value for its customer base of nearly 20 million Microsoft Office Project user licenses."[8] Microsoft is using the PMI's *PMBOK® Guide* as a foundation for supporting specific Project 2007 competencies. See Microsoft's Web site (*www.microsoft.com*) for more details.

- The International Project Management Association (IPMA) offers a four-level certification program. The main requirements for each level are derived from typical activities, responsibilities, and requirements from practice. The IPMA four-level certification system, in descending order, includes the Certified Project Director, Certified Senior Project Manager, Certified Project Manager, and Certified Project Management Associate. More than 50,000 people worldwide had earned IPMA certifications, according to IPMA's Web site in January 2007. See *www.ipma.ch* for more details.
- Certified IT Project Manager (CITPM): In 1998, the Singapore Computer Society collaborated with the Infocomm Development Authority of Singapore to establish an Information Technology Project Management Certification Program. PMI signed a Memorandum of Understanding with the Singapore Computer Society to support and advance the global credentialing of project management and information technology expertise. In January 2007, the Singapore Computer Society's Web site listed three levels of the CITPM Certification: CITPM (Senior), CITPM, and CITPM (Associate). For more details consult their Web site (*www.scs.org.sg/about_certprog.php*).
- PMI's additional certifications. As mentioned earlier, as of January 2009, PMI also offers certification as a Certified Associate in Project Management (CAPM), a PMI-Scheduling Professional (PMI-SP), a PMI-Risk Management Professional (PMI-RMP), and a Program Management Professional (PgMP). Check PMI's Web site for updated information on their certification programs.
- Many colleges, universities, and corporate training companies now provide their own certificate programs or entire degrees in project management. Typing "project management certificate" into *www.google.com* in January 2009 resulted in 58,400 hits. Some of the certificate courses apply toward bachelor's or advanced degrees, while many do not. Like any other educational program, it is important to research the quality of the program and find one that will meet your specific needs. The December 2006 issue of the Project Management Institute's PMI Today included a supplement that said they had identified over 284 project management degree programs at 235 institutions worldwide. See the author's Web site for a summary of more than 120 U.S. graduate programs created as part of a class project in 2006 (*www.kathyschwalbe.com*, under Project Management Info). Also see sites like *www.gradschools.com* to find more information on graduate programs in project management throughout the world.

## Discussion Questions

1. What is PMP certification, and why do you think the number of people earning it has grown so much in the past 10 years?
2. What do you need to do before you can take the PMP exam? What is the exam itself like? What do you need to do to maintain PMP certification? What do you need to do to take the Project+ exam? How does the Project+ exam differ from the PMP exam? Do you need to renew Project+ certification?
3. What is the difference between conceptual, application, and evaluative questions? Which project management process groups have the most questions on the PMP exam? What are the four domain areas tested on the Project+ exam?
4. Which tips for taking the PMP exam do you think would be most helpful for you?
5. If you plan to take the Project+ or PMP exam soon, what should you do to prepare?
6. Briefly describe project management certification programs other than the PMP or Project+ certifications.

## Exercises

1. Go to PMI's Web site and review the information about taking the PMP exam. Write a two-page paper summarizing what you found.
2. Go to CompTIA's Web site and review the information about taking the Project+ exam. Write a two-page paper summarizing what you found.
3. Answer the 20 sample PMP questions in this text or take another sample test, and then score your results. Summarize how you did and areas you would need to study before you could take the PMP exam.
4. Interview someone who has PMP or Project+ certification. Ask him or her why he or she earned the certification and how it has affected his or her career. Write your findings in a two-page paper.
5. Do an Internet search on earning PMP or Project+ certification. Be sure to search for Yahoo! groups or similar sites related to these topics. What are some of the options you found to help people prepare for either exam? If you were to take one of the exams, what do you think you would do to help study for it? Do you think you would need additional information beyond what is in this text to help you pass? Write a two-page paper describing your findings and opinions.
6. Read a recent issue of PMI's *PM Network* magazine. You can access a free sample copy from PMI's Web site. Summarize all of the ads you find in the magazine related to earning PMP or other project management related certification in a two-page paper. Also, include your opinion on which ad, course, book, CD-ROM, or other media appeals to you the most.

## Answers to Sample PMP Exam Questions

1. c
2. d
3. b
4. b
5. a
6. b
7. d
8. a
9. c
10. a
11. b
12. c
13. c
14. c
15. d
16. d
17. c
18. a
19. b
20. b

## Answers to Sample Project+ Exam Questions

1. c
2. c, d, a, b
3. a, b, d
4. b, d, a, c
5. d
6. d
7. a, c
8. d
9. a, b, c
10. c, b, a

## End Notes

[1] Project Management Institute, "PMI® Certification Programs," (*www.pmi.org/*) (January 2007).

[2] Project Management Institute Information Systems Specific Interest Group (ISSIG), *ISSIG Bits* (*www.pmi-issig.org*) (February 2003).

[3] Becky Nagel, "CertCities.com's 10 Hottest Certifications for 2006," *CertCities.com*, (December 14, 2005).

[4] Global Knowledge, "2003 Certification Salary by Certmag," *Global Knowledge E-Newsletter*, Issue #56 (March 2003).

[5] Adrienne Rewi, "The Rise of PMP," *PM Network*, (October 2004), p. 18.

[6] The Project Management Institute, "PMI Today," (February 2009).

[7] CompTIA Web site (*http://certification.comptia.org/project*) (January 2009).

[8] Microsoft, "Microsoft Advances Its Project Management Technology and the Project Management Profession," Microsoft PressPass (October 20, 2006).

# ADDITIONAL RUNNING CASES AND SIMULATION SOFTWARE

## INTRODUCTION

This appendix provides two additional running cases as well as information about using simulation software. The first case includes tasks ordered by each of the nine knowledge areas discussed in Chapters 4 through 12 of this text. The second case includes tasks based on the five project management process groups. There is also information about using Fissure's project management simulation software. Additional running cases, including the Video Game Delivery Project case that was in the fifth edition of this text, and suggestions for finding more cases or having students do real projects are available on the companion Web site.

The purpose of these cases is to help you practice and develop the project management skills learned from this text. Several of the tasks involve using templates provided on the companion Web site (*www.cengage.com/mis/*schwalbe) and the author's personal Web site (*www.kathyschwalbe.com*). Instructors can download the suggested solutions for these cases from the password-protected section on Course Technology's Web site. Contact a sales rep at *www.cengage.com/coursetechnology*.

## ADDITIONAL CASE 1: GREEN COMPUTING RESEARCH PROJECT

### Part 1: Project Integration Management

You are working for We Are Big, Inc., an international firm with over 100, 000 employees located in several different countries. A strategic goal is to help improve the environment while increasing revenues and reducing costs. The Environmental Technologies Program just started, and the VP of Operations, Natalie, is the program sponsor. Ito is the program manager, and there is a steering committee made up of ten senior executives, including

Natalie, overseeing the program. There are several projects underneath this program, one being the Green Computing Research Project. The CIO and project sponsor, Ben, has given this project high priority and plans to hold special interviews to hand-pick the project manager and team. Ben is also a member of the program steering committee. Before coming to We Are Big, Inc., Ben sponsored a project at a large computer firm to improve data center efficiency. This project, however, is much broader than that one was. The main purpose of the Green Computing Research Project is to research possible applications of green computing including:



- Data center and overall energy efficiency
- The disposal of electronic waste and recycling
- Telecommuting
- Virtualization of server resources
- Thin client solutions
- Use of open source software, and
- Development of new software to address green computing for internal use and potential sale to other organizations

The budget for the project was $500, 000, and the goal was to provide an extensive report, including detailed financial analysis and recommendations on what green computing technologies to implement. Official project request forms for the recommended solutions would also be created as part of the project.

Ben decided to have a small group of people, five to be exact, dedicated to working on this six-month project full-time and to call on people in other areas on an as-needed basis. He wanted to personally be involved in selecting the project manager and have that person help him to select the rest of the project team. Ben wanted to find people already working inside the company, but he was also open to reviewing applications for potential new employees to work specifically on this project as long as they could start quickly. Since many good people were located in different parts of the world, Ben thought it made sense to select the best people he could find and allow them to work virtually on the project. Ben also wanted the project manager to do more than just manage the project. He or she would also do some of the research, writing, editing, and the like required to produce the desired results. He was also open to paying expert consultants for their advice and purchasing books and related articles, as needed.

## Tasks

1. Research green computing and projects that have been done or are being done by large organizations such as IBM, Dell, HP, and Google. See *www.greener-computing.com* and similar sites provided on the companion Web site or that you find yourself. Include your definition of green computing to include all of the topics listed in the background scenario. Describe each of these areas of green computing, including a detailed example of how at least one organization has implemented each one, and investigate the return on investment. Summarize your results in a two- to three-page paper, citing at least three references.
2. Prepare a weighted decision matrix using the template from the companion Web site (wtd_decision_matrix.xls) for Ben to use to evaluate people applying to be the project manager for this important project. Develop at least five

criteria, assign weights to each criterion, assign scores, and then calculate the weighted scores for four fictitious people. Print the spreadsheet and bar chart with the results. Write a one-page paper describing this weighted decision matrix and summarize the results.

3. Prepare the financial section of a business case for the Green Computing Research Project. Assume this project will take six months to complete (done in Year 0) and cost $500,000, and costs to implement some of the technologies would be $2,000,000 for year one and $600,000 for years two and three. Estimated benefits are $500,000 the first year after implementation and $2.5 million the following two years. Use the business case spreadsheet template from the companion Web site (business_case_financials.xls) to help calculate the NPV, ROI, and the year in which payback occurs. Assume a 7 percent discount rate, but make sure it is an input that is easy to change.
4. Prepare a project charter for the Green Computing Research Project. Assume the project will take six months to complete and the budget is $500,000. Use the project charter template (charter.doc) and examples of project charters in Chapters 3 and 4 as guidelines. Assume that part of the approach is to select the project team as quickly as possible.
5. Since people will request changes to the project, you want to make sure you have a good integrated change control process in place. You will also want to address change requests as quickly as possible. Review the template for a change request form provided on the companion Web site (change_request.doc). Write a two-page paper describing how you plan to manage changes on this project in a timely manner. Address who will be involved in making change control decisions, what paperwork or electronic systems will be used to collect and respond to changes, and other related issues.



## Part 2: Project Scope Management

Congratulations! You have been selected as the project manager for the Green Computing Research Project. The company's CIO, Ben, is the project sponsor, and Ito is the program manager for the larger Environmental Technologies Program that this project is part of. Now you need to put together your project team and get to work on this high-visibility project. You will work with Ben to hand-pick your team. Ben had already worked with the HR department to advertise these openings internally as well as outside the company. Ben had also used his personal contacts to let people know about this important project. In addition, you are encouraged to use outside consultants and other resources, as appropriate. Initial estimates suggest that about $300,000 budgeted for this project will go to internal staffing and the rest to outside sources. The main products you'll produce will be a series of research reports—one for each green computing technology listed earlier plus one final report including all data—plus formal project proposals for at least four recommendations for implementing some of these technologies. Ben also suggested that the team come up with at least 20 different project ideas and then recommend the top four based on extensive analysis. Ben thought some type of decision support model would make sense to help collect and analyze the project ideas. You are expected to tap into resources available from the Environmental Technologies Program, but you will need to include some of those resources in your project budget. Ben mentioned that he knew there had already been some research done on increasing the use of telecommuting. Ben also showed you examples of what he considered to be good research reports. You notice that his examples are very professional, with a lot of charts and references, and most are 20–30 pages long, single-spaced. Ben has also shown you examples of good formal project proposals for We Are Big, Inc., and you are surprised to see how detailed they are, as well. They often reference other research and include a detailed business case.



### Tasks

1. Document requirements for your project so far, including a requirements traceability matrix. Use the template provided (reqs_matrix.xls). Also include a list of questions you would like to ask the sponsor about the scope.
2. Develop a scope statement for the project using the template provided (scope_statement.doc). Be as specific as possible in describing product characteristics and deliverables. Make assumptions as needed, assuming you got answers to the questions you had in Task 1.
3. Develop a work breakdown structure (WBS) for the project. Break down the work to level 3 or level 4, as appropriate. Use the template on the companion Web site (wbs.doc) and samples in the text as guides. Print the WBS in list form as a Word file. Be sure to base your WBS on the project scope statement, stakeholder requirements, and other relevant information. Remember to include the work involved in selecting the rest of your project team and outside resources as well as coordinating with the Environmental Technologies Program. Use the project management process groups as level 2 WBS items or include project management as a level 2 WBS item to make sure you include work related to managing the project.
4. Use the WBS you developed in Task 3 above to create a Gantt chart for the project in Microsoft Project 2007. Use the outline numbering feature to display the outline numbers (click Tools on the menu bar, click Options, and then click Show outline number). Do not enter any durations or dependencies. Print the resulting Gantt chart on one page, being sure to display the entire Task Name column.

## Part 3: Project Time Management

As project manager, you are actively leading the Green Computing Research Project team in developing a schedule. You and Ben found three internal people and one new hire to fill the positions on the project team as follows:

- Matt was a senior technical specialist in the corporate IT department located in the building next to yours and Ben's. He is an expert in collaboration technologies and volunteers in his community helping to organize ways for residents to dispose of computers, printers, and cell phones.
- Teresa was a senior systems analyst in the IT department in a city 500 miles away from your office. She just finished an analysis of virtualization of server resources for her office, which has responsibility for the company's data center.

- James was a senior consultant in the strategic research department in a city 1,000 miles away from your office. He has a great reputation as being a font of knowledge and excellent presenter. Although he is over 60, he has a lot of energy.
- Le was a new hire and former colleague of Ben's. She was working in Malaysia, but she planned to move to your location, starting work about four weeks after the project started. Le has a lot of theoretical knowledge in green computing, and her doctoral thesis was on that topic.

While waiting for everyone to start working on your project, you talked to several people working on other projects under the Environmental Technologies Program and did some research on green computing. You can use a fair amount of the work already done on telecommuting, and you have the name of a consulting firm to help with that part of your project, if needed. Ito and Ben both suggested that you get up to speed on available collaboration tools since much of your project work will be done virtually. They knew that Matt would be a tremendous asset for your team in that area. You have also contacted other IT staff to get detailed information on your company's needs and plans in other areas of green computing. You also found out that there is a big program meeting in England next month that you and one or two of your team members should attend. It is a three-day meeting, plus travel. Recall that the Green Computing Research Project is expected to be completed in six months, and you and your four team members are assigned full-time to this project. Your project sponsor, Ben, has made it clear that delivering a good product is most important, but he also thinks you should have no problem meeting your schedule goal. He can authorize additional funds, if needed. You have decided to hire a part-time editor/consultant, Deb, whom you know from a past job to help your team produce the final reports and project proposals. Your team has agreed to add a one-week buffer at the end of the project to ensure that you finish on time or early.

## Tasks

1. Review the WBS and Gantt chart you created for Tasks 3 and 4 in Part 2. Propose three to five additional activities you think should be added to help you estimate resources and durations. Write a one-page paper describing these new activities.
2. Identify at least four milestones for this project. Write a one-page paper describing each milestone using the SMART criteria.
3. Using the Gantt chart created for Task 4 in Part 2, and the new activities and milestones you proposed in Tasks 1 and 2 above, estimate the task durations and enter dependencies as appropriate. Remember that your schedule goal for the project is six months. Print the Gantt chart and network diagram.
4. Write a one-page paper summarizing how you would assign people to each activity. Include a table or matrix listing how many hours each person would work on each task. These resource assignments should make sense given the duration estimates made in Task 3 above.
5. Assume that your project team starts falling behind schedule. In several cases, it is difficult to find detailed information on some of the green computing technologies, especially financial data. You know that it is important to meet

or beat the six-month schedule goal, but quality is most important. Describe contingency strategies for making up lost time and avoiding schedule slips in the future.



## Part 4: Project Cost Management

Your project sponsor has asked you and your team to refine the existing cost estimate for the project so that there is a solid cost baseline for evaluating project performance. Recall that your schedule and cost goals are to complete the project in six months or less for under $500, 000. Initial estimates suggested that about $300,000 for this project would go toward internal labor. You mistakenly thought that travel costs would be included in that $300,000, but now you realize that it is a separate cost item. The one trip to England early in the project cost $6,000, which you had not expected.

### Tasks

1. Prepare and print a one-page cost estimate for the project, similar to the one provided in Chapter 7. Use the WBS categories you created earlier, and be sure to document assumptions you make in preparing the cost estimate. Assume a burdened labor rate of $100/hour for the project manager, $90 for Teresa, James, and Le, and $80/hour for Matt. Assume about $200/hour for outsourced labor.
2. Using the cost estimate you created in task 1, prepare a cost baseline by allocating the costs by WBS for each month of the project.
3. Assume you have completed three months of the project and have actual data. The BAC was $500,000 for this six-month project. Also assume the following:

   PV= $160,000
   EV= $150,000
   AC= $180,000

   Using this information, write a short report that answers the following questions.

   a. What is the cost variance, schedule variance, cost performance index (CPI), and schedule performance index (SPI) for the project?
   b. Use the CPI to calculate the estimate at completion (EAC) for this project. Use the SPI to estimate how long it will take to finish this project. Sketch an earned value chart using the above information, including the EAC point. See Figure 7-6 as a guide. Write a paragraph explaining what this chart shows.
   c. How is the project doing? Is it ahead of schedule or behind schedule? Is it under budget or over budget? Should you alert your sponsor or other senior management and ask for assistance?
4. You notice that several of the tasks that involve getting inputs from consultants outside of your own company have cost more and taken longer to complete than planned. You have talked to the consultants several times, but they say they are doing the best they can. You also underestimated travel costs for this project. Write a one-page paper describing corrective action you could take to address these problems.

## Part 5: Project Quality Management

The Green Computing Research Project team is working hard to ensure their work meets expectations. The team has a detailed project scope statement, schedule, and so on, but as the project manager, you want to make sure you'll satisfy key stakeholders, especially Ben, the project sponsor, and Ito, the program manager. You have seen how tough Ito can be on project managers after listening to his critiques of other project managers at the monthly program review meeting. He was adamant on having solid research and financial analysis and liked to see people use technology to make quick what-if projections. You were impressed to see that several other project teams had developed computer models to help them perform sensitivity analysis and make important decisions. Most of the models were done using Excel, which Ito preferred, and you were glad that you were an expert with Excel, as was Matt. Ito was pretty easy on you at your first monthly review because things were just getting started, but he did give you a list of items to report on next month. You had Ben there to help answer some of the tough questions, but you wanted to be able to hold your own at future monthly meetings.



### Tasks

1. Develop a list of at least five quality standards or requirements related to meeting the stakeholder expectations especially for Ben and Ito. Also provide a brief description of each standard or requirement. For example, a requirement might be related to the computer model (that the computer model you create to analyze the 20 or more technologies be done in Excel 2007). Other standards or requirements might be related to the quality of the financial analysis and research you use.
2. Review the Seven Basic Tools of Quality. Pick one and make up a scenario related to this project where it would be useful. Document the scenario and tool in a one-to-two page paper.
3. Find a high quality research report related to the green computing. Summarize the report and why you think it is of high quality in a one-to-two page paper.

## Part 6: Project Human Resource Management

You are five weeks into the Green Computing Research Project, and all of the full-time team members are together face-to-face for the first time. Recall that you, Ben, Matt, and Le all work in the same location, but Teresa and James are based out of town and will do most of their work virtually. Le is also new to the company and just moved to the U.S. She is currently staying in a hotel and looking for a place to live. She'd like to buy her first home, but she wants to make sure it's a good investment and somewhere she'd like to stay for at least five years. You get along very well with your project sponsor, Ben, and Matt is a great resource, even though he is extremely reserved. Le is also very quiet, and you quickly discover that she is an excellent researcher and writer, but she is not comfortable speaking in public. Teresa and James are much more talkative and are excited to be working on this project. However, James seems to be reluctant to use much technology to share ideas and really enjoys face-to-face meetings and discussions. You have made preliminary agreements with two outside consultants to assist you with editing and the teleconferencing topic for your research. You have to prepare a monthly progress report and presentation for Ito, the program manager. You also have short meetings as needed with Ben, your sponsor, and send him a weekly progress report.

C.8

## Tasks

1. Before this first face-to-face meeting, you asked everyone to send a brief introductory e-mail, including links to their personal Web sites, LinkedIn site, etc. You also asked everyone to take a short version of the Meyers-Briggs Type Indicator (MBTI) online and share their results with everyone else. Take this test yourself from *www.humanmetrics.com* and research how different MBTI types respond to work environments, especially for research projects and virtual teams. Summarize your findings in a one-to-two page paper. Also document what you would write in an e-mail to introduce yourself, assuming you are the project manager for this project. Be creative in your response.
2. Prepare a responsibility assignment matrix in RACI chart format based on the WBS you created earlier and the information you have on project team members and other stakeholders. Use the template (ram.xls) and samples in the text. Document key assumptions you made in preparing the chart.
3. Since everyone will be in town for most of the week, you want to make sure they develop good working relationships. You also want everyone to work together efficiently. You asked Matt to review collaboration tools he recommends the team use for this project. As Matt starts demonstrating some of these tools, including webcams and wikis, you notice that a couple of team members seem uncomfortable, especially James. He thought that he would be in charge of certain aspects of the research reports and didn't know how he'd feel about any team member being able to change what he wrote in a wiki. Le did not like the idea of using a Webcam. She'd rather not have her face on video when communicating virtually. Discuss these human resource-related concerns and others that you think would be common in this situation. Include strategies for addressing them as well.

## Part 7: Project Communications Management

Several communications issues have arisen on the Green Computing Research Project. Three months have passed since the project started. Your team had agreed to post all of their work on a shared site, but a couple of team members don't seem to like using that site and prefer to use e-mails and attachments. When they do that, other team members cannot easily see what work is done in one place or provide feedback using the wiki tools. It is also clear that some team members are better researchers and writers than others. When you have weekly conference calls with the Webcams, at least a couple of people don't have the Webcam working and just use the audio. You also find that these meetings rarely end on time as some team members get very talkative. You also got grilled by Ito at the last monthly program review meeting. He thought you'd be much further along in the project than you are and expects you to have one recommendation on a green computing project that looks very promising by next month. You haven't seen any great ideas yet. You want to start having face-to-face meetings at least twice a month, but you know it will make your project go over budget even more. At least the Excel model is going well. You and Matt have put a good deal of time into developing it. If only you had enough good data to put into it.

### Tasks



1. Create an issue log for the project using the template provided (issue_log.doc). List at least four issues and related information based on the scenario presented.
2. Research the use of wikis and address the concerns several team members have about using them, especially their fear of having others "mess up" their work. Document your findings in a one- to two-page paper.
3. Write a two-page paper describing how you might approach two of the conflicts described above.

## Part 8: Project Risk Management

Since several problems have been occurring on the Green Computing Research Project (see the case information in Part 7), you decide to be more proactive in managing risks. You also want to address positive and negative risks.

### Tasks

1. Create a risk register for the project, using the template (risk_register.xls). Identify six potential risks, including at least two positive risks.
2. Plot the six risks on a probability/impact matrix, using the template (prob_impact_matrix.ppt) and print it out. Assign a numeric value for the probability of each risk, and its impact on meeting the main project objectives. Use a scale of 1 to 10 to assign the values, with 1 being lowest and 10 being highest. For a simple risk factor calculation, multiply these two values (the probability score and the impact score). Document the results in a one-page paper, including your rationale for how you determined the scores for one of the negative risks and one of the positive risks.
3. Develop a response strategy for one of the negative risks and one of the positive risks. Enter the information in the risk register and print out your complete risk register. Also write a one-page paper describing what specific tasks would need to be done to implement these two strategies. Include time and cost estimates for each strategy.

## Part 9: Project Procurement Management

After a monthly program management review meeting four months into your project, Ito and Ben approved adding $100,000 and one additional month to the project. You provided strong rationale for the need for additional travel funds and more money for outside consultants to help you in finding good research information. You decided to have James go back to his old job since he didn't seem open to sharing ideas with others. It would be best to have a consulting firm, one you were already using, pick up the work he was supposed to do, even though it would cost a lot more. The lead consultant, Anne, had done a great analysis of improving overall energy efficiency for the company that would save millions of dollars each year. Ben, your project sponsor, was disappointed that you couldn't meet the original time and cost goals, but he wanted to make sure the final results were of high quality.

## Tasks

1. Draft a contract to have Anne's consulting firm perform the work that James was supposed to do for this project. Assume that the contract would last for three months and that Anne herself would be working about half-time, earning $200/hour. She would also have some other consultants do up to 100 hours of work at $150/hour. They would do most of the work virtually, but Anne would come in town at least once a month for face-to-face meetings. Limit the contract to two or three pages, and be sure to address specific personnel and travel requirements. Also make sure that all work produced would be owned and copyrighted by your company exclusively.
2. Deb, the editor you hired for this project, has asked for your assistance in organizing the final comprehensive research report. Draft a one-page executive summary and a table of contents for this report.
3. Although this is not really a procurement task, it is provided here for convenience purposes. Prepare a lessons-learned report for what you may have learned so far as project manager for this project. Use the template provided on the companion Web site (lessons_learned_report.doc) and be creative in your response.

# ADDITIONAL CASE 2: PROJECT MANAGEMENT VIDEOS PROJECT

## Part 1: Initiating

You and several of your classmates are taking a project management class, and your instructor suggested a project to find or create good video clips to illustrate various concepts related to the class. For example, the *Oceans 11*, *12*, and *13* movies all include great planning and execution clips. *Apollo 13* provides a great example of scope management and creative problem solving when the team has to figure out how to keep the astronauts alive. *The Office* television show includes many examples of poor motivation techniques. In addition to providing the clips on DVD, you will write a summary of them, including the length and source of the clip, introductions for each clip, discussion questions that you can pose before and after each clip, and suggested answers to those questions. Your instructor has suggested that teams find or create at least two good clips per team member. If several teams in your class work on this project, you will have to coordinate with them to avoid duplicating clips and to share resources. Everything your team creates for the project should fit on one DVD that will run on your instructor's computer. The DVD will be for educational use only, so there should not be any copyright issues.

## Tasks

1. To become more familiar with finding short video clips, do some preliminary research. Go to sites like *youtube.com* and search for videos related to project management concepts. Also search for articles related to project management in the movies, and visit sites such as *imdb.com* to see movie trailers. Find other

sites that have legitimate movie and television clips. Also discuss movies or television shows that you and your teammates are familiar with that could be used for this project. Write a two- to three-page paper with your findings, citing all references.

2. To become familiar with creating or editing short video clips, research how to take short segments of an existing DVD and put it on a computer. Also research the devices and software needed to create, edit, and post your own videos (such as *theFlip.com* and *youtube.com*). Summarize at least three options, including price information. Write a two- to three-page paper with your findings, citing all references.
3. Prepare a team contract for this project. Use the team_contract.doc template provided on the companion Web site, and review the sample in the text.
4. Prepare a draft project charter for the Project Management Videos Project. Assume the project will be completed by the last day of class, and costs will include an estimate of hours (unpaid) your team will work on this project plus the cost of any necessary hardware/software you would like for the project (such as DVDs, a camcorder, video editing software, etc.). Use the charter.doc template provided on the companion Web site, and review the sample in the text.
5. Prepare a draft schedule for completing all of the tasks for this project. Include columns that list each task by process group; estimated start and end dates for each task; who has the main responsibility for each task; estimated hours for each task by person; and actual hours for each task by person that you'll complete as you have the information.
6. Write a brief summary of your team's MBTI types and how they might affetct your team dynamics. You can take a version of the test from *www.humanmetrics.com.*
7. Prepare a 10-minute presentation that summarizes results from the above initiating tasks. Assume the presentation is for a review with your class and instructor. Be sure to document notes of any feedback received during the presentation and hand in hard copies of everything you produced.

## Part 2: Planning

Work with your teammates and instructor to perform several planning activities for this project.

## Tasks

1. Develop a scope statement for the project. Use the scope_statement.doc template provided on the companion Web site, and review the sample in the text. Be as specific as possible in describing product characteristics and requirements, as well as key deliverables. Determine which video clips your team will provide and what resources you think you will need (DVDs, camcorders, etc.). Be sure to coordinate the clips with your instructor and other teams and get feedback before handing in your scope statement.
2. Develop a WBS for the project. Use the wbs.doc template provided on the companion Web site, and review the samples in the text. Print the WBS in list form

as a Word file. Be sure the WBS is based on the project charter, scope statement, draft schedule, and other relevant information.

3. Create a milestone list for this project, and include at least ten milestones and estimated completion dates for them. Note that your instructor should have input for several of these milestones and completion dates. Use the milestone_report.doc template.
4. Develop a cost estimate for the project. Estimate hours needed to complete each task (including those already completed) and the costs of any items you would like to purchase for the project. Assume a rate of $10 per hour for all labor. Use the cost_estimate.xls template.
5. Use the WBS and milestone list you developed in numbers 2 and 3 above and the draft schedule you created earlier to create a Gantt chart and network diagram in Project 2007 for the project. Estimate task durations and enter dependencies, as appropriate. Print the Gantt chart and network diagram. Also update the draft schedule you created under Initiating, Task 5.
6. Create a quality checklist for ensuring that the project is completed successfully. Also define at least two quality metrics for the project.
7. Create a RACI chart for the main tasks and deliverables for the project.
8. Develop a communications management plan for the project. Use the comm_plan.doc template and sample provided.
9. Create a probability/impact matrix and list of prioritized risks for the project. Include at least ten risks. Use the prob_impact_matrix.ppt template and sample provided.
10. Prepare a ten-minute presentation that you would give to summarize results from the above planning tasks. Assume the presentation is for a review with your class and instructor. Be sure to document notes of any feedback received during the presentation and hand in hard copies of everything you produced. Plan to show one video clip along with the discussion questions to get feedback.

## Part 3: Executing

Work with your teammates and instructor to perform several executing activities for this project.

### Tasks

1. Find or create your video clips and put them on one DVD. Be sure it runs on your instructor's computer.
2. Write the clip summaries, introductions, discussion questions, and suggested answers to those questions.
3. Document any change requests you have during project execution and get sponsor approval, if needed.

## Part 4: Monitoring and Controlling

Work with your teammates and instructor to perform several monitoring and controlling activities for this project.

## Tasks

1. Review the Seven Basic Tools of Quality. Pick one of these tools and create a chart or diagram to help you solve problems you are facing. Use the available templates and samples provided. Note: There is only a template for the Pareto chart called pareto_chart.xls.
2. Create and update, as required, an issue log. Use the issue_log.doc template and sample provided.
3. As described in the last task for the initiating and planning sections, be ready to show progress you have made as part of a project review. Also be sure to document actual hours on each task in the draft schedule you first created for Task 5 under Initiating and updated for Task 5 under Planning.

## Part 5: Closing

Work with your teammates and instructor to perform several closing activities for this project.

## Tasks

1. Prepare a 20-minute final project presentation to summarize the results of the project. Describe the initial project goals, planned versus actual scope, time, and cost information, challenges faced, lessons-learned, and key products produced. Be sure to list all of the clips your team found and show at least two of them along with the discussion questions.
2. Prepare a final project report. Include a cover page and detailed table of contents, getting feedback from your instructor on information required. Be sure to include all of the documents and products you have prepared as appendices.
3. Get feedback from your sponsor in the form of a customer acceptance/project completion form (see the template called client_acceptance.doc) or in some other fashion. Also get feedback from your classmates.
4. If you are comfortable doing so, send a copy of your final project report and feedback on this case to the author of this text at *schwalbe@augsburg.edu*.

# FISSURE SIMULATION SOFTWARE

## Introduction

Another way to practice your project management skills is by using simulation software. This text can be purchased so that it is bundled with Fissure's project management simulation software, or you can buy it separately at *www.ichapters.com* for about $12. Search for Fissure from the site. The version provided is the correct one.

Fissure's project management simulation software is based on the SimProject Alliance Prototype project. The demo version of the simulation software includes a fairly simple, 11-week project consisting of only seven tasks and ten potential team members. Fissure estimates that it takes about three to four hours to run the demo simulation. To participate in this simulated project, you are expected to read about the company, project, and people available to work on this project. You plan your project and make typical project decisions each week, such as when to assign staff, when to hold meetings, and when to send staff to various training opportunities. You run your project a week at a time, analyzing your results each week, referring to your weekly reports, and making your decisions for the next week. As you run the simulation each week, you are presented with communications from people within the company, team members, or other stakeholders related to the project. You choose how to respond to these communications, and all the decisions you make impact how your project progresses. You can close the simulation at any time and save your work, if desired. You can also run the simulation as many times as you like, and the results vary based on your decisions. You can use the software for 120 days.

## Instructions

**NOTE**

It often works best if students can do this assignment as a small team or in pairs, if possible.

Run the Fissure simulation software at least two different times. Summarize key decisions and results from each week for each run, and print out the final earned value chart as well as any other information you think is valuable. Write a two- to three-page single-spaced report summarizing what you thought about using this simulation software. Be honest, specific, and thorough in your report.

# GLOSSARY

5 whys — A technique where you repeatedly ask the question "Why?" (five is a good rule of thumb) to help peel away the layers of symptoms that can lead to the root cause of a problem.

acceptance decisions — Decisions that determine if the products or services produced as part of the project will be accepted or rejected.

activity — An element of work, normally found on the WBS, that has an expected duration and cost, and expected resource requirements; also called a task.

activity attributes — Information about each activity, such as predecessors, successors, logical relationships, leads and lags, resource requirements, constraints, imposed dates, and assumptions related to the activity.

activity list — A tabulation of activities to be included on a project schedule.

activity-on-arrow (AOA) — A network diagramming technique in which activities are represented by arrows and connected at points called nodes to illustrate the sequence of activities; also called arrow diagramming method (ADM).

actual cost (AC) — The total of direct and indirect costs incurred in accomplishing work on an activity during a given period.

adaptive software development (ASD) — A software development approach used when requirements cannot be clearly expressed early in the life cycle.

agile software development — A method for software development that uses new approaches, focusing on close collaboration between programming teams and business experts.

analogous estimates — A cost estimating technique that uses the actual cost of a previous, similar project as the basis for estimating the cost of the current project, also called top-down estimates.

analogy approach — Creating a WBS by using a similar project's WBS as a starting point.

appraisal cost — The cost of evaluating processes and their outputs to ensure that a project is error-free or within an acceptable error range.

arrow diagramming method (ADM) — A network diagramming technique in which activities are represented by arrows and connected at points called nodes to illustrate the sequence of activities; also called activity-on-arrow (AOA).

backward pass — A project network diagramming technique that determines the late start and late finish dates for each activity in a similar fashion.

balanced scorecard — A methodology that converts an organization's value drivers to a series of defined metrics.

baseline — The original project plan plus approved changes.

baseline dates — The planned schedule dates for activities in a Tracking Gantt chart.

benchmarking — A technique used to generate ideas for quality improvements by comparing specific project practices or product characteristics to those of other projects or products within or outside the performing organization.

best practice — An optimal way recognized by industry to achieve a stated goal or objective.

bid — Also called a tender or quote (short for quotation), a document prepared by sellers providing pricing for standard items that have been clearly defined by the buyer.

blogs — Easy to use journals on the Web that allow users to write entries, create links, and upload pictures, while readers can post comments to journal entries.

bottom-up approach — Creating a WBS by having team members identify as many specific tasks related to the project as possible and then grouping them into higher level categories.

bottom-up estimates — A cost estimating technique based on estimating individual work items and summing them to get a project total.

brainstorming — A technique by which a group attempts to generate ideas or find a solution for a specific problem by amassing ideas spontaneously and without judgment.

budget at completion (BAC) — The original total budget for a project.

budgetary estimate — A cost estimate used to allocate money into an organization's budget.

buffer — Additional time to complete a task, added to an estimate to account for various factors.

burst — When a single node is followed by two or more activities on a network diagram.

business service management (BSM) tools — Tools that help track the execution of business process flows and expose how the state of supporting IT systems and resources is impacting end-to-end business process performance in real time.

Capability Maturity Model Integration (CMMI) — A process improvement approach that provides organizations with the essential elements of effective processes.

capitalization rate — The rate used in discounting future cash flow; also called the discount rate or opportunity cost of capital.

cash flow — Benefits minus costs or income minus expenses.

cash flow analysis — A method for determining the estimated annual costs and benefits for a project.

cause-and-effect diagram — A diagram that traces complaints about quality problems back to the responsible production operations to help find the root cause. Also known as fishbone diagram or Ishikawa diagram.

champion — A senior manager who acts as a key proponent for a project.

change control board (CCB) — A formal group of people responsible for approving or rejecting changes on a project.

change control system — A formal, documented process that describes when and how official project documents may be changed.

closing processes — Formalizing acceptance of the project or project phase and ending it efficiently.

coercive power — Using punishment, threats, or other negative approaches to get people to do things they do not want to do.

collaborating mode — A conflict-handling mode where decision makers incorporate different viewpoints and insights to develop consensus and commitment.

communications management plan — A document that guides project communications.

compromise mode — Using a give-and-take approach to resolving conflicts; bargaining and searching for solutions that bring some degree of satisfaction to all the parties in a dispute.

configuration management — A process that ensures that the descriptions of the project's products are correct and complete.

conformance — Delivering products that meet requirements and fitness for use.

conformance to requirements — The project processes and products meet written specifications.

confrontation mode — Directly facing a conflict using a problem-solving approach that allows affected parties to work through their disagreements.

constructive change orders — Oral or written acts or omissions by someone with actual or apparent authority that can be construed to have the same effect as a written change order.

contingency allowances — Provisions held by the project sponsor or organization to reduce the risk of cost or schedule overruns to an acceptable level; also called contingency reserves.

contingency plans — Predefined actions that the project team will take if an identified risk event occurs.

contingency reserves — Provisions held by the project sponsor or organization to reduce the risk of cost or schedule overruns to an acceptable level; also called contingency allowances.

contract — A mutually binding agreement that obligates the seller to provide the specified products or services, and obligates the buyer to pay for them.

control chart — A graphic display of data that illustrates the results of a process over time.

controlling costs — Controlling changes to the project budget.

cost management plan — A document that describes how cost variances will be managed on the project.

cost baseline — A time-phased budget that project managers use to measure and monitor cost performance.

cost of capital — The return available by investing the capital elsewhere.

cost of nonconformance — Taking responsibility for failures or not meeting quality expectations.

cost of quality — The cost of conformance plus the cost of nonconformance.

cost performance index (CPI) — The ratio of earned value to actual cost; can be used to estimate the projected cost to complete the project.

cost variance (CV) — The earned value minus the actual cost.

cost plus award fee (CPAF) contract — A contract in which the buyer pays the supplier for allowable performance costs plus an award fee based on the satisfaction of subjective performance criteria.

cost plus fixed fee (CPFF) contract — A contract in which the buyer pays the supplier for allowable performance costs plus a fixed fee payment usually based on a percentage of estimated costs.

cost plus incentive fee (CPIF) contract — A contract in which the buyer pays the supplier for allowable performance costs along with a predetermined fee and an incentive bonus.

cost plus percentage of costs (CPPC) contract — A contract in which the buyer pays the supplier for allowable performance costs along with a predetermined percentage based on total costs.

cost-reimbursable contracts — Contracts involving payment to the supplier for direct and indirect actual costs.

crashing — A technique for making cost and schedule trade-offs to obtain the greatest amount of schedule compression for the least incremental cost.

critical chain scheduling — A method of scheduling that takes limited resources into account when creating a project schedule and includes buffers to protect the project completion date.

critical path — The series of activities in a network diagram that determines the earliest completion of the project; it is the longest path through the network diagram and has the least amount of slack or float.

critical path method (CPM) or critical path analysis — A project network analysis technique used to predict total project duration.

decision tree — A diagramming analysis technique used to help select the best course of action in situations in which future outcomes are uncertain.

decomposition — Subdividing project deliverables into smaller pieces.

defect — Any instance where the product or service fails to meet customer requirements.

DMAIC (Define, Measure, Analyze, Improve, Control) — A systematic, closed-loop process for continued improvement that is scientific and fact based.

definitive estimate — A cost estimate that provides an accurate estimate of project costs.

deliverable — A product or service, such as a technical report, a training session, a piece of hardware, or a segment of software code, produced or provided as part of a project.

Delphi technique — An approach used to derive a consensus among a panel of experts, to make predictions about future developments.

dependency — The sequencing of project activities or tasks; also called a relationship.

deputy project managers — People who fill in for project managers in their absence and assist them as needed, similar to the role of a vice president.

design of experiments — A quality technique that helps identify which variables have the most influence on the overall outcome of a process.

determining the budget — Allocating the overall cost estimate to individual work items to establish a baseline for measuring performance.

direct costs — Costs that can be directly related to producing the products and services of the project.

directives — New requirements imposed by management, government, or some external influence.

discretionary dependencies — Sequencing of project activities or tasks defined by the project team and used with care since they may limit later scheduling options.

discount factor — A multiplier for each year based on the discount rate and year.

discount rate — The rate used in discounting future cash flow; also called the capitalization rate or opportunity cost of capital.

dummy activities — Activities with no duration and no resources used to show a logical relationship between two activities in the arrow diagramming method of project network diagrams.

duration — The actual amount of time worked on an activity *plus* elapsed time.

early finish date — The earliest possible time an activity can finish based on the project network logic.

early start date — The earliest possible time an activity can start based on the project network logic.

earned value (EV) — An estimate of the value of the physical work actually completed.

earned value management (EVM) — A project performance measurement technique that integrates scope, time, and cost data.

effort — The number of workdays or work hours required to complete a task.

empathic listening — Listening with the intent to understand.

enterprise or portfolio project management software — Software that integrates information from multiple projects to show the status of active, approved, and future projects across an entire organization.

estimate at completion (EAC) — An estimate of what it will cost to complete the project based on performance to date.

estimating costs — Developing an approximation or estimate of the costs of the resources needed to complete the project.

ethics — A set of principles that guide our decision making based on personal values of what is "right" and "wrong."

executing processes — Coordinating people and other resources to carry out the project plans and produce the products, services, or results of the project or project phase.

executive steering committee — A group of senior executives from various parts of the organization who regularly review important corporate projects and issues.

expectations management matrix — A tool to help understand unique measures of success for a particular project.

expected monetary value (EMV) — The product of the risk event probability and the risk event's monetary value.

expert power — Using one's personal knowledge and expertise to get people to change their behavior.

extrinsic motivation — Causes people to do something for a reward or to avoid a penalty.

external dependencies — Sequencing of project activities or tasks that involve relationships between project and non-project activities.

external failure cost — A cost related to all errors not detected and corrected before delivery to the customer.

fallback plans — Plans developed for risks that have a high impact on meeting project objectives, to be implemented if attempts to reduce the risk are not effective.

fast tracking — A schedule compression technique in which you do activities in parallel that you would normally do in sequence.

features — The special characteristics that appeal to users.

feeding buffers — Additional time added before tasks on the critical path that are preceded by non-critical-path tasks.

fishbone diagram — Diagram that traces complaints about quality problems back to the responsible production operations to help find the root cause. Also known as cause-and-effect diagram or Ishikawa diagram.

fitness for use — A product can be used as it was intended.

finish-to-start dependency — A relationship on a project network diagram where the "from" activity must be finished before the "to" activity can be started.

fixed-price contract — Contract with a fixed total price for a well-defined product or service; also called a lump-sum contract.

float — The amount of time a project activity may be delayed without delaying a succeeding activity or the project finish date; also called slack.

flowchart — Graphic display of the logic and flow of processes that helps you analyze how problems occur and how processes can be improved.

forecasts — Used to predict future project status and progress based on past information and trends.

forcing mode — Using a win-lose approach to conflict resolution to get one's way.

forward pass — A network diagramming technique that determines the early start and early finish dates for each activity.

free slack (free float) — The amount of time an activity can be delayed without delaying the early start of any immediately following activities.

functionality — The degree to which a system performs its intended function.

functional organizational structure — An organizational structure that groups people by functional areas such as information technology, manufacturing, engineering, and human resources.

Gantt chart — A standard format for displaying project schedule information by listing project activities and their corresponding start and finish dates in a calendar format; sometimes referred to as bar chart.

Google Docs — Online applications offered by Google that allow users to create, share, and edit documents, spreadsheets, and presentations online.

green IT or green computing — Developing and using computer resources in an efficient way to improve economic viability, social responsibility, and environmental impact.

groupthink — Conformance to the values or ethical standards of a group.

hierarchy of needs — A pyramid structure illustrating Maslow's theory that people's behaviors are guided or motivated by a sequence of needs.

histogram — A bar graph of a distribution of variables.

human resources frame — Focuses on producing harmony between the needs of the organization and the needs of people.

indirect costs — Costs that are not directly related to the products or services of the project, but are indirectly related to performing the project.

influence diagram — Diagram that represents decision problems by displaying essential elements, including decisions, uncertainties, and objectives, and how they influence each other.

initiating processes — Defining and authorizing a project or project phase.

intangible costs or benefits — Costs or benefits that are difficult to measure in monetary terms.

integrated change control — Identifying, evaluating, and managing changes throughout the project life cycle.

integration testing — Testing that occurs between unit and system testing to test functionally grouped components to ensure a subset(s) of the entire system works together.

interface management — Identifying and managing the points of interaction between various elements of a project.

internal failure cost — A cost incurred to correct an identified defect before the customer receives the product.

internal rate of return (IRR) — The discount rate that results in an NPV of zero for a project.

interviewing — A fact-finding technique that is normally done face-to-face, but can also occur through phone calls, e-mail, or instant messaging.

intrinsic motivation — Causes people to participate in an activity for their own enjoyment.

Ishikawa diagram — Diagram that traces complaints about quality problems back to the responsible production operations to help find the root cause. Also known as cause-and-effect diagram or fishbone diagram.

ISO 9000 — A quality system standard developed by the International Organization for Standardization (ISO) that includes a three-part, continuous cycle of planning, controlling, and documenting quality in an organization.

issue — A matter under question or dispute that could impede project success.

issue log — A tool to document and monitor the resolution of project issues.

IT governance — Addresses the authority and control for key IT activities in organizations, including IT infrastructure, IT use, and project management.

Joint Application Design (JAD) — Using highly organized and intensive workshops to bring together project stakeholders—the sponsor, users, business analysts, programmers, and so on—to jointly define and design information systems.

kick-off meeting — A meeting held at the beginning of a project so that stakeholders can meet each other, review the goals of the project, and discuss future plans.

known risks — Risks that the project team have identified and analyzed and can be managed proactively.

known unknowns — Dollars included in a cost estimate to allow for future situations that may be partially planned for (sometimes called contingency reserves) and are included in the project cost baseline.

late finish date — The latest possible time an activity can be completed without delaying the project finish date.

late start date — The latest possible time an activity may begin without delaying the project finish date.

leader — A person who focuses on long-term goals and big-picture objectives, while inspiring people to reach those goals.

learning curve theory — A theory that states that when many items are produced repetitively, the unit cost of those items normally decreases in a regular pattern as more units are produced.

legitimate power — Getting people to do things based on a position of authority.

lessons-learned report — Reflective statements written by project managers and their team members to document important things they have learned from working on the project.

life cycle costing — Considers the total cost of ownership, or development plus support costs, for a project.

lump-sum contract — Contract with a fixed total price for a well-defined product or service; also called a fixed-price contract.

maintainability — The ease of performing maintenance on a product.

make-or-buy decision — When an organization decides if it is in its best interests to make certain products or perform certain services inside the organization, or if it is better to buy them from an outside organization.

Malcolm Baldrige National Quality Award — An award started in 1987 to recognize companies that have achieved a level of world-class competition through quality management.

management reserves — Dollars included in a cost estimate to allow for future situations that are unpredictable (sometimes called unknown unknowns).

manager — A person who deals with the day-to-day details of meeting specific goals.

mandatory dependencies — Sequencing of project activities or tasks that are inherent in the nature of the work being done on the project.

matrix organizational structure — An organizational structure in which employees are assigned to both functional and project managers.

maturity model — A framework for helping organizations improve their processes and systems.

mean — The average value of a population.

measurement and test equipment costs — The capital cost of equipment used to perform prevention and appraisal activities.

merge — When two or more nodes precede a single node on a network diagram.

methodology — Describes *how* things should be done.

metric — A standard of measurement.

milestone — A significant event that normally has no duration on a project; serves as a marker to help in identifying necessary activities, setting schedule goals, and monitoring progress.

mind mapping — A technique that uses branches radiating out from a core idea to structure thoughts and ideas.

mirroring — Matching certain behaviors of the other person.

Monte Carlo analysis — A risk quantification technique that simulates a model's outcome many times, to provide a statistical distribution of the calculated results.

monitoring and controlling processes — Regularly measuring and monitoring progress to ensure that the project team meets the project objectives.

multitasking — When a resource works on more than one task at a time.

Murphy's Law — Principle that if something can go wrong, it will.

Myers-Briggs Type Indicator (MBTI) — A popular tool for determining personality preferences.

net present value (NPV) analysis — A method of calculating the expected net monetary gain or loss from a project by discounting all expected future cash inflows and outflows to the present point in time.

network diagram — A schematic display of the logical relationships or sequencing of project activities.

node — The starting and ending point of an activity on an activity-on-arrow diagram.

normal distribution — A bell-shaped curve that is symmetrical about the mean of the population.

offshoring — Outsourcing from another country.

organizational breakdown structure (OBS) — A specific type of organizational chart that shows which organizational units are responsible for which work items.

organizational culture — A set of shared assumptions, values, and behaviors that characterize the functioning of an organization.

organizational process assets — Formal and informal plans, policies, procedures, guidelines, information systems, financial systems, management systems, lessons learned, and historical information that can be used to influence a project's success.

opportunities — Chances to improve the organization.

opportunity cost of capital — The rate used in discounting future cash flow; also called the capitalization rate or discount rate.

outsourcing — When an organization acquires goods and/or sources from an outside source.

overallocation — When more resources than are available are assigned to perform work at a given time.

overrun — The additional percentage or dollar amount by which actual costs exceed estimates.

parametric modeling — A cost-estimating technique that uses project characteristics (parameters) in a mathematical model to estimate project costs.

Pareto analysis — Identifying the vital few contributors that account for most quality problems in a system.

Pareto chart — Histogram that helps identify and prioritize problem areas.

Parkinson's Law — Principle that work expands to fill the time allowed.

payback period — The amount of time it will take to recoup, in the form of net cash inflows, the total dollars invested in a project.

performance — How well a product or service performs the customer's intended use.

PERT weighted average — optimistic time + 4 * most likely time + pessimistic time

planned value (PV) — That portion of the approved total cost estimate planned to be spent on an activity during a given period.

planning processes — Devising and maintaining a workable scheme to ensure that the project addresses the organization's needs.

phase exit or kill point — Management review that should occur after each project phase to determine if projects should be continued, redirected, or terminated.

Point of Total Assumption (PTA) — The cost at which the contractor assumes total responsibility for each additional dollar of contract cost in a fixed price incentive fee contract.

political frame — Addresses organizational and personal politics.

politics — Competition between groups or individuals for power and leadership.

power — The potential ability to influence behavior to get people to do things they would not otherwise do.

Precedence Diagramming Method (PDM) — A network diagramming technique in which boxes represent activities.

predictive life cycle — A software development approach used when the scope of the project can be clearly articulated and the schedule and cost can be accurately predicted.

prevention cost — The cost of planning and executing a project so that it is error-free or within an acceptable error range.

probability/impact matrix or chart — A matrix or chart that lists the relative probability of a risk occurring on one side of a matrix or axis on a chart and the relative impact of the risk occurring on the other.

probabilistic time estimates — Duration estimates based on using optimistic, most likely, and pessimistic estimates of activity durations instead of using one specific or discrete estimate.

problems — Undesirable situations that prevent the organization from achieving its goals.

process — A series of actions directed toward a particular result.

process adjustments — Adjustments made to correct or prevent further quality problems based on quality control measurements.

procurement — Acquiring goods and/or services from an outside source.

profit margin — The ratio between revenues and profits.

profits — Revenues minus expenses.

program — A group of projects managed in a coordinated way to obtain benefits and control not available from managing them individually.

Program Evaluation and Review Technique (PERT) — A project network analysis technique used to estimate project duration when there is a high degree of uncertainty with the individual activity duration estimates.

program manager — A person who provides leadership and direction for the project managers heading the projects within a program.

progress reports — Reports that describe what the project team has accomplished during a certain period of time.

project — A temporary endeavor undertaken to create a unique product, service, or result.

project acquisition — The last two phases in a project (implementation and close-out) that focus on delivering the actual work.

project archives — A complete set of organized project records that provide an accurate history of the project.

project buffer — Additional time added before the project's due date.

project charter — A document that formally recognizes the existence of a project and provides direction on the project's objectives and management.

project cost management — The processes required to ensure that the project is completed within the approved budget.

project feasibility — The first two phases in a project (concept and development) that focus on planning.

project integration management — Processes that coordinate all project management knowledge areas throughout a project's life, including developing the project charter, developing the preliminary project scope statement, developing the project management plan, directing and managing the project, monitoring and controlling the project, providing integrated change control, and closing the project.

project life cycle — A collection of project phases, such as concept, development, implementation, and close-out.

project management — The application of knowledge, skills, tools, and techniques to project activities to meet project requirements.

Project Management Institute (PMI) — An international professional society for project managers.

project management knowledge areas — Project integration management, scope, time, cost, quality, human resource, communications, risk, and procurement management.

Project Management Office (PMO) — An organizational group responsible for coordinating the project management functions throughout an organization.

project management plan — A document used to coordinate all project planning documents and guide project execution and control.

project management process groups — The progression of project activities from initiation to planning, executing, monitoring and controlling, and closing.

Project Management Professional (PMP) — Certification provided by PMI that requires documenting project experience and education, agreeing to follow the PMI code of ethics, and passing a comprehensive exam.

project management tools and techniques — Methods available to assist project managers and their teams; some popular tools in the time management knowledge area include Gantt charts, network diagrams, and critical path analysis.

project manager — The person responsible for working with the project sponsor, the project team, and the other people involved in a project to meet project goals.

project organizational structure — An organizational structure that groups people by major projects, such as specific aircraft programs.

project procurement management — The processes required to acquire goods and services for a project from outside the performing organization.

project portfolio management (portfolio management) — When organizations group and manage projects as a portfolio of investments that contribute to the entire enterprise's success.

project quality management — Ensuring that a project will satisfy the needs for which it was undertaken.

project scope management — The processes involved in defining and controlling what work is or is not included in a project.

project scope statement — A document that includes, at a minimum, a description of the project, including its overall objectives and justification, detailed descriptions of all project deliverables, and the characteristics and requirements of products and services produced as part of the project.

PRojects IN Controlled Environments (PRINCE2) — A project management methodology developed in the U.K. that defines 45 separate sub-processes and organizes these into eight process groups.

project sponsor — The person who provides the direction and funding for a project.

project time management — The processes required to ensure timely completion of a project.

proposal — A document prepared by sellers when there are different approaches for meeting buyer needs.

prototyping — Developing a working replica of the system or some aspect of the system to help define user requirements.

quality — The totality of characteristics of an entity that bear on its ability to satisfy stated or implied needs or the degree to which a set of inherent characteristics fulfill requirements.

quality assurance — Periodically evaluating overall project performance to ensure that the project will satisfy the relevant quality standards.

quality audit — Structured review of specific quality management activities that helps identify lessons learned and can improve performance on current or future projects.

quality circles — Groups of nonsupervisors and work leaders in a single company department who volunteer to conduct group studies on how to improve the effectiveness of work in their department.

quality control — Monitoring specific project results to ensure that they comply with the relevant quality standards and identifying ways to improve overall quality.

quality planning — Identifying which quality standards are relevant to the project and how to satisfy them.

RACI charts — Charts that show Responsibility, Accountability, Consultation, and Informed roles for project stakeholders.

rapport — A relation of harmony, conformity, accord, or affinity.

rate of performance (RP) — The ratio of actual work completed to the percentage of work planned to have been completed at any given time during the life of the project or activity.

Rational Unified Process (RUP) — An iterative software development process that focuses on team productivity and delivers software best practices to all team members.

referent power — Getting people to do things based on an individual's personal charisma.

reliability — The ability of a product or service to perform as expected under normal conditions.

Request for Proposal (RFP) — A document used to solicit proposals from prospective suppliers.

Request for Quote (RFQ) — A document used to solicit quotes or bids from prospective suppliers.

required rate of return — The minimum acceptable rate of return on an investment.

requirement — A condition or capability that must be met or possessed by a system, product, service, result, or component to satisfy a contract, standard, specification, or other formal document.

requirements management plan — A plan that describes how project requirements will be analyzed, documented, and managed.

requirements traceability matrix — A table that lists requirements, various attributes of each requirement, and the status of the requirements to ensure that all requirements are addressed.

reserves — Dollars included in a cost estimate to mitigate cost risk by allowing for future situations that are difficult to predict.

residual risks — Risks that remain after all of the response strategies have been implemented.

resource breakdown structure — A hierarchical structure that identifies the project's resources by category and type.

resource histogram — A column chart that shows the number of resources assigned to a project over time.

resource leveling — A technique for resolving resource conflicts by delaying tasks.

resource loading — The amount of individual resources an existing schedule requires during specific time periods.

resources — People, equipment, and materials.

responsibility assignment matrix (RAM) — A matrix that maps the work of the project as described in the WBS to the people responsible for performing the work as described in the organizational breakdown structure (OBS).

return on investment (ROI) — (Benefits minus costs) divided by costs.

reward power — Using incentives to induce people to do things.

rework — Action taken to bring rejected items into compliance with product requirements or specifications or other stakeholder expectations.

risk — An uncertainty that can have a negative or positive effect on meeting project objectives.

risk acceptance — Accepting the consequences should a risk occur.

risk-averse — Having a low tolerance for risk.

risk avoidance — Eliminating a specific threat or risk, usually by eliminating its causes.

risk breakdown structure — A hierarchy of potential risk categories for a project.

risk enhancement — Changing the size of an opportunity by identifying and maximizing key drivers of the positive risk.

risk events — Specific uncertain events that may occur to the detriment or enhancement of the project.

risk exploitation — Doing whatever you can to make sure the positive risk happens.

risk factors — Numbers that represent overall risk of specific events, given their probability of occurring and the consequence to the project if they do occur.

risk management plan — A plan that documents the procedures for managing risk throughout a project.

risk mitigation — Reducing the impact of a risk event by reducing the probability of its occurrence.

risk-neutral — A balance between risk and payoff.

risk owner — The person who will take responsibility for a risk and its associated response strategies and tasks.

risk register — A document that contains results of various risk management processes, often displayed in a table or spreadsheet format.

risk-seeking — Having a high tolerance for risk.

risk sharing — Allocating ownership of the risk to another party.

risk tolerance — The amount of satisfaction or pleasure received from a potential payoff; also called risk utility.

risk transference — Shifting the consequence of a risk and responsibility for its management to a third party.

risk utility — The amount of satisfaction or pleasure received from a potential payoff; also called risk tolerance.

Robust Design methods — Methods that focus on eliminating defects by substituting scientific inquiry for trial-and-error methods.

rough order of magnitude (ROM) estimate — A cost estimate prepared very early in the life of a project to provide a rough idea of what a project will cost.

runaway projects — Projects that have significant cost or schedule overruns.

run chart — Chart that displays the history and pattern of variation of a process over time.

scatter diagram — Diagram that helps to show if there is a relationship between two variables; also called XY charts.

schedule baseline — The approved planned schedule for the project.

schedule performance index (SPI) — The ratio of earned value to planned value; can be used to estimate the projected time to complete a project.

schedule variance (SV) — The earned value minus the planned value.

scope — All the work involved in creating the products of the project and the processes used to create them.

scope baseline — The approved project scope statement and its associated WBS and WBS dictionary.

scope creep — The tendency for project scope to keep getting bigger.

secondary risks — Risks that are a direct result of implementing a risk response.

sellers — Contractors, suppliers, or providers who provide goods and services to other organizations.

sensitivity analysis — A technique used to show the effects of changing one or more variables on an outcome.

seven run rule — If seven data points in a row on a quality control chart are all below the mean, above the mean, or are all increasing or decreasing, then the process needs to be examined for nonrandom problems.

SharePoint portal — Allows users to create custom Web sites to access documents and applications stored on shared devices.

six 9s of quality — A measure of quality control equal to 1 fault in 1 million opportunities.

Six Sigma — A comprehensive and flexible system for achieving, sustaining, and maximizing business success that is uniquely driven by close understanding of customer needs, disciplined use of facts, data, statistical analysis, and diligent attention to managing, improving, and reinventing business processes.

Six Sigma methodologies — DMAIC (Define, Measure, Analyze, Improve, and Control) is used to improve an existing business process and DMADV (Define, Measure, Analyze, Design, and Verify) is used to create new product or process designs.

slack — The amount of time a project activity may be delayed without delaying a succeeding activity or the project finish date; also called float.

slipped milestone — A milestone activity that is completed later than planned.

SMART criteria — Guidelines to help define milestones that are **s**pecific, **m**easurable, **a**ssignable, **r**ealistic, and **t**ime-framed.

smoothing mode — Deemphasizing or avoiding areas of differences and emphasizing areas of agreements.

software defect — Anything that must be changed before delivery of the program.

Software Quality Function Deployment (SQFD) model — A maturity model that focuses on defining user requirements and planning software projects.

staffing management plan — A document that describes when and how people will be added to and taken off a project team.

stakeholder management strategy — An approach to help increase the support of stakeholders throughout the project.

stakeholder register — A document that includes details related to the identified project stakeholders.

stakeholders — People involved in or affected by project activities.

standard — Describes best practices for *what* should be done.

standard deviation — A measure of how much variation exists in a distribution of data.

start-to-finish dependency — A relationship on a project network diagram where the "from" activity cannot start before the "to" activity is finished.

start-to-start dependency — A relationship on a project network diagram in which the "from" activity cannot start.

statement of work (SOW) — A description of the work required for the procurement.

statistical sampling — Choosing part of a population of interest for inspection until the "to" activity starts.

status reports — Reports that describe where the project stands at a specific point in time.

strategic planning — Determining long-term objectives by analyzing the strengths and weaknesses of an organization, studying opportunities and threats in the business environment, predicting future trends, and projecting the need for new products and services.

structural frame — Deals with how the organization is structured (usually depicted in an organizational chart) and focuses on different groups' roles and responsibilities to meet the goals and policies set by top management.

subproject managers — People responsible for managing the subprojects that a large project might be broken into.

sunk cost — Money that has been spent in the past.

SWOT analysis — Analyzing **S**trengths, **W**eaknesses, **O**pportunities, and **T**hreats; used to aid in strategic planning.

symbolic frame — Focuses on the symbols, meanings, and culture of an organization.

synergy — An approach where the whole is greater than the sum of the parts.

system outputs — The screens and reports the system generates.

systems — Sets of interacting components working within an environment to fulfill some purpose.

systems analysis — A problem-solving approach that requires defining the scope of the system to be studied, and then dividing it into its component parts for identifying and evaluating its problems, opportunities, constraints, and needs.

systems approach — A holistic and analytical approach to solving complex problems that includes using a systems philosophy, systems analysis, and systems management.

systems development life cycle (SDLC) — A framework for describing the phases involved in developing and maintaining information systems.

systems management — Addressing the business, technological, and organizational issues associated with creating, maintaining, and making changes to a system.

systems philosophy — An overall model for thinking about things as systems.

system stakeholder register — A public document that includes details related to the identified project stakeholders.

systems thinking — Taking a holistic view of an organization to effectively handle complex situations.

tangible costs or benefits — Costs or benefits that can be easily measured in dollars.

task — An element of work, normally found on the WBS, that has an expected duration and cost, and expected resource requirements; also called an activity.

team development — Building individual and group skills to enhance project performance.

termination clause — A contract clause that allows the buyer or supplier to end the contract.

testing — Testing the entire system as one entity to ensure that it is working properly.

Theory of Constraints (TOC) — A management philosophy that states that any complex system at any point in time often has only one aspect or constraint that is limiting its ability to achieve more of its goal.

three-point estimate — An estimate that includes an optimistic, most likely, and pessimistic estimate.

time and material (T&M) contracts — A hybrid of both fixed-price and cost-reimbursable contracts.

top-down approach — Creating a WBS by starting with the largest items of the project and breaking them into their subordinate items.

top-down estimates — A cost estimating technique that uses the actual cost of a previous, similar project as the basis for estimating the cost of the current project, also called analogous estimates.

Top Ten Risk Item Tracking — A qualitative risk analysis tool for identifying risks and maintaining an awareness of risks throughout the life of a project.

total slack (total float) — The amount of time an activity may be delayed from its early start without delaying the planned project finish date.

Tracking Gantt chart — A Gantt chart that compares planned and actual project schedule information.

triggers — Indications for actual risk events.

triple constraint — Balancing scope, time, and cost goals.

Tuckman model — Describes five stages of team development: forming, storming, norming, performing, and adjourning.

unit pricing — An approach in which the buyer pays the supplier a predetermined amount per unit of service, and the total value of the contract is a function of the quantities needed to complete the work.

unit test — A test of each individual component (often a program) to ensure that it is as defect-free as possible.

unknown risks — Risks that have not been identified and analyzed so they cannot be managed proactively.

unknown unknowns — Dollars included in a cost estimate to allow for future situations that are unpredictable (sometimes called management reserves).

use case modeling — A process for identifying and modeling business events, who initiated them, and how the system should respond to them.

user acceptance testing — An independent test performed by end users prior to accepting the delivered system.

variance — The difference between planned and actual performance.

virtualization — Hiding the physical characteristics of computing resources from their users, such as making a single server, operating system, application, or storage device appear to function as multiple virtual resources.

virtual team — A group of individuals who work across time and space using communication technologies.

watch list — A list of risks that are low priority, but are still identified as potential risks.

WBS dictionary — A document that describes detailed information about each WBS item.

weighted scoring model — A technique that provides a systematic process for basing project selection on numerous criteria.

wiki — A Web site that has a page or pages designed to enable anyone who accesses it to contribute or modify content.

withdrawal mode — Retreating or withdrawing from an actual or potential disagreement.

workarounds — Unplanned responses to risk events when there are no contingency plans in place.

work breakdown structure (WBS) — A deliverable-oriented grouping of the work involved in a project that defines the total scope of the project.

work package — A task at the lowest level of the WBS.

yield — The number of units handled correctly through the development process.

# INDEX

Note: Page numbers in **boldface** type indicate key terms and their definitions.

## C

## D



## E

## F



## G



## H

## I

## J



## K



## L



## M

## N



## O



## P

## Q



## R

## S

## T

## U



## V



## W

## X



## Z